Radon Series on Computational and Applied Mathematics 8

Radon Series on Computational and Applied Mathematics

1 *Lectures on Advanced Computational Methods in Mechanics*
Johannes Kraus and Ulrich Langer (eds.), 2007

2 *Gröbner Bases in Symbolic Analysis*
Markus Rosenkranz and Dongming Wang (eds.), 2007

3 *Gröbner Bases in Control Theory and Signal Processing*
Hyungju Park and Georg Regensburger (eds.), 2007

4 *A Posteriori Estimates for Partial Differential Equations*
Sergey Repin, 2008

5 *Robust Algebraic Multilevel Methods and Algorithms*
Johannes Kraus and Svetozar Margenov, 2009

6 *Iterative Regularization Methods for Nonlinear Ill-Posed Problems*
Barbara Kaltenbacher, Andreas Neubauer and Otmar Scherzer, 2008

7 *Robust Static Super-Replication of Barrier Options*
Jan H. Maruhn, 2009

8 *Advanced Financial Modelling*
Hansjörg Albrecher, Wolfgang J. Runggaldier and Walter Schachermayer (eds.), 2009

Advanced Financial Modelling

Edited by

Hansjörg Albrecher
Wolfgang J. Runggaldier
Walter Schachermayer

Walter de Gruyter · Berlin · New York

Editors

Hansjörg Albrecher
Université de Lausanne
Quartier UNIL-Dorigny
Bâtiment Extranef
1015 Lausanne, Switzerland
E-mail: hansjoerg.albrecher@unil.ch

Wolfgang J. Runggaldier
Dipartimento di Matematica Pura ed Applicata
Università degli Studi di Padova
Via Trieste 63
35121 Padova, Italy
E-mail: runggal@math.unipd.it

Walter Schachermayer
Faculty of Mathematics
University of Vienna
Nordbergstraße 15
1090 Vienna, Austria
E-Mail: walter.schachermayer@univie.ac.at

Keywords
Mathematical finance, actuarial mathematics, stochastic differential equations, optimization, mathematical modelling, computational methods.

Mathematics Subject Classification 2000
91-02, 60G35, 60H35, 60J60, 62P05, 65C05, 91B16, 91B28, 91B70, 93E20.

∞ Printed on acid-free paper which falls within the guidelines of the ANSI to ensure permanence and durability.

ISBN 978-3-11-021313-3

Bibliographic information published by the Deutsche Nationalbibliothek

The Deutsche Nationalbibliothek lists this publication in the Deutsche Nationalbibliografie; detailed bibliographic data are available in the Internet at http://dnb.d-nb.de.

Printed in Germany
Cover design: Martin Zech, Bremen.
Typeset using the authors' LaTeX files: Jan Nitzschmann, Leipzig.
Printing and binding: Hubert & Co. GmbH & Co. KG, Göttingen.

Preface

This book is a collection of state-of-the-art surveys on various topics in mathematical finance, with an emphasis on recent modeling and computational approaches. The volume is related to a Special Semester on Stochastics with Emphasis on Finance that took place from September to December 2008 at the Johann Radon Institute for Computational and Applied Mathematics (RICAM) of the Austrian Academy of Sciences in Linz, Austria.

The Special Semester was built around a number of selected topics and each of these topics was the theme of an international workshop with about 20 invited speakers. Besides a Tutorial, a Kick-Off Workshop focusing also on "Academics meeting Practitioners" and a Concluding Workshop, the thematic workshops concerned the following topics:

Advanced Modelling in Finance and Insurance; Optimization and Optimal Control; Inverse and Partial Information Problems: Methodology and Applications; Computational Methods with Applications in Finance, Insurance and the Life Sciences; Stochastic Methods in Partial Differential Equations and Applications of Deterministic and Stochastic PDEs.

In addition to the workshops, the idea arose to collect surveys on important aspects and recent developments related to the topics of the Special Semester and this forms the contents of the present volume. The topics covered include the following (listed alphabetically and grouped according to their relation with the topics of the Special Semester in the above order):

- Affine diffusion processes in finance
- Default and prepayment modeling using Levy processes
- Volatility inference in models beyond semimartingales
- Optimal asset allocation
- Optimal consumption and investment in illiquid markets and with downside risk measures
- Multiperiod acceptability functionals
- Worst-case portfolio optimization
- Good deal bounds
- Optimal investment and hedging under partial and inside information
- Regularization of inverse problems and calibration of option price models
- Advanced simulation techniques
- Applications of Malliavin Calculus
- Probabilistic schemes for fully nonlinear PDE's

The contributions themselves are arranged in alphabetic order according to the first named author.

More details on the Special Semester and the full workshop program can be found at the RICAM Special Semester webpage at:

http://www.ricam.oeaw.ac.at/specsem/sef

We would like to take this opportunity to thank all those who have contributed scientifically to this Special Semester, in particular the authors of this volume and the speakers at the workshops as well as the (more than 250) participants in the workshops. Further thanks go to the Austrian Academy of Sciences and in particular the Johann Radon Institute of Computational and Applied Mathematics in Linz and its director Heinz W. Engl for making this Special Semester possible. We also thank Robert Plato from the publishing house de Gruyter for the professional editorial support during the preparation of this volume.

Lausanne, Padua and Vienna, June 2009

Hansjoerg Albrecher
Wolfgang Runggaldier
Walter Schachermayer

Contents

Radon Series Comp. Appl. Math **8**, 1–25 © de Gruyter 2009

Brownian semistationary processes and volatility/intermittency

Ole E. Barndorff-Nielsen and Jürgen Schmiegel

Abstract. A new class of stochastic processes, termed Brownian semistationary processes ($\mathcal{BSS}$), is introduced and discussed. This class has similarities to that of Brownian semimartingales ($\mathcal{BSM}$), but is mainly directed towards the study of stationary processes, and $\mathcal{BSS}$ processes are not in general of the semimartingale type. We focus on semimartingale - nonsemimartingale issues and on inference problems concerning the underlying volatility/intermittency process, in the nonsemimartingale case and based on normalised realised quadratic variation. The concept of $\mathcal{BSS}$ processes has arisen out of an ongoing study of turbulent velocity fields and is the purely temporal version of the general tempo-spatial framework of ambit processes. The latter, which may have applications also to the finance of energy markets, is briefly considered at the end of the paper, again with reference to the question of inference on the volatility/intermittency.

Key words. Ambit processes, intermittency, nonsemimartingales, stationary processes, realised quadratic variation, turbulence, volatility.

AMS classification. 60G10

1 Introduction

This paper discusses stochastic processes $Y = \{Y_t\}_{t\in\mathbb{R}}$ of the form

$$Y_t = \mu + \int_{-\infty}^{t} g(t-s)\sigma_s \mathrm{d}B_s + \int_{-\infty}^{t} q(t-s)a_s \mathrm{d}s \tag{1.1}$$

where μ is a constant, B is Brownian motion, g and q are nonnegative deterministic functions on $\mathbb{R}$, with $g(t) = q(t) = 0$ for $t \leq 0$, and σ and a are càdlàg processes. When σ and a are stationary then so is Y. Accordingly we shall refer to processes of this type as *Brownian semistationary* ($\mathcal{BSS}$) processes. It is sometimes convenient to indicate the formula for Y as

$$Y = \mu + g * \sigma \bullet B + q * a \bullet Leb, \tag{1.2}$$

where Leb denotes Lebesgue measure.

We consider the $\mathcal{BSS}$ processes to be the natural analogue, for stationarity related processes, of the class $\mathcal{BSM}$ of Brownian semimartingales

$$Y_t = \int_0^t \sigma_s \mathrm{d}B_s + \int_0^t a_s \mathrm{d}s. \tag{1.3}$$

In the present paper the processes σ and a will, unless otherwise stated, be taken to be stationary, and we then refer to σ as the *volatility* or *intermittency* process. The term intermittency comes from turbulence, and in that scientific field intermittency plays a key role, similar to that of (stochastic) volatility in finance.

In turbulence the basic notion of intermittency refers to the fact that the energy in a turbulent field is unevenly distributed in space and time. The present paper is part of a project with aim to construct a stochastic process model of the field of velocity vectors representing the fluid motion, conceiving of the intermittency as a positive random field with values $\sigma_t(x)$ at positions (x,t) in space-time. However, most extensive data sets on turbulent velocities only provide the time series of the main component (i.e. the component in the main direction of the fluid flow) of the velocity vector at a single location in space. In the present paper the focus is on this latter case, but in Sections 8 and 9 some discussion will be given on the further intriguing issues that arise when addressing tempo-spatial settings. For a detailed discussion of $\mathcal{BSS}$ and the more general concept of tempo-spatial ambit processes, in the context of turbulence modelling, we refer to Barndorff-Nielsen and Schmiegel (2004), Barndorff-Nielsen and Schmiegel (2007), Barndorff-Nielsen and Schmiegel (2008a), Barndorff-Nielsen and Schmiegel (2008b) and Barndorff-Nielsen and Schmiegel (2008c). There it is shown that such processes are able to reproduce main stylized facts of turbulent data.

In general, as we shall discuss in Section 3, models of the $\mathcal{BSS}$ form are not semimartingales. One consequence of this is that various useful techniques developed for semimartingales, such as the calculation of quadratic variation by Ito algebra and those of multipower variation, need extension or modification.

The recently established theory of multipower variation (Barndorff-Nielsen et al. (2006a), Barndorff-Nielsen et al. (2006b) and Jacod (2008a), cf. also Barndorff-Nielsen and Shephard (2003), Barndorff-Nielsen and Shephard (2004), Barndorff-Nielsen and Shephard (2006a), Barndorff-Nielsen and Shephard (2006b), Barndorff-Nielsen et al. (2006c) and Jacod (2008b)) was developed as a basis for inference on σ under $\mathcal{BSM}$ models and, more generally Ito semimartingales, with particular focus on inference about the integrated squared volatility σ^{2+} given by

$$\sigma_t^{2+} = \int_0^t \sigma_s^2 \mathrm{d}s. \tag{1.4}$$

In the present paper the focus is similarly on inference for σ_t^{2+}. Specifically we shall discuss to what extent (a suitable normalised version of) realised quadratic variation of Y can be used to estimate σ_t^{2+}.

It is important to realise that, as regards inference on σ^{2+}, there may be substantial differences between cases where g is positive on all of $(0,\infty)$ and those where $g(t) = 0$ for $t > l$ for some $l \in (0,\infty)$. This will be discussed in detail later.

In semimartingale theory the quadratic variation $[Y]$ of Y is defined in terms of the Ito integral $Y \bullet Y$, as $[Y] = Y^2 - 2Y \bullet Y$. In that setting $[Y]$ equals the limit in probability as $\delta \to 0$ of the *realised quadratic variation* $[Y_\delta]$ of Y defined by

$$[Y_\delta]_t = \sum_{j=1}^{\lfloor t/\delta \rfloor} \left(Y_{j\delta} - Y_{(j-1)\delta}\right)^2 \tag{1.5}$$

where $\lfloor t/\delta \rfloor$ is the largest integer smaller than or equal to t/δ.

However, the question of whether $[Y_\delta]$ has a limit in probability – and what that limit is – is of interest more broadly than for semimartingales, and in particular for $\mathcal{BSS}$ processes. For any process $Y = \{Y_t\}_{t\geq 0}$ we shall use $[Y.]$ to denote the limit, when it exists, i.e.

$$[Y.]_t = p\text{-}\lim_{\delta\to 0} [Y_\delta]_t \, .$$

Thus, in case $Y \in \mathcal{BSM}$ we have $[Y.] = [Y]$.[1]

We abbreviate realised quadratic variation to *RQV* and write *QV* for $[Y.]$.

The paper is organised as follows. Brownian semistationary processes are introduced in Section 2 and related non-semimartingale issues are considered in Section 3. Section 4 introduces a concept of q-orthogonality of stochastic processes and considers the computation of QV in some semimartingale difference cases. In Section 5 we turn to the increments of Brownian semistationary processes. Section 6 defines a normalised version $\overline{[Y_\delta]}$ of RQV, and Section 7 derives sufficient conditions for the convergence in probability of $\overline{[Y_\delta]}$ to σ^{2+}. Extensions to the tempo-spatial setting is discussed in Sections 8 and 9. Some indications of ongoing further work and open problems are given in the concluding Section 10.

2 $\mathcal{BSS}$ processes

We have defined the concept of Brownian semistationary processes ($\mathcal{BSS}$) as processes $Y = \{Y_t\}_{t\in\mathbb{R}}$ of the form

$$Y_t = \mu + \int_{-\infty}^{t} g(t-s)\sigma_s \mathrm{d}B_s + \int_{-\infty}^{t} q(t-s)a_s \mathrm{d}s \tag{2.1}$$

where, in the context of the present paper, the processes σ and a are taken to be stationary. The integrals in (2.1) are to be understood as limits in probability for $u \to -\infty$ of the integrals

$$\int_{u}^{t} g(t-s)\sigma_s \mathrm{d}B_s + \int_{u}^{t} q(t-s)a_s \mathrm{d}s$$

which are assumed to exist, the first defined for each fixed t as an Ito integral. This of course poses restrictions on which functions g and q are feasible, including square integrability of g.

The focus of the present paper is on inference about the integrated squared volatility σ^{2+} given by (1.4). In particular, we shall discuss to what extent realised quadratic variation of Y can be used to estimate σ_t^{2+}. Note that the relevant question here is whether a suitably normalised version of the realised quadratic variation, and not necessarily the realised quadratic variation itself, converges in probability and law.

[1] Of course, for semimartingales Y we have the more general result that

$$[Y]_t = p\text{-}\lim_{|\tau|\to 0} [Y_\tau]$$

where τ denotes a subdivision of $[0, t]$, $|\tau|$ is the maximal span in the subdivision, and Y_τ is the τ-discretisation of Y over the interval $[0, t]$.

As a modelling framework for continuous time stationary processes the specification (2.1) is quite general. In fact, the continuous time Wold–Karhunen decomposition says[1] that any second order stationary stochastic process, possibly complex valued, of mean 0 and continuous in quadratic mean can be represented as

$$Z_t = \int_{-\infty}^{t} \phi(t-s)\,\mathrm{d}\Xi_s + V_t$$

where

- the deterministic function ϕ is an, in general complex, deterministic square integrable function
- the process Ξ has orthogonal increments with $\mathrm{E}\{\,|\mathrm{d}\Xi_t|^2\,\} = \varpi \mathrm{d}t$ for some constant $\varpi > 0$
- the process V is nonregular (i.e. its future values can be predicted by linear operations on past values without error).

Under the further condition that $\cap_{t\in\mathbb{R}}\overline{\mathrm{sp}}\,\{Z_s : s \le t\} = \{0\}$, the function ϕ is real and uniquely determined up to a real constant of proportionality; and the same is therefore true of Ξ (up to an additive constant).

3 $\mathcal{BSS}$ and semi - nonsemimartingale issues

If $Y \in \mathcal{BSS}$ then Y may or may not be of the semimartingale type. This section discusses criteria for either of these cases.

3.1 Semimartingale cases

We begin by recalling a classical necessary and sufficient condition, due to Knight (1992), for the process Y to be a semimartingale, valid in the special simple situation where $\sigma = 1$ and $a = 0$, i.e. where the process is of the form

$$Y_t = \int_{-\infty}^{t} g(t-s)\,\mathrm{d}B_s. \tag{3.1}$$

Knight's theorem says that $(Y_t)_{t\ge 0}$ is a semimartingale in the $\left(\mathcal{F}_t^B\right)_{t\ge 0}$ filtration if and only if

$$g(t) = c + \int_0^t b(s)\,\mathrm{d}s \tag{3.2}$$

for some $c \in \mathbb{R}$ and a square integrable function b.

[1] See Doob (1953) and Karhunen (1950)

Example. An example of some particular interest is where

$$g(t) = t^{\alpha} e^{-\lambda t} \quad \text{for} \quad t \in (0, \infty)$$

and some $\lambda > 0$. In order for the integral (3.1) to exist, α is required to be greater than $-\frac{1}{2}$, and for g to be of the form (3.2) we must have $\alpha = 0$ or $\alpha > \frac{1}{2}$. In other words, the nonsemimartingale cases are $\alpha \in \left(-\frac{1}{2}, 0\right) \cup \left(0, \frac{1}{2}\right]$.

Generally, one may ask under what conditions moving average processes of the form

$$X_t = \int_{-\infty}^{\infty} \left(g(t-s) - h(-s) \right) \mathrm{d}B_s,$$

with g and h deterministic, are semimartingales. More specifically, when is $(X_t)_{t \geq 0}$ a $\left(\mathcal{F}_t^X\right)_{t \geq 0}$-seminartingale, where $\mathcal{F}_t^X$ is the σ-algebra generated by $\{X_s, s \leq t\}$. Constructive necessary and sufficient conditions for this are given in a recent paper by Basse, see Basse (2007a).

More generally still is the question of when a process X is a Gaussian semimartingale. Also for this case a necessary and sufficient criterion has been obtained by Basse, in Basse (2008), cf. also Basse (2007b) which discusses the spectral representation of Gaussian semimartingales.

At a further level of generalisation, Basse and Pedersen, in Basse and Pedersen (2008), consider processes X of the general form

$$X_t = \int_{-\infty}^{t} \left(\phi(t-s) - \psi(-s) \right) \mathrm{d}L_s$$

where L is a (two-sided) nondeterministic Lévy process with characteristic triplet $\left(\gamma, \sigma^2, \nu\right)$, ϕ and ψ are deterministic functions and the integral exists, in the sense of Rajput and Rosinski (1989). These authors establish various necessary conditions on $\left(\gamma, \sigma^2, \nu\right)$ and ϕ, ψ in order for $(X_t)_{t \geq 0}$ to be an $\left(\mathcal{F}_t^L\right)_{t \geq 0}$-semimartingale.

Now, turning to the general $\mathcal{BSS}$ case, we first argue formally, as if the differential of Y exists. From (2.1),

$$\mathrm{d}Y_t = g(0+) \sigma_t \mathrm{d}B_t + \{\dot{g} * \sigma \bullet B_t + q(0+) a_t + \dot{q} * a \bullet Leb_t\} \mathrm{d}t$$

suggesting that Y_t can be reexpressed as

$$Y_t = Y_0 + g(0+) \int_0^t \sigma_s \mathrm{d}B_s + \int_0^t A_s \mathrm{d}s$$

with

$$A = \dot{g} * \sigma \bullet B + q(0+) a + \dot{q} * a \bullet Leb.$$

This will indeed be the case provided the following conditions are satisfied (recall that we have assumed that σ and a are stationary):

(i) $g(0+)$ and $q(0+)$ exist and are finite.

(ii) g is absolutely continuous with square integrable derivative $\dot{g}$

(iii) The process $\dot{g}(-\cdot)\sigma_{\cdot}$ is square integrable

(iv) The process $\dot{q}(-\cdot)a_{\cdot}$ is integrable.

In view of the results by Knight and Basse, mentioned above, these conditions must be close to necessary as well.

We shall here not further discuss affirmative conditions for Y to be of the semimartingale type. Instead we turn to cases where Y can be written as a linear combination of semimartingales which are orthogonal, in a sense that will be specified, and have different filtrations.

4 RQV and linear combinations of semimartingales

While the focus will be on cases where a given $\mathcal{BSS}$ process Y can be rewritten as $Y^+ - Y^-$, where both Y^+ and Y^- are semimartingales, we begin by considering the broader issue of existence and calculation of $[Y_{\cdot}]$ when Y is a linear combination of q-orthogonal processes, q-orthogonality being defined below.

4.1 General considerations

Suppose that a process $Y = \{Y_t\}$ is representable in law as a linear combination $Y = Y' + Y''$ of some processes Y' and Y'' of interest, of semimartingale type or not. Then, defining $[Y'_\delta, Y''_\delta]$ and $[Y'_{\cdot}, Y''_{\cdot}]$ by

$$[Y'_\delta, Y''_\delta]_t = \sum_{j=1}^{\lfloor t/\delta \rfloor} \left(Y'_{j\delta} - Y'_{(j-1)\delta}\right)\left(Y''_{j\delta} - Y''_{(j-1)\delta}\right)$$

and

$$[Y'_{\cdot}, Y''_{\cdot}] = p\text{-}\lim_{\delta \to 0} [Y'_\delta, Y''_\delta]_t$$

we have

$$[Y_\delta] = [Y'_\delta] + [Y''_\delta] + 2\,[Y'_\delta, Y''_\delta]$$

and hence, provided the limit exists (in probability),

$$[Y_{\cdot}] = [Y'_{\cdot}] + [Y''_{\cdot}] + 2\,[Y'_{\cdot}, Y''_{\cdot}]\,.$$

We will write this symbolically as

$$d\,[Y_{\cdot}] = \mathrm{d}\,[Y'_{\cdot}] + \mathrm{d}\,[Y''_{\cdot}] + 2\mathrm{d}\,[Y'_{\cdot}, Y''_{\cdot}]\,.$$

In case $[Y'_{\cdot}, Y''_{\cdot}] = 0$ we say that Y' and Y'' are *q-orthogonal* and express this by writing

$$\mathrm{d}Y'\mathrm{d}Y'' = 0.$$

Then

$$[Y_{\cdot}] = [Y'_{\cdot}] + [Y''_{\cdot}] .$$

In particular, if Y' and Y'' are both semimartingales, in general with different own filtrations, and q-orthogonal then

$$[Y_{\cdot}] = [Y] = [Y'] + [Y'']$$

and $\mathrm{d}\,[Y_{\cdot}]$ may be calculated as

$$\mathrm{d}\,[Y_{\cdot}]_t = (\mathrm{d}Y'_t)^2 + (\mathrm{d}Y''_t)^2 .$$

In this case we may define $\mathrm{d}Y$ as $\mathrm{d}Y' + \mathrm{d}Y''$ and then, as in the usual semimartingale world, we have

$$[Y_{\cdot}]_t = \int_0^t (\mathrm{d}Y_s)^2 \,\mathrm{d}s.$$

An elementary instance of this $Y_t = Y'_t + Y''_t$ with $Y'_t = B_t$ and $Y''_t = -B_{t-1}$ and where $B = \{B_t\}_{\mathbb{R}}$ is Brownian motion on the real line.

These considerations are extendable to settings where Y is a linear combination $Y = \int Y_t^{(c)} M\,(\mathrm{d}c)$ of mutually q-orthogonal processes $Y^{(c)}$ and where M is a deterministic, possibly signed, measure. We shall not here discuss specific general conditions for this; however an example is given in the next subsection.

4.2 Some $\mathcal{BSS}$ cases

Let $\mathfrak{G}$ be the class of functions of the form (3.2). If $g \in \mathfrak{G}$ then for any $u > 0$ the function $h\,(\cdot) = g\,(\cdot + u)$ also belongs to $\mathfrak{G}$. This has the important consequence that if g is of the form $g^{\#} 1_A$ with $A = (0, l)$ for some $l > 0$ and $g^{\#} \in \mathfrak{G}$ then Y itself is not a semimartingale but it is the difference between two semimartingales, specifically

$$Y_t = Y_t^+ - Y_t^-$$

where

$$Y_t^+ = \mu + \int_{-\infty}^t g\,(t-s)\,\sigma_s \mathrm{d}B_s + q * a \bullet Leb$$

and

$$Y_t^- = \int_{-\infty}^t g\,(t-s+l)\,\sigma_{s-l} \mathrm{d}B_{s-l}.$$

Both Y^+ and Y^- are semimartingales but with different filtrations, namely $\{\mathcal{F}_t^B\}_{t\in\mathbb{R}}$ and $\{\mathcal{F}_{t-l}^B\}_{t\in\mathbb{R}}$. Moreover, Y^+ and Y^- are q-orthogonal, and hence

$$\mathrm{d}\,[Y_{\cdot}]_t = \left(\mathrm{d}Y_t^+\right)^2 + \left(\mathrm{d}Y_t^-\right)^2 .$$

More generally, suppose that g has the form

$$g\,(t) = \int_0^\infty g_0\,(t-c)\,\mathrm{d}M\,(c)$$

for a $g_0 \in \mathfrak{G}$ and where M is a function of bounded variation on $\mathbb{R}_+$. In this case we have

$$\begin{aligned} Y_t &= \int_{-\infty}^{t} g(t-s)\,\sigma_s \mathrm{d}B_s \\ &= \int_{-\infty}^{t}\int_{0}^{\infty} g_0(t-s-c)\,\mathrm{d}M(c)\,\sigma_s \mathrm{d}B_s \\ &= \int_{0}^{\infty}\int_{-\infty}^{t} g_0(t-s-c)\,\sigma_s \mathrm{d}B_s \mathrm{d}M(c) \\ &= \int_{0}^{\infty}\int_{-\infty}^{t-c} g_0(t-c-s)\,\sigma_s \mathrm{d}B_s \mathrm{d}M(c) \\ &= \int_{0}^{\infty} Y_t^{(c)} \mathrm{d}M(c) \end{aligned}$$

where

$$Y_t^{(c)} = \int_{-\infty}^{t} g_0(t-s)\,\sigma_{s-c} \mathrm{d}B_{s-c}$$

showing that Y is a linear combination of q-orthogonal semimartingales with different filtrations (namely, conditional on σ the filtration of $Y^{(c)}$ is $\{\mathcal{F}_{t-c}^B\}_{t\in\mathbb{R}}$).

5 Increment processes

Again, suppose that $g = g^{\#}1_{[0,l]}$ for some $l > 0$ and $g^{\#} \in \mathfrak{G}$. For any given t we define the increment process $\{X_{u|t}\}_{u\geq 0}$ by

$$\begin{aligned} X_{t+u|t} &= Y_{t+u} - Y_t \\ &= \int_{t}^{t+u} g(t+u-s)\,\sigma_s \mathrm{d}B_s + \int_{-\infty}^{t} \{g(t+u-s) - g(t-s)\}\,\sigma_s \mathrm{d}B_s \\ &\quad + \int_{t}^{t+u} q(t+u-s) a_s \mathrm{d}s + \int_{-\infty}^{t} \{q(t+u-s) - q(t-s)\} a_s \mathrm{d}s \end{aligned}$$

It will be convenient to rewrite $X_{t|t-u}$ as

$$X_{t|t-u} = \int_{-\infty}^{0} \phi_u(-v)\,\sigma_{v+t} \mathrm{d}B_{v+t} + \int_{-\infty}^{0} \chi_u(-v)\, a_{v+t} \mathrm{d}v \tag{5.1}$$

where ϕ_u and χ_u are defined by

$$\phi_u(v) = \begin{cases} g(v) & \text{for } 0 \leq v < u \\ g(v) - g(v-u) & \text{for } u \leq v < \infty \end{cases}$$

and

$$\chi_u(v) = \begin{cases} q(v) & \text{for } 0 \leq v < u \\ q(v) - q(v-u) & \text{for } u \leq v < \infty\,. \end{cases}$$

From now on we assume that (σ, a) is independent of B and that a is adapted to the filtration $\mathcal{F}^\sigma$. This, together with (5.1), implies in particular that the conditional variance of $Y_t - Y_{t-u}$ given the process σ takes the form

$$\mathrm{E}\left\{(Y_t - Y_{t-u})^2 \mid \sigma\right\} = \int_0^\infty \psi_u(v)\,\sigma_{t-v}^2 \mathrm{d}v + \left(\int_0^\infty \chi_u(v)\,a_{t-v}\mathrm{d}v\right)^2$$

where

$$\psi_u(v) = \begin{cases} g^2(v) & \text{for } 0 \leq v < u \\ \{g(v-u) - g(v)\}^2 & \text{for } u \leq v < \infty\,. \end{cases}$$

Remark 5.1. Note that $\phi_u(v) = \psi_u(v) = \chi_u(v) = 0$ for $v \geq l + u$ while for $l \leq v < l + u$ we have $\psi_u(v) = g(v-u)^2$ and $\chi_u(v) = q(v-u)$.

Let

$$c(u) = \int_0^\infty \psi_u(v)\,\mathrm{d}v. \tag{5.2}$$

Remark 5.2. Trivially,

$$c(\delta) \geq \int_0^\delta \psi_\delta(v)\,\mathrm{d}v = \int_0^\delta g^2(v)\,\mathrm{d}v$$

implying that if $g(0+) > 0$ then $c(\delta)$ cannot tend to 0 faster than δ.

Remark 5.3. We have

$$c(u) = 2\,\|g\|^2\,\bar{r}(u)$$

where $\bar{r} = 1 - r$ with r being the autocorrelation function of Y. Furthermore,

$$\begin{aligned} \mathrm{E}\left\{(Y_t - Y_{t-u})^2\right\} &= \mathrm{E}\left\{\sigma_0^2\right\} c(u) \\ &\quad + \mathrm{E}\left\{a_0^2\right\} \int_0^\infty \int_0^\infty \chi_u(v)\,\chi_u(w)\,\varrho(|v-w|)\,\mathrm{d}v\,\mathrm{d}w \end{aligned}$$

where ϱ is the autocorrelation function of a.

6 Normalised RQV

We now define the *normalised RQV* as

$$\overline{[Y_\delta]} = \frac{\delta}{c(\delta)}\,[Y_\delta]\,.$$

The question we wish to address here is whether and under what conditions $\overline{[Y_\delta]}$ converges in probability to σ^{2+}. Concerning the related question of a central limit theorem for $\overline{[Y_\delta]}$, see Section 10.

In the present paper we shall largely restrict the discussion to quite regular forms of the weight function g, assuming in particular that g is positive on a finite interval $(0,l)$ only. Specifically, we now assume that the function g is positive, continuously differentiable, convex and decreasing on an interval $(0,l)$ where $0<l<\infty$ and that $g(t)=0$ outside that interval. Also, we require that σ and a are stationary and càdlàg and, as before, that a is adapted to the natural filtration of σ. Without loss of generality we take t/δ to be an integer n so that $t=n\delta$. Below C denotes a constant that is independent of n but whose value may change with the context.

7 Consistency

To discuss the question of when $\overline{[Y_\delta]} \xrightarrow{p} \sigma^{2+}$ we first note that, by (5.1),

$$
\begin{aligned}
[Y_\delta]_t &= \sum_{k=1}^{n}\left(\int_{-\infty}^{0}\phi_\delta(-v)\,\sigma_{v+k\delta}\,\mathrm{d}B_{v+k\delta}\right)^2 \\
&\quad +2\sum_{k=1}^{n}\int_{-\infty}^{0}\phi_\delta(-v)\,\sigma_{v+k\delta}\,\mathrm{d}B_{v+k\delta}\int_{-\infty}^{0}\chi_\delta(-v)\,a_{v+k\delta}\,\mathrm{d}v \\
&\quad +\sum_{k=1}^{n}\left(\int_{-\infty}^{0}\chi_\delta(-v)\,a_{v+k\delta}\,\mathrm{d}v\right)^2.
\end{aligned}
$$

It follows that

$$
\mathrm{E}\left\{\overline{[Y_\delta]_t}\,|\,\sigma\right\}=\int_0^\infty\left\{\delta\sum_{k=1}^{n}\sigma_{k\delta-v}^2\right\}\pi_\delta(\mathrm{d}v)+c(\delta)^{-1}D_\delta(a) \tag{7.1}
$$

where

$$
\pi_\delta(\mathrm{d}v)=\frac{\psi_\delta(v)}{c(\delta)}\mathrm{d}v
$$

and

$$
D_\delta(a)=\int_0^\infty\int_0^\infty\chi_\delta(v)\,\chi_\delta(w)\left\{\delta\sum_{k=1}^{n}a_{k\delta-v}a_{k\delta-w}\right\}\mathrm{d}v\,\mathrm{d}w.
$$

Thus π_δ is an absolutely continuous probability measure on $(0,l+\delta)$. Furthermore,

$$
|D_\delta(a)|\le C\left(\int_0^\infty|\chi_\delta(v)|\,\mathrm{d}v\right)^2
$$

where the constant C depends on a, l and t.

This leads us to introduce

Condition A. $c(\delta)^{-1}\left(\int_0^\infty |\chi_\delta(v)|\,\mathrm{d}v\right)^2 \to 0.$

Remark 7.1. Note that in this connection if q is a positive decreasing function then

$$\int_0^\infty |\chi_\delta(v)|\,\mathrm{d}v = 2\int_0^\delta q(v)\,\mathrm{d}v. \tag{7.2}$$

Suppose that π_δ converges weakly, as $\delta \to 0$, to a probability measure π on $[0, l]$, i.e.

$$\pi_\delta \xrightarrow{w} \pi. \tag{7.3}$$

Then, if Condition A holds we obtain from (7.1) that

$$\mathrm{E}\left\{\overline{[Y_\delta]}_t \,|\, \sigma\right\} \to \int_0^\infty \left(\sigma_{t-v}^{2+} - \sigma_{-v}^{2+}\right)\pi(\mathrm{d}v). \tag{7.4}$$

In particular, if $\pi = \delta_0$, the delta measure at 0, then

$$\mathrm{E}\left\{\overline{[Y_\delta]}_t \,|\, \sigma\right\} \to \sigma_t^{2+} \tag{7.5}$$

where

$$\sigma_t^{2+} = \int_0^t \sigma_s^2\,\mathrm{d}s.$$

The following two subsections derive sufficient conditions for $\pi_\delta \xrightarrow{w} \delta_0$ and for $\mathrm{Var}\left\{\overline{[Y_\delta]}_t \,|\, \sigma\right\} \to 0$, respectively. These two relations together with Condition A imply that

$$\overline{[Y_\delta]} \xrightarrow{p} \sigma^{2+}. \tag{7.6}$$

We will refer to the case where (7.6) is satisfied by saying that the model for Y is *volatility memoryless*.

7.1 Pidelta to pi

Suppose that $l < \infty$ and let

$$\Psi_\delta(u) = \int_0^u \psi_\delta(v)\,\mathrm{d}v \quad \text{and} \quad \bar{\Psi}_\delta(u) = \int_{l+\delta-u}^{l+\delta} \psi_\delta(v)\,\mathrm{d}v,$$

so that $c(\delta)^{-1}\Psi_\delta$ is the distribution function, say Π_δ, of π_δ.

Next, for $k = 1, 2, \ldots$, let

$$\begin{aligned} c_k(\delta) &= \int_{(k-1)\delta}^{k\delta} \psi_\delta(u)\,\mathrm{d}u \\ &= \int_0^\delta \psi_\delta((k-1)\delta + u)\,\mathrm{d}u \\ &= \delta\int_0^1 \psi_\delta((k-1+u)\delta)\,\mathrm{d}u \end{aligned}$$

i.e.

$$c_k(\delta) = \delta \int_0^1 \{g((k-2+u)\delta) - g((k-1+u)\delta)\}^2 \mathrm{d}u. \tag{7.7}$$

We must now distinguish between the cases $t < l$ and $t \geq l$.

Suppose first that $t \geq l$. Let $k^* = \max\{k : k\delta \leq l\}$. Then, by (7.7), for $1 < k \leq k^*$

$$c_k(\delta) = \delta^3 \int_0^1 g'((k-2+u+\theta_k(u))\delta)^2 \mathrm{d}u$$

where the $\theta_k(u)$ satisfy $0 \leq \theta_k(u) \leq 1$. Since g is convex and decreasing this implies, provided $k_* \leq k \leq k^*$ where $k_* > 2$, that

$$c_k(\delta) \leq \delta^3 g'((k-2)\delta)^2 \leq \delta^3 g'((k_*-2)\delta)^2.$$

Therefore, for any $\varepsilon \in (2\delta, l)$ with $1 < \lfloor \varepsilon/\delta \rfloor < k^*$ we have

$$\begin{aligned}
\Psi_\delta(k^*\delta) - \Psi_\delta(\varepsilon) &\leq \sum_{k=\lfloor \varepsilon/\delta \rfloor + 1}^{k^*} c_k(\delta) \\
&\leq \delta^3 (k^* - \lfloor \varepsilon/\delta \rfloor) g'(\lfloor \varepsilon - 2\delta \rfloor)^2 \\
&\leq (l - \varepsilon + \delta) g'(\lfloor \varepsilon - 2\delta \rfloor)^2 \delta^2
\end{aligned}$$

so that

$$\Pi_\delta(k^*\delta) - \Pi_\delta(\varepsilon) \leq (l - \varepsilon + \delta) g'(\lfloor \varepsilon - 2\delta \rfloor)^2 \delta^2 c(\delta)^{-1}.$$

Consequently, as $\delta \to 0$,

$$\Pi_\delta(k^*\delta) - \Pi_\delta(\varepsilon) \to 0.$$

It follows that if π_δ converges to a probability measure π then π is necessarily a linear combination of the delta measures at 0 and l.

Furthermore,

$$\Psi_\delta(l+\delta) - \Psi_\delta(k^*\delta) = c_{k^*+1}(\delta) + c_{k^*+2}(\delta)$$

where

$$\begin{aligned}
c_{k^*+1}(\delta) &= \int_{k^*\delta}^{(k^*+1)\delta} \psi_\delta(v) \mathrm{d}v \\
&= \int_{k^*\delta}^{l} \{g(v-\delta) - g(v)\}^2 \mathrm{d}v \\
&\quad + \int_l^{(k^*+1)\delta} g^2(v-\delta) \mathrm{d}v
\end{aligned}$$

and

$$c_{k^*+2}(\delta) = \int_{(k^*+1)\delta}^{l+\delta} g^2(v-\delta) \mathrm{d}v.$$

So, combining, for $\pi_\delta \to \delta_0$ to hold we must require that

$$c(\delta)^{-1} \int_{k^*\delta}^{l} \{g(v-\delta) - g(v)\}^2 \, \mathrm{d}v \to 0$$

and

$$c(\delta)^{-1} \int_{l}^{l+\delta} g^2(v-\delta) \, \mathrm{d}v \to 0.$$

But the first relation follows from the smoothness of g, so to guarantee $\pi_\delta \xrightarrow{w} \delta_0$, when $t \geq l$, we therefore only need to add

Condition B. $c(\delta)^{-1} \int_l^{l+\delta} g^2(v-\delta) \, \mathrm{d}v \to 0 \quad$ as $\delta \downarrow 0$.

Remark 7.2. Condition B is equivalently to having

$$\frac{\int_{l-\delta}^{l} g^2(v)\mathrm{d}v}{\int_0^{\delta} g^2(v)\mathrm{d}v} \to 0,$$

as follows from the above discussion. In particular, it suffices to have $g(v) \to 0$ as $v \uparrow l$.

Remark 7.3. In case $c(\delta)^{-1} \int_l^{l+\delta} g^2(v-\delta) \, \mathrm{d}v \to \lambda \in (0,1)$ we obtain $\pi_\delta \to (1-\lambda)\delta_0 + \lambda\delta_1$.

When $t < l$, for any $\varepsilon \in (2\delta, t)$ with $1 < \lfloor \varepsilon/\delta \rfloor < n$,

$$\begin{aligned} \Psi_\delta(t+\delta) - \Psi_\delta(\varepsilon) &\leq \sum_{k=\lfloor \varepsilon/\delta \rfloor + 1}^{n} c_k(\delta) \\ &\leq (t-\varepsilon+\delta)\, g'(\lfloor \varepsilon - 2\delta \rfloor)^2 \delta^2 \end{aligned}$$

which tends to 0 at the order of δ^2. To obtain $\pi_\delta \xrightarrow{w} \delta_0$ we therefore only need to add the assumption that

$$\Psi_\delta(l+\delta) - \Psi_\delta(t+\delta) = o(c(\delta)). \tag{7.8}$$

Now,

$$\Psi_\delta(l+\delta) - \Psi_\delta(t+\delta) = \sum_{k=n+1}^{\infty} c_k(\delta).$$

Thus, letting

$$\bar{c}_k(\delta) = \frac{c_k(\delta)}{c(\delta)} \tag{7.9}$$

we have that (7.8) is implied by

Condition C. $\sum_{k=n+1}^{\infty} \bar{c}_k(\delta) \to 0 \quad \text{as} \quad \delta \downarrow 0.$

7.2 Conditional Var to 0

We now establish conditions under which the conditional variance of the normalised realised quadratic variation tends to 0 as $\delta \to 0$, i.e.

$$\operatorname{Var}\{\overline{[Y_\delta]_t} \,|\, \sigma\} \to 0. \tag{7.10}$$

Suppose first that $a = 0$.

Let $\Delta_j^n Y = Y_{j\delta} - Y_{(j-1)\delta}$. Then

$$\begin{aligned}\operatorname{Var}\{\overline{[Y_\delta]_t} \,|\, \sigma\} = \frac{\delta^2}{c(\delta)^2}\Bigg\{ & \sum_{j=1}^{n} \operatorname{Var}\{(\Delta_j^n Y)^2 \,|\, \sigma\} \\ & + 2\sum_{j=1}^{n}\sum_{k=j+1}^{n} \operatorname{Cov}\{(\Delta_j^n Y)^2, (\Delta_k^n Y)^2 \,|\, \sigma\}\Bigg\}\end{aligned}$$

where, for $j < k$,

$$\begin{aligned}\operatorname{Cov}\{\Delta_j^n Y \Delta_k^n Y \,|\, \sigma\} &= \mathrm{E}\left\{\left(Y_{j\delta} - Y_{(j-1)\delta}\right)\left(Y_{k\delta} - Y_{(k-1)\delta}\right) \,|\, \sigma\right\} \\ &= \int_0^\infty \phi_\delta\left((k-j)\delta + u\right)\phi_\delta(u)\,\sigma_{j\delta-u}^2 \,\mathrm{d}u.\end{aligned}$$

Let $K(\sigma) = \sup_{-l \le s \le t} \sigma_s^2$. As σ is assumed càdlàg, $K(\sigma) < \infty$ a.s.. Hence, by the Cauchy–Schwarz inequality,

$$\begin{aligned}&\left|\operatorname{Cov}\{Y_{j\delta} - Y_{(j-1)\delta}, Y_{k\delta} - Y_{(k-1)\delta} \,|\, \sigma\}\right| \\ &\le K(\sigma)\left(\int_0^\infty \psi_\delta(u)\,\mathrm{d}u\right)^{1/2}\left(\int_{(k-j)\delta}^\infty \psi_\delta(u)\,\mathrm{d}u\right)^{1/2}.\end{aligned}$$

Now, recall that for any pair X and Y of normal, mean zero random variables we have

$$\operatorname{Cov}\{X^2, Y^2\} = 2\operatorname{Cov}\{X, Y\}^2. \tag{7.11}$$

Therefore

$$\begin{aligned}\operatorname{Var}\{\overline{[Y_\delta]_t} \,|\, \sigma\} &\le 2K(\sigma)^2 \frac{\delta^2}{c(\delta)^2}\left(l\delta^{-1}c(\delta)^2 + 2c(\delta)\sum_{j=1}^{n-1}\sum_{i=j+1}^{n}\int_{(i-j)\delta}^\infty \psi_\delta(u)\,\mathrm{d}u\right) \\ &= 2K(\sigma)^2\delta\left(l + 2\frac{\delta}{c(\delta)}\sum_{j=1}^{n-1}\sum_{i=j+1}^{n}\int_{(i-j)\delta}^\infty \psi_\delta(u)\,\mathrm{d}u\right)\end{aligned}$$

Here

$$\begin{aligned}
\sum_{j=1}^{n-1}\sum_{i=j+1}^{n}\int_{(i-j)\delta}^{\infty}\psi_{\delta}(u)\,\mathrm{d}u &= \sum_{j=1}^{n-1}\sum_{i=1}^{n-j}\int_{i\delta}^{\infty}\psi_{\delta}(u)\,\mathrm{d}u \\
&= \sum_{\nu=1}^{n-1}\sum_{i=1}^{\nu}\sum_{k=i}^{\infty}c_{k+1}(\delta) \\
&= \sum_{\nu=1}^{n-1}\sum_{k=1}^{\infty}c_{k+1}(\delta)\sum_{i=1}^{\nu}1_{\leq k}(i) \\
&= \sum_{\nu=1}^{n-1}\left(\sum_{k=1}^{\nu}kc_{k+1}(\delta)+\nu\sum_{k=\nu}^{\infty}c_{k+2}(\delta)\right) \\
&= \sum_{k=1}^{n-1}(n-k)\,kc_{k+1}(\delta)+\sum_{k=1}^{\infty}c_{k+2}(\delta)\sum_{\nu=1}^{k\wedge(n-1)}\nu \\
&= \sum_{k=1}^{n-1}\left\{(n-k)\,kc_{k+1}(\delta)+\frac{1}{2}(k+1)\,kc_{k+2}\right\} \\
&\quad+\frac{(n-1)n}{2}\sum_{k=n}^{\infty}c_{k+2}(\delta).
\end{aligned}$$

With the notation (7.9) we thus have

$$\begin{aligned}
\mathrm{Var}\{\overline{[Y_{\delta}]}_t\,|\,\sigma\} &\leq 2K(\sigma)^2\,\delta\Bigg(l+2\delta\sum_{k=1}^{n-1}(n-k)\,k\bar{c}_{k+1}(\delta) \\
&\qquad +2\delta\sum_{k=1}^{\infty}\bar{c}_{k+2}(\delta)\sum_{\nu=1}^{k\wedge(n-1)}\nu\Bigg) \\
&= 2K(\sigma)^2\,l\delta \\
&\quad +2K(\sigma)^2\,\delta^2\sum_{k=1}^{n-1}\left\{(n-k)\,k\bar{c}_{k+1}(\delta)+\frac{1}{2}(k+1)\,k\bar{c}_{k+2}\right\} \\
&\quad +2\delta K(\sigma)^2\,\frac{(n-1)\,n}{2}\sum_{k=n}^{\infty}\bar{c}_{k+2}(\delta).
\end{aligned}$$

Here

$$\delta^2\sum_{k=1}^{n-1}\left\{(n-k)\,k\bar{c}_{k+1}(\delta)+\frac{1}{2}(k+1)\,k\bar{c}_{k+2}\right\}\leq C\delta\sum_{k=1}^{n}k\bar{c}_k(\delta)$$

and

$$\delta^2\frac{(n-1)\,n}{2}\sum_{k=n}^{\infty}\bar{c}_{k+2}(\delta)\leq C\sum_{k=n+1}^{\infty}\bar{c}_k(\delta).$$

Consequently, when $a = 0$, for (7.10) to be valid it suffices to have

$$\delta \sum_{k=1}^{n} k\bar{c}_k(\delta) \to 0 \quad \text{and} \quad \sum_{k=n+1}^{\infty} \bar{c}_k(\delta) \to 0. \tag{7.12}$$

Condition C will ensure the second limit result, and we now introduce

Condition D. $\delta \sum_{k=1}^{n} k\bar{c}_k(\delta) \to 0 \quad \text{as} \quad \delta \downarrow 0.$

Provided $a = 0$, for (7.10) to be valid it suffices that Conditions C and D to hold.

Next we show that the convergence $\mathrm{Var}\{\overline{[Y_\delta]_t} \mid \sigma\} \to 0$ also holds if a is not 0 provided Condition A is fulfilled too. In case $a \neq 0$, $\mathrm{Var}\{\overline{[Y_\delta]_t} \mid \sigma\}$ is a sum of two terms, one as above for $a = 0$ while the other is

$$4\frac{\delta^2}{c(\delta)^2} \sum_{k=1}^{n} \int_0^\infty \psi_\delta(v)\, \sigma_{k\delta-v}^2 \,\mathrm{d}v \left(\int_0^\infty \chi_\delta(v)\, a_{k\delta-v} \,\mathrm{d}v \right)^2 \tag{7.13}$$

which is bounded above by $4HK$ where

$$H = \limsup_{k,\delta} \frac{\delta}{c(\delta)} \int_0^\infty \psi_\delta(v)\, \sigma_{k\delta-v}^2 \,\mathrm{d}v$$

and

$$K = \frac{\delta}{c(\delta)} \sum_{k=1}^{n} \left(\int_0^\infty \chi_\delta(v)\, a_{k\delta-v} \,\mathrm{d}v \right)^2.$$

Here

$$\int_0^\infty \psi_\delta(v)\, \sigma_{k\delta-v}^2 \,\mathrm{d}v \leq Cc(\delta)$$

where the constant C depends on t and σ. Hence $H \to 0$. Furthermore,

$$\begin{aligned} \sum_{k=1}^{n} \left(\int_0^\infty \chi_\delta(v)\, a_{k\delta-v} \,\mathrm{d}v \right)^2 &\leq C \sum_{k=1}^{n} \left(\int_0^\infty |\chi_\delta(v)| \,\mathrm{d}v \right)^2 \\ &= C\delta^{-1} \left(\int_0^\infty |\chi_\delta(v)| \,\mathrm{d}v \right)^2 \end{aligned}$$

where C, again, depends on t and a. Hence Condition A implies $K \to 0$.

7.3 Summing up

Suppose first that $t < l < \infty$, which is the most interesting case from the viewpoint of turbulence modelling. If

$$c(\delta)^{-1} \left(\int_0^\infty |\chi_\delta(v)| \,\mathrm{d}v \right)^2 \to 0 \tag{7.14}$$

$$\sum_{k=n+1}^{\infty} \bar{c}_k(\delta) \to 0 \tag{7.15}$$

and

$$\delta \sum_{k=1}^{n-1} k\bar{c}_k(\delta) \to 0 \tag{7.16}$$

then

$$\pi_\delta \to \delta_0, \mathrm{Var}\{\overline{[Y_\delta]} \mid \sigma\} \to 0 \text{ and } \overline{[Y_\delta]} \xrightarrow{p} \sigma^{2+}. \tag{7.17}$$

If $l \leq t$ then the additional assumption that

$$\frac{\int_{l-\delta}^{l} g^2(v)\,\mathrm{d}v}{\int_0^\delta g^2(v)\,\mathrm{d}v} \to 0 \tag{7.18}$$

is required. The latter is, in particular, fulfilled if $g(v) \to 0$ for $v \uparrow l$. In case (7.18) is violated but (7.14), (7.15) and (7.16) hold and $\pi_\delta \xrightarrow{w} \pi$ for some π, necessarily of the form $\pi = \lambda\delta_0 + (1-\lambda)\delta_l$ for some $\lambda \in (0,1)$, then

$$\overline{[Y_\delta]}_t \xrightarrow{p} \lambda\sigma_t^{2+} + (1-\lambda)\left(\sigma_{t-l}^{2+} - \sigma_{-l}^{2+}\right). \tag{7.19}$$

7.4 Examples

Recall Conditions A–D:

$$c(\delta)^{-1}\left(\int_0^\infty |\chi_\delta(v)|\,\mathrm{d}v\right)^2 \to 0 \tag{7.20}$$

$$c(\delta)^{-1}\int_l^{l+\delta} g^2(v-\delta)\,\mathrm{d}v \to 0$$

$$\sum_{k=n+1}^{\infty} \bar{c}_k(\delta) \to 0 \tag{7.21}$$

$$\delta \sum_{k=1}^{n} k\bar{c}_k(\delta) \to 0. \tag{7.22}$$

In this section we suppose that $q = g$. Then Condition A has the form

$$c(\delta)^{-1} c_1(\delta)^2 \to 0. \tag{7.23}$$

Example. Suppose that $t = l$ and $g(v) = e^{-\lambda v} 1_{(0,l)}(v)$ (a non-semimartingale case). Then

$$\psi_\delta(v) = e^{-2\lambda v}\begin{cases} 1 & \text{for} \quad 0 \leq v < \delta \\ \left(e^{\lambda\delta}-1\right)^2 & \text{for} \quad \delta \leq v < l \\ e^{2\lambda\delta} & \text{for} \quad l \leq v < l+\delta \\ 0 & \text{for} \quad l+\delta \leq v. \end{cases}$$

Here we find

$$c_1(\delta) = \frac{1}{2\lambda}\left(1 - e^{-2\lambda\delta}\right) \sim \delta$$

while for $k = 2, \ldots, n$

$$c_k(\delta) = \frac{1}{2\lambda}\left(e^{\lambda\delta} - 1\right)^3 e^{-2k\lambda} \sim \frac{\lambda^2}{2} e^{-2k\lambda}\delta^3 .$$

Moreover we have

$$c_{n+1}(\delta) = \frac{1}{2\lambda}\left(1 - e^{-2\lambda\delta}\right) \sim e^{-2\lambda l}\delta ,$$

whereas $c_k(\delta) = 0$ for $k > n+1$. Finally, $c(\delta) \sim \delta(1 + e^{-2\lambda l})$ and

$$\overline{c}_{n+1}(\delta) c(\delta)^{-1} \int_l^{l+\delta} g^2(v - \delta)\,\mathrm{d}v \to \left(1 + e^{2\lambda l}\right)^{-1} .$$

So, Conditions A, C and D are met. But Condition B is not and we have that $\pi_\delta \to \pi$, where

$$\pi = \frac{1}{1 + e^{-2\lambda l}}\delta_0 + \frac{1}{1 + e^{2\lambda l}}\delta_1,$$

and thus

$$\overline{[Y_\delta]} \xrightarrow{p} \sigma_t^{2+} - \left(1 + e^{2\lambda l}\right)^{-1} \sigma_{-t}^{2+} .$$

Example. Let $g(v) = v^\alpha (1-v)^\beta 1_{(0,1)}(v)$ with $-\frac{1}{2} < \alpha$ and $\beta \geq 1$. The first inequality ensures existence of the stochastic integral $g * \sigma \bullet B$, and if $\alpha < 0$ then we are in the nonsemimartingale situation. In showing that $\pi_\delta \to \delta_0$ and $\overline{[Y_\delta]} \xrightarrow{p} \sigma^{2+}$ it suffices to consider the case where $-\frac{1}{2} < \alpha < 0$, $\beta = 1$ and $n\delta = t$. Let $\gamma = -\alpha$, and suppose $t < 1$.

We find

$$\begin{aligned} c_0(\delta) &= \int_0^\delta u^{-2\gamma}(1-u)^2\,\mathrm{d}u \\ &= (1-2\gamma)^{-1}\delta^{1-2\gamma}(1 + O(\delta)) \end{aligned}$$

and, for $k = 1, 2, \ldots, n-1$,

$$\begin{aligned} c_k(\delta) = \delta \int_0^1 \Big[&((k+u)\delta)^{-\gamma} - ((k+u)\delta)^{1-\gamma} \\ &- ((k+u-1)\delta)^{-\gamma} + ((k+u-1)\delta)^{1-\gamma} \Big]^2 \mathrm{d}u \\ = \delta^{1-2\gamma} \int_0^1 \Big[&(k+u)^{-\gamma} - (k+u-1)^{-\gamma} \\ &- \delta \left\{ (k+u)^{1-\gamma} - (k+u-1)^{1-\gamma} \right\} \Big]^2 \mathrm{d}u \end{aligned}$$

while

$$\begin{aligned} c_n(\delta) &= \delta^{3-2\gamma}\int_0^1\left[(n+u)^{1-\gamma}-(n+u-1)^{1-\gamma}\right]^2 du \\ &= \delta^{3-2\gamma}n^{2-2\gamma}\int_0^1\left[\left(1+\frac{u}{n}\right)^{1-\gamma}-\left(1+\frac{u-1}{n}\right)^{1-\gamma}\right]^2 du \\ &= O\left(\delta^2\right) \end{aligned}$$

and $c_k(\delta)=0$ when $k>n$. It follows, in particular, that

$$c_1(\delta)=O\left(\delta^{1-2\gamma}\right);$$

furthermore, since for $1<k<n$ and $0\le u\le 1$

$$\left|(k+u)^{-\gamma}-(k+u-1)^{-\gamma}\right|\le\gamma(k-1)^{-\gamma-1}$$

and

$$\left|(k+u)^{1-\gamma}-(k+u-1)^{1-\gamma}\right|\le(1-\gamma)(k-1)^{-\gamma}$$

we have (when $\delta<1$)

$$\begin{aligned} c_k(\delta) &\le \delta^{1-2\gamma}(k-1)^{-2\gamma-2}\left[\gamma+(1-\gamma)\delta(k-1)\right]^2 \\ &\le \delta^{1-2\gamma}k^{-2\gamma-2}\left(1-\frac{1}{k}\right)^{-2\gamma-2} \\ &\le C\delta^{1-2\gamma}k^{-2\gamma-2}. \end{aligned}$$

Consequently,

$$c(\delta)=O\left(\delta^{1-2\gamma}\right)$$

while for $1<k<n$

$$k\bar{c}_k(\delta)\le Ck^{-2\gamma-1}$$

so that

$$\sum_1^{n-1}k\bar{c}(\delta)\le C.$$

We conclude that the Conditions A–D are satisfied and hence that $\overline{[Y_\delta]}_t\xrightarrow{p}\sigma_t^{2+}$.

8 Tempo-spatial setting

Above only the case of time-wise behaviour at a single point in space was considered. In the real turbulence setting, space and the velocity vector are three dimensional. The general modelling framework specifies the velocity and intermittency fields as

$$
\begin{aligned}
Y_t(x) &= \mu + \int_{A_t(x)} g(t-s, |\xi - x|)\, \sigma_s(\xi)\, W(\mathrm{d}\xi, \mathrm{d}s) \\
&\quad + \int_{C_t(x)} q(t-s, |\xi - x|)\, a_s(\xi)\, \mathrm{d}\xi\, \mathrm{d}s
\end{aligned}
$$

and

$$
\sigma_t^2(x) = \int_{D_t(x)} h(t-s, |\xi - x|)\, L(\mathrm{d}\xi, \mathrm{d}s)
$$

Here Y_t is a vector process of dimension d ($d = 0, 1, 2$ or 3), g, q and h are deterministic matrices of dimension $d \times k$, $\sigma_s(\xi) \geq 0$ and $a_s(\xi)$ are random field matrices of dimension $k \times m$ on $\mathbb{R}^3 \times \mathbb{R}$, W is an m-dimensional white noise on $\mathbb{R}^3 \times \mathbb{R}$, L is an m-dimensional nonnegative Lévy basis or exponential of a Lévy basis on $\mathbb{R}^3 \times \mathbb{R}$, and $A_t(x)$, $C_t(x)$ and $D_t(x)$ are (homogeneous) *ambit sets*, i.e. $A_t(x)$ is of the form $A_t(x) = A + (x, t)$ where

$$
A = \left\{(\xi, s) : s \leq 0, c_s^- \leq \xi \leq c_s^+\right\}
$$

for some functions $c_\cdot^-$ and $c_\cdot^+$ with $c_s^- \leq 0$ and $c_s^+ \geq 0$; and similarly for $C_t(x)$ and $D_t(x)$.

In this space-time setting the key questions (analogous to those discussed above) are substantially more intricate, major differences occurring already for the case of a one-dimensional space component. Here only a particular aspect of this will be discussed.

For simplicity we consider the case where the spatial dimension is 1 and $Y_t(x)$ is one-dimensional, i.e. $d = k = m = 1$.

9 Ambit processes

Now, let $\tau = \{\tau(w) : w \in \mathbb{R}\}$, with $\tau(w) = (\xi(w), s(w))$, be a smooth curve in $\mathbb{R} \times \mathbb{R}$ such that $s(w)$ is increasing in w and $s(\mathbb{R}) = \mathbb{R}$, and let

$$
X_w = Y_{s(w)}(\xi(w))
$$

with Y defined as in Section 8. The process $X = \{X_w\}_{w \in \mathbb{R}}$ is said to be an *ambit process*.

Under the specific assumptions made earlier

$$
\begin{aligned}
X_w &= \int_{A+\tau(w)} g(t-s, x-\xi)\sigma_s(\xi)\, W(\mathrm{d}\xi\, \mathrm{d}s) \\
&\quad + \int_{D+\tau(w)} q(t-s, x-\xi)a_s(\xi)\, \mathrm{d}\xi\, \mathrm{d}s
\end{aligned}
$$

and we now consider the questions of whether the quadratic variation $[X.]$ exists, as the probability limit of the realised quadratic variation

$$[X_\delta]_w = \sum_{j=1}^{\lfloor w/\delta \rfloor} \left(X_{j\delta} - X_{(j-1)\delta}\right)^2,$$

and whether $[X.]_w = \int_0^w \sigma^2_{s(\phi)}(\xi(\phi))\,\mathrm{d}\phi$. A comprehensive treatment of these questions will not be attempted here, and we restrict the discussion to outlining a setting where the curve τ and the ambit set A are 'aligned' in a specified sense. A general formula is then available for the quadratic variation. Moreover, under certain conditions on g and A, X_w is representable as the difference $X_w = X_w^+ - X_w^-$ between two q-orthogonal semimartingales; however, such cases are not of prime interest in the context of turbulence and we shall not discuss them further here.

9.1 Alignment

Definition 9.1. The curve τ and the ambit set A, with rectifiable and parametrised boundary $C = \{c(\gamma) : \gamma \in \Gamma\}$, are said to be *aligned* if the following conditions are satisfied. Let $c^\perp$ denote the transversal of $\dot{c}$, i.e. $c^\perp = (\dot{c}_2, -\dot{c}_1)$.

(i) For all w there exists a partition of C into two sets C_w^+ and C_w^- such that $\dot{\tau}(w) \cdot c^\perp(\gamma) \geq 0$ for all γ with $c(\gamma) \in C_w^+$ while $\dot{\tau}(w) \cdot c^\perp(\gamma) \leq 0$ for all γ with $c(\gamma) \in C_w^-$.

(ii) The subsets Γ_w^+ and Γ_w^- of Γ corresponding to C_w^+ and C_w^- are connected.

(iii) For all w the curve lengths of C_w^+ and C_w^- are positive.

The sets C_w^+ and C_w^- constitute the 'front' and the 'rear' of $A_{t(w)}(x(w))$ as $(x(w), t(w))$ moves along the curve τ.

Figures 9.1 and 9.2 illustrate a case of nonalignment and one of alignment, respectively.

9.2 QV under alignment

Suppose the curve τ and the ambit set A are aligned, and that A is convex and bounded. Then, under suitable conditions, the quadratic variation $[X.]$ of X exists as the limit in probability of the realised quadratic variation $[X_\delta]$ and

$$[X.]_w - [X.]_{w_0} = \int_{w_0}^{w} \int_C g^2(-c_1(\gamma), -c_2(\gamma))\,\sigma^2(c(\gamma) + \tau(u))\,\dot{c}^\perp(\gamma) \cdot \dot{\tau}(u)\,\mathrm{d}\gamma\,\mathrm{d}u.$$

In other words:

$$\mathrm{d}[X.]_w = \int_C g^2(-c_1(\gamma), -c_2(\gamma))\,\sigma^2(c(\gamma) + \tau(u))\,\dot{c}^\perp(\gamma) \cdot \dot{\tau}(w)\,\mathrm{d}\gamma\,\mathrm{d}w$$

which can be rewritten as

$$\frac{\mathrm{d}\,[X.]_w}{\mathrm{d}w} = \oint_{A+\tau(w)} g^2\left(\tau\left(w\right) - \left(\xi, s\right)\right)^2 \sigma_s^2\left(\xi\right) \mathrm{d}\xi\,\mathrm{d}s.$$

A detailed discussion of the pertinent conditions will be given elsewhere. Here we just mention that a conceptually important ingredient for the proof is the following pure analysis result (which is likely to be known but to which we have not been able to find a reference).

Let $m = 2$ and let $\tau\left(w\right)$ be a curve in $\mathbb{R}^2$ as before, and assume that τ and the boundary curve c of the ambit set A are both continuously differentiable. Furthermore, suppressing w in the notation $\tau\left(w\right)$, let

$$x_w = y_\tau = \int_{A+\tau} H\left(\tau, v\right) \mathrm{d}v$$

where the function $H : \mathbb{R} \times \mathbb{R} \to \mathbb{R}$ is assumed to be integrable on all sets $A + \tau$ and such that $H\left(t, x\right)$ is continuously differentiable with respect to t for almost all x (with respect to Lebesgue measure).

Proposition. The differential of y_τ along τ is

$$\mathrm{d}y_\tau = \int_C H\left(\tau, c+\tau\right) \mathrm{d}c^\perp \cdot \mathrm{d}\tau + \int_{A+\tau} \mathrm{d}_\tau H\left(\tau, v\right) \mathrm{d}v \cdot \mathrm{d}\tau$$

where $\mathrm{d}c^\perp = \left(\mathrm{d}c_2, -\mathrm{d}c_1\right)$ is the transversal of $\mathrm{d}c$.

Sketch of proof. Suppose for simplicity that y_τ can be rewritten as

$$y_\tau = \int_{a+\tau_1(w)}^{b+\tau_1(w)} \int_{l(\xi)+\tau_2(w)}^{u(\xi)+\tau_2(w)} H\left(\tau, \xi, \eta\right) \mathrm{d}\eta\,\mathrm{d}\xi$$

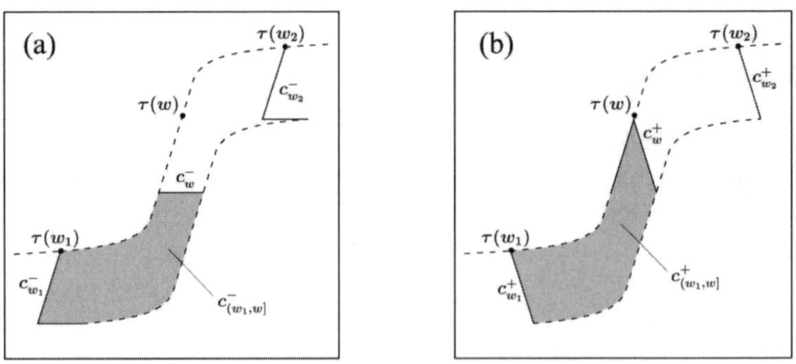

Figure 9.1. Illustration of the concept of alignment with a triangular ambit set. The curve τ and the triangular ambit set are not aligned.

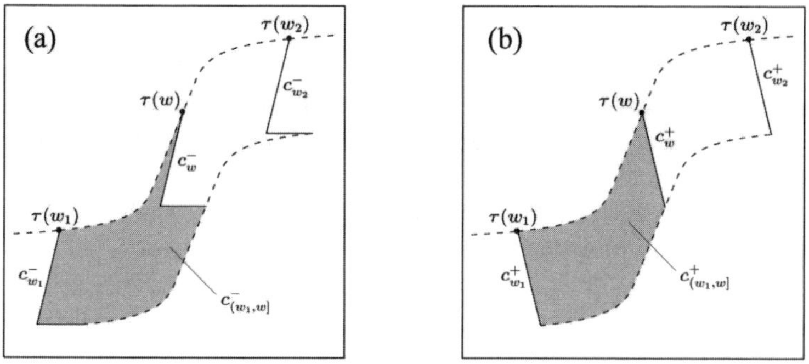

Figure 9.2. Illustration of the concept of alignment with a triangular ambit set. The curve τ and the triangular ambit set are aligned.

Then, by ordinary rules of calculus, and using anticlockwise orientation for curvilinear integrals, we find

$$\begin{aligned}
\mathrm{d}y_\tau &= \int_{a+\tau_1(w)}^{b+\tau_1(w)} \mathrm{d} \int_{l(\xi)+\tau_2(w)}^{u(\xi)+\tau_2(w)} H\left(\tau(w), \xi, \eta\right) \mathrm{d}\eta\, \mathrm{d}\xi \\
&= \int_{a+\tau_1(w)}^{b+\tau_1(w)} H\left(\tau, \xi, u(\xi) + \tau_2(w)\right) \mathrm{d}\tau_2\, \mathrm{d}\xi \\
&\quad - \int_{a+\tau_1(w)}^{b+\tau_1(w)} H\left(\tau, \xi, l(\xi) + \tau_2(w)\right) \mathrm{d}\tau_2\, \mathrm{d}\xi \\
&\quad + \int_{a+\tau_1(w)}^{b+\tau_1(w)} \int_{l(\xi)+\tau_2(w)}^{u(\xi)+\tau_2(w)} \mathrm{d}_\tau H\left(\tau, \xi, \eta\right) \mathrm{d}\eta\, \mathrm{d}\xi \cdot \mathrm{d}\tau \\
&= -\int_{C+\tau} H\left(\tau, \xi, \eta\right) \mathrm{d}\xi \mathrm{d}\tau_2 + \int_{A+\tau} \mathrm{d}_\tau H\left(\tau, v\right) \mathrm{d}v \cdot \mathrm{d}\tau \\
&= \int_C H\left(\tau, c + \tau\right) \mathrm{d}c^\perp \cdot \mathrm{d}\tau + \int_{A+\tau} \mathrm{d}_\tau H\left(\tau, v\right) \mathrm{d}v \cdot \mathrm{d}\tau.
\end{aligned}$$

10 Conclusion

In the purely temporal setting, so far we have assumed that $\sigma \perp\!\!\!\perp B$. In joint work with José Manuel Corcuera and Mark Podolskij (Barndorff-Nielsen et al. (2009)) this condition has been substantially weakened. This more refined analysis – which uses the theory of multipower variation and recent powerful results of Malliavin calculus due to Nualart, Peccati et al. – has shown:

- In wide generality, $\overline{[Y_\delta]} \xrightarrow{p} \sigma^{2+}$
- Under certain conditions a feasible CLT for $\overline{[Y_\delta]}$ can be established.

- The results can be further extended to consistency and feasible CLTs for multipower variations, in particular for bipower variation.

Extensions of these results to the tempo-spatial regimes will be of key interest but the inclusion of a spatial component makes the issues considerably more challenging, as the discussion in Sections 8 and 9 will have indicated.

We are indebted to Jose Manuel Corcuera for a careful reading of the manuscript and accompanying helpful comments.

Bibliography

[1] Barndorff-Nielsen, O.E., Corcuera, J.M. and Podolskij, M. (2009): Multipower variation of Brownian semistationary processes. (Submitted.)

[2] Barndorff-Nielsen, O.E., Graversen, S.E., Jacod, J., Podolskij, M. and Shephard, N. (2006a): A central limit theorem for realised power and bipower variations of continuous semimartingales. In Yu. Kabanov, R. Liptser and J. Stoyanov (Eds.): *From Stochastic Calculus to Mathematical Finance*. Festschrift in Honour of A.N. Shiryaev. Heidelberg: Springer. Pp. 33–68.

[3] Barndorff-Nielsen, O.E., Graversen, S.E., Jacod, J., and Shephard, N. (2006b): Limit theorems for bipower variation in financial econometrics. *Econometric Theory* **22**, 677–719.

[4] Barndorff-Nielsen, O.E. and Schmiegel, J. (2004): Lévy-based tempo-spatial modelling; with applications to turbulence. *Uspekhi Mat. NAUK* **59**, 65–91.

[5] Barndorff-Nielsen, O.E. and Schmiegel, J. (2007): Ambit processes; with applications to turbulence and cancer growth. In F.E. Benth, Nunno, G.D., Linstrøm, T., Øksendal, B. and Zhang, T. (Eds.): *Stochastic Analysis and Applications: The Abel Symposium 2005* . Heidelberg: Springer. Pp. 93–124.

[6] Barndorff-Nielsen, O.E. and Schmiegel, J. (2008a): A stochastic differential equation framework for the timewise dynamics of turbulent velocities. *Theory Prob. Its Appl.* **52**, 372–388.

[7] Barndorff-Nielsen, O.E. and Schmiegel, J. (2008b): Time change, volatility and turbulence. In A. Sarychev, A. Shiryaev, M. Guerra and M.d.R. Grossinho (Eds.): *Proceedings of the Workshop on Mathematical Control Theory and Finance*. Lisbon 2007. Berlin: Springer. Pp. 29–53.

[8] Barndorff-Nielsen, O.E. and Schmiegel, J. (2008c): Time change and universality in turbulence. Research Report 2007-8. Thiele Centre for Applied Mathematics in Natural Science. (Submitted.)

[9] Barndorff-Nielsen, O.E. and Shephard, N. (2003): Realised power variation and stochastic volatility models. *Bernoulli* **9**, 243–265.

[10] Barndorff-Nielsen, O.E. and Shephard, N. (2004): Power and bipower variation with stochastic volatility and jumps (with discussion). *J. Fin. Econometrics* **2**, 1–48.

[11] Barndorff-Nielsen, O.E. and Shephard, N. (2006a): Econometrrics of testing for jumps in financial economics using bipower variation. *J. Fin. Econometrics* **4**, 217–252.

[12] Barndorff-Nielsen, O.E. and Shephard, N. (2006b): Multipower variation and stochastic volatility. In M. do Rosário Grossinho, A.N. Shiryaev, M.L. Esquível and P.E. Oliveira: *Stochastic Finance*. New York: Springer. Pp. 73–82.

[13] Barndorff-Nielsen, O.E., Winkel, M. and Shephard, N. (2006c): Limit theorems for multi-power variation in the presence of jumps. *Stoch. Proc. Appl.* **116**, 796–806.

[14] Basse, A. (2007a): Spectral representation of Gaussian semimartingales. Research Report 2008-3. Thiele Centre for Applied Mathematics in Natural Science.

[15] Basse, A. (2007b): Representation of Gaussian semimartingales and applications to the co-variance function. Research Report 2008-5. Thiele Centre for Applied Mathematics in Natural Science.

[16] Basse, A. (2008): Gaussian moving averages and semimartingales. *Electron. J. Probab.* **13**, 1140–1165.

[17] Basse, A. and Pedersen, J. (2008): Lévy driven moving averages and semimartingales. (To appear in *Stoch. Proc. Appl.*.)

[18] Doob, J.L. (1953): *Stochastic Processes*. New York: Wiley.

[19] Jacod, J. (2008a): Asymptotic properties of realized power variations and related functionals of semimartingales. *Stoch. Proc. and their Appl.* **118**, 517–559.

[20] Jacod, J. (2008b): Statistics and high frequency data. Lecture notes.

[21] Karhunen, K. (1950): Über die Struktur stationärer zufälliger Funktionen. *Ark. Mat.* **1**, 141–160.

[22] Knight, F. (1992): *Foundations of the Prediction Process*. Oxford: Clarendon Press.

[23] Rajput, B.S. and Rosinski, J. (1989): Spectral representations of infinitely divisible distributions. *Probab. Theory Related Fields* **82**, 451–487.

Author information

Ole E. Barndorff-Nielsen, Thiele Centre for Applied Mathematics in Natural Science, Department of Mathematical Sciences, Ny Munkegade, 8000 Aarhus, Denmark.
Email: `oebn@imf.au.dk`

Jürgen Schmiegel, Thiele Centre for Applied Mathematics in Natural Science, Department of Mathematical Sciences, Ny Munkegade, 8000 Aarhus, Denmark.
Email: `schmiegl@imf.au.dk`

From bounds on optimal growth towards a theory of good-deal hedging

Dirk Becherer

Abstract. Good-deal bounds have been introduced as a way to obtain valuation bounds for derivative assets which are tighter than the arbitrage bounds. This is achieved by ruling out not only those prices that violate no-arbitrage restrictions but also trading opportunities that are 'too good'. We study dynamic good-deal valuation bounds that are derived from bounds on optimal expected growth rates. This leads naturally to restrictions on the set of pricing measure which are local in time, thereby inducing good dynamic properties for the good-deal valuation bounds. We study good-deal bounds by duality arguments in a general semimartingale setting. In a Wiener space setting where asset prices evolve as Itô processes, good-deal bounds are then conveniently described by backward SDEs. We show how the good-deal bounds arise as the value function for an optimal control problem, where a dynamic coherent a priori risk measure is minimised by the choice of a suitable hedging strategy. This demonstrates how the theory of no-good-deal valuations can be associated to an established concept of dynamic hedging in continuous time.

Key words. Good deal bounds, good deal hedging, incomplete markets, backward stochastic differential equations, dynamic coherent risk measures, logarithmic utility, optimal growth.

AMS classification. (MSC2000) 60G35, 60G44, 60H30, 91B16, 91B28

1 Introduction

When pricing a contingent claim solely based on no-arbitrage arguments, the range of possible arbitrage free prices can easily be too wide for practical purposes. This can happen generically if the financial market is not taken as complete or, more generally, if the payoff of the contingent claim cannot be perfectly synthesised by dynamical trading. The approach of good-deal bounds has been developed to derive a tighter range of no-good-deal prices, by using only a subset of martingale (pricing) measures to value contingent claims. This subset is chosen such that some notion of too favourable trading opportunities (good deals) is prevented, if the market is extended by additional price processes which are obtained by using those martingale measures for the pricing of contingent claims.

The most cited reference on the topic is probably [12], where no good deals are defined by bounds on Sharpe ratios. For comprehensive references on the topic we refer to [9, 18, 7, 20]. Let us just mention here, that in [7] the authors also consider bounds for Sharpe ratios in a continuous time model which permits for asset price jumps, thereby

Support by Deutsche Forschungsgemeinschaft through the DFG research center MATHEON *Mathematics for Key Technologies* is gratefully acknowledged. I like to thank Tomas Björk for bringing the hedging problem to my attention and Stephen Dodgson for advice.

extending [12]. The starting point of another stream of papers is given by a set of acceptable financial positions. From there, no-good-deal price bounds are then defined as the smallest amount which a seller of a claim must charge, respectively the largest amount a buyer can pay, in order to obtain an acceptable position in the market, see e.g. [9, 18]. The route taken in the present article is to look at bounds for expected optimal growth rates, which corresponds to bounds on expected logarithmic utility. In deriving dynamic good-deal bounds in continuous time from bounds on expected utility, it is maybe closest in spirit to and has been motivated by [20], where a rigorous study of such bounds has been presented in a Levy process setting for the case of exponential utility. For important conceptual earlier contributions in this direction, see [10, 11] and also references in [10, 20] on earlier work by S. Hodges. In [20] it has been pointed out that restrictions on the no-good-deal set of martingale (pricing) measures should be local in time, in order to achieve good dynamic properties of the good-deal valuation bounds. However, the authors have concentrated on the case of exponential utility from terminal wealth where additional, essentially technical, assumptions were needed to ensure such local restrictions. Apparently, those do not arises naturally from the good-deal approach in this case. To ensure that their local restrictions imply certain 'global' restrictions that have financial meaning in terms of bounds on utility, the authors have therefore imposed additional structural assumptions, like preservation of a Levy structure of the model under all no-good-deal pricing measures.

To arrive at local restrictions for the set of no-good-deal martingale measures in a more endogenous way, it appears natural to start from bounds on expected optimal growth, since this is linked to logarithmic utility which is know to be myopic. By this choice, one also avoids to have good-deal bounds that depend on an exogenously given time horizon, for which expected terminal utility is maximised. This is the starting point for us in Section 3, where we consider the problem in a general semimartingale model, after the model framework has been set in Section 2. To derive more explicit results we specialise in Section 4 to a Wiener space setting with asset prices that are Itô processes, but need not be Markovian diffusions. In this setting, we obtain a convenient dynamic description of good-deal bounds in terms of standard backward stochastic differential equations (BSDEs) with generators that satisfy a Lipschitz property in Section 5. It is well known that the good-deal approach is in essence an approach to pricing, but that a theory on notions of hedging that can be linked to it still needs to be further developed and, possibly even more so, constructive descriptions for respective hedging strategies are needed. This is emphasised in the concluding remarks in [7], where the authors write "the task of developing such a [good-deal hedging] theory constitutes a highly challenging open problem". The final Section 5 discusses such links to hedging, showing that a trading strategy, which corresponds to the (upper) good-deal valuation bound, is given by a strategy that minimises a suitable coherent risk measure ρ (of good-deal type). The hedging strategy can then be defined as the minimiser. In other words, the hedging strategy is the strategy that minimises (maximises) the upper (lower) a priori ρ-valuation bounds for the seller (buyer) of a contingent claim, such that the good-deal bounds arise as the tightest 'a priori' ρ-valuation bounds for the residual risk that can be obtained by optimal hedging in the market. For our mathematical analysis, BSDE theory provides convenient methods to solve the optimisation problems involved and to describe both the valuation bounds and the optimal hedging strategies. To our best knowledge, the connection between growth-optimal dynamic good-deal bounds and a corresponding notion of hedging with respect to dynamic coherent risk measures has not been elaborated so far. Mathematically, this connection fits well with the general theory for inf-convolutions of risk measures, see [4] for an excellent exposition and references.

The contributions of this article are threefold. First, we show that restrictions on optimal expected growth rates lead to good-deal valuation bounds with good dynamic properties in a general model. Secondly, we describe the dynamics of the valuation bounds and of a suitable hedging strategy in a Wiener space setting by BSDE solutions where, finally, the corresponding hedging strategy is obtained as the minimiser of a suitable a-priori risk measure of good-deal type.

2 General framework and preliminaries

Let $(\Omega, \mathcal{F}, \mathbb{F}, P)$ be a stochastic basis with fixed time horizon $\bar{T} < \infty$ and a filtration $\mathbb{F} = (\mathcal{F}_t)_{t\in[0,\bar{T}]}$ satisfying the usual conditions of right-continuity and completeness. All semimartingales are taken to have right-continuous paths with left limits (RCLL paths). For simplicity let $\mathcal{F}_0$ be trivial and $\mathcal{F} = \mathcal{F}_{\bar{T}}$. Conditional expectations with respect to $\mathcal{F}_t$ under P are denoted by $E_t[\cdot] = E_t^P[\cdot]$. For random variables X we define $E_t[X]$ as $E_t[X^+] - E_t[X^-]$ where the latter is well defined, and as $-\infty$ elsewhere. Inequalities between random variables (processes) are meant to hold almost surely (respectively $P\times dt$ almost everywhere). Equality between processes with RCLL paths then means indistinguability. By convention, $\frac{0}{0}$ is defined to be 0.

All prices of assets in our model for the financial market are expressed in terms of a numeraire asset, whose (discounted) price is thus constant at one. A common choice for the numeraire is a cash account where money can be deposited and lent from at the same rate of interest; for such a choice, asset prices are expressed in units of one Euro put into the cash account at time 0. Alternatively, if the numeraire is chosen to be the zero coupon bond with maturity $\bar{T}$, then asset prices are expressed in $\bar{T}$-forward units, i.e. in terms of Euros at time $\bar{T}$. We consider a financial market model with $d+1$ tradable assets, which comprise the numeraire asset and furthermore d risky assets, whose price processes $S = (S^i)_{i=1,\dots,d}$ are semimartingales. All S^i are assumed to be locally bounded from below. For example, a natural (globally uniform) lower bound for stocks would be zero.

As usual, a (self-financing) portfolio strategy is defined by some initial capital V_0 and a predictable and S-integrable $\mathbb{R}^d$-valued process ϑ (i.e. $\vartheta \in L(S)$), describing the numbers of risky assets to be held dynamically over time. The wealth process of such strategy is given by $V = V_0 + \vartheta\cdot S = V_0 + \int \vartheta dS$. Since we will be interested in the returns V_t/V_s $(s \le t \le \bar{T})$ of wealth processes that are strictly positive, it is convenient to normalise initial wealth to one by simple scaling. To this end, we define

$$\mathcal{N} := \mathcal{N}(S) := \{N \mid N = 1 + \theta\cdot S\,,\ \theta \in L(S),\ N > 0\} \tag{2.1}$$

as the family of all strictly positive wealth processes starting at one. These processes are called tradable numeraires (with respect to S).

We denote by $\mathcal{M}^e(\mathcal{N})$ the set of all probability measures equivalent to P ($Q \sim P$) such that any $N \in \mathcal{N}$ is a local Q-martingale. It is easy to see that $Q \sim P$ is a local martingale measure for S if and only if it is a local martingale measure for any $N \in \mathcal{N}$. Indeed, one inclusion of the identity $\mathcal{M}^e(S) = \mathcal{M}^e(\mathcal{N})$ follows from S being locally bounded from below by adding a suitable constant and a simple scaling argument, while the other inclusion follows from [1], Corollary 3.5.

Throughout the sequel, we assume that the market is free of arbitrage in the sense that

$$\mathcal{M}^e := \mathcal{M}^e(\mathcal{N}(S)) \neq \emptyset. \tag{2.2}$$

More precisely, condition (2.2) is equivalent to the property of 'no free lunch with vanishing risk' introduced in [14], cf. Proposition 2.3 in [5].

Let us recall some properties about the set of numeraires $\mathcal{N}(S)$.

Lemma 2.1. *1. $\mathcal{N}$ is convex.*

2. For $N^1, N^2 \in \mathcal{N}$ and for stopping times T, τ with $0 \le T \le \tau \le \bar{T}, A \in \mathcal{F}_T$, the process

$$\begin{aligned} N \quad := I_{[\![0,T]\!]} N^1 \quad & + I_{]\!]T,\tau]\!]} \left(I_A \frac{N_T^1}{N_T^2} N^2 + I_{A^c} N^1 \right) \\ & + I_{]\!]\tau,\bar{T}]\!]} \left(I_A \frac{N_T^1}{N_T^2} \frac{N_\tau^2}{N_\tau^1} + I_{A^c} \right) N^1 \end{aligned}$$

is an element of $\mathcal{N}$.

3. Every $N \in \mathcal{N}$ can be written as a stochastic exponential, and the set

$$\mathcal{L} := \left\{ L \,\middle|\, L \text{ semimartingale},\ L_0 = 0, \mathcal{E}(L) \in \mathcal{N} \right\}$$

contains 0 and is predictably convex, i.e., for any predictable $[0,1]$-valued process H and $L^1, L^2 \in \mathcal{L}$, the process $L := H \cdot L^1 + (1 - H) \cdot L^2$ is an element of $\mathcal{L}$.

4. Any $N \in \mathcal{N}$ is bounded away from 0 uniformly in t: $P[\inf_{t \le \bar{T}} N_t > 0] = 1$, and is a strictly positive supermartingale under Q for any $Q \in \mathcal{M}^e$.

Proof. Part 1 is immediate. Part 3 can be shown using that any $N \in \mathcal{N}$, being a strictly positive Q-martingale for $Q \in \mathcal{M}^e(\mathcal{N})$, can be written as a stochastic exponential $N = \mathcal{E}(L) \equiv 1 + N_- \cdot L$ for $L := (1/N_-) \cdot N$. Letting $L := H \cdot L^1 + (1 - H) \cdot L^2$ for $N^i = \mathcal{E}(L^i) \in \mathcal{N}, i = 1, 2$, it is straightforward to show that N is in $\mathcal{N}$. For part 2, let $H := I_{[\![0,T]\!]} + I_{]\!]T,\tau]\!]} I_{A^c} + I_{]\!]\tau,\bar{T}]\!]}$ and $L := H \cdot L^1 + (1 - H) \cdot L^2$. Then one can check that $N = \mathcal{E}(L)$ and the claim follows by part 3. Details on parts 2, 3 are given in [5], Lemma 2.5. For part 4, the second claim holds since any positive local martingale is a supermartingale by Fatou's lemma. The first claim then follows by the minimum principle for positive supermartingales, which is an application of the optional stopping theorem, see [23] II.3. □

Dynamic good-deal bounds: Preliminaries

For equivalent martingale measures $Q \in \mathcal{M}^e$ we will for convenience often identify the measure Q with its density process $Z_t = Z_t^Q = E_t[dQ/dP]$, omitting indices like Q when there is no ambiguity. To limit notations, we write compactly $Z \in \mathcal{M}^e$ for Z^Q of some $Q \in \mathcal{M}^e$.

We denote by $\mathcal{Q}$ the set

$$\mathcal{Q} := \mathcal{Q}(S) := \{Q \in \mathcal{M}^e(S) \,|\, E[-\log Z_{\bar{T}}] < \infty\} . \tag{2.3}$$

The integrability condition in (2.3) requires that the relative entropy $H(P|Q)$ of P with respect to Q, that is

$$H(P|Q) := E[-\log Z_{\bar{T}}] = E^Q\left[\frac{1}{Z_{\bar{T}}} \log \frac{1}{Z_{\bar{T}}}\right] , \tag{2.4}$$

is finite. For an equivalent martingale measure Q, it is usual to call $H(P|Q)$ the reverse relative entropy of Q, whereas $H(Q|P)$ is called the relative entropy of Q.

Let

$$h = (h_t) \geq 0$$

be a predictable process that is bounded, measurable and positive. For given h, let $\mathcal{Q}^{\mathrm{ngd}} := \mathcal{Q}^{\mathrm{ngd}}(S)$ denote the subset of those measures $Q \in \mathcal{M}^e(S)$ which satisfy that

$$E_T\left[-\log \frac{Z_\tau}{Z_T}\right] \;\leq\; \frac{1}{2} E_T\left[\int_T^\tau h_u^2 \, du\right] \quad \text{for all } T \leq \tau \leq \bar{T}, \tag{2.5}$$

where T, τ are stopping times. Let us note here, that we have not included the factor $1/2$ and the square into the function h to simplify formulae in the later sections, e.g. 5.

We shall write $h(t)$ instead of h_t $(0 \leq t \leq \bar{T})$ if h is deterministic, depending only on time but not on $\omega \in \Omega$. The simplest case is to take $h(t) = \bar{h} \geq 0$ to be constant. Permitting h to depend on time t and ω increases generality. In later sections, $\mathcal{Q}^{\mathrm{ngd}}$ will serve as a set of pricing measures which are taken into consideration to determine a range of no-good-deal valuations. If, for instance, market prices of risk in the distant future are considered to be of higher variability this could be captured by $t \mapsto h(t)$ being increasing.

Clearly, we have $\mathcal{Q}^{\mathrm{ngd}} \subset \mathcal{Q} \subset \mathcal{M}^e$. Going beyond condition (2.2), we will subsequently assume that $\mathcal{Q}^{\mathrm{ngd}}$ contains at least one element $\widehat{Q}$, i.e. it is non-empty,

$$\mathcal{Q}^{\mathrm{ngd}} \neq \emptyset. \tag{2.6}$$

The next result collects properties of the density process Z for measures Q in $\mathcal{Q}$.

Proposition 2.2. *For any $Q \sim P$ with $-\log Z_{\bar{T}} \in L^1(P)$ the following holds.*

1. *The process $(-\log Z_t)_{t \leq \bar{T}}$ is a submartingale of class (D), that is the family $\{-\log Z_T\}$ is uniformly P-integrable, with T ranging over the set of all stopping times $T \leq \bar{T}$.*

2. *The process* $-\log Z$ *has a unique Doob–Meyer decomposition*

$$-\log Z_t = M_t + A_t\,, \quad t \le \bar{T}\,, \tag{2.7}$$

with M *being a uniformly integrable martingale and* A *being an integrable increasing predictable process with* $A_0 = 0$.

3. *For any stopping times* $T \le \tau \le \bar{T}$, *it holds*

$$E_T\left[-\log\frac{Z_\tau}{Z_T}\right] = E_T[A_\tau - A_T]\,. \tag{2.8}$$

Proof. 1. By assumption we have $-\log Z_{\bar{T}} \in L^1(P)$. For any stopping time $T \le \bar{T}$,

$$-E_T[Z_{\bar{T}}] + 1 \le -Z_T + 1 \le -\log Z_T \le E_T[-\log Z_{\bar{T}}]$$

holds by Jensen's inequality. Since the upper and lower bounds are uniformly integrable martingales stopped at T, the family $\{-\log Z_T\}$ is of class (D). Using again Jensen's inequality, we obtain that $-\log Z$ is a supermartingale.

2. This follows by a version of the Doob–Meyer decomposition, see Theorem 8 in Ch. III of [22]. That A is integrable, means that $E[A_{\bar{T}}] < \infty$.
3. This follows from part 2 since martingale increments vanish in expectation.

□

For any Q as in Proposition 2.2 with density process Z and corresponding A from (2.7), we can define a finite measure $\mu = \mu^Q$ on the predictable σ-field $\mathcal{P}$ by

$$\mu(B) := E\left[\int 1_B\, dA_t\right], \quad B \in \mathcal{P}\,. \tag{2.9}$$

The next result shows that μ corresponding $Q \in \mathcal{Q}$ is dominated by the measure

$$\nu(B) := \frac{1}{2}E\left[\int 1_B h_t^2\, dt\right], \quad B \in \mathcal{P}\,, \tag{2.10}$$

and thereby provides simpler criteria (2.11), (2.13) for Q being in $\mathcal{Q}^{\mathrm{ngd}}$, which are formulated with respect to deterministic times only instead of stopping times as (2.5).

Proposition 2.3. *If the measure* $Q \sim P$ *satisfies*

$$E_s\left[-\log\frac{Z_t}{Z_s}\right] \le \frac{1}{2}E_s\left[\int_s^t h_u^2 du\right] \quad \text{for all } s \le t \le \bar{T} \tag{2.11}$$

with s, t *ranging over the set of all deterministic times, then*

$$\mu(B) \le \nu(B) \tag{2.12}$$

holds for all predictable sets $B \in \mathcal{P}$. *In particular, the conditions (2.11) and (2.5) are equivalent, and for* $Q \in \mathcal{M}^e(S)$ *condition (2.11) implies that* Q *is in* $\mathcal{Q}^{\mathrm{ngd}}$.

In the case where $h_t = h(t)$ is deterministic, (2.11) simplifies to

$$E_s\left[-\log\frac{Z_t}{Z_s}\right] \leq \frac{1}{2}\int_s^t h^2(u)du \quad \text{for all } s \leq t \leq \bar{T}. \tag{2.13}$$

Proof. By assumption (2.11), inequality (2.12) holds for predictable sets of the form $B = A_s \times (s,t]$ with $s < t \leq \bar{T}$ and $A_s \in \mathcal{F}_s$. The class of all such sets is a semi-ring that generates the predictable σ-field $\mathcal{P}$. Since both measures are finite and (2.12) holds on the semi-ring, it follows that (2.12) also holds on the predictable σ-field $\mathcal{P}$ generated by it. Indeed, the inequality extends directly from the semi-ring to $\mathcal{P}$ for the outer measures that are generated from the restrictions of μ, ν onto the semi-ring. Since these outer measures coincide with μ respectively ν on $\mathcal{P}$ by the extension theorems of measure theory, (2.12) holds for $B \in \mathcal{P}$. Hence, condition (2.11) implies (2.5) since sets of the form $B = A_T \times \rrbracket T, \tau \rrbracket$ are in $\mathcal{P}$. Necessity is obvious. The last claim follows from the definition of $\mathcal{Q}$. □

Definition 2.4. A set $\mathcal{S}$ of probability measures, all elements of which are equivalent to P, is called multiplicativity stable (m-stable) if for all elements $Q^1, Q^2 \in \mathcal{S}$ with density processes Z^1, Z^2 and for all stopping times $T \leq \bar{T}$, it holds that $Z_{\bar{T}} := Z^1_T \frac{Z^2_{\bar{T}}}{Z^2_T}$ is the density of some $Q \in \mathcal{S}$.

This definition follows [13], where m-stable sets of measures are studied in a general framework. Examples for m-stable sets that play a role in the sequel are given by

Proposition 2.5. *The sets $\mathcal{M}^e$, $\mathcal{Q}$ and $\mathcal{Q}^{\mathrm{ngd}}$ are m-stable.*

Proof. The arguments for Proposition 5 in [13] show that $\mathcal{M}^e$ is m-stable. This holds also without S being locally bounded. The m-stability $\mathcal{Q}$ follows then by part 3 of Proposition 2.2, which implies

$$E\left[-\log\left(Z^1_T\frac{Z^2_{\bar{T}}}{Z^2_T}\right)\right] \leq E[A^1_T + (A^2_{\bar{T}} - A^2_T)] < \infty.$$

To show that the property (2.5) defining $\mathcal{Q}^{\mathrm{ngd}}$ is consistent with m-stability of $\mathcal{Q}^{\mathrm{ngd}}$, let $\tau' \leq \tau$ be stopping times. Letting $T' = (T \vee \tau') \wedge \tau$ for T and Z as in Definition 2.4, we obtain, using again part 3 of Proposition 2.2, that

$$\begin{aligned} E_{\tau'}\left[-\log\frac{Z_\tau}{Z_{\tau'}}\right] &= E_{\tau'}\left[A^2_\tau - A^2_{T'} + A^1_{T'} - A^1_{\tau'}\right] \\ &\leq \frac{1}{2}E_{\tau'}\left[\int_{\tau'}^{\tau} h^2_u\,du\right]. \end{aligned}$$

□

With respect to a given m-stable set $\mathcal{S}$ of equivalent measures, we define for $X \in L^\infty$ the upper and lower valuation bounds by

$$\pi^u_t(X;\mathcal{S}) = \operatorname*{ess\,sup}_{Q\in\mathcal{S}} E^Q_t[X], \tag{2.14}$$

$$\pi^\ell_t(X;\mathcal{S}) = \operatorname*{ess\,inf}_{Q\in\mathcal{S}} E^Q_t[X]. \tag{2.15}$$

Clearly, $\pi^\ell(X;\mathcal{S}) = -\pi^u(-X;\mathcal{S})$ holds. Therefore, we can restrict the analysis to the upper bounds π^u in the sequel, without loss of generality.

The next results recalls that the family $\{\pi_t^u(X)\}_{t\le\bar{T}}$ satisfies the properties of a dynamic coherent risk measure (or a dynamic monetary coherent utility functional), and can be represented by a process with good path properties. We will always choose such a version in the sequel without further notice. Moreover, it shows that the family of mappings $X \mapsto \pi_T^u(X)$ where T ranges over all stopping times $T \le \bar{T}$, exhibits good dynamic consistency properties.

Proposition 2.6. *Assume $\mathcal{S} \neq \emptyset$ (e.g. $\mathcal{S} = \mathcal{Q}^{\mathrm{ngd}}$). As mappings from L^∞ to $L^\infty(\mathcal{F}_t)$ the family $X \mapsto \pi_t^u(X;\mathcal{S})$ ($t \le \bar{T}$) has the following properties.*

1. *(Path properties) For any $X \in L^\infty$, there is a version of $(\pi_t^u(X))_{t\le\bar{T}}$ having RCLL paths and such that*

$$\pi_T^u(X) = \operatorname*{ess\,sup}_{Q\in\mathcal{S}} E_T^Q[X] \quad \textit{for all stopping times } T \le \bar{T}.$$

2. *(Recursiveness) For any stopping times $T \le \tau \le \bar{T}$, it holds that*

$$\pi_T^u(X) = \pi_T^u(\pi_\tau^u(X)).$$

3. *(Stopping-time consistency) For stopping times $T \le \tau \le \bar{T}$, the inequality $\pi_\tau^u(X^1) \ge \pi_\tau^u(X^2)$ implies $\pi_T^u(X^1) \ge \pi_T^u(X^2)$.*
4. *(Supermartingale property) $(\pi_t^u(X))$ is a supermartingale under any $Q \in \mathcal{S}$.*
5. *(Coherent risk measure) For any stopping time $T \le \bar{T}$ and $m_T, \alpha_T, \lambda_T \in L^\infty(\mathcal{F}_T)$ with $0 \le \alpha_T \le 1$, $\lambda_T \ge 0$, the mapping $X \mapsto \pi_T^u(X)$ satisfies the properties:*
    - *monotonicity: $X^1 \ge X^2$ implies $\pi_T^u(X^1) \ge \pi_T^u(X^2)$*
    - *translation invariance: $\pi_T^u(X + m_T) = \pi_T^u(X) + m_T$*
    - *convexity: $\pi_T^u(\alpha_T X^1 + (1-\alpha_T)X^2) \le \alpha_T \pi_T^u(X^1) + (1-\alpha_T)\pi_T^u(X^2)$*
    - *positive homogeneity: $\pi_T^u(\lambda_T X) = \lambda_T \pi_T^u(X)$*
6. *No arbitrage consistency: If moreover $\mathcal{S} \subset \mathcal{M}^e$, then $\pi_T^u(X) = x + \vartheta \cdot S_T$ holds for any $X = x + \vartheta \cdot S_{\bar{T}}$ with $((\vartheta \cdot S_t)_{t\le\bar{T}})$ being uniformly bounded.*

Proposition 2.6 is analogous to Theorem 2.7 in [20] and a direct consequence of results by Delbaen [13]. As the proof shows, it follows essentially from m-stability of $\mathcal{S}$.

Remark 2.7. In the literature on risk measures, those are often applied to the net value Y of a position. Under such convention, one would rather call $Y \mapsto \pi^u(-Y) = -\pi^\ell(Y)$ a dynamic coherent risk measure and call $Y \mapsto \pi^\ell(Y)$ a monetary coherent utility functional. Clearly, the difference is only a matter of sign conventions, where we take $X = -Y$ as a liability instead of a net value. This should not cause confusion.

Proof of Proposition 2.6. Since $\mathcal{S}$ is m-stable by assumption, part 1 follows by Lemmata 22 and 23 from [13], and using his Theorem 12 yields the claims 2–4. Finally, part 5 is immediate from part 1. Also part 6 follows from 1, since bounded local martingales are uniformly integrable. □

In the sequel, the aim is to study good-deal bounds with respect to optimal growth. In the next section, those bounds are shown to be valuations bounds as above for $\mathcal{S} = \mathcal{Q}^{\mathrm{ngd}}(S)$. This explains the terminology of the next definition.

Definition 2.8. The upper and lower good-deal bounds for $X \in L^\infty$ are defined by (2.14) respectively (2.15) for $\mathcal{S} = \mathcal{Q}^{\mathrm{ngd}}(S)$. For brevity of notation, they are denoted in the sequel by

$$\pi_t^u(X) := \pi_t^u(X; \mathcal{Q}^{\mathrm{ngd}}(S)) \quad \text{and} \quad \pi_t^\ell(X) := \pi_t^\ell(X; \mathcal{Q}^{\mathrm{ngd}}(S)). \tag{2.16}$$

3 Good-deal bounds for optimal growth rates: duality results

To motivate the subsequent results, let us suppose that the market is to be extended by adding further tradable risky assets. From the general theory of no-arbitrage pricing, it is known that each arbitrage-free price process of contingent claims with (discounted) payoffs $X \in L^0(\mathcal{F}_{\bar{T}}, \mathbb{R}^{d'})$ should in principle be of the form

$$S'_t = S'_t(X, Q) = E_t^Q[X], \quad t \le \bar{T}, \tag{3.1}$$

for some $Q \in \mathcal{M}^e$. Indeed, this ensures 'no arbitrage' for the extended market with risky asset price processes

$$\bar{S} = (S, S') = (S, S'(X, Q)) \tag{3.2}$$

evolving in $\mathbb{R}^{d+d'}$, in the sense that $Q \in \mathcal{M}^e(\bar{S}) \neq \emptyset$, subject to suitable integrability assumptions. A sufficient condition to ensure that $\bar{S}$ meets the same assumptions that our general setting imposes on S is obviously that $X \in L^\infty$. A more general condition is that all coordinates of $X = (X^i)_{1 \le i \le d'}$ are bounded from below and integrable with respect to the $Q \in \mathcal{M}^e$ chosen in (3.1), i.e.

$$(X^i)^- \in L^\infty \text{ and } X^i \in L^1(Q) \text{ for all } i = 1, \ldots, d'. \tag{3.3}$$

Then $S'(X, Q)$ is finite and bounded from below in each coordinate.

However, in incomplete markets the set $\mathcal{M}^e$ generically contains not one unique martingale measures but infinitely many. The price range, which would be obtained from (3.1) by letting Q range over all $\mathcal{M}^e$ is typically too wide for practical purposes. In other words, the restrictions on prices that result from no-arbitrage arguments alone will often lead to price bounds that are too wide. The leads naturally to the idea to let

Q range over a suitable smaller set (say $\mathcal{Q}^{\mathrm{ngd}}$) to obtain tighter price bounds. To give financial meaning to the desired tighter bounds, the general idea of good-deal bounds is to let Q range over a subset of $\mathcal{M}^e$ which is taken such that price processes $S'(X,Q)$ do not permit trading opportunities that are 'too good' to be realistic. Prices within the resulting range are then considered to be consistent with 'no good deals' while prices outside the range might be interpreted as 'good deals' for either the seller or buyer, depending whether they are above π^u or below π^ℓ.

There are different possibilities for defining good deals, which can result in different subsets of $\mathcal{M}^e$. We are going to show that our definition of $\mathcal{Q}^{\mathrm{ngd}}$ is such that *any* market extension $\bar{S} = (S, S')$ by price processes of the form (3.1) for $Q \in \mathcal{Q}^{\mathrm{ngd}}$ does only permit for (conditional) expected optimal growth of returns at a rate not exceeding the bound $h_t^2/2$ specified via h in the definition (2.5) of $\mathcal{Q}^{\mathrm{ngd}}$.

To show this, we are going to apply convex duality arguments. Recall that the conjugate function of $U(x) = \log x$ with $x > 0$ is

$$V(y) := \sup_{x>0} U(x) - xy = -\log y - 1\,, \quad y > 0\,. \tag{3.4}$$

It follows that

$$U(x) \le V(y) + xy \quad \text{for all } x, y > 0\,, \tag{3.5}$$

with equality holding for $x = 1/y$.

Any local martingale $\bar{N} > 0$ with respect to some $Q \in \mathcal{M}^e$ is a Q-supermartingale. Hence, $\bar{N}Z$ is a P-supermartingale. Letting $x = \bar{N}_\tau/\bar{N}_T$ and $y = Z_\tau/Z_T$ in (3.5) for stopping times $T \le \tau \le \bar{T}$, we obtain by taking conditional expectations that

$$\begin{aligned} & E_T\left[\log \frac{\bar{N}_\tau}{\bar{N}_T}\right] \\ \le\;& E_T\left[-\log \frac{Z_\tau}{Z_T}\right] + E_T\left[\frac{Z_\tau \bar{N}_\tau}{Z_T \bar{N}_T}\right] - 1 \\ \le\;& E_T\left[-\log \frac{Z_\tau}{Z_T}\right] = E_T^Q\left[\frac{Z_T}{Z_\tau}\log\frac{Z_T}{Z_\tau}\right], \end{aligned} \tag{3.6}$$

and these inequalities become equalities for $\bar{N} = 1/Z$.

Theorem 3.1. *1. For any $Q \in \mathcal{Q}^{\mathrm{ngd}}$ with density process Z and any X satisfying (3.3), it holds that any tradable numeraire $\bar{N} = 1 + \bar{\vartheta} \cdot \bar{S} \in \mathcal{N}(\bar{S})$ in the extended market (3.2) satisfies*

$$E_T\left[\log \frac{\bar{N}_\tau}{\bar{N}_T}\right] \le E_T\left[-\log \frac{Z_\tau}{Z_T}\right] \le \frac{1}{2} E_T\left[\int_T^\tau h_u^2\, du\right] \tag{3.7}$$

for all stopping times $T \le \tau \le \bar{T}$.

2. The former inequality in (3.7) is sharp: For any $Q \in \mathcal{Q}^{\mathrm{ngd}}$ with density process Z, there exists a real-valued X satisfying (3.3) (with $d' = 1$) such that $\bar{N} :=$

$S'(X,Q)/S'_0(X,Q)$ in $\mathcal{N}(\bar{S})$ is a tradable numeraire in the extended market (3.2), satisfying

$$E_T\left[\log \frac{\bar{N}_\tau}{\bar{N}_T}\right] = E_T\left[-\log \frac{Z_\tau}{Z_T}\right] \tag{3.8}$$

for all stopping times $T \le \tau \le \bar{T}$.

Proof. Equation (3.6) and the remark following it imply the claim. □

The first part of the theorem shows that $\mathcal{Q}^{\mathrm{ngd}}$ is defined such that all market extensions (3.2) based on $Q \in \mathcal{Q}^{\mathrm{ngd}}$ will respect the good-deal bounds on expected growth rates specified via h. The second part shows that the duality relation is sharp, in that the duality bound (3.6) is attained by a suitable market extension with respect to a given Q. Whether the second inequality in (3.7) is sharp and might be attained for some Q with equality, in general depends on the market model under consideration. For instance, if $\mathcal{M}^e$ were a singelton $\{\widehat{Q}\}$ with $\widehat{Q} \in \mathcal{Q}^{\mathrm{ngd}}$, the second inequality in (3.7) easily becomes strict if h is taken somewhat larger.

4 A financial market model with Itô processes

In the next sections, we are going to obtain more explicit results on good-deal valuation bounds arising from restrictions on expected growth rates, in comparison to the convex duality results for general semimartingale models in the previous section. To this end, we study a model where the dynamics of financial market prices S are described by Itô processes on a Wiener space. This permits to use in Section 5 the well-developed theory of backward stochastic differential equations (BSDEs) to describe the dynamics of the good-deal valuation bounds and a corresponding notion of hedging more explicitly.

An Itô process model

From the remainder of the paper, we strengthen the assumptions from Section 2 on the underlying model, by assuming that the filtration $\mathbb{F}$ of our stochastic basis $(\Omega, \mathcal{F}, \mathbb{F}, P)$ is the filtration generated by an n-dimensional Brownian motion (Wiener process) $(W_t)_{t\in[0,\bar{T}]}$, completed by nullsets. It is well known that $\mathbb{F}$ is then right-continuous. Furthermore, we assume there are $d \le n$ risky assets whose price processes $S = (S^i_t)_{1\le i\le d}$ are described by the unique solution to the stochastic differential equation (SDE)

$$dS^i_t = S^i_t dR^i_t\,, \quad t \le \bar{T},\ 1 \le i \le d\,, \tag{4.1}$$

with $S^i_0 > 0$, where the return process $R = (R^i_t)_{1\le i\le d}$ is given by the solution to the SDE

$$dR_t = \gamma_t\, dt + \sigma_t\, dW_t\,. \tag{4.2}$$

We assume that γ and σ are predictable processes taking values in $\mathbb{R}^d$ and $\mathbb{R}^{d\times n}$. The volatility process σ is taken to have full rank d, in the sense that

$$\det \sigma\sigma^{\mathrm{tr}} \neq 0 \quad (P \times dt - a.e.). \tag{4.3}$$

Denoting the market price of risk process by

$$\xi := \sigma^{\mathrm{tr}}(\sigma\sigma^{\mathrm{tr}})^{-1}\gamma \tag{4.4}$$

one can write the SDE describing R compactly as

$$R_0 = 0\,, \quad dR_t = \sigma_t(\xi_t dt + dW_t) =: \sigma_t d\widehat{W}_t\,, \quad t \le \bar{T}\,. \tag{4.5}$$

We assume that there exists some $\varepsilon \in (0, \infty)$ such that, for h from (2.5), it holds

$$h - |\xi| > \varepsilon \quad (dP \times dt - a.e.)\,. \tag{4.6}$$

By boundedness of h, this implies in particular that

the market price of risk process ξ is bounded. (4.7)

Each $S^i = S^i_0 \mathcal{E}(R^i)$ is a stochastic exponential and can be written explicitly as

$$S^i_t = S^i_0 \exp\left(\int_0^t (\sigma_u)^i dW_u + \int_0^t (\sigma_u \xi_u)^i - \frac{1}{2}(\sigma_u \sigma_u^{\mathrm{tr}})^{ii} du \right)\,.$$

For subsequent analysis, it turns out to be more convenient to describe trading strategies not in terms of numbers $\vartheta = (\vartheta^i_t)$ of risky assets held, but instead by amounts of wealth $\varphi = (\varphi^i_t)$ invested in each of the risky assets. To this end, we define the set Φ_φ of permitted trading strategies to consist of those predictable processes $\varphi = (\varphi^i_t)$ which satisfy

$$E\left[\int_0^{\bar{T}} |\varphi_t^{\mathrm{tr}} \sigma_t|^2 dt \right] < \infty\,. \tag{4.8}$$

The wealth process V that is obtained from initial wealth V_0 by investing according to $\varphi \in \Phi$ is given by the solution to the SDE

$$dV_t = \varphi_t^{\mathrm{tr}} dR_t = \varphi_t^{\mathrm{tr}} \sigma_t(\xi_t dt + dW_t) = \varphi_t^{\mathrm{tr}} \sigma_t d\widehat{W}_t\,, \tag{4.9}$$

where all occurring integral terms are being well defined thanks to

$$\int_0^{\bar{T}} |\xi_t^{\mathrm{tr}} \sigma_t^{\mathrm{tr}} \varphi_t| dt \le \left(\int_0^{\bar{T}} |\sigma_t^{\mathrm{tr}} \varphi_t|^2 dt \right)^{\frac{1}{2}} \left(\int_0^{\bar{T}} |\xi_t|^2 dt \right)^{\frac{1}{2}} < \infty\,.$$

Remark 4.1. In the Itô process framework for the financial market S in this section the risky asset prices are modeled as continuous processes. This is like in the continuous time models in [12] or [10], and different from [7] where asset prices can also have jumps. On the other hand, differently from [12, 10, 7] our dynamics (4.1) do not impose any Markov structure on the evolution of S, neither alone nor jointly with additional factor processes.

Parameterisation of strategies

Strategies φ in Φ_φ have as clear financial meaning in terms of wealth invested in the risky assets. Still, it will be technically convenient to re-parameterise strategies in terms of integrands with respect to $d\widehat{W} = \xi_t dt + dW_t$. To this end, let

$$C_t := \operatorname{Im} \sigma_t^{\mathrm{tr}}\,, \quad t \le \bar{T},$$

denote the image (range) of $\sigma_t^{\mathrm{tr}} \in \mathbb{R}^{n\times d}$. With any $\varphi \in \Phi$ one can associate the image of φ under σ^{tr}, i.e.

$$\phi_t = \sigma_t^{\mathrm{tr}} \varphi_t \in C_t\,, \quad t \le \bar{T}\,,$$

and write the evolution of wealth (4.9) conveniently as

$$dV = \phi^{\mathrm{tr}}(\xi dt + dW) = \phi^{\mathrm{tr}} d\widehat{W}\,. \tag{4.10}$$

We let $\Phi := \Phi_\phi := \left\{\phi \,\middle|\, \phi = \sigma_t^{\mathrm{tr}}\varphi_t\,, \varphi \in \Phi_\varphi\right\}$. Then, by definition of Φ_φ,

$$\Phi = \Phi_\phi = \left\{\phi \,\middle|\, \phi \text{ is predictable}, \ \phi_t \in C_t\ \forall t,\ E\left[\int_0^{\bar{T}} |\phi_t|^2\, dt\right] < \infty\right\}.$$

By applying the pseudo-inverse $\sigma^{\mathrm{tr}+} := (\sigma\sigma^{\mathrm{tr}})^{-1}\sigma$ to $\phi = \sigma^{\mathrm{tr}}\varphi$, one re-obtains $\varphi = \sigma^{\mathrm{tr}+}\phi$. Hence, the relations $\varphi = \sigma^{\mathrm{tr}+}\phi$ and $\phi = \sigma^{\mathrm{tr}}\varphi$ provide a bijection between $\Phi = \Phi_\phi$ and Φ_φ. This allows to consider Φ as the set of permitted trading strategies.

Let $\Pi_t = \Pi_{C_t}$ and $\Pi_t^\perp = \Pi_{C_t^\perp}$ denote the orthogonal projections onto $C_t = \operatorname{Im} \sigma_t^{\mathrm{tr}} = (\operatorname{Ker} \sigma_t)^\perp$ and $C_t^\perp = \operatorname{Ker} \sigma_t = (\operatorname{Im} \sigma_t^{\mathrm{tr}})^\perp$, respectively, that are given by

$$\Pi_t : \mathbb{R}^n \to C_t\,, \qquad z \mapsto \Pi_t(z) := \sigma^{\mathrm{tr}}(\sigma_t\sigma_t^{\mathrm{tr}})^{-1}\sigma_t z\,, \tag{4.11}$$

$$\Pi_t^\perp : \mathbb{R}^n \to C_t^\perp\,, \qquad z \mapsto \Pi_t^\perp(z) := (\mathrm{Id} - \Pi_t)(z)\,. \tag{4.12}$$

Clearly, any z in $\mathbb{R}^n = C_t \oplus C_t^\perp = \operatorname{Im} \sigma_t^{\mathrm{tr}} \oplus \operatorname{Ker} \sigma_t$ has a unique orthogonal decomposition

$$z = \Pi_t(z) + \Pi_t^\perp(z) = \Pi_{\operatorname{Im} \sigma_t^{\mathrm{tr}}}(z) + \Pi_{\operatorname{Ker} \sigma_t}(z) \quad \text{in } C_t \oplus C_t^\perp\,. \tag{4.13}$$

Equivalent martingale measures

In the current setting of a Brownian filtration, the equivalent martingale measures $Q \in \mathcal{M}^e$ for S can be parameterised quite explicitly.

In order to construct a martingale measure, one can use the Girsanov transformation to eliminate the drift of S. For instance, since ξ is bounded,

$$\frac{d\widehat{Q}}{dP} := \mathcal{E}\left(-\int \xi dW\right)_{\bar{T}}$$

clearly defines a probability measure $\widehat{Q}$, which is known as the minimal martingale measure, see e.g. [25]. Under $\widehat{Q}$, the process

$$\widehat{W}_0 = 0\,, \quad d\widehat{W}_t = dW_t + \xi_t\, dt \quad \text{for } t \le \bar{T}, \tag{4.14}$$

is Brownian motion.

More generally, the density process Z of any equivalent measure $Q \sim P$ must be a stochastic exponential

$$Z_t = \frac{dQ}{dP}\Big|_{\mathcal{F}_t} \quad = \quad \mathcal{E}(L)_t = \mathcal{E}\left(\int \lambda dW\right)_t, \quad t \le \bar{T},$$

with $dL := \frac{1}{Z} dZ$ being a local martingale of the form $L = \int \lambda dW$ for some predictable $\lambda = \lambda^Q$ with $\int_0^{\bar{T}} |\lambda|^2 \, dt < \infty$. By Girsanov's theorem and Lévy's characterisation

$$W^Q = W - \int \lambda dt \text{ is a } Q\text{-Brownian Motion.}$$

Since

$$\sigma_t d\widehat{W}_t = \sigma_t((\xi_t + \lambda_t)dt + dW_t^Q)$$

holds, Q is in $\mathcal{M}^e$ if and only if $\sigma_t(\xi_t + \lambda_t) = 0$ holds ($P \times dt$-a.e.), that is $\lambda_t^Q = -\xi_t + \eta_t$ with $\eta_t = \eta_t^Q \in \operatorname{Ker} \sigma_t = C_t^\perp$. Hence, any equivalent martingale measure Q for S must have a density process of the form

$$Z_t := \frac{dQ}{dP}\Big|_t = \mathcal{E}\left(\int \lambda dW\right)_t = \mathcal{E}\left(-\int \xi \, dW\right)_t \mathcal{E}\left(\int \eta \, dW\right)_t, \quad t \le \bar{T}, \quad (4.15)$$

with $\lambda = -\xi + \eta$ satisfying $-\xi = \Pi_\cdot(\lambda)$ and $\eta = \Pi_\cdot^\perp(\lambda)$ ($P \times dt$-a.e.). Since $\sigma\eta = 0$ implies $\eta^{\text{tr}}\xi = 0$, the second equality in (4.15) holds by Yor's formula.

If $d = n$ (as many risky assets as sources of noise), it thus holds that $\eta = \sigma^{-1}0 = 0$, hence $\widehat{Q}$ is the unique equivalent local martingale measure for S. In that case, the market is complete by the strong predictable representation theorem.

The next result summaries the convenient parameterisation of $\mathcal{Q}$ and $\mathcal{Q}^{\text{ngd}}$.

Proposition 4.2. *1. Any $Q \in \mathcal{Q}$ has a density process $Z = Z^Q$ of the form (4.15) with a predictable process $\lambda = -\xi + \eta$ with $\Pi_t(\lambda_t) = -\xi_t$ and $\Pi_t^\perp(\lambda_t) = \eta_t$, satisfying $\int_0^T |\lambda|_t^2 dt < \infty$. In particular, $\xi^{\text{tr}}\eta = 0$. The processes λ, η, ξ are unique ($P \times dt$-a.e.). For $Q \in \mathcal{Q}^{\text{ngd}} \subset \mathcal{Q}$, it holds in addition that $|\lambda|^2 = |\xi|^2 + |\eta|^2 \le h^2$.*

2. In turn, any predictable λ with $|\lambda|^2 \le h^2$ and $\Pi_t(\lambda_t) = -\xi_t$ ($P \times dt$-a.e.) defines a density process Z of the form (4.15) for some $Q \in \mathcal{Q}^{\text{ngd}}$ with $\eta = \Pi^\perp(\lambda)$.

3. For $Q \in \mathcal{Q}$, the Doob–Meyer decomposition (2.7) for $-\log Z$ is given by

$$M_t \quad = \quad \int_0^t \xi - \eta \, dW, \quad (4.16)$$

$$A_t \quad = \quad \frac{1}{2}\int_0^t |\lambda_s|^2 \, ds, \quad t \le \bar{T}. \quad (4.17)$$

Proof. Part 3 and the first claims of part 1 follows from the foregoing discussion, and $|\lambda| \le h$ ($P \times dt$-a-e.) then holds by Proposition 2.3, which implies that the predictable set $B = \{|\lambda| > h\}$ is a nullset with respect to $P \times dt$ for $\lambda = \lambda^Q$ from $Q \in \mathcal{Q}^{\mathrm{ngd}}$. Concerning part 2, assumption $|\lambda|^2 \leq h^2$ and boundedness of h imply that Z is a martingale and defines a measure $Q \sim P$, which is martingale measure by the foregoing discussion. The integrability condition (2.5) defining $\mathcal{Q}^{\mathrm{ngd}}$ is readily verified for Q from $|\lambda|^2 = |\xi|^2 + |\eta|^2 \leq h^2$. □

Backward SDEs

This section introduces the notion and recalls some classical results on standard BSDEs whose generator satisfies a Lipschitz condition, as stated in [15].

For $p \in (1, \infty)$, we denote by $\mathcal{S}^p_{\bar{T}} = \mathcal{S}^p_{\bar{T}}(P)$ the space of real valued adapted RCLL processes Y with the norm $\|Y\|_{\mathcal{S}^p_{\bar{T}}} := E\left[\sup_{t \leq \bar{T}} |Y_t|^p\right]^{1/p} < \infty$. Let $\mathcal{H}^p_{\bar{T}} = \mathcal{H}^p_{\bar{T}}(P)$ denote the space of predictable $\mathbb{R}^n$-valued processes Z with the norm $\|Z\|_{\mathcal{H}^p_{\bar{T}}} := E\left[(\int_0^{\bar{T}} |Z_t|^2\, dt)^{p/2}\right]^{1/p} < \infty$. Let $BMO(P)$ denote the subspace of those $Z \in \mathcal{H}^2_{\bar{T}}(P)$ which satisfy that there is some $c \in \mathbb{R}_+$ such that $E_T\left[\int_T^{\bar{T}} |Z_t|^2\, dt\right] < c$ for all stopping times T. For any Z in $BMO(P)$, the process $Z \cdot W$ is called a P-BMO martingale. The abbreviation BMO stands for 'bounded mean oscillation', see Chapter X in [17].

A (simplified) standard generator of a BSDE is a measurable function

$$f : (\Omega \times [0, \bar{T}] \times \mathbb{R}^n, \mathcal{P} \times \mathcal{B}(\mathbb{R}^n)) \;\to\; (\mathbb{R}, \mathcal{B}(\mathbb{R}))$$

which is such that $f_t(z) = f(\omega, t, z)$ satisfies $(f_t(0))_{t \leq \bar{T}} \in \mathcal{H}^2_{\bar{T}}$, and $P \times dt$-a.e.

$$|f_t(z) - f_t(z')| \leq L_f |z - z'| \quad \text{for all } z, z' \in \mathbb{R}^n\,,$$

with some Lipschitz constant $L_f < \infty$. For given BSDE standard parameters (f, X), which we take to be given by a standard generator f and a terminal condition $X \in L^2(P)$, a solution to the BSDE

$$Y_{\bar{T}} = X \quad \text{and} \quad -\,dY_t = f_t(Z_t)dt - Z_t\, dW_t\,, \quad t \leq \bar{T}, \tag{4.18}$$

is a tuple (Y, Z) of processes in $\mathcal{S}^2_{\bar{T}} \times \mathcal{H}^2_{\bar{T}}$, satisfying

$$Y_t = X + \int_t^{\bar{T}} f_u(Z_u)\, du - \int_t^{\bar{T}} Z\, dW \quad \text{for all } t \leq \bar{T}\,. \tag{4.19}$$

Since it is sufficient for our purposes, we consider simplified generators f, in comparison to [15], that do not depend on Y.

Proposition 4.3. *For given standard BSDE parameters (f, X), there exists a unique solution $(Y, Z) \in \mathcal{S}^2_{\bar{T}} \times \mathcal{H}^2_{\bar{T}}$ to the BSDE (4.18).*

This result holds by Theorems 2.1 and 5.1 in [15]. It ensures that unique solutions exists for the BSDEs in subsequent sections. Let us note that, more generally, there is even a unique (Y, Z) in $\mathcal{S}^p_{\bar{T}} \times \mathcal{H}^p_{\bar{T}}$ $(1 < p < \infty)$ satisfying (4.19) if the BSDE data satisfy $X \in L^p(P)$ and condition $(f_t(0))_{t\le\bar T} \in \mathcal{H}^2_{\bar T}$ is replaced by $(f_t(0))_{t\le\bar T} \in \mathcal{H}^p_{\bar T}$, see [15], Theorem 5.1.

The main argument for identifying solutions to optimal control problems in the sequel in terms of BSDE solutions is provided by the next result, which is a simplified version of Proposition 3.1 in [15].

Proposition 4.4. *For a family of standard parameters (f, X) and $\{(f^\alpha, X)\}$, with α from an arbitrary index set, let (Y, Z) and (Y^α, Z^α) denote the solution to the corresponding BSDEs. If there exists $\bar\alpha$ such that*

$$f_t(Z_t) = f_t^{\bar\alpha}(Z_t) = \operatorname*{ess\,inf}_{\alpha} f_t^\alpha(Z_t)\,, \quad P\times dt - a.e.,$$

then $Y_t = \operatorname*{ess\,inf}_{\alpha} Y_t^\alpha = Y_t^{\bar\alpha}$ holds for all $t \le \bar T$, P-a.s..

5 Good-deal valuation and hedging via BSDEs

In this section, we obtain a dynamic description for the good-deal valuation bounds, that arise from no-good-deal restrictions on optimal expected growth rates. The valuation bounds are given by the solutions to standard non-linear backward SDEs, whose generator satisfies a Lipschitz condition. Moreover, we develop a corresponding notion of hedging and show that also the hedging strategy is described by a BSDE.

Dynamic good-deal valuation bounds

In extension to Definition 2.8, let us define for $X \in L^2(P) \supset L^\infty$

$$\pi_t^u(X) := \operatorname*{ess\,sup}_{Q\in\mathcal{Q}^{\mathrm{ngd}}} E_t^Q[X]\,, \quad t \le \bar T\,. \tag{5.1}$$

Using an L^2 space for X fits conveniently with the present BSDE setting. As results in the present section show, (5.1) induces a mapping $L^2(P) \to L^2(P, \mathcal{F}_t)$. Using results from standard BSDE theory one can check that properties 1.-4. of Proposition 2.6 are maintained for $X \mapsto \pi_t^u(X)$ ($t \le \bar T$) with $X \in L^2(P)$ and that properties 5.-6. follow from (5.1).

Lemma 5.1. *Let $Q \sim P$ with density process $\frac{dQ}{dP}|_{\mathbb{F}} =: D = \mathcal{E}(\int \lambda dW)$ for a predictable and bounded process λ. Then there exists a unique solution $(Y, Z) \in \mathcal{S}^2_{\bar T} \times \mathcal{H}^2_{\bar T}$ to the linear BSDE*

$$Y_{\bar T} = X\,, \quad -dY_t = -\lambda_t^{\mathrm{tr}} Z_t dt - Z_t\, dW_t\,, \quad t \le \bar T\,, \tag{5.2}$$

for $X \in L^2(P)$. Moreover Y is a Q-martingale and $W^\lambda := W - \int \lambda dt$ is a Q Brownian motion, satisfying

$$Y_t = E_t^Q[X] = Y_0 + \int_0^t Z\, dW^\lambda\,, \quad t \le \bar T\,.$$

If $X \in L^\infty$, then Y is bounded and Z is in $BMO(P)$, i.e. $Z \cdot W$ is a P-BMO martingale.

Proof. Since λ is bounded, the parameters of the linear BSDE (5.2) are standard and it has a unique solution (Y, Z) in $\mathcal{S}^2_{\bar{T}} \times \mathcal{H}^2_{\bar{T}}$. By application of Itô's formula, the process DY is seen to be a local P-martingale. Because D is in $\mathcal{S}^p_{\bar{T}}(P)$ for any $p < \infty$, DY is in $\mathcal{S}^{2-\varepsilon}_{\bar{T}}(P)$ for any $\varepsilon > 0$ and thus a P-martingale. Hence, Y is a Q-martingale with $Y_t = E^Q_t[X]$. If $X \in L^\infty$, $Y - Y_0 = Z \cdot W^\lambda$ is bounded, so $Z \cdot W^\lambda$ is a Q-BMO martingale. This implies that $Z \cdot W$ is a P-BMO martingale by Theorem 3.6 from [19], noting that $dP/dQ|_{\mathbb{F}} = \mathcal{E}(-\lambda \cdot W^\lambda)$ holds for $\lambda \cdot W^\lambda$ being a Q-BMO martingale due to the boundedness of λ. □

Recall that the density process of any $Q \in \mathcal{Q}^{\text{ngd}}$ is determined by the predictable process $\eta = \eta^Q = \Pi^\perp(\lambda^Q)$ from Proposition 4.2. For any $\eta = \eta^Q$ let $(Y^\eta, Z^\eta) \in \mathcal{S}^2_{\bar{T}} \times \mathcal{H}^2_{\bar{T}}$ denote the solution to the BSDE

$$\begin{aligned} -dY_t &= \left(-\xi_t^{\text{tr}}\Pi_t(Z_t) + \eta_t^{\text{tr}}\Pi_t^\perp(Z_t)\right) dt - Z_t\, dW_t\,, \quad t \le \bar{T}\,, \\ Y_{\bar{T}} &= X\,. \end{aligned} \tag{5.3}$$

We are going to demonstrate that the good-deal bound $\pi^u_t(X)$ is described by the solution to the BSDE

$$\begin{aligned} -dY_t &= \left(-\xi_t^{\text{tr}}\Pi_t(Z_t) + \sqrt{h_t^2 - |\xi_t|^2}\,\left|\Pi_t^\perp(Z_t)\right|\right) dt - Z_t\, dW_t\,, \quad t \le \bar{T}\,, \\ Y_{\bar{T}} &= X\,. \end{aligned} \tag{5.4}$$

Theorem 5.2. *For $X \in L^2(P)$, let (Y, Z) and the family (Y^η, Z^η) (for $\eta = \eta^Q$ with $Q \in \mathcal{Q}^{\text{ngd}}$) be the solutions to the standard BSDEs (5.4) and (5.3), respectively. Then, there exists $\bar{Q} = Q^{\bar{\eta}} \in \mathcal{Q}^{\text{ngd}}$ corresponding (by (4.15)) to*

$$\bar{\eta} = \eta^{\bar{Q}} = \frac{\sqrt{h^2 - |\xi_t|^2}}{|\Pi_t^\perp(Z)|}\Pi_t^\perp(Z)$$

such that

$$Y_t = \operatorname*{ess\,sup}_\eta Y_t^\eta = Y_t^{\bar{\eta}}\,, \quad t \le \bar{T}\,, \tag{5.5}$$

holds, with η ranging over all $\eta = \eta^Q$ for $Q \in \mathcal{Q}^{\text{ngd}}$. Moreover, the upper good-deal bound for X is given by

$$\pi_t^u(X) = \operatorname*{ess\,sup}_{Q \in \mathcal{Q}^{\text{ngd}}} E_t^Q[X] = E_t^{\bar{Q}}[X] = Y_t\,, \quad t \le \bar{T}\,. \tag{5.6}$$

This result not surprising. Noting the definition of the good-deal bound as a supremum of conditional expectation, the equality (5.6) is basically a special case of the dual representation for g-conditional risk measures (respectively, for non-linear g-expectations), which are defined by solutions to BSDEs with suitable generators g, see [4]. We give a short direct proof in our setting, to show how Proposition 4.4 provides the essential argument for the equality (5.6) and to identify $\bar{\eta}$ and $\bar{Q}$ explicitly for π^u.

Proof. Comparing the BSDE generators in (5.4) and (5.3), evaluated at Z, one sees that for any η the inequality

$$-\xi_t^{\mathrm{tr}}\Pi_t(Z_t) + \eta_t^{\mathrm{tr}}\Pi_t^{\perp}(Z_t) \leq -\xi_t^{\mathrm{tr}}\Pi_t(Z_t) + \sqrt{h^2 - |\xi_t|^2}|\Pi_t^{\perp}(Z_t)|$$

holds ($P \times dt$-a.e.) by the definition of $\mathcal{Q}^{\mathrm{ngd}}$, since $|-\xi + \eta|^2 = |\xi|^2 + |\eta|^2 \leq h^2$ with equality holding for $\bar{\eta} := \sqrt{h^2 - |\xi_t|^2}\Pi_t^{\perp}(Z)/|\Pi_t^{\perp}(Z)|$. By Proposition 4.4 thus follows (5.5). By Lemma 5.1, we have $Y_t^{\eta} = E_t^Q[X]$ for $\eta = \eta^Q$ and $Y_t^{\bar{\eta}} = E_t^{\bar{Q}}[X]$. Hence, (5.5) and the definition of $\pi^u(X)$ imply (5.6). □

Dynamic good-deal hedging

We are going to show in this section, that a hedging strategy which minimises a suitable dynamic coherent risk measure is naturally linked to the good-deal valuation bounds.

To this end, let $\mathcal{P}^{\mathrm{ngd}}$ denote the set of those equivalent probability measures Q whose density process is of the form $\mathcal{E}(\int \lambda dW)$ for a predictable bounded process λ with $|\lambda| \leq h$, that is

$$\mathcal{P}^{\mathrm{ngd}} := \left\{ Q \sim P \;\Big|\; \frac{dQ}{dP}\Big|_{\mathbb{F}} = \mathcal{E}\left(\int \lambda dW\right) \text{ for } \lambda \text{ predictable with } |\lambda| \leq h \right\}. \tag{5.7}$$

The notation $\mathcal{P}^{\mathrm{ngd}}$ is motivated by the following observation. In Section 3, it has been shown that the defining properties for $\mathcal{Q}^{\mathrm{ngd}}(S)$ ensure two properties for any market $\bar{S} = (S, S')$ (3.2) that is enlarged by an additional price process S' which is obtained by some pricing measure Q from $\mathcal{Q}^{\mathrm{ngd}}$. Firstly, the enlarged market is free of arbitrage since $Q \in \mathcal{M}^e(S)$. Secondly, it does not permit investment opportunities whose expected growth rates exceed the good-deal restrictions. How would the situation be, if we would take a step back and reduce instead of increase the number of risky assets? Suppose we start from a trivial initial market without risky assets S where only the riskless numeraire asset with price 1 is tradable. Since $\mathcal{M}^e(1) = \{Q | Q \sim P\}$, the integrability condition in the definition (2.5) of $\mathcal{Q}^{\mathrm{ngd}}(1)$ yields $\mathcal{Q}^{\mathrm{ngd}}(1) = \mathcal{P}^{\mathrm{ngd}}$. That means, $\mathcal{P}^{\mathrm{ngd}}$ is such that any market $(1, S')$, whose price processes S' are given by (3.1) for some $Q \in \mathcal{P}^{\mathrm{ngd}}$, does not permit trading opportunities that are 'too good'.

In analogy to the definition of $\pi^u(X) = \pi^u(X; \mathcal{Q}^{\mathrm{ngd}})$ in (2.16), we define

$$\rho_t(X) := \operatorname*{ess\,sup}_{Q \in \mathcal{P}^{\mathrm{ngd}}} E_t^Q[X], \quad t \leq \bar{T}, \tag{5.8}$$

for $X \in L^2(P)$. That is, $\rho(X) = \pi^u(X; \mathcal{Q}^{\mathrm{ngd}}(1))$ is of the same 'good-deal'-type as $\pi^u(X) = \pi^u(X; \mathcal{Q}^{\mathrm{ngd}}(S))$ but defined with respect to 1 instead of S.

Due to m-stability of $\mathcal{P}^{\mathrm{ngd}}$, the mapping ρ satisfies by Proposition 2.6 the same good dynamic properties (on L^∞) as π^u and is thus a dynamic coherent risk measure. By Proposition 5.3, $X \mapsto \rho_t(X)$ induces a mapping $L^2(P) \to L^2(P, \mathcal{F}_t)$ and, in the same way as with π^u, one can show that for ρ the properties stated in Proposition 2.6 for $X \in L^\infty$ even hold on $L^2(P)$. By $\mathcal{P}^{\mathrm{ngd}} \supset \mathcal{Q}^{\mathrm{ngd}} = \mathcal{Q}^{\mathrm{ngd}}(S)$, it is clear that $\rho(X) \geq \pi^u(X)$. We are going to show that $\pi_t^u(X)$ is obtained from $\rho_t(X - \int_t^{\bar{T}} \phi \, d\widehat{W})$ by minimising over all permitted trading strategies $\phi \in \Phi$.

To show firstly that $\rho_t(X)$ is described by the solution to the BSDE

$$\begin{aligned} -dY_t &= h|Z_t|\,dt - Z_t\,dW_t\,, \quad t \le \bar{T}\,, \\ Y_{\bar{T}} &= X\,, \end{aligned} \tag{5.9}$$

we consider for any $Q \in \mathcal{P}^{\mathrm{ngd}}$ the BSDE

$$\begin{aligned} -dY_t &= \lambda_t^{\mathrm{tr}} Z_t\,dt - Z_t\,dW_t\,, \quad t \le \bar{T}\,, \\ Y_{\bar{T}} &= X\,, \end{aligned} \tag{5.10}$$

where $\lambda = \lambda^Q$ denotes the process in (5.7) that determines the density of Q.

Proposition 5.3. *For $X \in L^2(P)$, let (Y, Z) and the family (Y^λ, Z^λ) (for $\lambda = \lambda^Q$ with $Q \in \mathcal{P}^{\mathrm{ngd}}$) be the solutions to the standard BSDEs (5.9) and (5.10), respectively. Then, there exists $\widehat{Q} = Q^{\widehat{\lambda}} \in \mathcal{P}^{\mathrm{ngd}}$ corresponding (by (5.7)) to $\widehat{\lambda} = \lambda^{\widehat{Q}} = hZ/|Z|$, such that*

$$Y_t = \operatorname*{ess\,sup}_{\lambda} Y_t^\lambda = Y_t^{\widehat{\lambda}}\,, \quad t \le \bar{T}\,, \tag{5.11}$$

holds, with λ ranging over all $\lambda = \lambda^Q$ for $Q \in \mathcal{P}^{\mathrm{ngd}}$. Moreover,

$$\rho_t(X) = \operatorname*{ess\,sup}_{Q \in \mathcal{P}^{\mathrm{ngd}}} E_t^Q[X] = E_t^{\widehat{Q}}[X] = Y_t\,, \quad t \le \bar{T}\,. \tag{5.12}$$

Since the proof for this result is very similar to the one of Theorem 5.2, we leave the details to the reader.

To motivate the next result on hedging, consider an investor who holds a contingent claim and is obliged to pay the liability X at maturity $\bar{T}$. If he measures his risk by the 'a priori' dynamic coherent risk measure ρ_t, he would assign at time t the monetary risk $\rho_t(X)$ to his liability if he had no access to the financial market S. By dynamic trading over the remaining period $(t, \bar{T}]$ according to some strategy $\phi \in \Phi$, he can transform his liability to $X - \int_t^{\bar{T}} \phi\,d\widehat{W}$. Accessing his risk in terms of the risk measure ρ_t, he should thus aim to trade according to some ϕ^* which minimises $\rho_t(X - \int_t^{\bar{T}} \phi\,d\widehat{W})$ at any time t. We are going to show that this is a well-posed optimal control problem whose value function turns out to be the good-deal bound $\pi_t^u(X)$ and whose optimal strategy ϕ^* can be obtained from the solution to the BSDE (5.4).

To this end, we consider for any permitted trading strategy $\phi \in \Phi$ the solution (Y^ϕ, Z^ϕ) to the BSDE

$$\begin{aligned} -dY_t &= \left(-\xi_t^{\mathrm{tr}} \phi_t + h_t|\phi_t - Z_t|\right)\,dt - Z_t\,dW_t\,, \quad t \le \bar{T}\,, \\ Y_{\bar{T}} &= X\,. \end{aligned} \tag{5.13}$$

This BSDE has a standard generator since h and ξ are bounded and $\phi \in \Phi = \mathcal{H}^2_{\bar{T}}$.

Theorem 5.4. *For $X \in L^2(P)$, let (Y, Z) and the family (Y^ϕ, Z^ϕ) (for $\phi \in \Phi$) be the solutions to the standard BSDEs (5.4) and (5.13), respectively. Then the strategy*

$$\phi^* = \frac{|\Pi^\perp(Z)|}{\sqrt{h^2 - |\xi|^2}}\xi + \Pi(Z) \tag{5.14}$$

is in Φ *and satisfies*

$$Y_t = \operatorname*{ess\,inf}_{\phi\in\Phi} Y_t^{\phi} = Y_t^{\phi^*}, \quad t \le \bar{T}. \tag{5.15}$$

Moreover, the upper good-deal bound for X *is given by*

$$\pi_t^u(X) = Y_t = \operatorname*{ess\,inf}_{\phi\in\Phi} \rho_t\left(X - \int_t^{\bar{T}} \phi\, d\widehat{W}\right) = \rho_t\left(X - \int_t^{\bar{T}} \phi^*\, d\widehat{W}\right) \tag{5.16}$$

for $t \le \bar{T}$*, and* ϕ^* *is the unique* $(P \times dt)$ *minimiser from* Φ *for the infimum in (5.16).*

Remark 5.5. 1. Equation 5.16 shows, that the hedging strategy (5.14) minimises the a priori risk measure ρ_t of the residual risk simultaneously for all $t \le \bar{T}$. Being coherent, ρ is a monetary risk measure (see [2]) and the good deal bound can be interpreted as the minimal capital required to make the position acceptable, after optimal hedging according to ϕ^*.

2. By the relation $\pi^\ell(X) = -\pi^u(-X)$ between lower and upper good-deal bounds, the result also yields the lower good-deal bound and the corresponding hedging strategy. Since the BSDE (5.4) is non-linear, upper and lower bounds as well as the respective hedging strategies are different, in general.

3. By (5.16) and (5.14), not only the good deal bounds but also the hedging strategies are given explicitly in terms of solutions (Y, Z) to the BSDE (5.4).

4. If $X \in L^2(P)$ is replicable by some $\phi^X \in \Phi$, in the sense that $X = c + \phi^X \cdot \widehat{W}_{\bar{T}}$ with $c \in \mathbb{R}$, then the solution (Y, Z) to the linear BSDE $-dY_t = -\xi_t^{\mathrm{tr}} Z_t\, dt - Z_t\, dW_t$ with $Y_{\bar{T}} = X$ (see Lemma 5.1) satisfies $Z = \phi^X$ and $Y_t = E_t^Q[X] = c + \phi^X \cdot \widehat{W}_t$ for any $Q \in \mathcal{Q}^{\mathrm{ngd}}$. Since $\phi_t^X \in C_t = \operatorname{Ker} \Pi_t^\perp$ holds, uniqueness of the BSDE solution and (5.14) imply that $\phi^* = \phi^X$.

5. Since good-deal bounds (5.16) and the corresponding hedging strategies (5.14) are given in terms of solutions to standard BSDEs, they can be computed by available numerical methods for BSDEs. Monte Carlo simulation methods for BSDEs are of particular relevance for problems in higher dimensions. We refer to [8, 16, 6] and references therein for advances in this field.

Proof of Theorem 5.4. That ϕ^* from (5.14) is indeed in Φ follows since Z is in $\mathcal{H}_{\bar{T}}^2$, ξ is bounded (4.7), and $|h| - |\xi| > \varepsilon$ holds by (4.6). Comparing the BSDE generators evaluated at Z, we obtain by Lemma 6.1 that for any $\phi \in \Phi$ the inequality

$$-\xi_t^{\mathrm{tr}} \phi_t + h_t |\phi_t - Z_t| \ge -\xi_t^{\mathrm{tr}} \Pi_t(Z_t) + \sqrt{h_t^2 - |\xi_t|^2}\, |\Pi_t^\perp(Z_t)|$$

holds ($P \times dt$-a.e.) with equality holding for ϕ^*. Hence, equation (5.15) follows by Proposition 4.4. The first equality in (5.16) thus holds by Theorem 5.2 since the BSDE solution is unique. To show the remaining equalities in (5.16), let $\widetilde{Y}_t := Y_t - \phi^* \cdot \widehat{W}_t$ $(t \le \bar{T})$. Noting that by definition of ϕ^* and Lemma 6.1

$$-\xi_t^{\mathrm{tr}} \Pi(Z_t) + \sqrt{h_t^2 - |\xi_t|^2} |\Pi_t(Z_t)| + \xi_t^{\mathrm{tr}} \phi_t^* = h_t |Z_t - \phi_t^*|$$

holds, it follows that

$$-d\widetilde{Y}_t = -dY_t + \xi_t^{\mathrm{tr}} \phi_t^* dt + \phi_t^* dW_t = h_t |Z_t - \phi_t^*| dt - (Z_t - \phi_t^*)\, dW_t\, .$$

Hence, equation $\widetilde{Y}_t = \rho_t(\widetilde{Y}_{\bar{T}})$ holds and adding $\phi^* \cdot \widehat{W}_t$ on both sides we obtain $Y_t = \rho_t(Y_{\bar{T}} - \int_t^{\bar{T}} \phi^* \, d\widehat{W}t)$. By analogous arguments one obtains the inequality $Y_t \leq \rho_t(Y_{\bar{T}} - \int_t^{\bar{T}} \phi \, d\widehat{W})$ for any $\phi \in \Phi$. This yields the last equality for (5.16). □

Equation (5.16) justifies to define the unique minimiser ϕ^* as the good-deal hedging strategy for X corresponding to the good-deal valuation bounds π^u, with respect to the dynamic coherent risk measure ρ and the strategy set Φ.

It is a natural idea to define a hedging strategy for a contingent claim as the optimal strategy which minimises some risk measure. Contributions in this direction can be found in several papers, including [3, 4, 9, 18, 21], and the result of Theorem 5.4 with respect to ρ belongs to the same family. Mathematically, the result of Theorem 5.4 fits well within the general theory for inf-convolutions of (BSDE-induced) risk measures, see [3, 4]. Indeed, Theorem 5.4 proves via Lemma 6.1 that the generator in the BSDE (5.4) for π^u is equal to the inf-convolution of the generator in the BSDE (5.9) for ρ with the (formal) generator $-\xi_t Z_t$, restricted to $Z \in C_t$, for the risk measure that is induced by (super-)hedging opportunities in the market; see Section 3.8 in [4]. In this sense, we have worked out a concrete solution to a dynamic inf-convolution problem. In sections 3-4 of [21], a BSDE similar to (5.4) has been obtained for a prototypical model where the martingale component of the risky asset prices is given by independent Brownian motions. As with [3, 4], the focus of [21] is on the minimisation of risk measures but proofs make less use of BSDE theory.

Despite these close relations, the perspective for our problem is in the following aspect opposite to that of the literature cited. For us, the starting point has been not a given a-priori risk measure from which a so-called market-consistent risk measure is to be found (see [4]) by optimal risk-sharing (hedging) with the market. Instead, we have started from the good-deal bounds π^u, which are already market consistent, and have constructed a suitable 'a priori risk' measure ρ in order to find a dynamic notion of good-deal hedging that can be associated to π^u. By this complementary perspective, we address the problem raised in the final conclusions of [7] about linking good-deal valuation to a suitable theory of hedging, what seems to have not been elaborated in the literature so far to our best knowledge.

Since $\pi_t^u(X)$ is the minimal risk (with respect to ρ_t) that is obtainable by optimal hedging when holding the (liability) position X, the position $X - Y_t$ ($Y_t \in L^2(P, \mathcal{F}_t)$) just becomes acceptable at t, in the sense that $\pi_t^u(X - Y_t) \leq 0$, for $Y_t = \pi_t^u(X)$. Considering $Y_t = \pi_t^u(X)$ as the minimal capital required at t (with respect to ρ_t) to hold position X, the process $\pi_0^u - \pi_t^u + \phi^* \cdot \widehat{W}_t$ can be interpreted as the tracking error of the hedging strategy ϕ^* and it is of interest to study its properties, following Remark 4.1 in [21]. In our setting, Theorem 5.4 readily yields

Corollary 5.6. *Under the assumptions and notations of Theorem 5.4, the tracking error* $\pi_0^u(X) - \pi_t^u(X) + \int_0^t \phi_s^* \, d\widehat{W}_s$ $(t \leq \bar{T})$ *of the good-deal hedging strategy* ϕ^* *is a submartingale under any* $Q \in \mathcal{P}^{\text{ngd}}$ *and a martingale unter the measure* $Q^\lambda \in \mathcal{P}^{\text{ngd}}$

corresponding (by (5.7)) to the bounded process

$$\lambda_t = -h_t \frac{\frac{|\Pi_t^\perp(Z_t)|}{\sqrt{h_t^2-|\xi_t|^2}}\xi_t - \Pi_t^\perp(Z_t)}{\left|\frac{|\Pi_t^\perp(Z_t)|}{\sqrt{h_t^2-|\xi_t|^2}}\xi_t - \Pi_t^\perp(Z_t)\right|}, \quad t \le \bar{T}. \tag{5.17}$$

Noting that positive signs of the tracking error correspond to gains and negative signs to losses, the result can be interpreted as a robustness property of the hedging strategy with respect to the family $\mathcal{P}^{\mathrm{ngd}}$ of probability measures as generalised scenarios (cf. [2]), with Q^λ being a worst case scenario in terms of the (conditional) expectation of additional funding needed to maintain the capital requirements when holding on to position X.

In the special case where the contingent claim X is replicable in the sense of part 4 of Remark 5.5, the tracking error vanishes (hence is a martingale under any measure) and equation (5.17) yields $\lambda = 0$ and thus $Q^\lambda = P$.

Proof of Corollary 5.6. First, note that the tracking error process is in $\mathcal{S}_{\bar{T}}^{2-\varepsilon}(Q)$ $(\varepsilon > 0)$ for any $Q \in \mathcal{P}^{\mathrm{ngd}}$ since it is in $\mathcal{S}_{\bar{T}}^2(P)$ and the density dQ/dP is in L^q for any $q < \infty$. Using Theorem 5.4, it follows from (5.14) and (5.4) with $Y_t = \pi_t^u(X)$ that

$$-dY_t + \phi_t^* d\widehat{W}_t = \frac{h_t^2}{\sqrt{h_t^2 - |\xi_t|^2}} |\Pi_t^\perp(Z_t)| dt + \left(\frac{|\Pi_t^\perp(Z_t)|}{\sqrt{h_t^2 - |\xi_t|^2}} \xi_t - \Pi_t^\perp(Z_t) \right) dW_t \tag{5.18}$$

for $t \le \bar{T}$. Since the minimum

$$\min_{\lambda_t} \lambda_t^{\mathrm{tr}} \left(\frac{|\Pi_t^\perp(Z_t)|}{\sqrt{h_t^2 - |\xi_t|^2}} \xi_t - \Pi_t^\perp(Z_t) \right) = -\frac{h_t^2}{\sqrt{h_t^2 - |\xi_t|^2}} |\Pi_t^\perp(Z_t)|,$$

taken over all λ_t in $\mathbb{R}^n$ with $|\lambda_t| \le h_t$, is attained ($P \times dt$-a.e.) by the predictable process λ from (5.17), the claim follows from (5.18) by a change of measure argument. □

6 Appendix

This section states a result on a deterministic convex optimisation problem, which is needed for the proof of Theorem 5.4. We recall our convention that $\frac{0}{0} = 0$.

Lemma 6.1. *Assume $h > 0$ and $\xi, z \in \mathbb{R}^n$. With $d \le n$, let $\sigma \in \mathbb{R}^{d \times n}$ be a matrix with full rank d, and let Π and $\Pi^\perp$ be the orthogonal projections on the linear subspaces $C := \operatorname{Im} \sigma^{\mathrm{tr}}$ respectively $C^\perp = (\operatorname{Im} \sigma^{\mathrm{tr}})^\perp = \operatorname{Ker} \sigma$ in $\mathbb{R}^n$. Let $\xi \in C$ and assume $h > |\xi|$.*

1. *For* $f_0(\phi) := -\xi^{\mathrm{tr}}\phi + h|\phi - z|$*, the ordinary convex minimisation program*

$$\min_{\phi} f_0(\phi) \quad \text{with } \phi \in \mathbb{R}^n \text{ and linear constraint } \Pi^{\perp}(\phi) = 0 \tag{6.1}$$

attains its minimum value

$$f_0(\phi^*) = -\xi^{\mathrm{tr}}\Pi(z) + \sqrt{h^2 - |\xi|^2}\,|\Pi^{\perp}(z)| \tag{6.2}$$

at the unique minimum

$$\phi^* = \frac{|\Pi^{\perp}(z)|}{\sqrt{h^2 - |\xi|^2}}\xi + \Pi(z) \quad \text{in } C\,. \tag{6.3}$$

2. *The maximisation problem* $\max_\lambda \lambda^{\mathrm{tr}} z$ *over* $\lambda \in \mathbb{R}^n$ *with constraint* $|\lambda| \le h$ *attains its maximum at* $\lambda^* = h\frac{z}{|z|}$ *with* $(\lambda^*)^{\mathrm{tr}} z = h|z|$.

Proof. 1. We prove that the Kuhn–Tucker conditions are satisfied. Since the convex function f_0 is differentiable for $\phi \neq z$, its subgradient $\partial f_0(\phi)$ at ϕ is simply the gradient

$$\partial f_0(\phi) = -\xi + \frac{h}{|\phi - z|}(\phi - z) \quad \text{for } \phi \neq z.$$

Noting that

$$|\phi^* - z| = \left| \frac{|\Pi^{\perp}(z)|}{\sqrt{h^2 - |\xi|^2}}\xi - \Pi^{\perp}(z) \right| = \frac{h}{\sqrt{h^2 - |\xi|^2}}|\Pi^{\perp}(z)|\,,$$

it follows that

$$\partial f_0(\phi^*) = -\frac{\sqrt{h^2 - |\xi|^2}}{|\Pi^{\perp}(z)|}\Pi^{\perp}(z) \quad \text{when } \phi^* \neq z\,. \tag{6.4}$$

At $\phi = z$, the subgradient of f_0 includes the closed ball in $\mathbb{R}^n$ with radius $h - |\xi| > 0$ around the origin, hence

$$\{g \in \mathbb{R}^n \,\big|\, |g| \le h - |\xi|\} \subset \partial f_0(\phi^*) \quad \text{when } \phi^* = z\,. \tag{6.5}$$

In either case, there exists by (6.4) and (6.5) a vector of Lagrange multipliers $\lambda^* \in \mathbb{R}^n$ such that

$$\Pi^{\perp}(\lambda^*) \in \partial f_0(\phi^*)\,. \tag{6.6}$$

The constraint $\Pi^{\perp}(\phi^*) = 0$ is satisfied by ϕ^* since ξ and $\Pi(z)$ are in $C = \operatorname{Ker}\Pi^{\perp}$. Since (ϕ^*, λ^*) satisfies the Kuhn–Tucker conditions, optimality of ϕ^* in (6.3) is ensured by the Kuhn–Tucker theorem, see [24]. Direct computation yields (6.2). If $\Pi^{\perp}(z) = 0$, then f_0 is strictly convex at $\phi^* = z$ since $h > |\xi|$. Otherwise, if $\Pi^{\perp}(z) \neq 0$, the restriction of f_0 onto $C = \operatorname{Im}\Pi = \operatorname{Ker}\Pi^{\perp}$ is strictly convex, in particular at ϕ^* from (6.3). Overall, this implies uniqueness of the minimum.

2. This is obvious from $|\lambda^{\mathrm{tr}} z| \le |\lambda||z| \le h|z|$. □

Bibliography

[1] J. P. Ansel and C. Stricker, *Couverture des Actifs Contingents*, Annales de l'Institut Henri Poincaré 30 (1994), pp. 303–315.

[2] P. Artzner, F. Delbaen, J.-M. Eber, and D. Heath, *Coherent Measures of Risk*, Mathematical Finance 9 (1999), pp. 203–228.

[3] P. Barrieu and N. El Karoui, *Inf-Convolution of Risk Measures and Optimal Risk Transfer*, Finance and Stochastics 9 (2005), pp. 269–298.

[4] ______, *Pricing, Hedging and Designing Derivatives with Risk Measures*, Indifference Pricing, Theory and Applications (R. Carmona, ed.), Princeton University Press, 2009, pp. 77–146.

[5] D. Becherer, *The Numeraire Portfolio for Unbounded Semimartingales*, Finance and Stochastics 5 (2001), pp. 327–341.

[6] C. Bender and R. Denk, *A Forward Scheme for Backward SDEs*, Stochastic Processes and their Applications 117 (2007), pp. 1793–1812.

[7] T. Björk and I. Slimko, *Towards a General Theory of Good-Deal Bounds*, Review of Finance 10 (2006), pp. 221–260.

[8] B. Bouchard and N. Touzi, *Discrete Time Approximation and Monte Carlo Simulation for Backward Stochastic Differential Equations*, Stochastic Processes and their Applications 111 (2004), pp. 175–2006.

[9] P. Carr, H. Geman, and D. B. Madan, *Pricing and Hedging in Incomplete Markets*, Journal of Financial Economics 62 (2001), pp. 131–167.

[10] A. Cerny, *Generalized Sharpe Ratios and Asset Pricing in Incomplete Markets*, European Finance Review 7 (2003), pp. 191–233.

[11] A. Cerny and S.D. Hodges, *The Theory of Good-Deal Pricing in Financial Markets*, Mathematical Finance – Bachelier Congress 2000 (H. Geman, Madan D.P., S.R. Plinska, and T. Vorst, eds.), Springer, Berlin,, 2002, pp. 175–202.

[12] J. Cochrane and J. Saá Requejo, *Beyond Arbitrage: Good Deal Asset Price Bounds in Incomplete Markets*, Journal of Political Economy 108 (2000), pp. 79–119.

[13] F. Delbaen, *The Structure of m-Stable Sets and in particular of the Set of Risk Neutral Measures*, Séminaire de Probabilités XXXIX, Lecture Notes in Mathematics 1874, Springer, Berlin, 2006, pp. 215–258.

[14] F. Delbaen and W. Schachermayer, *The Fundamental Theorem of Asset Pricing for Unbounded Stochastic Processes*, Mathematische Annalen 312 (1998), pp. 215–250.

[15] N. El Karoui, S. Peng, and M. C. Quenez, *Backward Stochastic Differential Equations in Finance*, Mathematical Finance 1 (1997), pp. 1–71.

[16] E. Gobet, J.P. Lemor, and X. Warin, *A Regression-Based Monte Carlo Method to Solve Backward Stochastic Differential Equations*, Annals of Applied Probability 15 (2005), pp. 2172–2202.

[17] S. He, J. Wang, and J. Yan, *Semimartingale Theory and Stochastic Calculus*, Science Press, CRC Press, New York, 1992.

[18] S. Jaschke and U. Küchler, *Coherent Risk Measures and Good-Deal Bounds*, Finance and Stochastics 5 (2001), pp. 181–200.

[19] N. Kazamaki, *Continuous Exponential Martingales and BMO*, Lecture Notes in Mathematics 1579, Springer, Berlin, 1994.

[20] S. Klöppel and M. Schweizer, *Dynamic Utility-Based Good-Deal Bounds*, Statistics and Decisions 25 (2007), pp. 285–309.

[21] J. Leitner, *Pricing and Hedging with Globally and Instantaneously Vanishing Risk*, Statistics and Decisions 25 (2007), pp. 311–332.

[22] P. Protter, *Stochastic Integration and Differential Equations*, Springer, Berlin, 2004.

[23] D. Revuz and M. Yor, *Continuous Martingales and Brownian Motion.*, Springer, Berlin, 1994.

[24] R. T. Rockafellar, *Convex Analysis*, Princeton University Press, Princeton, 1970.

[25] M. Schweizer, *A Guided Tour through Quadratic Hedging Approaches*, Option Pricing, Interest Rates and Risk Management (E. Jouini, J. Cvitanić, and M. Musiela, eds.), Cambridge University Press, Cambridge, 2001, pp. 538–574.

Author information

Dirk Becherer, Institut für Mathematik
Humboldt Universität zu Berlin
Unter den Linden 6
D-10099 Berlin, Germany.
Email: `becherer[at]math.hu-berlin.de`

Radon Series Comp. Appl. Math **8**, 53–89

Viscosity solutions to optimal portfolio allocation problems in models with random time changes and transaction costs

Christophette Blanchet-Scalliet, Rajna Gibson Brandon, Benoîte de Saporta, Denis Talay, and Etienne Tanré

Abstract. We consider a risky asset whose instantaneous rate of return takes two different values and changes from one to the other one at random times which are neither known, nor directly observable. We study the optimal allocation strategy of traders who, in the presence of cost of transactions, invest in this risky asset or in a non risky asset according to their belief on the current state of the instantaneous rate of return.

We prove that the related trader's value function is the unique viscosity solution of a system of HJB inequalities. We carefully prove the Dynamic Programming Principle and show results of numerical experiments.

Key words. Stochastic control, HJB inequalities, viscosity solutions, dynamic programming principle, portfolio allocation, transaction costs.

AMS classification. 49L,93E20,91B28

1 Introduction

The practitioners use various rules to rebalance their portfolios. These rules usually come, either from fundamental economic principles, or from mathematical approaches derived from mathematical models, or technical analysis approaches. Technical analysis, which provides decision rules based on past prices behaviour, avoids model specification and thus model risk (for a survey, see, e.g., Achelis [1]). Let us describe the technical analysis methodology in the framework of the detection of changes in stock returns.

There has been a considerable literature over the last three decades emphasising predictability in stock returns. Researchers – such as Jegadeesh and Titman [9] and Lakonishok et al [12] have shown that stock returns are characterised by short term momentum or price continuation patterns. But the latter are only temporary in nature since stock returns also display long term reversals.The recent financial market crisis also clearly demonstrates that observed trends in stock or real estate prices eventually can be subject to sharp interruptions and thus detecting such interruptions or changes of regimes in modelling stock return distributions is of the highest importance. Speculative price bubbles, such as the one that was observed in the US equity housing markets until the subprime crisis or in the dot.com stock market segment at the turn of this Century further suggest that financial markets can be subject to market over-reactions driven by a subset of agents who can temporarily drive asset prices far away form their fundamental values. Once again, such tendencies will eventually correct and lead to sharp corrections that can be quite painful for agents who missed the "turning point". In light of these historically observed price patterns and irrespectively of their- rational or irrational -origins, it is not surprising that technical or chartist methods have always been popular at least among a segment of the trading population and of investors who thereby attempt to identify how long they can "surf the wave" and even more crucially at what time a given price trend or pattern is likely to reverse and thus commands a reversal of their specific transaction. While the origins of price regime changes are certainly debatable within the efficient market paradigm, the pursuit of scientific methods to potentially detect them is worthy of an academic analysis. This statement is reinforced by the fact that with access to intra - daily financial data and with the presence of a growing population of short term traders in pursuit of "quick trades", chartist and scientific methods of price regime changes detection have gained renewed interest.

Chartist methodologies have not been intensively studied from a mathematical point of view. Pastukhov [16] has studied mathematical properties of volatility indicators. Shiryaev and Novikov [18] exhibit an optimal one-time rebalancing strategy in the Black–Scholes model when the drift term of the stock may change its value spontaneously at some random non-observable (hidden) time. Blanchet et al. [4] propose a framework allowing one to compare the performances obtained by various strategies derived from erroneously calibrated mathematical models and the performances obtained by technical analysis techniques, and compare such strategies when the exact model is a diffusion model with one and only one change of stock returns at a random time.

In this paper, motivated by the extension of the analysis made in [4] to models with a random number of changes of stock returns at random times and including transactions costs, we study the corresponding optimal allocation problem: obviously, the value function of this optimal allocation problem with a perfect model calibration is the benchmark for strategies derived from statistical procedures or misspecified mathematical models, as well as for the chartist strategies.

We therefore had to solve a stochastic control problem which, to the best of our knowledge, had not been solved in the literature so far. Related works actually concern other dynamics. For instance, Tang and Yong [20] study optimal switching and impulse controls. Brekke and Øksendal [6] consider optimal switching in an economic activity; Pham [17], Ly Vath and Pham [13] and Ly Vath et al. [14] obtained results on the optimal switching problems and the regularity of the related value functions for families of particular models which do not include our model.

Our paper is organised as follows. We first introduce our model and some notation in Section 2. In Section 3, we list a few useful estimates. In Section 4, we prove the continuity of the value functions of our optimal allocation problem. In Section 5, we rigorously prove the Dynamic Programming Principle. In Section 6 we prove that the value function is a viscosity solution to a system of Hamilton–Jacobi–Bellman (HJB) inequalities; uniqueness is proved in Section 7. Finally, in Section 8, we present numerical approximations of the value function, compare performances of several strategies, and briefly discuss misspecification issues.

2 Description of the model and notation

Consider a market with a deterministic short term rate r, a non risky asset with price process S^0, and a stock with price process S^1 whose instantaneous trend may only take two values μ_1 and μ_2 with $\mu_1 < r < \mu_2$. The changes of trend may occur at random times τ_n defined as follows:

$$\tau_0 = 0,\ \tau_n := \nu_1 + \cdots + \nu_n,$$

where the time intervals ν_j between changes of trend are independent. The ν_{2n+1} (respectively, ν_{2n}) are identically distributed; their common law is exponential with parameter λ_1 (respectively, λ_2). Thus the trend process is

$$\mu(\theta) := \begin{cases} \mu_1 \text{ if } \tau_{2n} \leq \theta < \tau_{2n+1}, \\ \mu_2 \text{ if } \tau_{2n+1} \leq \theta < \tau_{2n+2}. \end{cases}$$

We suppose that the dynamics of the stock price obeys the stochastic differential equation

$$\frac{dS_\theta}{S_\theta} = \mu(\theta)d\theta + \sigma dB_\theta,$$

where (B_θ) is a Brownian motion under the historical probability and $\sigma > 0$ is the constant and deterministic volatility of the stock.

Obviously the trader should totally rebalance his/her portfolio at each change of the trend. We actually consider that he/she will do it at certain decision times which should ideally be equal to τ_n. However the times τ_n cannot be detected exactly and the trader's strategy needs to be assumed progressively measurable w.r.t. the filtration generated by the observed prices, that is, the filtration $\mathcal{F}^S := (\mathcal{F}^S_\theta,\ 0 \leq \theta \leq T)$ which is strictly smaller than the filtration generated by the Brownian motion and the τ_n's. This leads us to introduce the following definition of the admissible strategies.

Definition 2.1. Let T be the investment time period. Denote by π_θ the proportion of wealth invested at time θ in the risky asset and by U a given function (utility function).

Given any time $t \in [0, T]$, an investment strategy (π_θ) over $[t, T]$ is said admissible if it is a piecewise constant càdlàg process taking values in the pair $\{0; 1\}$ which is progressively measurable w.r.t. the filtration $\mathcal{F}^S$ and satisfies

$$\mathbb{E}|U(W_T^\pi)| < +\infty,$$

where W^π denotes the wealth process resulting from the strategy π.

The set of such admissible strategies is denoted by $\mathcal{A}_t$.

We now introduce the Optional Projection process

$$F_\theta := \mathbb{P}(\mu(\theta) = \mu_1 \mid \mathcal{F}^S_\theta).$$

It is a classical result in filtering theory (see, e.g., Kurtz and Ocone [11]) that the process

$$\overline{B}_\theta := \frac{1}{\sigma}\left(\log\frac{S_\theta}{S_0} - \int_0^\theta \left(\mu_1 F_s + \mu_2(1 - F_s) - \frac{\sigma^2}{2}\right) ds\right) \tag{2.1}$$

is a $\mathcal{F}^S$ Brownian motion, and that

$$dF_\theta = (-\lambda_1 F_\theta + \lambda_2(1 - F_\theta))d\theta + \frac{\mu_1 - \mu_2}{\sigma} F_\theta(1 - F_\theta)d\overline{B}_\theta. \tag{2.2}$$

Notice that Feller's criterion ensures that the solution of the preceding SDE takes values in $[0, 1]$ when $0 \leq F_0 \leq 1$. Equation (2.1) clearly yields

$$\frac{dS_\theta}{S_\theta} = (\mu_1 F_\theta + \mu_2(1 - F_\theta))d\theta + \sigma d\overline{B}_\theta, \tag{2.3}$$

from which $\mathcal{F}^S = \mathcal{F}^{\overline{B}}$.

We consider the situation where the trader needs to face proportional transaction costs: given an amount W to transfer from the bank account to the stock, the cost is $g_{01}W$; if W is transfered from the stock to the bank account, then the cost is $g_{10}W$. In view of (2.3), we have, for all $\theta > 0$,

$$\frac{dW_\theta^\pi}{W_{\theta-}^\pi} = (\pi_\theta(\mu_1 F_\theta + \mu_2(1 - F_\theta) - r) + r)\, d\theta + \pi_\theta \sigma d\overline{B}_\theta - g_{01}\mathbb{I}_{\Delta\pi_\theta = 1} - g_{10}\mathbb{I}_{\Delta\pi_\theta = -1}, \tag{2.4}$$

where $\Delta\pi_\theta := \pi_\theta - \pi_{\theta-}$. In addition, if (Z_θ^π) is the continuous part of the process $(\log(W_\theta^\pi))$, we have

$$dZ_\theta^\pi = \left(\pi_\theta(\mu_1 F_\theta + \mu_2(1 - F_\theta) - \frac{\sigma^2}{2} - r) + r\right) d\theta + \pi_\theta \sigma d\overline{B}_\theta. \tag{2.5}$$

We need to introduce some notation.

Notation 2.2. Given a time t in $[0, T]$, $i \in \{0, 1\}$ and an admissible strategy π in $\mathcal{A}_t$, we denote by $(F^{t,f}, Z^{t,z,f,\pi}, W^{t,x,f,i,\pi})$ the solution to (2.2), (2.5), and (2.4) respectively issued at time t from $f \in [0, 1]$, $z \in \mathbb{R}$, and $x > 0$ if $\pi_t = i$ and $x(1 - g_{ij})$ if $\pi_t = j = 1 - i$.

For all $t \leq \theta \leq T$ we also set

$$\begin{aligned} \xi_\theta^{t,i,\pi} &= -\log(1 - g_{01})\mathbb{I}_{\pi_t - i = 1} - \log(1 - g_{10})\mathbb{I}_{\pi_t - i = -1} \\ &\quad - \sum_{t < s \leq \theta} [\log(1 - g_{01})\mathbb{I}_{\Delta\pi_s = 1} + \log(1 - g_{10})\mathbb{I}_{\Delta\pi_s = -1}]. \end{aligned} \tag{2.6}$$

Notice that $\xi^{t,i,\pi}$ is a positive process and that given a time t in $[0, T]$, $i \in \{0, 1\}$ and an admissible strategy π in $\mathcal{A}_t$, the process

$$W_\theta^{t,x,f,i,\pi} = x \exp(Z_\theta^{t,0,f,\pi} - \xi_\theta^{t,i,\pi}) \tag{2.7}$$

is the unique solution of (2.4) issued from $x > 0$ if $\pi_t = i$ and $x(1 - g_{ij})$ if $\pi_t = j = 1 - i$.

We consider a utility function U which is, either the logarithmic utility function, or an element of the set $\mathcal{U}$ of the increasing and concave functions of class $\mathcal{C}^1((0, +\infty); \mathbb{R})$ which satisfy: $U(0) = 0$, and there exist real numbers $C > 0$ and $0 \leq \alpha \leq 1$ such that

$$0 < U'(x) \leq C(1 + x^{-\alpha}) \text{ for all } x > 0. \tag{2.8}$$

Notice that HARA utilities belong to the class $\mathcal{U}$.

For all $i \in \{0; 1\}$, $t \in (0, T]$, $x > 0$, $0 \le f \le 1$, we set

$$\forall \pi \in \mathcal{A}_t, \quad J^i(t, x, f, \pi) := \mathbb{E}[U(W_T^{t,x,f,i,\pi})].$$

Define the value functions as

$$V^i(t, x, f) := \sup_{\pi \in \mathcal{A}_t} J^i(t, x, f, \pi). \tag{2.9}$$

We aim to show that the functions V^i are continuous, satisfy the dynamic programming principle, and that the pair (V^0, V^1) is the unique viscosity solution to a system of Hamilton–Jacobi–Bellman inequalities.

We end this section with an elementary inequality which we will often use in the sequel:

Proposition 2.3. *Under the above assumptions on the utility function U, there exists $C > 0$ such that, for all real numbers z and $\tilde{z}$ and all positive real numbers x, $\tilde{x}$, and ζ,*

$$\begin{aligned} &\left|U\left(xe^{z-\zeta}\right) - U\left(\tilde{x}e^{\tilde{z}-\zeta}\right)\right| \\ &\qquad \le C\left(1 + x^{-\alpha}e^{-\alpha z} + \tilde{x}^{-\alpha}e^{-\alpha\tilde{z}}\right)\left(|x - \tilde{x}| + (x + \tilde{x})\,|z - \tilde{z}|\right)\left(e^z + e^{\tilde{z}}\right), \end{aligned} \tag{2.10}$$

where $\alpha = 1$ if $U(x) = \log(x)$.

Proof. If $U(x) = \log(x)$, as $\log(u) \le u - 1$ for all $u > 0$ we have

$$\begin{aligned} \log\left(\frac{x}{\tilde{x}}e^{(z-\tilde{z})}\right) &\le \frac{x}{\tilde{x}}e^{(z-\tilde{z})} - 1 \\ &\le \frac{1}{\tilde{x}e^{\tilde{z}}}\left((x - \tilde{x})\,e^z + \tilde{x}\left(e^z - e^{\tilde{z}}\right)\right) \\ &\le \frac{1}{\tilde{x}e^{\tilde{z}}}\left(|x - \tilde{x}|\,e^z + \tilde{x}\,|z - \tilde{z}|\left(e^z + e^{\tilde{z}}\right)\right), \end{aligned}$$

and, similarly,

$$\log\left(\frac{\tilde{x}}{x}e^{(\tilde{z}-z)}\right) \le \frac{1}{xe^z}\left(|x - \tilde{x}|\,e^{\tilde{z}} + x\,|z - \tilde{z}|\left(e^z + e^{\tilde{z}}\right)\right).$$

If the function U belongs to the class $\mathcal{U}$, using the monotonicity of U' and (2.8) we get

$$\begin{aligned} \left|U(xe^{z-\zeta}) - U(\tilde{x}e^{\tilde{z}-\zeta})\right| &\le |xe^z - \tilde{x}e^{\tilde{z}}|\,e^{-\zeta}\left(U'(xe^{z-\zeta}) + U'(\tilde{x}e^{\tilde{z}-\zeta})\right) \\ &\le C|xe^z - \tilde{x}e^{\tilde{z}}|\,e^{-(1-\alpha)\zeta}\left(2 + x^{-\alpha}e^{-\alpha z} + \tilde{x}^{-\alpha}e^{-\alpha\tilde{z}}\right). \end{aligned}$$

As $0 \le \alpha \le 1$ and $\zeta \ge 0$, the result follows from the obvious inequality

$$|xe^z - \tilde{x}e^{\tilde{z}}| \le |x - \tilde{x}|e^z + \tilde{x}|z - \tilde{z}|(e^z + e^{\tilde{z}}).$$

□

2.1 Our main result

Consider the system

$$\begin{cases} \min\left\{-\dfrac{\partial V^0}{\partial t} - \mathcal{L}^0 V^0;\ V^0(t,x,f) - V^1(t,x(1-g_{01}),f)\right\} = 0, \\ \min\left\{-\dfrac{\partial V^1}{\partial t} - \mathcal{L}^1 V^1;\ V^1(t,x,f) - V^0(t,x(1-g_{10}),f)\right\} = 0, \end{cases} \tag{2.11}$$

with the boundary condition $V^0(T,x,f) = V^1(T,x,f) = U(x)$, where

$$\begin{aligned} \mathcal{L}^0\varphi(t,x,f) := {} & xr\frac{\partial\varphi}{\partial x}(t,x,f) + (-\lambda_1 f + \lambda_2(1-f))\frac{\partial\varphi}{\partial f}(t,x,f) \\ & + \frac{1}{2}\left(\frac{\mu_1-\mu_2}{\sigma}\right)^2 f^2(1-f)^2\frac{\partial^2\varphi}{\partial f^2}(t,x,f), \end{aligned}$$

and

$$\begin{aligned} \mathcal{L}^1\varphi(t,x,f) := {} & x(\mu_1 f + \mu_2(1-f) - r)\frac{\partial\varphi}{\partial x}(t,x,f) + (-\lambda_1 f + \lambda_2(1-f))\frac{\partial\varphi}{\partial f}(t,x,f) \\ & + \frac{1}{2}x^2\sigma^2\frac{\partial^2\varphi}{\partial x^2}(t,x,f) \\ & + \frac{1}{2}\left(\frac{\mu_1-\mu_2}{\sigma}\right)^2 f^2(1-f)^2\frac{\partial^2\varphi}{\partial f^2}(t,x,f) \\ & + x(\mu_1-\mu_2)f(1-f)\frac{\partial^2\varphi}{\partial x\partial f}(t,x,f). \end{aligned}$$

Let us comment the system (2.11). If there would exist smooth value functions V^0 and V^1 and an optimal control π^*, on the time intervals where π^* is equal to i, the classical PDE would be satisfied:

$$-\frac{\partial V^i}{\partial t} - \mathcal{L}^i V^i = 0.$$

When π^* could switch from $\pi^* = i$ to $j = 1-i$, we would have the boundary condition

$$V^i(t,x,f) = V^j(t,x(1-g_{ij}),f).$$

In general, the value functions are not smooth and an optimal control does not exist. In Sec. 6, we rigorously prove that V^0, V^1 are viscosity solutions of the system (2.11).

Moreover, the system (2.11) combines the usual specificities of HJB equations for impulse and switching controls. The impulse part, due to the transaction costs, gives rise to the comparison between the *PDE* term and the boundary term. The switching part is due to the change of dynamics of the portfolio at each transaction. The originality of this system is that it is neither a classical switching nor classical impulse problem.

In addition, system (2.11) allows one to develop a numerical procedure to approximate numerically the value functions (see Sec. 8).

We define viscosity solutions for (2.11) as follows.

Definition 2.4. A pair of continuous functions (V^0, V^1) from $[0,T] \times (0,+\infty) \times [0,1]$ to $\mathbb{R}$ is a viscosity upper solution to the system (2.11) if $V^0(T,x,f) = V^1(T,x,f) = U(x)$ and if, for all $i \neq j$ in $\{0,1\}$, all bounded function ϕ of class $\mathcal{C}^{1,2}([0,T] \times \mathbb{R}_+ \times [0,1])$ with bounded derivatives, and all local minimum $(\hat{t},\hat{x},\hat{f})$ of $V^i - \phi$, one has

$$\min\left\{-\frac{\partial\phi}{\partial t}(\hat{t},\hat{x},\hat{f}) - \mathcal{L}^i\phi(\hat{t},\hat{x},\hat{f});\ V^i(\hat{t},\hat{x},\hat{f}) - V^j(\hat{t},\hat{x}(1-g_{ij}),\hat{f})\right\} \geq 0.$$

A viscosity lower solution is defined analogously: for all local maximum $(\hat{t},\hat{x},\hat{f})$ of $V^i - \phi$, for $j \neq i$,

$$\min\left\{-\frac{\partial\phi}{\partial t}(\hat{t},\hat{x},\hat{f}) - \mathcal{L}^i\phi(\hat{t},\hat{x},\hat{f});\ V^i(\hat{t},\hat{x},\hat{f}) - V^j(\hat{t},\hat{x}(1-g_{ij}),\hat{f})\right\} \leq 0.$$

Finally, a viscosity solution is both a upper and lower viscosity solution.

Theorem 2.5. *Let $\mathcal{V}_\alpha$ be the class of functions Υ which are continuous on $[0,T] \times [0,+\infty) \times [0,1]$ and satisfy: for all $(t,f) \in [0,T] \times [0,1]$, $\Upsilon(t,0,f) = 0$ and there exists $C > 0$ such that*

$$|\Upsilon(t,x,f)| \leq C(1 + x^{-\alpha} + x) \textit{ for all } (t,x,f) \in [0,T] \times (0,+\infty) \times [0,1].$$

Suppose that the utility function belongs to the class $\mathcal{U}$ defined above. Then the pair of value functions (V^0,V^1) is the unique viscosity solution of (2.11) in $\mathcal{V}_\alpha$ satisfying the boundary condition $V^0(T,x,f) = V^1(T,x,f) = U(x)$ for all (x,f) in $[0,+\infty) \times [0,1]$.

If U is logarithmic, (V^0,V^1) is the unique viscosity solution of (2.11) in the set of function $\{\log(x) + \bar{V}(t,f)\}$ where $\bar{V}$ is continuous on $[0,T] \times [0,1]$.

Remark 2.6. The Theorem 2.5 allows one to use the numerical solution of the system of inequalities (2.11) in order to construct Markov allocation strategies $\bar{\pi}$ such that $J^i(t,x,f,\bar{\pi})$ is close to $V^i(t,x,f)$. To implement such a strategy, the investor needs to estimate F_t at each time t from the observation of the prices $(S_\theta;\ \theta \leq t)$. From (2.2) and (2.3), we have for some smooth functions α_1 and α_2,

$$dF_\theta = \alpha_1(F_\theta)d\theta + \alpha_2(F_\theta)\frac{dS_\theta}{S_\theta}.$$

One can discretise this equation by using, e.g., the Euler scheme. In [15] the authors construct an approximation of F based on filtering theory which is more accurate than the Euler approximation.

3 Preliminary estimates

Notation 3.1. In this section we are given arbitrary times $0 \leq s \leq t \leq \hat{t} \leq T$, admissible strategies π, $\mathcal{F}^S_s$-measurable random variables F and $\hat{F}$ taking values in $[0,1]$, and $\mathcal{F}^S_s$-measurable random variables Z, $\hat{Z}$.

It must be understood that the various constants C in the inequalities below only depend on some of the constants m, μ_1, μ_2, λ_1, λ_2, σ and T.

Elementary calculations show that we have, for all integer $m \geq 1$ and all time $t \leq \theta \leq T$,

$$\mathbb{E}|Z_\theta^{t,0,F,\pi}|^{2m} \leq C(\theta - t)^m, \tag{3.1}$$

$$\mathbb{E}|Z_T^{\hat{t},\hat{Z},\hat{F},\pi} - Z_T^{t,Z,F,\pi}|^{2m} \leq C\left(\mathbb{E}|Z - \hat{Z}|^{2m} + \mathbb{E}|F - \hat{F}|^{2m} + |\hat{t} - t|^m\right), \tag{3.2}$$

$$\mathbb{E}|F_\theta^{t,F} - F_\theta^{t,\hat{F}}|^{2m} \leq C\mathbb{E}|F - \hat{F}|^m, \tag{3.3}$$

and, for all $t \leq \theta \leq \theta^* \leq T$,

$$\mathbb{E}|F_{\theta^*}^{t,F} - F_\theta^{t,F}|^{2m} \leq C(\theta^* - \theta)^m. \tag{3.4}$$

As well, one has, for all $\beta > 0$,

$$\mathbb{E}\exp\left(\beta \int_t^\theta \pi_s \sigma d\overline{B}_s\right) \leq C \exp\left(\frac{\beta^2\sigma^2}{2}(\theta - t)\right),$$

and, for all $\mathcal{F}_t^S$-measurable random variables Z such that $\mathbb{E}\exp(\beta Z) < \infty$,

$$\mathbb{E}\exp(\beta Z_\theta^{t,Z,F,\pi}) \leq \mathbb{E}\exp(\beta Z)\exp(C(\theta - t)). \tag{3.5}$$

We end this part by the following result which is easy to deduce from (2.7), (2.10), (3.2) and (3.5)

Proposition 3.2. *There exists $C > 0$ such that, for all $0 \leq t \leq \hat{t} \leq T$, all $(x, \hat{x}, f, \hat{f}) \in [0, +\infty)^2 \times [0, 1]^2$, for all admissible controls $\pi \in \mathcal{A}_t$ such that $\pi_\theta = i$ for $\theta \in [t, \hat{t})$,*

$$\begin{aligned}|J^i(\hat{t}, \hat{x}, \hat{f}, \pi) &- J^i(t, x, f, \pi)| \\ &\leq C(1 + x^{-\alpha} + \hat{x}^{-\alpha})\left(|\hat{x} - x| + (x + \hat{x})(|\hat{f} - f| + |\hat{t} - t|^{1/2})\right).\end{aligned}$$

4 Continuity of the value functions

Theorem 4.1. *There exists $C > 0$ such that, for all $i \in \{0; 1\}$, $0 \leq t \leq \hat{t} \leq T$, x and $\hat{x}$ in $(0, +\infty)$, f and $\hat{f}$ in $[0, 1]$, one has*

$$\begin{aligned}&\left|V^i(\hat{t}, \hat{x}, \hat{f}) - V^i(t, x, f)\right| \\ &\qquad \leq C(1 + x^{-\alpha} + \hat{x}^{-\alpha})\left(|\hat{x} - x| + (x + \hat{x})(|\hat{f} - f| + |\hat{t} - t|^{1/2})\right).\end{aligned}$$

Proof. We successively consider $|V^i(\hat{t},\hat{x},\hat{f})-V^i(\hat{t},x,f)|$ and $|V^i(\hat{t},x,f)-V^i(t,x,f)|$.
For the first term, choose a control $\pi \in \mathcal{A}_{\hat{t}}$. In view of Proposition 3.2:

$$|J^i(\hat{t},\hat{x},\hat{f},\pi) - J^i(\hat{t},x,f,\pi)| \leq C(1+x^{-\alpha}+\hat{x}^{-\alpha})\left(|\hat{x}-x|+(x+\hat{x})|\hat{f}-f|\right).$$

Taking the supremum over all admissible control $\pi \in \mathcal{A}_{\hat{t}}$ yields the desired inequality. We now examine $V^i(\hat{t},x,f)-V^i(t,x,f)$. For all admissible control $\hat{\pi} \in \mathcal{A}_{\hat{t}}$, define the new admissible control $\pi \in \mathcal{A}_t$ on $[t,T]$ by $\pi := \hat{\pi}$ on $[\hat{t},T]$ and $\pi = i$ on $[t,\hat{t})$. Then $\xi_T^{\hat{t},i,\pi} = \xi_T^{t,i,\pi}$. Therefore, using (2.10) again, we have

$$\begin{aligned}
J^i(\hat{t},x,f,\hat{\pi}) - V^i(t,x,f) &\leq J^i(\hat{t},x,f,\hat{\pi}) - J^i(t,x,f,\pi) \\
&\leq Cx\mathbb{E}[(1+x^{-\alpha}\exp(-\alpha Z_T^{\hat{t},0,f,\hat{\pi}}) + x^{-\alpha}\exp(-\alpha Z_T^{t,0,f,\pi})) \\
&\qquad |Z_T^{\hat{t},0,f,\hat{\pi}} - Z_T^{t,0,f,\pi}|(\exp(Z_T^{\hat{t},0,f,\hat{\pi}}) + \exp(Z_T^{t,0,f,\pi}))] \\
&\leq Cx(1+x^{-\alpha})\left(\mathbb{E}|Z_T^{\hat{t},0,f,\hat{\pi}} - Z_T^{t,0,f,\pi}|^2\right)^{1/2}.
\end{aligned}$$

As $\pi = \hat{\pi}$ on $[\hat{t},T]$, the inequalities (3.1), (3.4) and (3.2) imply

$$\begin{aligned}
\mathbb{E}|Z_T^{\hat{t},0,f,\hat{\pi}} - Z_T^{t,0,f,\pi}|^2 &\leq C\left(\mathbb{E}|Z_{\hat{t}}^{t,0,f,\pi}|^2 + \mathbb{E}|Z_T^{\hat{t},0,f,\hat{\pi}} - Z_T^{\hat{t},0,F_{\hat{t}}^{t,f},\hat{\pi}}|^2\right) \\
&\leq C\left(\mathbb{E}|Z_{\hat{t}}^{t,0,f,\pi}|^2 + \mathbb{E}|F_{\hat{t}}^{t,f} - f|^2\right) \\
&\leq C(\hat{t}-t).
\end{aligned}$$

Taking the supremum over all admissible controls $\hat{\pi} \in \mathcal{A}_{\hat{t}}$, we obtain

$$V^i(\hat{t},x,f) - V^i(t,x,f) \leq Cx(1+x^{-\alpha})|\hat{t}-t|^{1/2}.$$

We finally consider $V^i(t,x,f) - V^i(\hat{t},x,f)$. Fix $\varepsilon > 0$ and an admissible control $\pi \in \mathcal{A}_t$ such that $V^i(t,x,f) \leq J^i(t,x,f,\pi) + \varepsilon$. Then, for all admissible controls $\hat{\pi} \in \mathcal{A}_{\hat{t}}$, one has

$$V^i(t,x,f) - V^i(\hat{t},x,f) \leq J^i(t,x,f,\pi) - J^i(\hat{t},x,f,\hat{\pi}) + \varepsilon.$$

We now aim to choose an admissible control $\hat{\pi} \in \mathcal{A}_{\hat{t}}$ on $[\hat{t},T]$ which is close to π and satisfies $\xi_T^{\hat{t},i,\hat{\pi}} = \xi_T^{t,i,\pi}$. The difficulty comes from the possible jumps of π before $\hat{t}$. This leads us to choose

$$\hat{\pi}_s := \begin{cases} \pi_{as+b} & \text{if } \hat{t} \leq s < \hat{t}+\varepsilon, \\ \pi_s & \text{if } s \geq \hat{t}+\varepsilon, \end{cases}$$

where $a := (\hat{t}+\varepsilon-t)/\varepsilon$, $b := (\hat{t}+\varepsilon)(t-\hat{t})/\varepsilon$. Notice that $a\hat{t}+b = t$, and $a(\hat{t}+\varepsilon)+b = \hat{t}+\varepsilon$. As $as+b \leq s$ for all $\hat{t} \leq s \leq \hat{t}+\varepsilon$, the control $\hat{\pi}$ is progressively measurable. In addition, the costs at time T due to the jumps of π and $\hat{\pi}$ are equal, and thus $\xi_T^{\hat{t},i,\hat{\pi}} = \xi_T^{t,i,\pi}$. As one also has $\hat{\pi} = \pi$ on $(\hat{t}+\varepsilon, T)$, in view of (2.10) and (3.5) it comes

$$\begin{aligned} V^i(t,x,f) - V^i(\hat{t},x,f) &\le \varepsilon + |J^i(t,x,f,\pi) - J^i(\hat{t},x,f,\hat{\pi})| \\ &\le \varepsilon + Cx(1+x^{-\alpha})\left(\mathbb{E}|Z_T^{\hat{t},0,f,\hat{\pi}} - Z_T^{t,0,f,\pi}|^2\right)^{1/2}. \end{aligned}$$

Using again inequalities (3.1), (3.4) and (3.2),

$$\begin{aligned} &\mathbb{E}|Z_T^{t,0,f,\pi} - Z_T^{\hat{t},0,f,\hat{\pi}}|^2 \\ &\quad \le C\left(\mathbb{E}|Z_{\hat{t}+\varepsilon}^{t,0,f,\pi}|^2 + \mathbb{E}|Z_{\hat{t}+\varepsilon}^{\hat{t},0,f,\hat{\pi}}|^2 + \mathbb{E}\left|Z_T^{\hat{t}+\varepsilon,0,F_{\hat{t}+\varepsilon}^{t,f},\pi} - Z_T^{\hat{t}+\varepsilon,0,F_{\hat{t}+\varepsilon}^{\hat{t},f},\pi}\right|^2\right) \\ &\quad \le C\left(\mathbb{E}|Z_{\hat{t}+\varepsilon}^{t,0,f,\pi}|^2 + \mathbb{E}|Z_{\hat{t}+\varepsilon}^{\hat{t},0,f,\hat{\pi}}|^2 + \mathbb{E}|F_{\hat{t}+\varepsilon}^{t,f} - F_{\hat{t}+\varepsilon}^{\hat{t},f}|^2\right) \\ &\quad \le C\left(\hat{t}+\varepsilon - t + \hat{t}+\varepsilon - \hat{t} + \mathbb{E}|F_{\hat{t}+\varepsilon}^{t,f} - f|^2 + \mathbb{E}|F_{\hat{t}+\varepsilon}^{\hat{t},f} - f|^2\right) \\ &\quad \le C((\hat{t}-t)+\varepsilon). \end{aligned}$$

Therefore

$$V^i(t,x,f) - V^i(\hat{t},x,f) \le \varepsilon + Cx(1+x^{-\alpha})(\varepsilon^{1/2} + (\hat{t}-t)^{1/2}),$$

and the desired result follows. □

Corollary 4.2. *For all $\beta \ge \alpha$, $0 \le s \le t \le T$, and i, x, f, for all admissible control $\pi \in \mathcal{A}_t$, one has*

$$\mathbb{E}\left|V^i(t, W_t^{s,x,f,i,\pi}, F_t^{s,f}) - V^i(s, xe^{-\xi_t^{s,i,\pi}}, f)\right|^\beta < C(t-s)^{\beta/2}.$$

We conclude this section by showing that the functions V^0 and V^1 are continuous.

Corollary 4.3. *If the utility function belongs to the class $\mathcal{U}$, then V^0 and V^1 are continuous on $[0,T]\times[0,+\infty)\times[0,1]$.*

Proof. As U is increasing and concave, for all i, t, x, f and all admissible controls π one has, in view of (3.5),

$$\begin{aligned} V^i(t,x,f) &= \sup_{\pi\in\mathcal{A}_t} \mathbb{E}[U(W_T^{t,x,f,i,\pi})] \\ &\le U(x \sup_{\pi\in\mathcal{A}_t} \mathbb{E}\exp(Z_T^{t,0,f,\pi} - \xi_T^{t,i,\pi})) \\ &\le U(x \sup_{\pi\in\mathcal{A}_t} \mathbb{E}\exp(Z_T^{t,0,f,\pi})) \\ &\le U(Cx), \end{aligned}$$

where C is a constant independent of t, x, f, π. As U is continuous at 0, $V^i(t,x,f)$ tends to 0 with x, and the convergence is uniform w.r.t. t, f. In addition, $W_T^{t,0,f,i,\pi} = 0$ for all t, f, π, which ends the proof. □

5 The dynamic programming principle

Proposition 5.1. *For all bounded continuous functions φ on $\mathbb{R}^+$, all stopping times τ such that $t \leq \tau \leq T$, all $x > 0$, $0 \leq f \leq 1$, $\pi \in \mathcal{A}_t$, one has*

$$\mathbb{E}[\varphi(W_T^{t,x,f,i,\pi}) \mid \mathcal{F}_\tau^S] = \mathbb{E}[\varphi(W_T^{\tau, W_\tau^{t,x,f,i,\pi}, F_\tau^{t,f}, \pi_\tau, \pi|_{[\tau,T]}})], \ \mathbb{P}_\tau - a.s.$$

Proof. We first remark that the filtration $\mathcal{F}^S$ is continuous ($\mathcal{F}^S = \mathcal{F}^{\bar{B}}$). So, we can approximate the stopping time τ by a sequence of stopping times τ_n with countably many values. Furthermore, the equality (2.7) shows that for all $\pi \in \mathcal{A}_t$, all $i \in \{0,1\}$, all Borel subset A of the set of càdlàg trajectories from $[t,T]$ to $[0,+\infty) \times [0,1]$, the mapping $(t,x) \mapsto \mathbb{P}\left((W_\cdot^{t,x,f,i,\pi}, F_\cdot^{t,F}) \in A\right)$ is measurable. Then, we apply Theorem 6.1.2 of [19] and obtain the result for each τ_n. Letting n go to infinity ends the proof.□

The Dynamic Programming Principle is a key step to establish the relationship between value functions and Hamilton–Jacobi–Bellman equations. This principle in the framework of processes with jumps has often been just assumed in the literature. A rigorous proof can be found in Ishikawa [8] for a model with jumps which substantially differs from ours. We thus carefully prove the following theorem.

Theorem 5.2 (Dynamic Programming Principle). *Let $\mathcal{T}_{t,T}$ denote the set of all $\mathcal{F}^S$ stopping times taking values in $[t,T]$. For all $0 \leq t \leq T$ and $x > 0$, $0 < f < 1$, $i \in \{0;1\}$, one has*

$$V^i(t,x,f) = \sup_{\pi \in \mathcal{A}_t} \sup_{\tau \in \mathcal{T}_{t,T}} \mathbb{E}[V^{\pi_\tau}(\tau, W_\tau^{t,x,f,i,\pi}, F_\tau^{t,f})]$$

$$V^i(t,x,f) = \sup_{\pi \in \mathcal{A}_t} \inf_{\tau \in \mathcal{T}_{t,T}} \mathbb{E}[V^{\pi_\tau}(\tau, W_\tau^{t,x,f,i,\pi}, F_\tau^{t,f})].$$

Proof. The proof is divided in two parts: we first prove the upper bound $V^i(t,x,f) \leq \sup_{\pi \in \mathcal{A}_t} \inf_{\tau \in \mathcal{T}_{t,T}} \mathbb{E}[V^{\pi_\tau}(\tau, W_\tau^{t,x,f,i,\pi}, F_\tau^{t,f})]$, and then the lower bound $V^i(t,x,f) \geq \sup_{\pi \in \mathcal{A}_t} \sup_{\tau \in \mathcal{T}_{t,T}} \mathbb{E}[V^{\pi_\tau}(\tau, W_\tau^{t,x,f,i,\pi}, F_\tau^{t,f})]$ (which is less immediate to get than the upper bound).

The upper bound

In view of Proposition 5.1 one has: for all admissible control π in $\mathcal{A}_t$, and all stopping times $\tau \in \mathcal{T}_{t,T}$

$$\begin{aligned} J^i(t,x,f,\pi) &= \mathbb{E}[U(W_T^{t,x,f,i,\pi})] \\ &= \mathbb{E}\left[\mathbb{E}[U(W_T^{t,x,f,i,\pi}) \mid \mathcal{F}_{t,\tau}^S]\right] \\ &= \mathbb{E}U(W_T^{\tau,W_\tau^{t,x,f,\pi},F_\tau^{t,f},\pi_\tau,\pi|_{[\tau,T]}}) \\ &= \sum_{k\in\{0,1\}} \mathbb{E}U(W_T^{\tau,W_\tau^{t,x,f,i,\pi},F_\tau^{t,f},k,\pi|_{[\tau,T]}})\mathbb{I}_{\pi_\tau=k} \\ &= \sum_{k\in\{0,1\}} \mathbb{E}\left[J^k(\tau,W_\tau^{t,x,f,i,\pi},F_\tau^{t,f},\pi|_{[\tau,T]})\mathbb{I}_{\pi_\tau=k}\right] \\ &\le \sum_{k\in\{0,1\}} \mathbb{E}\left[V^k(\tau,W_\tau^{t,x,f,i,\pi},F_\tau^{t,f})\mathbb{I}_{\pi_\tau=k}\right] \\ &\le \mathbb{E}[V^{\pi_\tau}(\tau,W_\tau^{t,x,f,i,\pi},F_\tau^{t,f})]. \end{aligned}$$

It then remains to take the infimum over all stopping times $\tau \in \mathcal{T}_{t,T}$ and the supremum over all admissible controls $\pi \in \mathcal{A}_t$ to obtain:

$$V^i(t,x,f) \le \sup_{\pi\in\mathcal{A}_t} \inf_{\tau\in\mathcal{T}_{t,T}} \mathbb{E}[V^{\pi_\tau}(\tau,W_\tau^{t,x,f,i,\pi},F_\tau^{t,f})].$$

The lower bound

In view of Theorem 4.1, for all $\varepsilon > 0$, one can find a countable partition of $[t,T] \times (0,+\infty) \times [0,1]$ with Borel subsets $\mathcal{B}_p$ such that, for all p, all (t,x,f) and $(\hat{t},\hat{x},\hat{f})$ in $\mathcal{B}_p$, all i,

$$|V^i(t,x,f) - V^i(\hat{t},\hat{x},\hat{f})| \le \varepsilon. \tag{5.1}$$

In addition, if $t \le \hat{t}$, thanks to Proposition 3.2, for all i, for all admissible controls π in $\mathcal{A}_t$ such that $\pi_\theta = i$ for $\theta \in [t,\hat{t})$, one has also for all p, all (t,x,f) and $(\hat{t},\hat{x},\hat{f})$ in $\mathcal{B}_p$,

$$|J^i(t,x,f,\pi) - J^i(\hat{t},\hat{x},\hat{f},\pi)| \le \varepsilon. \tag{5.2}$$

We now set

$$\rho := \sup_{\pi\in\mathcal{A}_t} \sup_{\tau\in\mathcal{T}_{t,T}} \mathbb{E}[V^{\pi_\tau}(\tau,W_\tau^{t,x,f,i,\pi},F_\tau^{t,f})].$$

Choose π in $\mathcal{A}_t$ and τ in $\mathcal{T}_{t,T}$ such that

$$\rho \le \varepsilon + \mathbb{E}[V^{\pi_\tau}(\tau,W_\tau^{t,x,f,i,\pi},F_\tau^{t,f})].$$

By definition of the partition $\{\mathcal{B}_p\}$,

$$\rho \le \varepsilon + \mathbb{E}\left[\sum_{p=0}^{\infty} V^{\pi_\tau}(\tau,W_\tau^{t,x,f,i,\pi},F_\tau^{t,f})\,\mathbb{I}_{(\tau,W_\tau^{t,x,f,i,\pi},F_\tau^{t,f})\in\mathcal{B}_p}\right].$$

Now, for all p choose a triple (t_p, x_p, f_p) in the closure $\bar{\mathcal{B}}_p$ of $\mathcal{B}_p$, where t_p is the largest time in the trace of $\bar{\mathcal{B}}_p$ in $[t,T]$. In view of (5.1) one thus has

$$\rho \leq 2\varepsilon + \mathbb{E}\left[\sum_{p=0}^{\infty} V^{\pi_\tau}(t_p, x_p, f_p)\, \mathbb{I}_{(\tau, W_\tau^{t,x,f,i,\pi}, F_\tau^{t,f}) \in \mathcal{B}_p}\right].$$

For all p, i, choose a control $\pi^{p,i}$ in $\mathcal{A}_{t_p}$ such that $V^i(t_p, x_p, f_p) \leq \varepsilon + J^i(t_p, x_p, f_p, \pi^{p,i})$. Then

$$\rho \leq 3\varepsilon + \mathbb{E}\left[\sum_{p=0}^{\infty} J^{\pi_\tau}(t_p, x_p, f_p, \pi^{p,\pi_\tau})\, \mathbb{I}_{(\tau, W_\tau^{t,x,f,i,\pi}, F_\tau^{t,f}) \in \mathcal{B}_p}\right].$$

Next, $p(\omega)$ being the integer s.t. $(\tau, W_\tau^{t,x,f,i,\pi}, F_\tau^{t,f})(\omega) \in \mathcal{B}_{p(\omega)}$, define the control $\hat{\pi}$ in $\mathcal{A}_t$ by

$$\hat{\pi}_s(\omega) := \begin{cases} \pi_s(\omega) & \text{if } t \leq s \leq \tau(\omega), \\ \pi_\tau(\omega) & \text{if } \tau(\omega) \leq s < t_{p(\omega)}, \\ \pi_s^{p(\omega),j}(\omega) & \text{if } s \geq t_{p(\omega)} \text{ and } \pi_\tau(\omega) = j. \end{cases}$$

From now on, we write $\hat{\pi}$ for $\hat{\pi}|_{[t_p,T]}$ and $\hat{\pi}|_{[\tau,T]}$. We have

$$\rho \leq 3\varepsilon + \mathbb{E}\left[\sum_{p=0}^{\infty} J^{\pi_\tau}(t_p, x_p, f_p, \hat{\pi})\, \mathbb{I}_{(\tau, W_\tau^{t,x,f,i,\pi}, F_\tau^{t,f}) \in \mathcal{B}_p}\right].$$

As the control $\hat{\pi}$ is constant on $[\tau, t_p)$, the inequality (5.2) leads to

$$\begin{aligned}
\rho &\leq 4\varepsilon + \mathbb{E}\left[\sum_{p=0}^{\infty} J^{\pi_\tau}(\tau, W_\tau^{t,x,f,i,\pi}, F_\tau^{t,f}, \hat{\pi}) \mathbb{I}_{(\tau, W_\tau^{t,x,f,i,\pi}, F_\tau^{t,f}) \in \mathcal{B}_p}\right] \\
&\leq 4\varepsilon + \mathbb{E}[J^{\hat{\pi}_\tau}(\tau, W_\tau^{t,x,f,i,\hat{\pi}}, F_\tau^{t,f}, \hat{\pi})] \\
&= 4\varepsilon + \mathbb{E}\left[U(W_T^{\tau, W_\tau^{t,x,f,i,\hat{\pi}}, F_\tau^{t,f}, \hat{\pi}_\tau, \hat{\pi}})\right].
\end{aligned}$$

Therefore, in view of Proposition 5.1,

$$\begin{aligned}
\rho &\leq 4\varepsilon + \mathbb{E}\left[U(W_T^{t,x,f,i,\hat{\pi}})\right] \\
&= 4\varepsilon + J^i(t, x, f, \hat{\pi}) \\
&\leq 4\varepsilon + V^i(t, x, f).
\end{aligned}$$

It now remains to let ε tend to 0. □

6 Existence of a viscosity solution

6.1 Existence of a viscosity upper solution

In this section we show that the pair (V^0, V^1) of value functions is a upper viscosity solution on $[0,T] \times (0,+\infty) \times [0,1]$ of the system (2.11).

Fix $i \in \{0;1\}$. Let $(\hat{t}, \hat{x}, \hat{f})$ be a local minimum of $V^i - \varphi$, where φ is a function of class $\mathcal{C}^{1,2}$ defined on a neighbourhood $[\hat{t}, \hat{t}+\varepsilon] \times \mathcal{B}_\varepsilon$ of $(\hat{t}, \hat{x}, \hat{f})$, where

$$\mathcal{B}_\varepsilon := \left[\hat{x}(1+\varepsilon)^{-1}, \hat{x}(1+\varepsilon)\right] \times \left[[\hat{f}-\varepsilon, \hat{f}+\varepsilon] \cap [0,1]\right],$$

and such that $V^i(\hat{t}, \hat{x}, \hat{f}) = \varphi(\hat{t}, \hat{x}, \hat{f})$.

For all controls $\pi \in \mathcal{A}_{\hat{t}}$, we have $W_{\hat{t}}^{\hat{t},(1-g_{ij})\hat{x},\hat{f},j,\pi} \leq W_{\hat{t}}^{\hat{t},\hat{x},\hat{f},i,\pi}$. Then, for all $\theta \geq \hat{t}$, we also have $W_\theta^{\hat{t},(1-g_{ij})\hat{x},\hat{f},j,\pi} \leq W_0^{\hat{t},\hat{x},\hat{f},i,\pi}$. Therefore

$$\begin{aligned}\mathbb{E}U(W_\theta^{\hat{t},(1-g_{ij})\hat{x},\hat{f},j,\pi}) &\leq \mathbb{E}U(W_\theta^{\hat{t},\hat{x},\hat{f},i,\pi}),\\ V^j(\hat{t}, \hat{x}(1-g_{ij}), \hat{f}) &\leq V^i(\hat{t}, \hat{x}, \hat{f}).\end{aligned}$$

We now aim to prove

$$-\left(\frac{\partial\varphi}{\partial t} + \mathcal{L}^i\varphi\right)(\hat{t}, \hat{x}, \hat{f}) \geq 0.$$

Fix $0 < h < \varepsilon$ and choose a control π which takes the value i on the time interval $[\hat{t}, \hat{t}+h]$. The Dynamic Programming Principle with the constant stopping time $\hat{t}+h$ leads to

$$\begin{aligned}\varphi(\hat{t}, \hat{x}, \hat{f}) &= V^i(\hat{t}, \hat{x}, \hat{f})\\ &\geq \mathbb{E}[V^i(\hat{t}+h, W_{\hat{t}+h}^{\hat{t},\hat{x},\hat{f},i,\pi}, F_{\hat{t}+h}^{\hat{t},\hat{f}})]\\ &= \mathbb{E}[V^i(\hat{t}+h, W_{\hat{t}+h}^{\hat{t},\hat{x},\hat{f},i,\pi}, F_{\hat{t}+h}^{\hat{t},\hat{f}})\mathbb{I}_{(W_{\hat{t}+h}^{\hat{t},\hat{x},\hat{f},i,\pi}, F_{\hat{t}+h}^{\hat{t},\hat{f}}) \in \mathcal{B}_\varepsilon}]\\ &\quad + \mathbb{E}[V^i(\hat{t}+h, W_{\hat{t}+h}^{\hat{t},\hat{x},\hat{f},i,\pi}, F_{\hat{t}+h}^{\hat{t},\hat{f}})\mathbb{I}_{(W_{\hat{t}+h}^{\hat{t},\hat{x},\hat{f},i,\pi}, F_{\hat{t}+h}^{\hat{t},\hat{f}}) \notin \mathcal{B}_\varepsilon}]\\ &\geq \mathbb{E}[\varphi(\hat{t}+h, W_{\hat{t}+h}^{\hat{t},\hat{x},\hat{f},i,\pi}, F_{\hat{t}+h}^{\hat{t},\hat{f}})] + \mathbb{E}[(V_i - \varphi)(\hat{t}+h, W_{\hat{t}+h}^{\hat{t},\hat{x},\hat{f},i,\pi}, F_{\hat{t}+h}^{\hat{t},\hat{f}})\\ &\qquad \mathbb{I}_{(W_{\hat{t}+h}^{\hat{t},\hat{x},\hat{f},i,\pi}, F_{\hat{t}+h}^{\hat{t},\hat{f}}) \notin \mathcal{B}_\varepsilon}].\end{aligned}$$

By Itô's formula to $\varphi(t, W_t, F_t)$,

$$\mathbb{E}[\varphi(\hat{t}+h, W_{\hat{t}+h}^{\hat{t},\hat{x},\hat{f},i,\pi}, F_{\hat{t}+h}^{\hat{t},\hat{f}})] = \varphi(\hat{t}, \hat{x}, \hat{f}) + \mathbb{E}\int_{\hat{t}}^{\hat{t}+h}\left(\frac{\partial\varphi}{\partial t} + \mathcal{L}^i\varphi\right)(s, W_s^{\hat{t},\hat{x},\hat{f},i,\pi}, F_s^{\hat{t},\hat{f}})ds,$$

since $\pi = i$ on $[\hat{t}, \hat{t}+h]$. Therefore

$$\begin{aligned}0 &\geq \frac{1}{h}\mathbb{E}\left[\int_{\hat{t}}^{\hat{t}+h}\left(\frac{\partial\varphi}{\partial t} + \mathcal{L}^i\varphi\right)(s, W_s^{\hat{t},\hat{x},\hat{f},i,\pi}, F_s^{\hat{t},\hat{f}})ds\right]\\ &\quad + \frac{1}{h}\mathbb{E}\left[(V_i - \varphi)(\hat{t}+h, W_{\hat{t}+h}^{\hat{t},\hat{x},\hat{f},i,\pi}, F_{\hat{t}+h}^{\hat{t},\hat{f}})\mathbb{I}_{(W_{\hat{t}+h}^{\hat{t},\hat{x},\hat{f},i,\pi}, F_{\hat{t}+h}^{\hat{t},\hat{f}}) \notin \mathcal{B}_\varepsilon}\right].\end{aligned} \tag{6.1}$$

When h tends to 0, the first term of the right-hand side tends to $\left(\frac{\partial\varphi}{\partial t} + \mathcal{L}^i\varphi\right)(\hat{t}, \hat{x}, \hat{f})$. For the second term, in view of Corollary 4.2,

$$\left|\mathbb{E}\left[(V_i - \varphi)(\hat{t} + h, W_{\hat{t}+h}^{\hat{t},\hat{x},\hat{f},i,\pi}, F_{\hat{t}+h}^{\hat{t},\hat{f}})\mathbb{I}_{(W_{\hat{t}+h}^{\hat{t},\hat{x},\hat{f},i,\pi}, F_{\hat{t}+h}^{\hat{t},\hat{f}}) \notin \mathcal{B}_\varepsilon}\right]\right|$$

$$\leq \left(\mathbb{E}\left|(V_i - \varphi)(\hat{t} + h, W_{\hat{t}+h}^{\hat{t},\hat{x},\hat{f},i,\pi}, F_{\hat{t}+h}^{\hat{t},\hat{f}})\right|^2\right)^{1/2} \left[\mathbb{P}(W_{\hat{t}+h}^{\hat{t},\hat{x},\hat{f},i,\pi}, F_{\hat{t}+h}^{\hat{t},\hat{f}}) \notin \mathcal{B}_\varepsilon\right]^{1/2}$$

$$\leq C\mathbb{P}[(W_{\hat{t}+h}^{\hat{t},\hat{x},\hat{f},i,\pi}, F_{\hat{t}+h}^{\hat{t},\hat{f}}) \notin \mathcal{B}_\varepsilon]^{1/2}.$$

One now uses the inequalities (3.1) and (3.4): as W is continuous on $[\hat{t}, \hat{t} + h))$,

$$\begin{aligned}
\mathbb{P}[(W_{\hat{t}+h}^{\hat{t},\hat{x},\hat{f},i,\pi}, F_{\hat{t}+h}^{\hat{t},\hat{f}}) \notin \mathcal{B}_\varepsilon] &\leq \mathbb{P}[\log(\hat{x}^{-1}W_{\hat{t}+h}^{\hat{t},\hat{x},\hat{f},i,\pi}) > \log(1+\varepsilon)] + \mathbb{P}[|F_{\hat{t}+h}^{\hat{t},\hat{f}} - \hat{f}| > \varepsilon] \\
&\quad + \mathbb{P}[\log(\hat{x}^{-1}W_{\hat{t}+h}^{\hat{t},\hat{x},\hat{f},i,\pi}) < -\log(1+\varepsilon)] \\
&\leq 2\frac{\mathbb{E}|Z_{\hat{t}+h}^{\hat{t},0,\hat{f}}|^8}{(\log(1+\varepsilon))^8} + \frac{\mathbb{E}|F_{\hat{t}+h}^{\hat{t},\hat{f}} - \hat{f}|^8}{\varepsilon^8} \\
&\leq Ch^4.
\end{aligned}$$

Then, the second term in (6.1) is of order h and it remains to let h tend to 0.

6.2 Existence of a viscosity lower solution

Suppose that (V^0, V^1) is not a viscosity lower solution. Then there exist $i \in \{0; 1\}$, a smooth function φ with bounded derivatives, $\epsilon > 0$, a local maximum $(\hat{t}, \hat{x}, \hat{f})$ of $V^i - \varphi$ on $[\hat{t}, \hat{t} + \varepsilon) \times \mathcal{B}_\varepsilon$, and $\gamma > 0$ such that, for all t, x, f in $[\hat{t}, \hat{t} + \varepsilon) \times \mathcal{B}_\varepsilon$,

$$-\left(\frac{\partial\varphi}{\partial t} + \mathcal{L}^i\varphi\right)(t, x, f) \geq \gamma,$$

$$V^i(t, x, f) - V^j(t, x(1 - g_{ij}), f) \geq \gamma,\ j \neq i.$$

We aim to exhibit a contradiction.

For the rest of this subsection we set

$$W^{i,\pi} \equiv W^{\hat{t},\hat{x},\hat{f},i,\pi} \text{ and } F \equiv F^{\hat{t},\hat{f}}.$$

For all controls π such that $\pi_{\hat{t}} = i$, let

$$\begin{aligned}
\tau_1 &:= \inf\{t \geq \hat{t};\ (W_t^{i,\pi}, F_t) \notin \mathcal{B}_\varepsilon\}, \\
\tau_2 &:= \inf\{t \geq \hat{t},\ \pi_t \neq i\}, \\
\tau &:= \tau_1 \wedge \tau_2 \wedge (\hat{t} + \tfrac{\varepsilon}{2}).
\end{aligned}$$

From the definition of admissible controls, we have $\tau_2 > \hat{t}$ *a.s.*. It is also clear that $\tau_1 > \hat{t}$ and thus $\tau > \hat{t}$. One has

$$\begin{aligned} V^i(\hat{t},\hat{x},\hat{f}) &= \varphi(\hat{t},\hat{x},\hat{f}) \\ &= \mathbb{E}\left[\varphi(\tau, W^{i,\pi}_{\tau-}, F_\tau) - \int_{\hat{t}}^{\tau} \left(\frac{\partial\varphi}{\partial s} + \mathcal{L}^{\pi_s}\varphi\right)(s, W^{i,\pi}_s, F_s)ds\right]. \end{aligned}$$

As $(\tau, W^{i,\pi}_{\tau-}, F_\tau)$ belongs to $[\hat{t}, \hat{t}+\varepsilon) \times \mathcal{B}_\varepsilon$ we have

$$V^i(\hat{t},\hat{x},\hat{f}) = \varphi(\hat{t},\hat{x},\hat{f}) \geq \mathbb{E}\left[V^i(\tau, W^{i,\pi}_{\tau-}, F_\tau) - \int_{\hat{t}}^{\tau} \left(\frac{\partial\varphi}{\partial s} + \mathcal{L}^{\pi_s}\varphi\right)(s, W^{i,\pi}_s, F_s)ds\right].$$

By definition of τ, one can substitute $\mathcal{L}^i$ to $\mathcal{L}^{\pi_s}$ in the preceding inequality, from which

$$V^i(\hat{t},\hat{x},\hat{f}) \geq \mathbb{E}\left[V^i(\tau, W^{i,\pi}_{\tau-}, F_\tau) + \gamma(\tau - \hat{t})\right].$$

For all (i,t,x,f), $j = 1-i$ and for all controls π as above, either one has $\pi_\tau = i$ and $W^{i,\pi}_{\tau-} = W^{i,\pi}_\tau$, or $\pi_\tau \neq i$ and $W^{i,\pi}_\tau = (1-g_{ij})W^{i,\pi}_{\tau-}$. In addition $V^i(t,x,f) \geq V^j(t,(1-g_{ij})x,f)$, hence

$$V^i(\tau, W^{i,\pi}_{\tau-}, F_\tau) \geq V^{\pi_\tau}(\tau, W^{i,\pi}_\tau, F_\tau).$$

The Dynamic Programming Principle then implies

$$V^i(\hat{t},\hat{x},\hat{f}) = \sup_{\pi\in\mathcal{A}_{\hat{t}}} \inf_{\tilde{\tau}\in\mathcal{T}_{t,T}} \mathbb{E}V^{\pi_{\tilde{\tau}}}(\tilde{\tau}, W^{i,\pi}_{\tilde{\tau}}, F_{\tilde{\tau}}).$$

Thus, for all $h > 0$ one can find a control π^h such that, for all stopping times $\tilde{\tau}$ in $\mathcal{T}_{t,T}$, one has

$$h \geq V^i(\hat{t},\hat{x},\hat{f}) - \mathbb{E}V^{\pi^h_{\tilde{\tau}}}(\tilde{\tau}, W^{i,\pi^h}_{\tilde{\tau}}, F_{\tilde{\tau}}).$$

We now define τ^h and τ_2^h in an obvious way, and we obtain

$$\begin{aligned} h &\geq \mathbb{E}\left[V^i(\tau, W^{i,\pi^h}_{\tau^h-}, F_{\tau^h}) - V^{\pi^h_{\tau^h}}(\tau^h, W^{i,\pi^h}_{\tau^h}, F_{\tau^h}) + \gamma(\tau^h - \hat{t})\right] \\ &\geq \gamma\mathbb{P}(\tau^h = \tau_2^h) + \gamma\mathbb{E}(\tau^h - \hat{t}). \end{aligned} \tag{6.2}$$

We now choose a sequence (h_n) which decreases to 0, and we distinguish two cases.

On the one hand, if there exists $0 < \beta < 1$ such that, for all n, $\mathbb{P}(\tau^{h_n} = \tau_2^{h_n}) \geq \beta$, then, for all n, $h_n \geq \gamma\beta$, and we have exhibited the desired contradiction.

On the other hand, there exists a subsequence of (h_n), still denoted by (h_n), such that $\mathbb{P}(\tau^{h_n} = \tau_2^{h_n}) \leq \frac{1}{n}$, then, for all n,

$$\mathbb{E}(\tau^{h_n} - \hat{t}) \geq \frac{\varepsilon}{2}\mathbb{P}(\tau^{h_n} = \hat{t} + \tfrac{\varepsilon}{2}).$$

Denote by $(W^{i,i}_t)$ the wealth process corresponding to the constant regime $\pi_t \equiv i$ and

$$\mathcal{E}^i := \left\{\inf\{t \geq \hat{t};\ (W^{i,i}_t, F_t) \notin \mathcal{B}_\varepsilon\} > \hat{t} + \frac{\varepsilon}{2}\right\}.$$

In view of (6.2) we have

$$\begin{aligned} h_n &\geq \gamma \frac{\varepsilon}{2} \mathbb{P}(\tau^{h_n} = \hat{t} + \tfrac{\varepsilon}{2}) \\ &\geq \gamma \frac{\varepsilon}{2} \mathbb{P}\left([\tau_1^{h_n} > \hat{t} + \tfrac{\varepsilon}{2}] \bigcap [\tau_2^{h_n} > \hat{t} + \tfrac{\varepsilon}{2}] \right) \\ &= \gamma \frac{\varepsilon}{2} \mathbb{P}\left(\mathcal{E}^i \bigcap [\tau_2^{h_n} > \hat{t} + \tfrac{\varepsilon}{2}] \right). \end{aligned}$$

Notice that the event $\mathcal{E}^i$ does not depend on n and that, by hypothesis,

$$\mathbb{P}[\tau_2^{h_n} > \hat{t} + \tfrac{\varepsilon}{2}] \geq 1 - \frac{1}{n}.$$

Therefore, letting n tend to 0 we get $0 \geq \mathbb{P}(\mathcal{E}^i)$, which again provides the desired contradiction.

7 Uniqueness of the viscosity solution

The aim of this section is to prove the uniqueness of a viscosity solution of HJB system (2.11), that is the uniqueness part of Theorem 2.5.

For technical reasons we need to distinguish the logarithmic utility case from the other cases.

7.1 Logarithmic utility function

In the logarithmic utility case we have

$$V^i(t,x,f) = \log x + \sup_{\pi \in \mathcal{A}_t} \mathbb{E}[Z_T^{t,0,f,\pi} - \xi_T^{t,i,\pi}].$$

Set

$$\overline{V}^i(t,f) = \sup_{\pi \in \mathcal{A}_t} \mathbb{E}[Z_T^{t,0,f,\pi} - \xi_T^{t,i,\pi}].$$

In the preceding section we have shown that the functions $\overline{V}^i$ are viscosity solutions of the system

$$\begin{cases} \min\left\{ -\dfrac{\partial \overline{V}^0}{\partial t} - \overline{\mathcal{L}}^0 \overline{V}^0;\ \overline{V}^0(t,f) - \log(1-g_{01}) - \overline{V}^1(t,f) \right\} = 0, \\ \min\left\{ -\dfrac{\partial \overline{V}^1}{\partial t} - \overline{\mathcal{L}}^1 \overline{V}^1;\ \overline{V}^1(t,f) - \log(1-g_{10}) - \overline{V}^0(t,f) \right\} = 0, \end{cases} \tag{7.1}$$

with boundary condition $\overline{V}^0(T,f) = \overline{V}^1(T,f) = 0$, where

$$\overline{\mathcal{L}}^0 \varphi(t,f) = r + (-\lambda_1 f + \lambda_2(1-f)) \frac{\partial \varphi}{\partial f}(t,f) + \frac{1}{2} \left(\frac{\mu_1 - \mu_2}{\sigma} \right)^2 f^2(1-f)^2 \frac{\partial^2 \varphi}{\partial f^2}(t,f),$$

and

$$\overline{\mathcal{L}}^1\varphi(t,f) = \mu_1 f + \mu_2(1-f) - \frac{\sigma^2}{2} + (-\lambda_1 f + \lambda_2(1-f))\frac{\partial\varphi}{\partial f}(t,f)$$
$$+ \frac{1}{2}\left(\frac{\mu_1-\mu_2}{\sigma}\right)^2 f^2(1-f)^2\frac{\partial^2\varphi}{\partial f^2}(t,f).$$

Suppose that (Υ^0, Υ^1) and (ψ^0, ψ^1) are two distinct viscosity solutions of (7.1) on $[0,T]\times[0,1]$ with boundary condition $\Upsilon^i(T,f) = \psi^i(T,f) = 0$ for all $f \in [0,1]$ and all $i \in \{0,1\}$. Then there would exist $(\hat{i},\hat{t},\hat{f})$ in $\{0;1\}\times]0,T[\times[0,1]$ such that

$$\eta := \Upsilon^{\hat{i}}(\hat{t},\hat{f}) - \psi^{\hat{i}}(\hat{t},\hat{f}) > 0.$$

Set

$$C^* := \max_{(t,f)\in[0,T]\times[0,1]} (|\Upsilon^1(t,f)| + |\Upsilon^2(t,f)| + |\psi^1(t,f)| + |\psi^2(t,f)|),$$

and, for all $\varepsilon > 0$,

$$\Phi^i(t,f,f') := (1+\gamma)\Upsilon^i(t,f) - \psi^i(t,f') - \frac{1}{2\varepsilon}(|f-f'|^2) + \beta t - \frac{\lambda}{t},$$

where

$$0 < \lambda < \frac{\eta\hat{t}T}{4(T-\hat{t})}, \qquad 0 < \beta < \frac{\eta}{4(T-\hat{t})}, \qquad 0 < \gamma < \frac{\eta}{4C^*}.$$

Thus $\Phi^i(t,f,f')$ tends to $-\infty$ uniformly in f, f' when t tends to 0, from which there exists $i_\varepsilon, t_\varepsilon, f_\varepsilon, f'_\varepsilon$ in $\{0;1\}\times]0,T]\times[0,1]^2$ such that

$$\Phi^{i_\varepsilon}(t_\varepsilon, f_\varepsilon, f'_\varepsilon) = \sup_{(i,t,f,f')\in\{0,1\}\times[0,T]\times[0,1]^2} \Phi^i(t,f,f'). \tag{7.2}$$

7.1.1 Auxiliary lemmae

Lemma 7.1. *For all $\varepsilon > 0$ one has $0 < t_\varepsilon < T$.*

Proof. Suppose that $t_\varepsilon = T$. Then we would have

$$\Phi^{i_\varepsilon}(T,f_\varepsilon,f'_\varepsilon) = -\frac{1}{2\varepsilon}(|f_\varepsilon - f'_\varepsilon|^2) + \beta T - \frac{\lambda}{T} \geq \Phi^{\hat{i}}(\hat{t},\hat{f},\hat{f}) = \eta + \gamma\Upsilon^{\hat{i}}(\hat{t},\hat{f}) + \beta\hat{t} - \frac{\lambda}{\hat{t}},$$

from which

$$0 \geq \eta - \gamma C^* + \beta(\hat{t}-T) - \lambda\left(\frac{1}{\hat{t}} - \frac{1}{T}\right) \geq \frac{\eta}{4} > 0,$$

which implies a contradiction. □

Proposition 7.2. *The function Υ^{i_ε} is a viscosity lower solution of*

$$-\frac{\partial\Upsilon}{\partial t}(t,f) - \overline{\mathcal{L}}^{i_\varepsilon}\Upsilon(t,f) = 0,$$

and the function ψ^{i_ε} is a viscosity upper solution of

$$-\frac{\partial\psi}{\partial t}(t,f) - \overline{\mathcal{L}}^{i_\varepsilon}\psi(t,f) = 0,$$

where i_ε is defined by (7.2).

Proof. In view of (7.1), it suffices to prove that, for all $\varepsilon > 0$,

$$\Upsilon^{i_\varepsilon}(t_\varepsilon, f_\varepsilon) > \log(1 - g_{i_\varepsilon j_\varepsilon}) + \Upsilon^{j_\varepsilon}(t_\epsilon, f_\varepsilon),$$

where $i_\varepsilon + j_\varepsilon = 1$. We have

$$\Phi^{j_\varepsilon}(t_\varepsilon, f_\varepsilon, f'_\varepsilon) \leq \Phi^{i_\varepsilon}(t_\varepsilon, f_\varepsilon, f'_\varepsilon).$$

As Υ and ψ are viscosity solutions of (7.1) we also have

$$\Upsilon^{i_\varepsilon}(t_\varepsilon, f_\varepsilon) \geq \log(1 - g_{i_\varepsilon j_\varepsilon}) + \Upsilon^{j_\varepsilon}(t_\varepsilon, f_\varepsilon),$$
$$\psi^{i_\varepsilon}(t_\varepsilon, f'_\varepsilon) \geq \log(1 - g_{i_\varepsilon j_\varepsilon}) + \psi^{j_\varepsilon}(t_\varepsilon, f'_\varepsilon).$$

Therefore, if the desired result were not true we would have

$$\log(1 - g_{i_\varepsilon j_\varepsilon}) \leq \log(1 - g_{i_\varepsilon j_\varepsilon})(1+\gamma),$$

which is impossible. □

7.1.2 Application of Ishii's lemma

We are in a position to apply Ishii's lemma (see Theorem 8.3 in [7] for this lemma and the definition of the sets $\overline{\mathcal{P}^{2+}}((1+\gamma)\Upsilon^{i_\varepsilon})(t_\varepsilon, f_\varepsilon)$ and $\overline{\mathcal{P}^{2-}}\psi^{i_\varepsilon}(t_\varepsilon, f'_\varepsilon)$ below). Consider the function Ψ defined on $[0,T] \times [0,1]^2$ as follows:

$$\Psi(t,f,f') = \frac{1}{2\varepsilon}(|f - f'|^2) - \beta t + \frac{\lambda}{t}.$$

Notice that $\Phi^i(t,f,f') = (1+\gamma)\Upsilon^i(t,f) - \psi^i(t,f') - \Psi(t,f,f')$. For all $\varepsilon > 0$, Ishii's lemma implies that there exist two real numbers d and d' and two positive numbers X and X' such that

$$\left(d, \frac{\partial\Psi}{\partial f}(t_\varepsilon, f_\varepsilon, f'_\varepsilon), X\right) \in \overline{\mathcal{P}^{2+}}((1+\gamma)\Upsilon^{i_\varepsilon})(t_\varepsilon, f_\varepsilon),$$
$$-\left(d', \frac{\partial\Psi}{\partial f'}(t_\varepsilon, f_\varepsilon, f'_\varepsilon), X'\right) \in \overline{\mathcal{P}^{2-}}\psi^{i_\varepsilon}(t_\varepsilon, f'_\varepsilon),$$
$$d + d' = \frac{\partial\Psi}{\partial t}(t_\varepsilon, f_\varepsilon, f'_\varepsilon),$$
$$-\left(\frac{1}{\varepsilon} + \|A\|\right) I \leq \begin{pmatrix} X & 0 \\ 0 & X' \end{pmatrix} \leq A + \varepsilon A^2,$$

where A is the Hessian matrix of Ψ in (f, f'), that is,

$$A = \begin{pmatrix} \frac{1}{\varepsilon} & -\frac{1}{\varepsilon} \\ -\frac{1}{\varepsilon} & \frac{1}{\varepsilon} \end{pmatrix}.$$

We now use the Proposition 7.2. We have:

$$\begin{aligned}
&-d - \left(i_\varepsilon(\mu_1 f_\varepsilon + \mu_2(1-f_\varepsilon) - r - \frac{\sigma^2}{2}) + r\right) \\
&\quad - (-\lambda_1 f_\varepsilon + \lambda_2(1-f_\varepsilon)) \frac{\partial \Psi}{\partial f}(t_\varepsilon, f_\varepsilon, f'_\varepsilon) - \frac{1}{2}\left(\frac{\mu_1-\mu_2}{\sigma}\right)^2 f_\varepsilon^2 (1-f_\varepsilon)^2 X \\
&\le 0 \\
&\le d' - \left(i_\varepsilon(\mu_1 f'_\varepsilon + \mu_2(1-f'_\varepsilon) - r - \frac{\sigma^2}{2}) + r\right) \\
&\quad + (-\lambda_1 f'_\varepsilon + \lambda_2(1-f'_\varepsilon)) \frac{\partial \Psi}{\partial f'}(t_\varepsilon, f_\varepsilon, f'_\varepsilon) + \frac{1}{2}\left(\frac{\mu_1-\mu_2}{\sigma}\right)^2 (f'_\varepsilon)^2 (1-f'_\varepsilon)^2 X'.
\end{aligned}$$

In view of the condition on $d + d'$ we deduce

$$\begin{aligned}
\beta + \frac{\lambda}{t_\varepsilon^2} &\le i_\varepsilon(\mu_1 - \mu_2)(f_\varepsilon - f'_\varepsilon) \\
&\quad + (-\lambda_1 f_\varepsilon + \lambda_2(1-f_\varepsilon)) \frac{1}{\varepsilon}(f_\varepsilon - f'_\varepsilon) + (-\lambda_1 f'_\varepsilon + \lambda_2(1-f'_\varepsilon)) \frac{1}{\varepsilon}(f'_\varepsilon - f_\varepsilon) \\
&\quad + \begin{pmatrix} \frac{\mu_1-\mu_2}{\sigma} f_\varepsilon(1-f_\varepsilon) \\ \frac{\mu_1-\mu_2}{\sigma} f'_\varepsilon(1-f'_\varepsilon) \end{pmatrix}^t \begin{pmatrix} X & 0 \\ 0 & X' \end{pmatrix} \begin{pmatrix} \frac{\mu_1-\mu_2}{\sigma} f_\varepsilon(1-f_\varepsilon) \\ \frac{\mu_1-\mu_2}{\sigma} f'_\varepsilon(1-f'_\varepsilon) \end{pmatrix}.
\end{aligned}$$

Notice that

$$A + \varepsilon A^2 = \begin{pmatrix} \frac{3}{\varepsilon} & -\frac{3}{\varepsilon} \\ -\frac{3}{\varepsilon} & \frac{3}{\varepsilon} \end{pmatrix}.$$

Therefore

$$\begin{aligned}
\beta + \frac{\lambda}{t_\varepsilon^2} &\le i_\varepsilon(\mu_1 f_\varepsilon + \mu_2(1-f_\varepsilon) - r - \frac{\sigma^2}{2}) + r + i_\varepsilon(\mu_1 f'_\varepsilon + \mu_2(1-f'_\varepsilon) - r - \frac{\sigma^2}{2}) + r \\
&\quad + (-\lambda_1 f_\varepsilon + \lambda_2(1-f_\varepsilon)) \frac{1}{\varepsilon}(f_\varepsilon - f'_\varepsilon) + (-\lambda_1 f'_\varepsilon + \lambda_2(1-f'_\varepsilon)) \frac{1}{\varepsilon}(f'_\varepsilon - f_\varepsilon) \\
&\quad + \frac{3}{\varepsilon}\left(\frac{\mu_1-\mu_2}{\sigma}\right)^2 (f_\varepsilon(1-f_\varepsilon) - f'_\varepsilon(1-f'_\varepsilon))^2 \\
&=: K_1 + \frac{K_2}{\varepsilon}.
\end{aligned}$$

We first estimate K_2:

$$K_2 \leq -(\lambda_1+\lambda_2)(f_\varepsilon - f'_\varepsilon)^2 + 3\left(\frac{\mu_1-\mu_2}{\sigma}\right)^2 (f_\varepsilon - f'_\varepsilon)^2(1+f_\varepsilon+f'_\varepsilon)^2$$
$$\leq 27\left(\frac{\mu_1-\mu_2}{\sigma}\right)^2 (f_\varepsilon - f'_\varepsilon)^2,$$

from which, owing to the Lemma 7.3 below,

$$\frac{K_2}{\varepsilon} \xrightarrow[\varepsilon\to 0]{} 0.$$

Lemma 7.3. *One has*

$$\lim_{\varepsilon\to 0} \frac{(f_\varepsilon - f'_\varepsilon)^2}{\varepsilon} = 0.$$

Proof. For all i, t, f, f' in $\{0; 1\} \times [0, T] \times [0, 1]^2$, by definition of the constant C^* one has

$$\Phi^i(t, f, f') \leq 2C^* - \frac{1}{2\varepsilon}(|f - f'|^2) + \beta T.$$

Set

$$H := \Phi^{\hat{i}}(\hat{t}, \hat{f}, \hat{f}) = \eta + \gamma \Upsilon^{\hat{i}}(\hat{t}, \hat{f}) + \beta \hat{t} - \frac{\lambda}{\hat{t}}.$$

If $|f - f'|^2 \geq 2\varepsilon(2C^* + \beta T - H + 1)$, one has $\Phi^i(t, f, f') \leq H - 1$; therefore

$$|f_\varepsilon - f'_\varepsilon|^2 \leq 2\varepsilon(2C^* + \beta T - H + 1).$$

In particular, $|f_\varepsilon - f'_\varepsilon|$ tends to 0 with ε. One also has

$$2\Phi^{i_\varepsilon}(t_\varepsilon, f_\varepsilon, f'_\varepsilon) \geq \Phi^{i_\varepsilon}(t_\varepsilon, f_\varepsilon, f_\varepsilon) + \Phi^{i_\varepsilon}(t_\varepsilon, f'_\varepsilon, f'_\varepsilon),$$

and thus

$$\frac{1}{2\varepsilon}(f_\varepsilon - f'_\varepsilon)^2 \leq \Upsilon^{i_\varepsilon}(t_\varepsilon, f_\varepsilon) - \Upsilon^{i_\varepsilon}(t_\varepsilon, f'_\varepsilon) + \psi^{i_\varepsilon}(t_\varepsilon, f'_\varepsilon) - \psi^{i_\varepsilon}(t_\varepsilon, f_\varepsilon).$$

The right-hand side tends to 0 with ε since Υ and ψ are uniformly continuous on $[0, 1]$, and since $|f_\varepsilon - f'_\varepsilon|$ tends to 0 with ε. □

It remains to estimate K_1. Notice that

$$K_1 = i_\varepsilon(\mu_1 - \mu_2)(f_\varepsilon - f'_\varepsilon),$$

and thus tends to 0 with ε. When ε tends to 0, one obtains $\beta \leq 0$. We thus have exhibited a contradiction, and proven the uniqueness of the viscosity solution.

7.2 Utility function in the class $\mathcal{U}$

In this subsection, we consider the case where the utility function U belongs to the class $\mathcal{U}$.

Suppose that Υ and ψ are two viscosity solutions of (2.11) in $\mathcal{V}_\alpha$. Then there would exist $(\hat{i}, \hat{t}, \hat{x}, \hat{f})$ in $\{0;1\}\times]0,T)\times]0,+\infty)\times[0,1]$ such that

$$\eta := \Upsilon^{\hat{i}}(\hat{t}, \hat{x}, \hat{f}) - \psi^{\hat{i}}(\hat{t}, \hat{x}, \hat{f}) > 0.$$

As Υ and ψ are null and continuous at $x = 0$, there also would exist $m > 0$ such that, for all i, t, f, f and $x, x' \le m$,

$$|\Upsilon^i(t, x, f) - \psi^i(t, x', f')| < \frac{\eta}{5}. \tag{7.3}$$

In addition, as Υ and ψ are in $\mathcal{V}_\alpha$, there exists $C > 0$ such that, for all i, t, x, f,

$$|\Upsilon^i(t, x, f)| + |\psi^i(t, x, f)| \le C(1 + x). \tag{7.4}$$

Define the functions Φ^0 and Φ^1 on $[0,T] \times [0,+\infty)^2 \times [0,1]^2$ by

$$\begin{aligned}\Phi^i(t, x, x', f, f') := \Upsilon^i(t, x, f) - \psi^i(t, x', f') - \nu e^{-Dt}(x^2 + x'^2)\\ - \frac{1}{2\varepsilon}(|x - x'|^2 + |f - f'|^2) + \beta t - \frac{\lambda}{t},\end{aligned}$$

where, C being as in (7.4),

$$\begin{aligned}\varepsilon > 0,\ D > (|\mu_1| + |\mu_2| + r + 3\sigma^2);\ 0 < \lambda < \frac{\eta \hat{t} T}{5(T - \hat{t})};\\ 0 < \beta < \frac{\eta}{5(T - \hat{t})}; 0 < \nu < \min\left\{ C e^{DT}, \frac{\eta e^{D\hat{t}}}{10\hat{x}^2} \right\}.\end{aligned} \tag{7.5}$$

Set

$$H := \Phi^{\hat{i}}(\hat{t}, \hat{x}, \hat{x}, \hat{f}, \hat{f}) = \eta - 2\nu e^{-D\hat{t}}\hat{x}^2 + \beta\hat{t} - \frac{\lambda}{\hat{t}}.$$

7.2.1 Auxiliary lemmae

Lemma 7.4. *Set*

$$M := \frac{e^{DT}}{2\nu}\left(C + \sqrt{C^2 + 4\nu e^{-DT}(2C + \beta T + C^2\nu^{-1}e^{DT} + 1 - H)} \right).$$

If $x > M$ or $x' > M$ then, for all i, t, x, x', f, f',

$$\Phi^i(t, x, x', f, f') \le H - 1 < \Phi^{\hat{i}}(\hat{t}, \hat{x}, \hat{x}, \hat{f}, \hat{f}).$$

Proof. For all i, t, x, x', f, f' one has

$$\Phi^i(t, x, x', f, f') \le 2C + \beta T + C(x + x') - \nu e^{-DT}(x^2 + x'^2).$$

On the one hand, if $x' \le C\nu^{-1}e^{DT}$ then

$$\Phi^i(t, x, x', f, f') \le 2C + \beta T + C^2\nu^{-1}e^{DT} + Cx - \nu e^{-DT}x^2.$$

Therefore $\Phi^i(t, x, x', f, f') \le H - 1$ when $x \ge M$.

On the other hand, if $x' > C\nu^{-1}e^{DT}$, then $Cx' - \nu e^{-DT}x'^2 < 0$ and

$$\begin{aligned}\Phi^i(t, x, x', f, f') &\le 2C + \beta T + Cx - \nu e^{-DT}x^2 \\ &\le 2C + \beta T + C^2\nu^{-1}e^{DT} + Cx - \nu e^{-DT}x^2.\end{aligned}$$

Therefore we also have $\Phi^i(t, x, x', f, f') \le H - 1$ when $x \ge M$. □

As Φ^0 and Φ^1 are continuous, we deduce that there exists $(i_\varepsilon, t_\varepsilon, x_\varepsilon, x'_\varepsilon, f_\varepsilon, f'_\varepsilon)$ in $\{0; 1\} \times [0, T] \times [0, M]^2 \times [0, 1]^2$ such that

$$\Phi^{i_\varepsilon}(t_\varepsilon, x_\varepsilon, x'_\varepsilon, f_\varepsilon, f'_\varepsilon) \ge \Phi^i(t, x, x', f, f')$$

for all i, t, x, x', f, f' in $\{0; 1\} \times [0, T] \times [0, +\infty)^2 \times [0, 1]^2$.

Lemma 7.5. *One has*

$$|x_\varepsilon - x'_\varepsilon|^2 + |f_\varepsilon - f'_\varepsilon|^2 \le 2\varepsilon(2C(1 + M) + \beta T - H + 1).$$

Proof. The constant C being defined as in (7.4), for all i, t, x, x', f, f' in $\{0; 1\} \times [0, T] \times [0, M]^2 \times [0, 1]^2$,

$$\Phi^i(t, x, x', f, f') \le 2C + 2CM - \frac{1}{2\varepsilon}(|x - x'|^2 + |f - f'|^2) + \beta T.$$

Therefore, if

$$|x - x'|^2 + |f - f'|^2 \ge 2\varepsilon(2C(1 + M) + \beta T - H + 1)$$

then $\Phi^i(t, x, x', f, f') \le H - 1$, from which the result follows. □

The preceding lemma implies that $|x_\varepsilon - x'_\varepsilon|^2 + |f_\varepsilon - f'_\varepsilon|^2$ tends to 0 with ε. We now prove that x_ε and x'_ε cannot be in a small neighbourhood of 0.

Lemma 7.6. *For all ε small enough, $x_\varepsilon \ge \frac{m}{2}$ and $x'_\varepsilon \ge \frac{m}{2}$, where m is defined by (7.3).*

Proof. Suppose that $x_\varepsilon < \frac{m}{2}$ or $x'_\varepsilon < \frac{m}{2}$. In view of the preceding lemma, there exists $\varepsilon_0 > 0$ such that, for all $0 < \varepsilon < \varepsilon_0$, $|x_\varepsilon - x'_\varepsilon| < \frac{m}{2}$. Therefore $x_\varepsilon \le m$ and $x'_\varepsilon \le m$.

By definition of m, η, β and λ, one thus has

$$\begin{aligned}\Phi^{i_\varepsilon}(t_\varepsilon, x_\varepsilon, x'_\varepsilon, f_\varepsilon, f'_\varepsilon) &= \Upsilon^{i_\varepsilon}(t_\varepsilon, x_\varepsilon, f_\varepsilon) - \psi^{i_\varepsilon}(t_\varepsilon, x'_\varepsilon, f'_\varepsilon) - \nu e^{-Dt_\varepsilon}(x_\varepsilon^2 + (x'_\varepsilon)^2) \\ &\quad - \frac{1}{2\varepsilon}(|x_\varepsilon - x'_\varepsilon|^2 + |f_\varepsilon - f'_\varepsilon|^2) + \beta t_\varepsilon - \frac{\lambda}{t_\varepsilon} \\ &\le \frac{\eta}{5} + \Phi^{\hat{i}}(\hat{t}, \hat{x}, \hat{x}, \hat{f}, \hat{f}) - \eta + 2\nu e^{-D\hat{t}}\hat{x}^2 - \beta(\hat{t} - T) + \frac{\lambda}{\hat{t}} - \frac{\lambda}{T} \\ &\le \Phi^{\hat{i}}(\hat{t}, \hat{x}, \hat{x}, \hat{f}, \hat{f}) - \frac{\eta}{5},\end{aligned}$$

which is impossible. □

Lemma 7.7. *One has* $0 < t_\varepsilon < T$.

Proof. In view of the definition of Φ and the preceding lemma, for all i, t, x, x', f, f' in $\{0; 1\} \times [0, T] \times [m/2, M]^2 \times [0, 1]^2$ one has

$$\Phi^i(t, x, x', f, f') \le 2C + 2CM + \beta T - \frac{\lambda}{t}.$$

Therefore, if $t \le \lambda(2C(1 + M) + \beta T - H + 1)^{-1}$, then $\Phi^i(t, x, x', f, f') \le H - 1$, which implies that $t_\varepsilon > 0$. Suppose that $t_\varepsilon = T$. Then

$$\begin{aligned}\Phi^{i_\varepsilon}(T, x_\varepsilon, x'_\varepsilon, f_\varepsilon, f'_\varepsilon) &= U(x_\varepsilon) - U(x'_\varepsilon) - \nu e^{-DT}(x_\varepsilon^2 + (x'_\varepsilon)^2) \\ &\quad - \frac{1}{2\varepsilon}(|x_\varepsilon - x'_\varepsilon|^2 + |f_\varepsilon - f'_\varepsilon|^2) + \beta T - \frac{\lambda}{T} \\ &\ge \Phi^{\hat{i}}(\hat{t}, \hat{x}, \hat{x}, \hat{f}, \hat{f}) \\ &= \eta - 2\nu e^{-D\hat{t}}\hat{x}^2 + \beta\hat{t} - \frac{\lambda}{\hat{t}}.\end{aligned}$$

Our choice (7.5) of ν, β, and λ implies

$$\begin{aligned}U'(m/2)|x_\varepsilon - x'_\varepsilon| &\ge U(x_\varepsilon) - U(x'_\varepsilon) \\ &\ge \eta - 2\nu e^{-D\hat{t}}\hat{x}^2 + \beta(\hat{t} - T) - \lambda\left(\frac{1}{\hat{t}} - \frac{1}{T}\right) \\ &\ge \frac{2\eta}{5}.\end{aligned}$$

Thanks to Lemma 7.5, we now choose ε small enough in order that $U'(m/2)|x_\varepsilon - x'_\varepsilon| \le \frac{\eta}{5}$ and obtain a contradiction. □

Lemma 7.8. *One has*

$$\Upsilon^{i_\varepsilon}(t_\varepsilon, x_\varepsilon, f_\varepsilon) > \Upsilon^{j_\varepsilon}(t_\varepsilon, x_\varepsilon(1 - g_{i_\varepsilon j_\varepsilon}), f_\varepsilon).$$

Proof. We already know:

$$\Upsilon^{i_\varepsilon}(t_\varepsilon, x_\varepsilon, f_\varepsilon) \geq \Upsilon^{j_\varepsilon}(t_\varepsilon, x_\varepsilon(1-g_{i_\varepsilon j_\varepsilon}), f_\varepsilon),$$
$$\psi^{i_\varepsilon}(t_\varepsilon, x'_\varepsilon, f'_\varepsilon) \geq \psi^{j_\varepsilon}(t_\varepsilon, x'_\varepsilon(1-g_{i_\varepsilon j_\varepsilon}), f'_\varepsilon),$$

where $i_\varepsilon + j_\varepsilon = 1$. By definition of $(i_\varepsilon, t_\varepsilon, x_\varepsilon, x'_\varepsilon, f_\varepsilon, f'_\varepsilon)$, we have

$$\Phi^{j_\varepsilon}(t_\varepsilon, x_\varepsilon(1-g_{i_\varepsilon j_\varepsilon}), x'_\varepsilon(1-g_{i_\varepsilon j_\varepsilon}), f_\varepsilon, f'_\varepsilon) \leq \Phi^{i_\varepsilon}(t_\varepsilon, x_\varepsilon, x'_\varepsilon, f_\varepsilon, f'_\varepsilon).$$

Suppose that the desired result does not hold true. Then we would have

$$0 \leq \nu \mathrm{e}^{-Dt_\varepsilon}(x_\varepsilon^2 + (x'_\varepsilon)^2)((1-g_{i_\varepsilon j_\varepsilon})^2 - 1) + \frac{1}{2\varepsilon}|x_\varepsilon - x'_\varepsilon|^2((1-g_{i_\varepsilon j_\varepsilon})^2 - 1) < 0,$$

which is impossible since $x_\varepsilon > 0$, $x'_\varepsilon > 0$, g_{10} and g_{01} are in $(0,1)$. □

Notice that the preceding lemma implies that

$$\begin{cases} -\dfrac{\partial \Upsilon^{i_\varepsilon}}{\partial t}(t_\varepsilon, x_\varepsilon, f_\varepsilon) - \mathcal{L}^{i_\varepsilon}\Upsilon^{i_\varepsilon}(t_\varepsilon, x_\varepsilon, f_\varepsilon) = 0, \\ -\dfrac{\partial \psi^{i_\varepsilon}}{\partial t}(t_\varepsilon, x'_\varepsilon, f'_\varepsilon) - \mathcal{L}^{i_\varepsilon}\psi^{i_\varepsilon}(t_\varepsilon, x'_\varepsilon, f'_\varepsilon) \geq 0. \end{cases}$$

7.2.2 Application of Ishii's lemma

Define the function Ψ on $[0,T] \times [0,+\infty)^2 \times [0,1]^2$ as follows:

$$\Psi(t,x,x',f,f') = \nu \mathrm{e}^{-Dt}(x^2 + x'^2) + \frac{1}{2\varepsilon}(|x-x'|^2 + |f-f'|^2) - \beta t + \frac{\lambda}{t}.$$

Notice that $\Phi^i(t,x,x',f,f') = \Upsilon^i(t,x,f) - \psi^i(t,x',f') - \Psi(t,x,x',f,f')$. For all $\varepsilon > 0$, Ishii's lemma implies that there exist two real numbers d and d' and two symmetric matrices X and X' such that

$$\left(d, \left(\frac{\partial \Psi}{\partial x}, \frac{\partial \Psi}{\partial f}\right)(t_\varepsilon, x_\varepsilon, x'_\varepsilon, f_\varepsilon, f'_\varepsilon), X\right) \in \overline{\mathcal{P}^{2+}}\Upsilon^{i_\varepsilon}(t_\varepsilon, x_\varepsilon, f_\varepsilon)$$
$$-\left(d', \left(\frac{\partial \Psi}{\partial x'}, \frac{\partial \Psi}{\partial f'}\right)(t_\varepsilon, x_\varepsilon, x'_\varepsilon, f_\varepsilon, f'_\varepsilon), X'\right) \in \overline{\mathcal{P}^{2-}}\psi^{i_\varepsilon}(t_\varepsilon, x'_\varepsilon, f'_\varepsilon),$$
$$d + d' = \frac{\partial \Psi}{\partial t}(t_\varepsilon, x_\varepsilon, x'_\varepsilon, f_\varepsilon, f'_\varepsilon),$$
$$-\left(\frac{1}{\varepsilon} + \|A\|\right) I \leq \begin{pmatrix} X & 0 \\ 0 & X' \end{pmatrix} \leq A + \varepsilon A^2,$$

where A is the Hessian matrix of Φ in x, x', f, f', that is,

$$A := \begin{pmatrix} a + \frac{1}{\varepsilon} & 0 & -\frac{1}{\varepsilon} & 0 \\ 0 & \frac{1}{\varepsilon} & 0 & -\frac{1}{\varepsilon} \\ -\frac{1}{\varepsilon} & 0 & a + \frac{1}{\varepsilon} & 0 \\ 0 & -\frac{1}{\varepsilon} & 0 & \frac{1}{\varepsilon} \end{pmatrix} = \begin{pmatrix} a & 0 & 0 & 0 \\ 0 & 0 & 0 & 0 \\ 0 & 0 & a & 0 \\ 0 & 0 & 0 & 0 \end{pmatrix} + \frac{1}{\varepsilon}\begin{pmatrix} 1 & 0 & -1 & 0 \\ 0 & 1 & 0 & -1 \\ -1 & 0 & 1 & 0 \\ 0 & -1 & 0 & 1 \end{pmatrix},$$

where $a := 2\nu e^{-Dt_\varepsilon}$.

We now use that we deal with viscosity solutions and get

$$\begin{aligned}
&-d - x_\varepsilon \left(i_\varepsilon(\mu_1 f_\varepsilon + \mu_2(1-f_\varepsilon) - r) + r\right) \frac{\partial \Psi}{\partial x}(t_\varepsilon, x_\varepsilon, x'_\varepsilon, f_\varepsilon, f'_\varepsilon) \\
&\quad - (-\lambda_1 f_\varepsilon + \lambda_2(1-f_\varepsilon)) \frac{\partial \Psi}{\partial f}(t_\varepsilon, x_\varepsilon, x'_\varepsilon, f_\varepsilon, f'_\varepsilon) - \frac{i_\varepsilon}{2} x_\varepsilon^2 \sigma^2 X_{11} \\
&\quad - \frac{1}{2}\left(\frac{\mu_1 - \mu_2}{\sigma}\right)^2 f_\varepsilon^2 (1-f_\varepsilon)^2 X_{22} \\
&\quad - i_\varepsilon x_\varepsilon (\mu_1 - \mu_2) f_\varepsilon (1-f_\varepsilon) X_{12} \\
&\le 0 \\
&\le d' + x'_\varepsilon \left(i_\varepsilon(\mu_1 f'_\varepsilon + \mu_2(1-f'_\varepsilon) - r) + r\right) \frac{\partial \Psi}{\partial x'}(t_\varepsilon, x_\varepsilon, x'_\varepsilon, f_\varepsilon, f'_\varepsilon) \\
&\quad + (-\lambda_1 f'_\varepsilon + \lambda_2(1-f'_\varepsilon)) \frac{\partial \Psi}{\partial f'}(t_\varepsilon, x_\varepsilon, x'_\varepsilon, f_\varepsilon, f'_\varepsilon) + \frac{i_\varepsilon}{2} (x'_\varepsilon)^2 \sigma^2 X'_{11} \\
&\quad + \frac{1}{2}\left(\frac{\mu_1 - \mu_2}{\sigma}\right)^2 (f'_\varepsilon)^2 (1-f'_\varepsilon)^2 X'_{22} + i_\varepsilon x'_\varepsilon (\mu_1 - \mu_2) f'_\varepsilon (1-f'_\varepsilon) X'_{12}.
\end{aligned}$$

In view of the condition on $d + d'$ we deduce

$$\begin{aligned}
&D e^{-Dt_\varepsilon} \nu (x_\varepsilon^2 + (x'_\varepsilon)^2) + \beta + \frac{\lambda}{t_\varepsilon^2} \\
&\le x_\varepsilon \left(i_\varepsilon(\mu_1 f_\varepsilon + \mu_2(1-f_\varepsilon) - r) + r\right) \left(2\nu e^{-Dt_\varepsilon} x_\varepsilon + \frac{1}{\varepsilon}(x_\varepsilon - x'_\varepsilon)\right) \\
&\quad + x'_\varepsilon \left(i_\varepsilon(\mu_1 f'_\varepsilon + \mu_2(1-f'_\varepsilon) - r) + r\right) \left(2\nu e^{-Dt_\varepsilon} x'_\varepsilon + \frac{1}{\varepsilon}(x'_\varepsilon - x_\varepsilon)\right) \\
&\quad + (-\lambda_1 f_\varepsilon + \lambda_2(1-f_\varepsilon)) \left(\frac{1}{\varepsilon}(f_\varepsilon - f'_\varepsilon)\right) + (-\lambda_1 f'_\varepsilon + \lambda_2(1-f'_\varepsilon)) \left(\frac{1}{\varepsilon}(f'_\varepsilon - f_\varepsilon)\right) \\
&\quad + \begin{pmatrix} \frac{i_\varepsilon}{\sqrt{2}} x_\varepsilon \sigma \\ \frac{\mu_1 - \mu_2}{\sigma} f_\varepsilon (1-f_\varepsilon) \\ \frac{i_\varepsilon}{\sqrt{2}} x'_\varepsilon \sigma \\ \frac{\mu_1 - \mu_2}{\sigma} f'_\varepsilon (1-f'_\varepsilon) \end{pmatrix}^t \begin{pmatrix} X & 0 \\ 0 & X' \end{pmatrix} \begin{pmatrix} \frac{i_\varepsilon}{\sqrt{2}} x_\varepsilon \sigma \\ \frac{\mu_1 - \mu_2}{\sigma} f_\varepsilon (1-f_\varepsilon) \\ \frac{i_\varepsilon}{\sqrt{2}} x'_\varepsilon \sigma \\ \frac{\mu_1 - \mu_2}{\sigma} f'_\varepsilon (1-f'_\varepsilon) \end{pmatrix}.
\end{aligned}$$

Notice that

$$A+\varepsilon A^2=\begin{pmatrix} 3a+\varepsilon a^2+\dfrac{3}{\varepsilon} & 0 & -2a-\dfrac{3}{\varepsilon} & 0\\ 0 & \dfrac{3}{\varepsilon} & 0 & -\dfrac{3}{\varepsilon}\\ -2a-\dfrac{3}{\varepsilon} & 0 & 3a+\varepsilon a^2+\dfrac{3}{\varepsilon} & 0\\ 0 & -\dfrac{3}{\varepsilon} & 0 & \dfrac{3}{\varepsilon}\end{pmatrix}.$$

In addition,

$$\begin{aligned}\beta+\frac{\lambda}{t_\varepsilon^2}\le\frac{1}{\varepsilon}\Bigg[&(x_\varepsilon-x'_\varepsilon)x_\varepsilon\left(i_\varepsilon(\mu_1 f_\varepsilon+\mu_2(1-f_\varepsilon)-r)+r\right)\\ &-x'_\varepsilon\left(i_\varepsilon(\mu_1 f'_\varepsilon+\mu_2(1-f'_\varepsilon)-r)+r\right)+3i_\varepsilon\frac{\sigma^2}{2}(x_\varepsilon-x'_\varepsilon)^2\\ &-(\lambda_1+\lambda_2)(f_\varepsilon-f'_\varepsilon)^2+3\left(\frac{\mu_1-\mu_2}{\sigma}\right)^2(f_\varepsilon(1-f_\varepsilon)-f'_\varepsilon(1-f'_\varepsilon))^2\Bigg]\\ &+\varepsilon i_\varepsilon a^2\frac{\sigma^2}{2}\left(x_\varepsilon^2+(x'_\varepsilon)^2\right)\\ &+\nu\left(-D\mathrm{e}^{-Dt_\varepsilon}(x_\varepsilon^2+(x'_\varepsilon)^2)+2\mathrm{e}^{-Dt_\varepsilon}\right)\left(x_\varepsilon^2\left(i_\varepsilon(\mu_1 f_\varepsilon+\mu_2(1-f_\varepsilon)-r)+r\right.\right.\\ &\left.\left.+(x'_\varepsilon)^2\left(i_\varepsilon(\mu_1 f'_\varepsilon+\mu_2(1-f'_\varepsilon)-r)+r\right)\right)\right)\\ &+i_\varepsilon a\frac{\sigma^2}{2}\left(3x_\varepsilon^2+3(x'_\varepsilon)^2-4x_\varepsilon x'_\varepsilon\right)\\ =&\frac{K_1}{\varepsilon}+\varepsilon K_2+K_3.\end{aligned}$$

We now estimate K_3. As $a>0$,

$$\begin{aligned}K_3\le&\,\mathrm{e}^{-Dt_\varepsilon}\nu\left(x_\varepsilon^2\left(-D+i_\varepsilon(\mu_1 f_\varepsilon+\mu_2(1-f_\varepsilon)-r+3\sigma^2)+r\right)\right.\\ &\left.+(x'_\varepsilon)^2\left(-D+i_\varepsilon(\mu_1 f'_\varepsilon+\mu_2(1-f'_\varepsilon)-r+3\sigma^2)+r\right)\right)\\ \le&\,\mathrm{e}^{-Dt_\varepsilon}\nu(x_\varepsilon^2+(x'_\varepsilon)^2)(-D+|\mu_1|+|\mu_2|+r+3\sigma^2).\end{aligned}$$

As $D>|\mu_1|+|\mu_2|+r+3\sigma$, we obtain $K_3\le 0$.

Notice that

$$K_2\le 4i_\varepsilon\sigma^2\nu^2M^2,$$

from which, since M and ν do not depend on ε,

$$\varepsilon K_2\xrightarrow[\varepsilon\to 0]{}0.$$

It remains to estimate K_1. One has

$$\begin{aligned}
K_1 &\le (x_\varepsilon - x'_\varepsilon)\left((x_\varepsilon - x'_\varepsilon)(|\mu_1| + |\mu_2| + r) + x'_\varepsilon(\mu_1 - \mu_2)(f_\varepsilon - f'_\varepsilon)\right) \\
&\quad + 3i_\varepsilon \frac{\sigma^2}{2}(x_\varepsilon - x'_\varepsilon)^2 + 12\left(\frac{\mu_1 - \mu_2}{\sigma}\right)^2 (f_\varepsilon - f'_\varepsilon)^2 \\
&\le (x_\varepsilon - x'_\varepsilon)^2 \left(3i_\varepsilon \frac{\sigma^2}{2} + |\mu_1| + |\mu_2| + r\right) + x'_\varepsilon(\mu_1 - \mu_2)(f_\varepsilon - f'_\varepsilon)(x_\varepsilon - x'_\varepsilon) \\
&\quad + 12\left(\frac{\mu_1 - \mu_2}{\sigma}\right)^2 (f_\varepsilon - f'_\varepsilon)^2 \\
&\le (x_\varepsilon - x'_\varepsilon)^2 \left(3i_\varepsilon \frac{\sigma^2}{2} + |\mu_1| + |\mu_2| + r\right) \\
&\quad + \frac{M}{2}|\mu_1 - \mu_2|\left((f_\varepsilon - f'_\varepsilon)^2 + (x_\varepsilon - x'_\varepsilon)^2\right) + 12\left(\frac{\mu_1 - \mu_2}{\sigma}\right)^2 (f_\varepsilon - f'_\varepsilon)^2 \\
&\le (x_\varepsilon - x'_\varepsilon)^2 \left(3\frac{\sigma^2}{2} + |\mu_1| + |\mu_2| + r + \frac{M}{2}|\mu_1 - \mu_2|\right) \\
&\quad + 12\left(\frac{\mu_1 - \mu_2}{\sigma} + \frac{M}{2}|\mu_1 - \mu_2|\right)^2 (f_\varepsilon - f'_\varepsilon)^2.
\end{aligned}$$

Thus, in view of the Lemma 7.9 below,

$$\frac{K_1}{\varepsilon} \xrightarrow[\varepsilon \to 0]{} 0.$$

As ε tends to 0, we obtain $\beta \le 0$, which exhibits a contradiction.

Lemma 7.9. *One has*

$$\lim_{\varepsilon \to 0} \frac{(x_\varepsilon - x'_\varepsilon)^2 + (f_\varepsilon - f'_\varepsilon)^2}{\varepsilon} = 0.$$

Proof. Observe that

$$2\Phi^{i_\varepsilon}(t_\varepsilon, x_\varepsilon, x'_\varepsilon, f_\varepsilon, f'_\varepsilon) \ge \Phi^{i_\varepsilon}(t_\varepsilon, x_\varepsilon, x_\varepsilon, f_\varepsilon, f_\varepsilon) + \Phi^{i_\varepsilon}(t_\varepsilon, x'_\varepsilon, x'_\varepsilon, f'_\varepsilon, f'_\varepsilon).$$

As the function Υ and ψ are in $\mathcal{V}_\alpha$, in view of Lemma 7.5,

$$\begin{aligned}
&\frac{1}{2\varepsilon}\left((x_\varepsilon - x'_\varepsilon)^2 + (f_\varepsilon - f'_\varepsilon)^2\right) \\
&\qquad \le \Upsilon^{i_\varepsilon}(t_\varepsilon, x_\varepsilon, f_\varepsilon) - \Upsilon^{i_\varepsilon}(t_\varepsilon, x'_\varepsilon, f'_\varepsilon) + \psi^{i_\varepsilon}(t_\varepsilon, x'_\varepsilon, f'_\varepsilon) - \psi^{i_\varepsilon}(t_\varepsilon, x_\varepsilon, f_\varepsilon) \\
&\qquad \le K(1 + x_\varepsilon^{-\alpha} + (x'_\varepsilon)^{-\alpha})(|x_\varepsilon - x'_\varepsilon| + (x_\varepsilon + x'_\varepsilon)|f_\varepsilon - f'_\varepsilon|) \\
&\qquad \le K(1 + 2(m/2)^{-\alpha} + 2M)2\varepsilon(2C(1 + M) + \beta T - H + 1).
\end{aligned}$$

The desired result follows. □

8 Numerical illustrations

8.1 Numerical scheme

The characterisation of (V^0, V^1) as the unique solution of (2.11) enables us to propose a numerical scheme to approximate the value functions. We only consider here the special case where the utility is a power function: $U(x) = x^\alpha$ with $0 < \alpha < 1$. Hence, for all t, x, f, i and π one has $U(W_T^{t,x,f,i,\pi}) = U(x)U(W_T^{t,1,f,i,\pi})$, and therefore, for $i \in \{0,1\}$, $V^i(t,x,f) = U(x)V^i(t,1,f)$. With a slight abuse of notation, set $V^i(t,f) = V^i(t,1,f)$. The system (2.11) can now be simplified as follows:

$$\begin{cases} \min\big\{ -\frac{\partial V^0}{\partial t}(t,f) - \mathcal{L}^0 V^0(t,f);\ V^0(t,f) - (1-g_{01})^\alpha V^1(t,f)\big\} = 0 \\ \min\big\{ -\frac{\partial V^1}{\partial t}(t,f) - \mathcal{L}^1 V^1(t,f);\ V^1(t,f) - (1-g_{10})^\alpha V^0(t,f)\big\} = 0 \end{cases}$$

with the boundary conditions $V^0(T,f) = 1$ and $V^1(T,f) = 1$ for all $0 \le f \le 1$. Here, we have for all $0 \le t \le T$, $0 \le f \le 1$ and $i \in \{0,1\}$

$$\mathcal{L}^i \varphi(t,f) = c(f,i)\varphi(t,f) + b(f,i)\frac{\partial \varphi}{\partial f}(t,f) + a(f)\frac{\partial^2 \varphi}{\partial f^2}(t,f),$$

with

$$\begin{aligned} a(f) &= \frac{1}{2}\Big(\frac{\mu_1-\mu_2}{\sigma}\Big)^2 f^2(1-f)^2, \\ b(f,0) &= -\lambda_1 f + \lambda_2(1-f), \\ b(f,1) &= -\lambda_1 f + \lambda_2(1-f) + \alpha(\mu_1-\mu_2)f(1-f), \\ c(f,0) &= \alpha r, \\ c(f,1) &= \alpha\Big(\mu_1 f + \mu_2(1-f) - (1-\alpha)\frac{\sigma^2}{2}\Big). \end{aligned}$$

For a time discretisation step δt and a space discretisation step δf, set:

$$\begin{aligned} S^i\varphi(t,f) &= \frac{\varphi(t,f)-\varphi(t-\delta t,f)}{\delta t} + \hat{\mathcal{L}}^i\varphi(t,f) \\ &= \frac{\varphi(t,f)-\varphi(t-\delta t,f)}{\delta t} + c(f,i)\varphi(t,f) \\ &\quad + b(f,i)^+\frac{\varphi(t,f+\delta f)-\varphi(t,f)}{\delta f} - b(f,i)^-\frac{\varphi(t,f)-\varphi(t,f-\delta f)}{\delta f} \\ &\quad + a(f)\frac{\varphi(t,f+\delta f)-2\varphi(t,\delta f)+\varphi(t,f-\delta f)}{\delta f^2}, \end{aligned}$$

where $x^+ = \frac{|x|+x}{2}$ and $x^- = \frac{|x|-x}{2}$. Note that the first-order term in f depends on the sign of b, and the second-order term $a(f)$ is positive except at the boundaries $f = 0$ and $f = 1$.

We construct the approximation $(\hat{V}^0, \hat{V}^1)$ of the value functions (V^0, V^1) as follows:

- Set $\hat{V}^0(T,\cdot) = \hat{V}^1(T,\cdot) = 1$.
- At each time step:
  - set $\bar{V}^0(t,\cdot) = \hat{V}^0(t,\cdot)$ and $\bar{V}^1(t,\cdot) = \hat{V}^1(t,\cdot)$.
  - compute $\bar{V}^i(t-\delta t, f)$ solution of $S^i\bar{V}^i(t,f) = 0$.
  - set $\hat{V}^0(t-\delta t, f) = \max\{\bar{V}^0(t-\delta t, f); (1-g_{01})^\alpha \bar{V}^1(t-\delta t, f)\}$, and $\hat{V}^1(t-\delta t, f) = \max\{\bar{V}^1(t-\delta t, f); (1-g_{10})^\alpha \bar{V}^0(t-\delta t, f)\}$.

One can easily show by induction the following result.

Lemma 8.1. $(\hat{V}^0, \hat{V}^1)$ *is the unique solution of the system*

$$\begin{cases} \min\{-S^0\varphi^0(t,f);\ \varphi^0(t-\delta t, f) - (1-g_{01})^\alpha \varphi^1(t-\delta t, f)\} = 0, \\ \min\{-S^1\varphi^1(t,f);\ \varphi^1(t-\delta t, f) - (1-g_{10})^\alpha \varphi^0(t-\delta t, f)\} = 0, \\ \hat{V}^0(T,\cdot) = \hat{V}^1(T,\cdot) = 1. \end{cases}$$

Remark 8.2. So far, we have not proven the convergence of this scheme to the value functions (V^0, V^1). Key results in that direction are in [10, 2, 5].

8.2 Approximate value function

We implemented the above-mentioned numerical scheme for $U(x) = \sqrt{x}$ ($\alpha = 1/2$), $T = 3$, $\mu_1 = -0.2$, $\mu_2 = 0.21$, $\lambda_1 = \lambda_2 = 2$, $\sigma = 0.15$, $g_{01} = g_{10} = 0.01$, $r = 0$ and the discretisation steps $\delta t = 10^{-6}$ and $\delta f = 10^{-3}$.

Figure 8.1 shows the approximate value function $\hat{V}^0$ as a function of time and f. Note that here $\mu_2 > \mu_1$, hence the value function is larger when f is close to 0, which means when $\mu(t)$ is likely to equal to μ_2. Theorem 4.1 shows that the value functions are Lipschitz-continuous in f and Hölder continuous with index $1/2$ in time. Figure 8.2 is a zoom of Figure 8.1 for $2.5 \le t \le 3$. Figure 8.3 shows $\hat{V}^0(t, 0.05)$ for $0 \le t \le 3$. It exhibits that the time derivative is discontinuous. Figure 8.4 shows $\hat{V}^0(t,f)$ for $t = 2.9$ (highest curve), $t = 2.91$, $t = 2.92$, $t = 2.93$, $t = 2.94$, $t = 2.95$ (flat curve) respectively.

8.3 Efficient strategy

As mentioned above, here $\mu_2 > \mu_1$, hence the investor should invest in the stock when $\mu(t) = \mu_2$, i.e. when f is close to 0, and sell when $\mu(t) = \mu_1$, i.e. when f is close to 1. One has to decide what *close to* means. We propose the following so called *efficient* strategy. It corresponds to the discrete Dynamic Programming Principle for $(\hat{V}^0, \hat{V}^1)$.

- Compute $(\hat{V}^0, \hat{V}^1)$ for all t and f in the discretisation grid.
- At time t in the grid, compute an estimate $\hat{F}_t$ of F_t from the observation of the stock (using classical filtering theory) (see [15]).

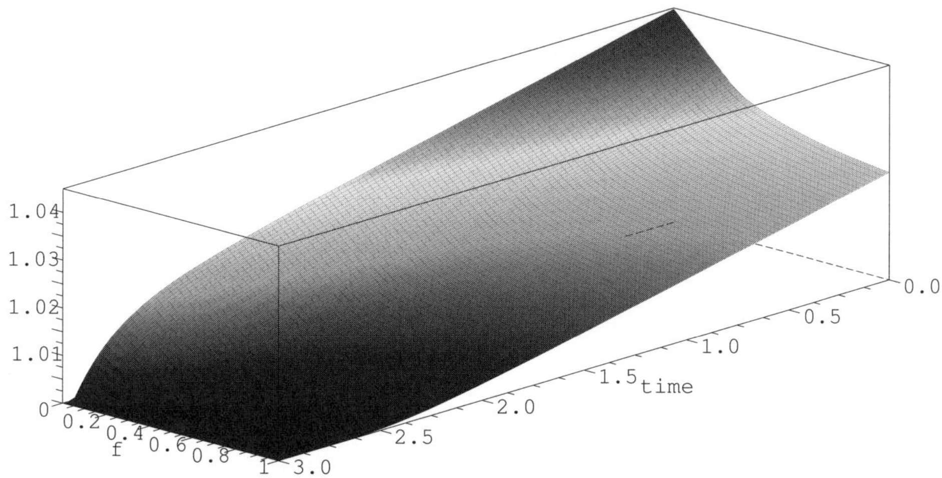

Figure 8.1. Approximated value function $\hat{V}^0$

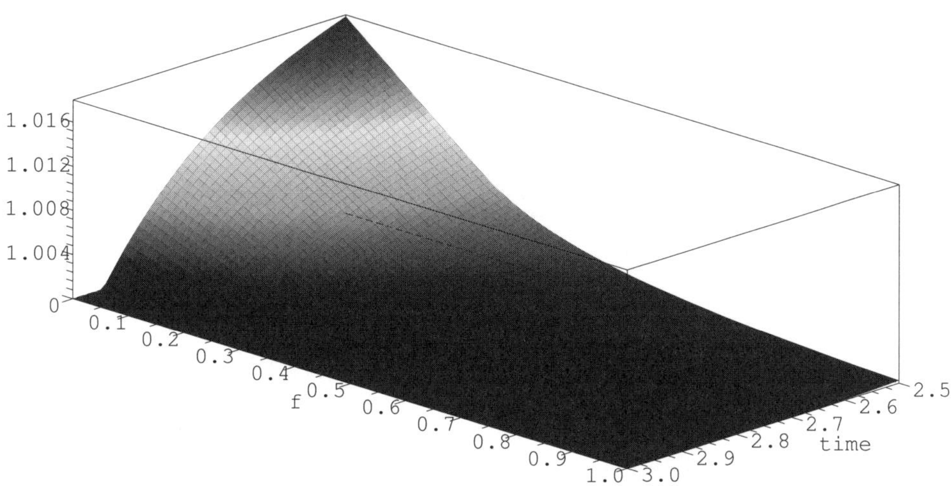

Figure 8.2. Zoom of $\hat{V}^0$ close to the horizon

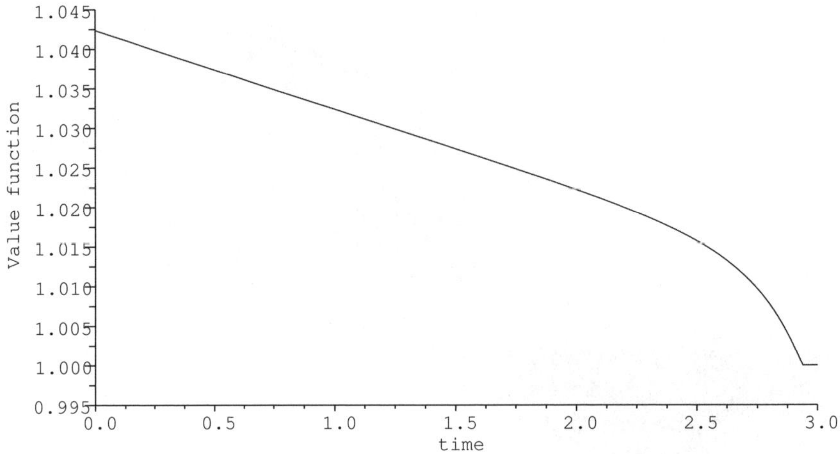

Figure 8.3. Regularity of $\hat{V}^0$ in time

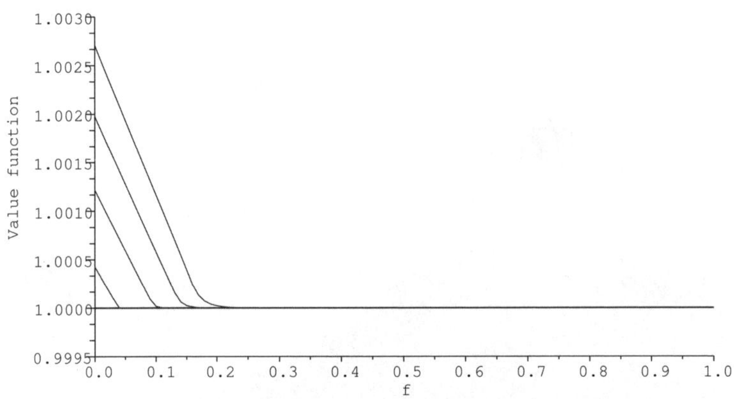

Figure 8.4. Regularity of $\hat{V}^0(t,\cdot)$ for $t = 2.9$ (highest curve), $t = 2.91$, $t = 2.92$, $t = 2.93$, $t = 2.94$, $t = 2.95$ (flat curve) respectively

- Compare $\hat{V}^0(t, \hat{F}_t)$ and $\hat{V}^1(t, \hat{F}_t)$: buy if $\hat{V}^0(t, \hat{F}_t) = (1 - g_{01})^\alpha \hat{V}^1(t, \hat{F}_t)$, sell if $\hat{V}^1(t, \hat{F}_t) = (1 - g_{10})^\alpha \hat{V}^0(t, \hat{F}_t)$.

Note that to apply this strategy, one only needs to know the areas where $\hat{V}^0 = (1 - g_{01})^\alpha \hat{V}^1$ or $\hat{V}^1 = (1 - g_{10})^\alpha \hat{V}^0$. It is not necessary to keep in memory the values of $\hat{V}^0$ and $\hat{V}^1$ at each point of the grid. Figure 8.5 illustrates this strategy. It shows the buying area where $\hat{V}^0 = (1 - g_{01})^\alpha \hat{V}^1$ (lower area), which means that F is close enough to 0

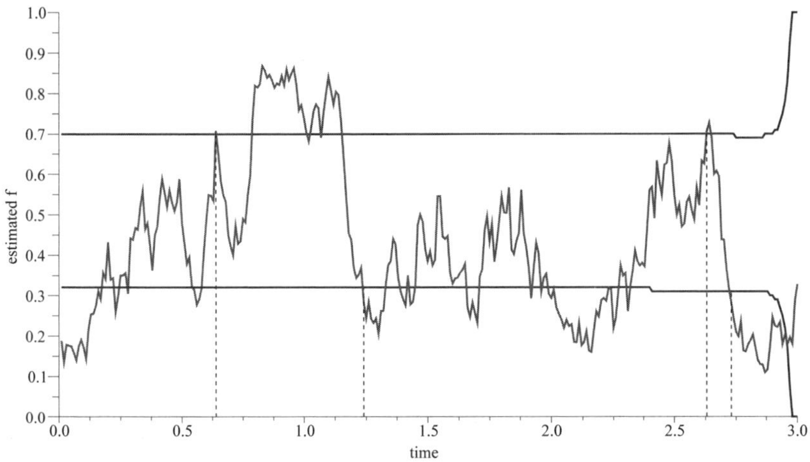

Figure 8.5. The efficient strategy

to buy the stock and the selling area where $\hat{V}^1 = (1 - g_{10})^\alpha \hat{V}^0$ (upper area), where F is close enough to 1 to sell the stock. The last area is a no-transaction zone: it means that the investor has to keep his/her position. This area is due to the transaction costs. On the Figure 8.5, we plot also the process $\hat{F}_t$ estimated from the stock. At time $t = 0$, $\hat{F}_0 \simeq 0.2$: the investor buys the stock. At time $t = 0.64$, $\hat{F}$ enters the selling zone, so he/she invests in the bond. At time $t = 1.24$, the process $\hat{F}$ reenters the buying zone, etc.

Note that all transactions should stop at a certain time before the time horizon T. This is due to the transaction costs: there is not enough time left to regain the price of the transaction. Far from the horizon, we can see that, approximately, $\hat{F}_t$ is small enough to buy when $\hat{F}_t \leq 0.3$ and is large enough to sell when $\hat{F}_t \geq 0.7$.

By Monte Carlo simulations, we can evaluate the expectation of the utility of the wealth when the efficient strategy is run, and compare the result to the approximate value function. Table 8.1 shows the results for 10^5 Monte Carlo simulations. One can see that this strategy is close to be optimal.

F_0	0	0.1	0.2	0.3	0.4	0.5	0.6	0.7	0.8	0.9	1
$\hat{V}^0$	1.061	1.057	1.053	1.049	1.045	1.043	1.041	1.039	1.038	1.037	1.036
$\tilde{V}^0$	1.061	1.056	1.052	1.049	1.045	1.043	1.040	1.039	1.038	1.037	1.036

Table 8.1. Optimality of the efficient strategy

8.4 Misspecifications

Our last result illustrates the critical effect of calibration. We compare two strategies: the first one, called misspecified strategy is the efficient strategy with miscalibrated coefficients: the stock follows the set of parameters mentionned above: $\mu_1 = -0.2$,

$\mu_2 = 0.21$, $\lambda_1 = \lambda_2 = 2$, $\sigma = 0.15$; but the agent computes the approximation of the value functions and the approximation of F_t with a set of different (misspecified) coefficients. The second strategy is a classical allocation procedure issued from technical analysis which does not require any knowledge of the dynamics of the stock or parameter estimation: the moving average strategy (with windowing size $\delta = 0.8$). The trader estimates the moving average of the prices

$$M_T^{(\delta)} = \frac{1}{\delta} \int_{t-\delta}^{t} S_u du,$$

and he/she decides to invest in the risky asset S if $S_t \geq M_t^{(\delta)}$ and to invest in the non risky asset S^0 otherwise. So, his/her strategy is $\pi_t = \mathbb{I}_{S_t \geq M_t^{(\delta)}}$. See [3] for more details.

As benchmarks, we use the efficient strategy (upper curve) with the right parameters and the *buy and hold* strategy (lower curve). For each strategy, we compute the utility

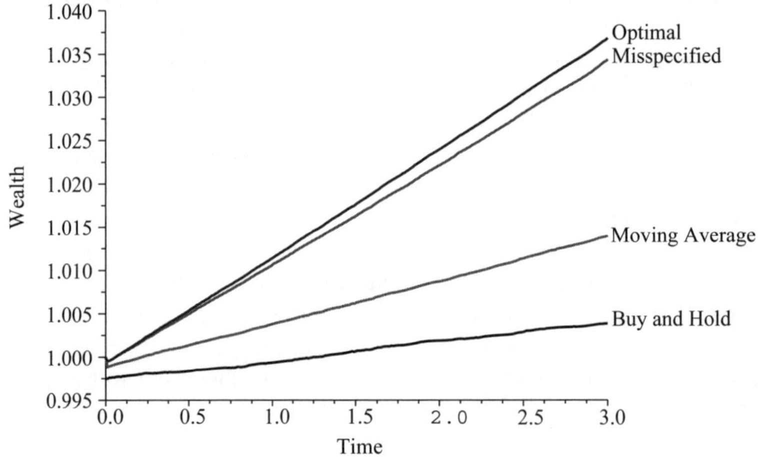

Figure 8.6. Comparison of strategies 1

of the corresponding wealth at each time and run 10^5 Monte Carlo simulations to estimate its expectation. Here, $g_{01} = g_{10} = 0.005$. Figure 8.6 shows the results for the set of misspecified parameters: $\mu_1 = -0.2$, $\mu_2 = 0.21$, $\sigma = 0.3$, $\lambda_1 = 0.5$, $\lambda_2 = 1$. Here the miscalibration mainly concerns the mean times of change of trend and the volatility. One can see that the miscalibrated strategy is still better than the moving average one. Figure 8.7 shows another set of miscalibrated parameters: $\mu_1 = -0.3$, $\mu_2 = 0.17$, $\sigma = 0.3$, $\lambda_1 = 2$, $\lambda_2 = 2$. Here the misscalibration mainly concerns the trends and the volatility. The moving average strategy is now better than the miscalibrated one. Further mathematical studies are necessary to understand these misspecification effects.

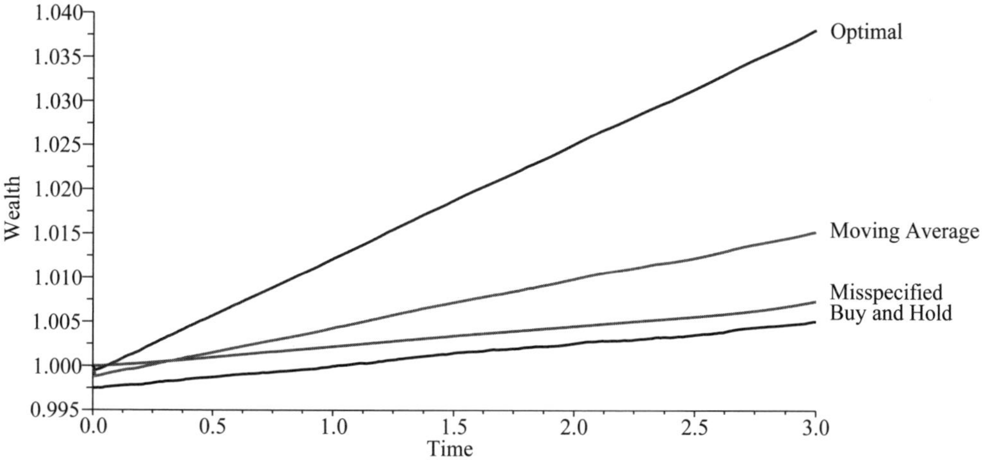

Figure 8.7. Comparison of strategies 2

Acknowledgement

This research has been carried out within the NCCR FINRISK project on "Credit Risk and Non-Standard Sources of Risk in Finance". Financial support by the National Centre of Competence in Research Financial valuation and Risk Management (NCCR FINRISK) is gratefully acknowledged. NCCR-FINRISK is a research program supported by the Swiss National Science Foundation.

Bibliography

[1] S. Achelis, *Technical analysis from A to Z*, McGraw Hill, 2000.

[2] G. Barles and E. R. Jakobsen, *On the convergence rate of approximation schemes for Hamilton-Jacobi-Bellman equations*, M2AN Math. Model. Numer. Anal. 36 (2002), pp. 33–54.

[3] C. Blanchet-Scalliet, A. Diop, R. Gibson, D. Talay, and E. Tanré, *Technical analysis techniques versus mathematical models: boundaries of their validity domains*, Monte Carlo and quasi-Monte Carlo methods 2004, Springer, Berlin, 2006, pp. 15–30.

[4] ———, *Technical analysis compared to Mathematical Models Based Methods Under Misspecification*, Journal of Banking & Finance 31 (2007), pp. 1351–1373.

[5] F. Bonnans, S. Maroso, and H. Zidani, *Error estimates for a stochastic impulse control problem*, Appl. Math. Optim. 55 (2007), pp. 327–357.

[6] K. A. Brekke and B. Øksendal, *Optimal switching in an economic activity under uncertainty*, SIAM J. Control Optim. 32 (1994), pp. 1021–1036.

[7] M. G. Crandall, H. Ishii, and P.-L. Lions, *User's guide to viscosity solutions of second order partial differential equations*, Bull. Amer. Math. Soc. (N.S.) 27 (1992), pp. 1–67.

[8] Y. Ishikawa, *Optimal control problem associated with jump processes*, Appl. Math. Optim. 50 (2004), pp. 21–65.

[9] N. Jegadeesh and S. Titman, *Returns to buying winners and selling losers: implications for stock market efficiency*, Journal of Finance 48 (1993), pp. 65–91.

[10] N. V. Krylov, *On the rate of convergence of finite-difference approximations for Bellman's equations with variable coefficients*, Probab. Theory Related Fields 117 (2000), pp. 1–16.

[11] T. G. Kurtz and D. L. Ocone, *Unique characterization of conditional distributions in nonlinear filtering*, Ann. Probab. 16 (1988), pp. 80–107.

[12] J. Lakonishok, L.K. Chan, and N. Jegadeesh, *Momentum strategies*, Journal of Finance 51 (1996), pp. 1681–1713.

[13] V. Ly Vath and H. Pham, *Explicit solution to an optimal switching problem in the two-regime case*, SIAM J. Control Optim. 46 (2007), pp. 395–426.

[14] V. Ly Vath, H. Pham, and S. Villeneuve, *A mixed singular/switching control problem for a dividend policy with reversible technology investment*, Ann. Appl. Probab. 18 (2008), pp. 1164–1200.

[15] M. Martinez, S. Rubenthaler, and E. Tanré, *Approximations of a Continuous Time Filter. Application to Optimal Allocation Problems in Finance*, Stoch. Anal. Appl. 27 (2009), pp. 270–296.

[16] S. V. Pastukhov, *On some probabilistic-statistical methods in technical analysis*, Teor. Veroyatn. Primen. 49 (2004), pp. 297–316.

[17] H. Pham, *On the smooth-fit property for one-dimensional optimal switching problem*, Séminaire de Probabilités XL, Lecture Notes in Math., vol. 1899, Springer, Berlin, 2007, pp. 187–199.

[18] A. N. Shiryaev and A.A. Novikov, *On a stochastic version of the trading rule "Buy and Hold"*, submitted for publication, 2009.

[19] D. W. Stroock and S. R. S. Varadhan, *Multidimensional diffusion processes*, Grundlehren der Mathematischen Wissenschaften [Fundamental Principles of Mathematical Sciences], vol. 233, Springer-Verlag, Berlin, 1979.

[20] S. J. Tang and J. M. Yong, *Finite horizon stochastic optimal switching and impulse controls with a viscosity solution approach*, Stochastics Stochastics Rep. 45 (1993), pp. 145–176.

Author information

Christophette Blanchet-Scalliet, Ecole Centrale de Lyon, ICJ CNRS UMR 5208, 36 avenue Guy de Collongue, 69134 Ecully, France.
Email: Christophette.Blanchet@ec-lyon.fr

Rajna Gibson Brandon, University of Geneva and Swiss Finance Institute, UNIMAIL, 40 Bd du Pont d'Arve, 1211 Genève 4, Switzerland.
Email: Rajna.Gibson@unige.ch

Benoîte de Saporta, University of Bordeaux, IMB, CNRS, UMR 5251, INRIA Bordeaux and Gretha CNRS UMR 5113, 351 Cours de le Libération, 33405 Talence Cedex, France.
Email: Benoite.deSaporta@math.u-bordeaux1.fr

Denis Talay, INRIA, 2004 Route des Lucioles. B.P. 93, 06902 Sophia-Antipolis, France.
Email: Denis.Talay@inria.fr

Etienne Tanré, INRIA, 2004 Route des Lucioles. B.P. 93, 06902 Sophia-Antipolis, France.
Email: Etienne.Tanre@inria.fr

Radon Series Comp. Appl. Math **8**, 91–124

Discrete-time approximation of BSDEs and probabilistic schemes for fully nonlinear PDEs

Bruno Bouchard, Romuald Elie, and Nizar Touzi

Abstract. The aim of this paper is to provide a survey on recent advances on probabilistic numerical methods for nonlinear PDEs, which serve as an alternative to classical deterministic schemes and allow to handle a large class of multidimensional nonlinear problems. These probabilistic schemes are based on the stochastic representation of semilinear PDEs by means of backward SDEs, which can be viewed as an extension of the well-known Feynman–Kac formula to the semilinear case. In this context, we first explain how smoothness properties can be obtained for non-reflected BSDEs set in the whole domain by using purely probabilistic techniques introduced in Ma and Zhang [59], and how they can be exploited to provide convergence rates of discrete time Euler-type approximation schemes. We then present some recent extensions to BSDEs with jumps, reflected BSDEs, BSDEs with horizon given by a finite stopping time, and second order BSDEs. The extension to fully nonlinear PDEs requires to enlarge the definition of backward SDEs. However, a natural probabilistic numerical scheme can be introduced formally by evaluating the nonlinear PDE along the trajectory of some chosen underlying diffusion. This point of view shows an intimate connection between the probabilistic numerical schemes reviewed in this paper and the standard finite differences methods. In particular, convergence and error estimates are provided by using the monotone schemes methods from the theory of viscosity solutions.

Key words. Backward stochastic differential equations, numerical resolution, conditional expectation, monte carlo methods.

AMS classification. 65C05, 60H35, 60G20

1 Introduction

The application field offered by the resolution of BSDEs is wide and diverse. As noticed by Bismut [12], Backward Stochastic Differential Equations (hereafter BSDEs) cover in particular the field of stochastic optimal control problems and the corresponding Hamilton–Jacobi–Bellman equations. See also [67], [72] for optimal control of diffusions with jumps, and [15] and [51] for optimal switching problems. Consequently, these techniques are of great interest for applications in economics. This is particularly true in finance, where their use was popularised by El Karoui, Peng and Quenez [48]. In incomplete markets, it can be used to solve optimal management problems by working along the lines presented in, e.g., El Kaouri, Peng and Quenez [49] for recursive utilities, El Karoui and Rouge [50] for indifference pricing, see also the more recent

Third author: This research is part of the Chair *Financial Risks* of the *Risk Foundation* sponsored by Société Générale, the Chair *Derivatives of the Future* sponsored by the Fédération Bancaire Française, the Chair *Finance and Sustainable Development* sponsored by EDF and Calyon.

paper [46] and the references therein, and Peng [68] for risk measures. Adding jumps to financial assets dynamics can be done as in [10] or [36]. BSDEs with jumps are also used for the valuation of financial derivatives with default risk. The valuation of American or game options follows from the resolution of reflected BSDEs, see [47]. The list of potential applications is long.

More generally, it is well known that the so-called Feynman Kac representation extends to second order semilinear PDEs which can be interpreted in terms of decoupled Forward-Backward SDEs, see Pardoux and Peng [63, 64] and Pardoux [62]. The term "decoupled" refers to the fact that the dynamic of the forward process does not depend on the solution of the Backward SDE. Hence, solving a BSDE or the related semilinear PDE is essentially the same. It was actually first proposed by Ma, Protter and Yong [58], see also [34], to use an estimation of the solution of the semilinear PDE to provide an approximation of the solution of Forward-Backward SDEs. However, solving PDEs in high dimension by finite-difference or finite-element schemes is a challenging task and any deterministic procedure suffers from the so-called "curse of dimensionality" which, very quickly, prevents from a good approximation of the solution of the PDE.

As a powerful alternative to classical purely deterministic techniques, more probabilistic approaches, directly based on the resolution of BSDEs, have been introduced these last years. We can essentially classify them in two categories. The first one concerns the approximation of the driving Brownian motion by a random walk on a finite tree. Such techniques have been widely studied, see e.g. [1], [28], [29] or [57], however they can be efficient only in low dimension since the size of the tree explodes exponentially with the dimension. The second one is based on Monte-Carlo type techniques. The main idea consists in simulating the Forward process by using standard Euler schemes and then computing backward the solution of the BSDE associated to the simulated forward paths by: 1. Discretising in time the BSDE and writing an Euler-type backward scheme which involves conditional expectations, 2. Approximating the conditional expectations by using non-parametric regression methods, see [22], [56], [27], [43] or integration by parts based estimators, see [55], [17] and [20]. The main advantage of such methods is that, like other Monte-Carlo type methods, the convergence rate does not depend a-priori on the dimension of the problem and they should therefore less suffer from the curse of dimensionality. In practice, the variance and/or the convergence rate of the estimators of the conditional expectations usually depends on the complexity/dimension of the problem, and the above comment has to be tempered. There is actually a third class of methods which lies in between the two previous ones and which were developed in a series of papers by Bally, Pagès and Printems, see e.g. [3], [4] and the references therein. In this approach, the original forward process is replaced by a quantised version in discrete time, i.e. a projection on a finite grid, whose associated Backward SDE can be computed by backward induction. The difference with the random walk technique is that the grid is computed in some optimal way, which allows to consider high dimension problems with a limited number of points, see the above quoted papers.

We shall not detail the pure numerical parts in this paper but rather insist on the first step of all these algorithms: the discrete time approximation of the BSDE which leads to a Backward Euler-type scheme.

Not surprisingly, the discrete time approximation error depends heavily on the regularity of the solution itself. The good news is that it essentially only depends on it. Therefore, all the analysis breaks down to that of the regularity of the solution of the BSDE. This has been first observed by Zhang [73, 74] and Bouchard and Touzi [20], who exploited the regularity results of Ma and Zhang [59] to provide a $|\pi|^{1/2}$ converge rate for decoupled Forward-Backward SDEs with Lipschitz coefficients, for a time-grid modulus $|\pi|$. Various extensions have been then developed in the literature and are reviewed in this paper: reflected BSDEs, the presence of a jump component, and the corresponding Dirichlet problem.

More recently, the backward schemes suggested by the discretisation of BSDEs have also been extended to the case of fully nonlinear PDEs by Fahim, Touzi and Warin [37]. The connection with an extended class of BSDEs, namely second order BSDEs, was outlined in [25]. The last section of this paper provides a presentation of the probabilistic numerical schemes which avoids BSDEs and only requires to evaluate the nonlinear PDE along the trajectory of some chosen underlying diffusion. In particular, this point of view highlights the relationship of the probabilistic schemes studied in this paper with the corresponding standard deterministic schemes. The convergence results and the corresponding error estimates in the fully nonlinear case are reviewed in the final section of this paper, and are based on the monotone schemes method initiated within the theory of viscosity solutions by Barles and Souganidis [9], Krylov [52, 53, 54], and Barles and Jakobsen [6, 7, 8].

To conclude, we should note that the case of coupled Forward-Backward SDEs has been considered in e.g. [31], [32], see also the references therein. The nature of the analysis being very different from the ones carried out here, we shall not discuss this case here (although it is important). See also [11] for Picard type iterations techniques.

Notations: Any element of $\mathbb{R}^d$ is viewed as a column vector and we denote by $'$ the transposition. $\mathbb{M}^d$ (resp. $\mathcal{S}_d$) denotes the set of $d \times d$ (resp. symmetric) matrices. For a smooth function $(t,x) \mapsto \varphi(t,x)$, we denote by $\partial_t\varphi$ its first derivative with respect to t, and by $D\varphi$ and $D^2\varphi$ its Jacobian and Hessian matrix with respect to x. It the function depends on more than one space variable, we use D_x, D_y, etc ... to denote the partial Jacobian with respect to a specific argument.

2 Numerical approximation of decoupled forward backward SDE

Let us first consider the simplest case of decoupled Forward-Backward SDE on the whole domain:

$$X_t = x + \int_0^t b(X_s)ds + \int_0^t \sigma(X_s)dW_s \tag{2.1}$$

$$Y_t = g(X_1) + \int_t^1 h(X_s,Y_s,Z_s)ds - \int_t^1 Z_s' dW_s\,, \quad t \le 1 \tag{2.2}$$

where W is d-dimensional Brownian Motion on some probability space $(\Omega, \mathcal{F}, \mathbb{P})$, whose associated complete and right-continuous filtration is denoted by $\mathbb{F} = (\mathcal{F}_t)_{t \le 1}$. Here, $b : \mathbb{R}^d \to \mathbb{R}^d$, $\sigma : \mathbb{R}^d \to \mathbb{M}^d$, $g : \mathbb{R}^d \to \mathbb{R}$ and $h : \mathbb{R}^d \times \mathbb{R} \times \mathbb{R}^d \to \mathbb{R}$ are some Lipschitz continuous functions.

2.1 Discrete time approximation

The discrete-time approximation is constructed as follows.

Euler scheme for the Forward process. Given a time-grid $\pi := \{0 = t_0 < t_1 < \cdots < t_n = 1\}$ of $[0,1]$ with modulus $|\pi|$, we first approximate the Forward process X by its Euler scheme X^π defined as

$$X_0^\pi := x\,, \quad \text{and} \quad X_{t_i+1}^\pi := b(X_{t_i}^\pi)\Delta t_i + \sigma(X_{t_i}^\pi)\Delta W_{t_i} \;\text{ for } i < n\,, \tag{2.3}$$

where $\Delta t_i := t_{i+1} - t_i$ and $\Delta W_{t_i} := W_{t_{i+1}} - W_{t_i}$.

Euler scheme for the Backward process. As for the solution (Y, Z) of the BSDE, we first approximate the terminal condition $Y_1 = g(X_1)$ by simply replacing X by its Euler scheme : $Y_1 \simeq g(X_1^\pi)$. Then, motivated by the formal Euler discretisation:

$$\begin{aligned} Y_{t_i} &= Y_{t_{i+1}} + \int_{t_i}^{t_{i+1}} h(X_s, Y_s, Z_s)ds - \int_{t_i}^{t_{i+1}} Z_s' dW_s \\ &\simeq Y_{t_{i+1}} + h(X_{t_i}^\pi, Y_{t_i}, Z_{t_i})\Delta t_i - Z_{t_i}'\Delta W_{t_i}, \end{aligned}$$

we define the discrete-time approximation of the BSDE as follows:
1. Taking expectation conditionally to $\mathcal{F}_{t_i}$ on both sides leads to

$$Y_{t_i} \simeq \mathbb{E}\left[Y_{t_{i+1}} \mid \mathcal{F}_{t_i}\right] + h(X_{t_i}^\pi, Y_{t_i}, Z_{t_i})\Delta t_i\,.$$

2. Pre-multiplying by ΔW_{t_i} and then taking conditional provides

$$0 \simeq \mathbb{E}\left[Y_{t_{i+1}}\Delta W_{t_i} \mid \mathcal{F}_{t_i}\right] - Z_{t_i}\Delta t_i\,.$$

This formal approximation argument leads to a Backward Euler scheme $(Y_{t_i}^\pi, \bar{Z}_{t_i}^\pi)$ of the form

$$\left\{ \begin{aligned} \bar{Z}_{t_i}^\pi &= \tfrac{1}{\Delta t_i}\,\mathbb{E}\left[Y_{t_{i+1}}^\pi \Delta W_{t_i} \mid \mathcal{F}_{t_i}\right] \\ Y_{t_i}^\pi &= \mathbb{E}\left[Y_{t_{i+1}}^\pi \mid \mathcal{F}_{t_i}\right] + h\left(X_{t_i}^\pi, Y_{t_i}^\pi, \bar{Z}_{t_i}^\pi\right)\Delta t_i \end{aligned} \right. \tag{2.4}$$

with terminal condition given by $Y_1^\pi = g(X_1^\pi)$. The reason why we denote the Z-component by $\bar{Z}^\pi$, and not simply Z^π, will become clear in Remark 2.1.2 below.

Remark 2.1.1 (Explicit or implicit schemes). Note that the above scheme is implicit as $Y_{t_i}^\pi$ appears in both sides of the equation. Since h is assumed to be Lipschitz and since it is multiplied by Δt_i, which is intended to be small, the equation can however be solved numerically very quickly by standard fixed point methods. Nevertheless, we could also consider an explicit version of this scheme by replacing the second equation in (2.4) by

$$Y^\pi_{t_i} = \mathbb{E}\left[Y^\pi_{t_{i+1}} + h\left(X^\pi_{t_i}, Y^\pi_{t_{i+1}}, \bar{Z}^\pi_{t_i}\right)\Delta t_i \mid \mathcal{F}_{t_i}\right].$$

This will not change the convergence rate, see [18]. Both can thus be used indifferently depending on the exact nature of the coefficients.

Remark 2.1.2 (Continuous time version of the Euler scheme). Note that the martingale representation theorem implies that we can find a square integrable process Z^π such that

$$Y^\pi_{t_i} = Y^\pi_{t_{i+1}} + h\left(X^\pi_{t_i}, Y^\pi_{t_i}, \bar{Z}^\pi_{t_i}\right)\Delta t_i - \int_{t_i}^{t_{i+1}} (Z^\pi_s)' dW_s \,.$$

This allows to consider a continuous version of the Backward Euler scheme by setting

$$Y^\pi_t := Y^\pi_{t_{i+1}} + \int_t^{t_{i+1}} h\left(X^\pi_{t_i}, Y^\pi_{t_i}, \bar{Z}^\pi_{t_i}\right) - \int_t^{t_{i+1}} (Z^\pi_s)' dW_s \ \text{ for } t \in [t_i, t_{i+1})\,. \tag{2.5}$$

It then follows from the Itô isometry that

$$\bar{Z}^\pi_{t_i} = \frac{1}{\Delta t_i}\mathbb{E}\left[\int_{t_i}^{t_{i+1}} Z^\pi_s ds \mid \mathcal{F}_{t_i}\right]. \tag{2.6}$$

In other words, the step constant process $\bar{Z}^\pi$, defined as $\bar{Z}^\pi_t := \bar{Z}^\pi_{t_i}$ for $t \in [t_i, t_{i+1})$, can be seen as the best $L^2(\Omega \times [0,1], d\mathbb{P} \otimes dt)$ approximation of Z^π in the class of adapted processes which are constant on each interval $[t_i, t_{i+1})$.

2.2 Error decomposition

In order to compare (Y, Z) and (Y^π, Z^π), let us introduce

$$\bar{Z}_{t_i} := \frac{1}{\Delta t_i}\mathbb{E}\left[\int_{t_i}^{t_{i+1}} Z_s ds \mid \mathcal{F}_{t_i}\right]$$

and observe that, since h is Lipschitz continuous,

$$\begin{aligned} -dY_t &= h(X_t, Y_t, Z_t)dt - Z_t' dW_t \\ &\simeq h(X^\pi_t, Y_{t_i}, \bar{Z}_{t_i})dt - Z_t' dW_t \quad \text{for } t \in [t_i, t_{i+1}) \end{aligned} \tag{2.7}$$

up to an error term of order $C\left(|X_t - X^\pi_{t_i}| + |Y_t - Y_{t_i}| + |Z_t - \bar{Z}_{t_i}|\right)dt$ for some constant $C > 0$. Next observe from the definition of $\bar{Z}_{t_i}$ and $\bar{Z}^\pi_{t_i}$ and Jensen's inequality that

$$\mathbb{E}\left[|\bar{Z}_{t_i} - \bar{Z}^\pi_{t_i}|^2\right] \le \mathbb{E}\left[\frac{1}{\Delta t_i}\int_{t_i}^{t_{i+1}} |Z_t - Z^\pi_t|^2 dt\right]. \tag{2.8}$$

It then follows from (2.5)–(2.7), recalling the involved error terms, and standard techniques for BSDEs with Lipschitz coefficients that

$$\max_{0\le i<n} \mathbb{E}\left[|Y_{t_i} - Y^\pi_{t_i}|^2 + \int_{t_i}^{t_{i+1}} |Z_t - Z^\pi_t|^2 dt\right]$$
$$\le C\,\mathbb{E}\left[\sup_{t_i\le t\le t_{i+1}} |X_t - X^\pi_{t_i}|^2 + \sup_{t_i\le t\le t_{i+1}} |Y_t - Y_{t_i}|^2 + \int_{t_i}^{t_{i+1}} |Z_t - \bar{Z}_{t_i}|^2 dt\right]$$

for some $C > 0$. Using (2.8) again and standard estimates for Forward Euler schemes, it follows that

$$\begin{aligned}
\mathrm{Err}(\pi)^2 &:= \max_{0\le i<n} \mathbb{E}\left[\sup_{t_i\le t\le t_{i+1}} |Y_t - Y^\pi_{t_i}|^2\right] + \mathbb{E}\left[\sum_{i=0}^{n-1}\int_{t_i}^{t_{i+1}} |Z_t - \bar{Z}^\pi_{t_i}|^2 dt\right] \\
&\le 2\max_{0\le i<n} \mathbb{E}\left[\sup_{t_i\le t\le t_{i+1}} \left(|Y_{t_i} - Y^\pi_{t_i}|^2 + |Y_t - Y_{t_i}|^2\right)\right] \\
&\quad +2\left(\mathbb{E}\left[\sum_{i=0}^{n-1}\int_{t_i}^{t_{i+1}} |Z_t - \bar{Z}_{t_i}|^2 + |\bar{Z}_{t_i} - \bar{Z}^\pi_{t_i}|^2 dt\right]\right) \\
&\le C\left(|\pi| + \mathcal{R}^2_Y(\pi) + \mathcal{R}^2_Z(\pi)\right)
\end{aligned}$$

for some $C > 0$, where

$$\begin{aligned}
\mathcal{R}^2_Y(\pi) &:= \max_{0\le i<n} \mathbb{E}\left[\sup_{t_i\le t\le t_{i+1}} |Y_t - Y_{t_i}|^2\right] \quad \text{and} \\
\mathcal{R}^2_Z(\pi) &:= \mathbb{E}\left[\sum_{i=0}^{n-1}\int_{t_i}^{t_{i+1}} |Z_t - \bar{Z}_{t_i}|^2 dt\right]
\end{aligned}$$

can be seen as (squared) modulus of regularity of Y and Z.

Proposition 2.2.1 ([20], [74]). *There is a constant $C > 0$, which is independent on π such that:*

$$\mathrm{Err}(\pi) \;\le\; C\left(|\pi| + \mathcal{R}^2_Y(\pi) + \mathcal{R}^2_Z(\pi)\right)^{\frac{1}{2}} .$$

Remark 2.2.2. 1. Note that, by continuity of the process Y, the quantity $\sup_{t_i\le t\le t_{i+1}} |Y_t - Y_{t_i}|^2$ converges $\mathbb{P}$ – a.s. to 0 as $|\pi| \to 0$. Since the coefficients are Lipschitz, we also have $\mathbb{E}\left[\sup_{t\le 1} |Y_t|^2\right] < \infty$. It thus follows from dominated convergence that

$$\mathcal{R}^2_Y(\pi) \longrightarrow 0 \text{ as } |\pi| \to 0 .$$

2. Also note that the step constant process $\bar{Z}$, defined as $\bar{Z}_t := \bar{Z}_{t_i}$ for $t \in [t_i, t_{i+1})$, can be seen as the best $L^2(\Omega\times[0,1], d\mathbb{P}\otimes dt)$ approximation of Z in the class of adapted processes which are constant on each interval $[t_i, t_{i+1})$. Since such processes can be used to approximate Z in $L^2(\Omega\times[0,1], d\mathbb{P}\otimes dt)$, it follows that

$$\mathcal{R}^2_Z(\pi) \longrightarrow 0 \text{ as } |\pi| \to 0 .$$

Combining Proposition 2.2.1 and the last remark, implies the first converge result.

Corollary 2.2.3. *We have:* $\mathrm{Err}(\pi)^2 \longrightarrow 0$ *as* $|\pi| \to 0$.

Remark 2.2.4. It follows from the interpretation of $\bar{Z}$ in terms of the best approximation of Z by step constant processes that

$$\mathbb{E}\left[\sum_{i=0}^{n-1} \int_{t_i}^{t_{i+1}} |Z_t - \bar{Z}^\pi_{t_i}|^2 dt\right] \geq \mathcal{R}^2_Z(\pi) .$$

In the case where $h \equiv 0$, then Y is a martingale and Y_{t_i} is therefore the best $L^2(\Omega, d\mathbb{P})$ approximation of Y_t for $t \in [t_i, t_{i+1}]$ by a $\mathcal{F}_{t_i}$-measurable random variable. The Burkholder–Davis–Gundy's inequality then implies that

$$\mathbb{E}\left[\sup_{t_i \leq t \leq t_{i+1}} |Y_t - Y^\pi_{t_i}|^2\right] \geq c\,\mathbb{E}\left[\sup_{t_i \leq t \leq t_{i+1}} |Y_t - Y_{t_i}|^2\right] ,$$

for some $c > 0$. In view of Proposition 2.2.1, this shows that

$$c\,\left(\mathcal{R}^2_Y(\pi) + \mathcal{R}^2_Z(\pi)\right) \leq \mathrm{Err}(\pi)^2 \leq C\,\left(|\pi| + \mathcal{R}^2_Y(\pi) + \mathcal{R}^2_Z(\pi)\right) .$$

Up to a $|\pi|$ term this can actually be shown even if $h \not\equiv 0$. This result is not surprising. As usual, the convergence rate should depend on the regularity of the functions/processes to be approximated.

2.3 Path regularity and convergence rate

In view of Proposition 2.2.1, the whole analysis of the convergence rate of the discrete-time approximation breaks down to the analysis of the (squared) modulus of continuity $\mathcal{R}^2_Y(\pi)$ and $\mathcal{R}^2_Z(\pi)$. The regularity of Y is standard in our context where the coefficients are Lipschitz continuous.

Proposition 2.3.1. *There exists* $C > 0$ *independent on* π *such that* $\mathcal{R}^2_Y(\pi) \leq C\,|\pi|$.

Note that this implies some regularity on the (unique viscosity) solution v of the semilinear PDE

$$\begin{aligned} -\mathcal{L}v - h(\cdot, v, \sigma' Dv) &= 0 \text{ on } [0,1) \times \mathbb{R}^d \\ v(T-, \cdot) &= g \text{ on } \mathbb{R}^d , \end{aligned}$$

where $\mathcal{L}$ is the Dynkin operator associated to X,

$$\mathcal{L}v := \partial_t v + \langle b, Dv \rangle + \frac{1}{2}\mathrm{Tr}\left[\sigma\sigma' D^2 v\right] ,$$

since it is related to (Y, Z) by the mapping $Y = v(\cdot, X)$ and (at least formally) $Z = \sigma(X)' Dv(\cdot, X)$, see e.g. [64]. As a matter of fact, it follows from our Lipschitz continuity assumption that this solution is Lipschitz continuous in space and $1/2$-Hölder in time. Since X is itself $1/2$-Hölder in time in L^2, this actually implies the regularity statement of Proposition 2.3.1.

It is however less standard to study the regularity of the Z-component. If Dv, whenever it is well-defined, were Lipschitz continuous in space and $1/2$-Hölder in time, we would easily deduce that $\mathcal{R}^2_Z(\pi)$ is of order $|\pi|$. But such a result is not true in general, and indeed much stronger than what we need.

A pure probabilistic analysis of the modulus of regularity of Z has been carried out by Ma and Zhang [59] and Zhang [74], actually in a more general setting. It is based on the following analysis ($d = 1$ to avoid notational complexities):

Step 1. The first step consists in assuming that g and σ are C^1_b and using the fact that (Y, Z) are differentiable in the Malliavin sense and that there exists a version of their Malliavin derivative (DY, DZ) which solves

$$D_s Y_t = Dg(X_1) D_s X_1 + \int_t^1 Dh(\Theta_r) D_s(X_r, Y_r, Z_r) dr - \int_t^1 D_s Z_r dW_r \ , \ s \leq 1 \ ,$$

where $\Theta := (X, Y, Z)$, see e.g. [61]. Moreover, Itô's formula and uniqueness of solutions of BSDEs with Lipschitz coefficients imply that it is related to the solution $(\nabla Y, \nabla Z)$ of

$$\nabla Y_t = Dg(X_1) \nabla X_1 + \int_t^1 Dh(\Theta_r) \nabla (X_r, Y_r, Z_r) dr - \int_t^1 \nabla Z_r dW_r \ , \ s \leq 1 \ ,$$

where ∇X denotes the first variation process of X, by the relation $D_s(X, Y, Z) = \mathbf{1}_{[s,1]} \nabla (X, Y, Z)(\nabla X_s)^{-1} \sigma(X_s)$.

Step 2. On the other hand, the identity

$$Y_t = Y_0 - \int_0^t h(\Theta_r) dr + \int_0^t Z_r dW_r$$

implies that $D_t Y_t = Z_t$, up to passing to a suitable version of these processes. It thus follows that

$$Z_t = \left(Dg(X_1) \nabla X_1 + \int_t^1 Dh(\Theta_r) \nabla \Theta_r dr - \int_t^1 \nabla Z_r dW_r \right) (\nabla X_t)^{-1} \sigma(X_t) \, .$$

Setting

$$\eta := \nabla X \exp \left\{ \int_0^{\cdot} D_z f(\Theta_r) dW_r - \int_0^{\cdot} \left(\frac{1}{2} |D_z f(\Theta_r)|^2 + D_y f(\Theta_r) \right) dr \right\} \ , \qquad (2.9)$$

we then deduce from Itô's formula that

$$Z_t = \mathbb{E} \left[Dg(X_1) \eta_1 + \int_t^1 D_x h(\Theta_r) \eta_r dr \mid \mathcal{F}_t \right] \eta_t^{-1} \sigma(X_t) \, . \qquad (2.10)$$

Step 3. The rest of the proof is based on a martingale argument. For sake of simplicity, let us consider the case $Dh \equiv 0$, $\nabla X \equiv 1$ and σ is constant. Then

$$Z_t = \mathbb{E}\left[Dg(X_1) \mid \mathcal{F}_t\right]\sigma ,$$

and, for $t \in [t_i, t_{i+1}]$,

$$\mathbb{E}\left[|Z_t - Z_{t_i}|^2\right] \leq \mathbb{E}\left[|Z_{t_{i+1}}|^2 - |Z_{t_i}|^2\right]$$

so that, by 2. of Remark 2.2.2,

$$\mathcal{R}_Z^2(\pi) \leq \sum_{i=0}^{n-1} \int_{t_i}^{t_{i+1}} \mathbb{E}\left[|Z_t - Z_{t_i}|^2\right] dt \leq |\pi| \mathbb{E}\left[|Z_1|^2\right] = |\pi| \mathbb{E}\left[|Dg(X_1)|^2\right] .$$

Also more technical, the general case is essentially treated similarly, by using the Lipschitz continuity of the coefficients. The above proof is performed for C_b^1 coefficients, but the final bound depends only on the Lipschitz constants. It can then be extended by standard approximation techniques.

Theorem 2.3.2 ([74]). *If the coefficients are Lipschitz, then* $\mathcal{R}_Z^2(\pi) = \mathcal{O}(|\pi|)$.

Combining the last theorem with Propositions 2.2.1 and 2.3.1 provides a $|\pi|^{\frac{1}{2}}$ convergence rate for the Backward Euler scheme.

Corollary 2.3.3. *If the coefficients are Lipschitz, then* $\mathrm{Err}(\pi) \leq \mathcal{O}(|\pi|^{\frac{1}{2}})$.

Remark 2.3.4. Importantly, all these results are obtained under the single assumption that the coefficients are Lipschitz continuous. In particular, no ellipticity condition is imposed on σ.

Remark 2.3.5. The $|\pi|^{\frac{1}{2}}$ convergence rate is clearly optimal and is similar to the one obtained for Forward SDEs.

Remark 2.3.6. Since no ellipticity assumption is made on σ, one component of X can be chosen so as to coincide with time. It is therefore not a restriction to consider a-priori time homogeneous dynamics. Note however, that a similar analysis could be performed under the assumption that the b, σ and f are only $1/2$-Hölder in the time component, as opposed to Lipschitz continuous.

Remark 2.3.7. Similar results are obtained in [74] in the case where the terminal condition is of the form $f(X)$ where f is a function of the whole path of X on $[0, 1]$, under suitable Lipschitz continuity assumptions on the space of continuous paths. The case of non-regular terminal conditions has been discussed in [44]. The grid has then to be refined near the terminal time $t = 1$ in order to compensate for the explosion of the gradient Dv. However, their proof relies on PDE arguments and requires strong smoothness assumptions on the coefficients as well as a uniform ellipticity condition on σ, a condition that was not required so far.

2.4 Weak expansion in the smooth case

So far, we have only discussed the strong convergence rate. But, as for linear problems, which correspond to the case where h does not depend on Y and Z, we can also study the weak approximation error. Such an analysis as been performed recently by Gobet and Labart [42], thus extending previous results of Chevance [26] who considered the case where h does not depend on Z.

Their result is of the form:

$$\begin{aligned} Y_t - Y_t^\pi &= Dv(t,X_t)(X_t - X_t^\pi) + \mathcal{O}(|\pi|) + \mathcal{O}\left(|X_t - X_t^\pi|^2\right) \\ Z_t - \bar{Z}_t^\pi &= D\left[Dv(t,X_t)\sigma\right]'(X_t - X_t^\pi) + \mathcal{O}(|\pi|) + \mathcal{O}\left(|X_t - X_t^\pi|^2\right) , \end{aligned}$$

where $(X_t^\pi)_{t\le 1}$ is the continuous version of the Euler scheme of X and (Y^π, Z^π) are defined consistently with the explicit Backward Euler scheme by

$$\begin{aligned} Y_t^\pi &= \mathbb{E}\left[Y_{t_{i+1}}^\pi + h\left(X_{t_i}^\pi, Y_{t_{i+1}}^\pi, \bar{Z}_{t_i}^\pi\right)(t_{i+1} - t) \mid \mathcal{F}_t\right] \\ \bar{Z}_t^\pi &= \frac{1}{t_{i+1} - t}\,\mathbb{E}\left[Y_{t_{i+1}}^\pi (W_{t_{i+1}} - W_t) \mid \mathcal{F}_t\right] , \; t \in [t_i, t_{i+1}) . \end{aligned}$$

In particular, this implies that $v(0, X_0) - Y_0^\pi = Y_0 - Y_0^\pi = \mathcal{O}(|\pi|)$. Moreover, when the grid is uniform, this allows to deduce from the weak convergence of the process $|\pi|^{-\frac{1}{2}}(X - X^\pi)$ that the process $|\pi|^{-\frac{1}{2}}(X - X^\pi, Y - Y^\pi, Z - \bar{Z}^\pi)$ weakly converges too.

As usual, this requires much stronger assumptions than only our previous Lipschitz continuity conditions, in particular they need to assume that σ is uniformly elliptic.

3 Extensions to other semilinear cases

The analysis of the previous Sections 2.2 and 2.3 consists in two parts:
1. Control the convergence rate of the discrete time approximation in terms of the modulus of regularity of Y and Z.
2. Study these modulus of regularity. Here, the difficult task consists in controlling $\mathcal{R}_Z^2(\pi)$. In the above classical case, it is obtained by applying martingale-type techniques to a suitable representation of Z.

In the following, we show how these techniques can be adapted to more general situations corresponding to BSDEs with jumps, reflected BSDEs or BSDEs with random time horizon.

3.1 BSDEs with jumps

The analysis of the previous section can be extended in a very natural way to decoupled Forward Backward SDEs with jumps. Such equations appear naturally in optimal control for jump diffusion processes, see [72], hedging problems, see [36], or portfolio management, see [10], in finance. From the PDE point of view, they have two important applications.

Example 3.1.1 (Semilinear PDEs with integral term). *Let us consider the solution* (X, Y, Z, U) *of*

$$\begin{aligned} X_t &= X_0 + \int_0^t b(X_r)dr + \int_0^t \sigma(X_r)dW_r + \int_0^t \int_E \beta(X_{r-}, e)\bar{\mu}(de, dr) , \\ Y_t &= g(X_1) + \int_t^1 h(\Theta_r)\, dr - \int_t^1 Z_r' dW_r - \int_t^1 \int_E U_r(e)\bar{\mu}(de, dr) \end{aligned} \tag{3.1}$$

where $\bar{\mu}(de, dt) = \mu(de, dt) - \lambda(de)dt$ *is a compensated compound point measure, independent on* W*, with mark-space* $E \subset \mathbb{R}^\kappa$*, and* $\Theta := (X, Y, Z, \Gamma)$ *with*

$$\Gamma := \int_E \rho(e)U(e)\lambda(de)$$

for some bounded map $\rho : E \mapsto \mathbb{R}^\kappa$*. Then, in the case where all the coefficients are Lipschitz continuous, uniformly in the e-component,* (Y, Z, U) *can be associated to the (unique viscosity) solution* v *of*

$$-\mathcal{L}v - h\left(\cdot, v, \sigma' Dv, \mathcal{I}[v]\right) = 0 \quad , \quad v(1-, \cdot) = g\,, \tag{3.2}$$

where $\mathcal{L}$ *is the Dynkin operator associated to* X *and*

$$\mathcal{I}[v](t, x) := \int_E \{v(t, x + \beta(x, e)) - v(t, x)\}\, \rho(e)\, \lambda(de)\,,$$

though the mapping, $Y_t = v(t, X_t)$*,* $U_t(e) = v(t, X_{t-} + \beta(X_{t-}, e)) - v(t, X_{t-})$ *and* $Z_t = \sigma(X_{t-})' Dv(t, X_{t-})$*, whenever it is well defined, see [5] and [18].*

Example 3.1.2 (Systems of semilinear PDEs). *More interestingly, it was shown by [65] and [71] that BSDEs of the form*

$$\begin{aligned} X_t &= X_0 + \int_0^t b(M_r, X_r)dr + \int_0^t \sigma(M_r, X_r)dW_r \,, \\ M_t &= 1 + \left(\int_0^t \sum_{j=1}^{\kappa-1} j\mu(\{j\}, dt)\right) [\text{modulo } \kappa] \\ Y_t &= g(M_1, X_1) + \int_t^1 h\left(M_r, X_r, Y_r, Z_r, U_r\right) dr - \int_t^1 Z_r dW_r \\ &- \int_t^1 \int_E U_r(e)\bar{\mu}(de, dr) \end{aligned} \tag{3.3}$$

allow to provide a probabilistic representation for systems of semilinear PDEs. Here the mark space is finite $E = \{1, \ldots, \kappa - 1\}$ *and* $\lambda(de) \equiv \lambda \sum_{k=1}^{\kappa-1} \delta_k(e)$*, where* $\lambda > 0$

is a constant. Namely, given a family $(\tilde{b}_m, \tilde{\sigma}_m, \tilde{h}_m, \tilde{g}_m)_{m\leq\kappa}$ *of Lipschitz continuous functions, if we set*

$$(b(m,x), \sigma(m,x), g(m,x)) = (\tilde{b}_m(x), \tilde{\sigma}_m(x), \tilde{g}_m(x))$$

and

$$h(m,x,y,z,\gamma) = \tilde{h}_m\Big(x, \big[y+\gamma^{\kappa-m+1}, \ldots, y+\gamma^{\kappa}, \underbrace{y}_{m}, y+\gamma^1, \ldots\big], z\Big) - \lambda\sum_{j=1}^{\kappa}\gamma^j\,,$$

then one can show that $Y_t = v^{M_t}(t, X_t)$ *where* $v = (v^1, \ldots, v^\kappa)$ *is the unique viscosity solution of the system*

$$-\mathcal{L}^m v^m - h_m\left(\cdot, v, Dv^m\sigma_m\right) = 0 \quad , \quad v^m(1,\cdot) = g_m(\cdot)\,,$$

with $\mathcal{L}^m$ *denoting the Dynkin operator associated to the Forward SDE with drift* $\tilde{b}_m$ *and volatility* $\tilde{\sigma}_m$.

Obviously, the formulations (3.1) and (3.3) can be combined so as to provide systems of semilinear PDEs with integral term.

The Backward Euler scheme for BSDEs with jumps of the form (3.1) was introduced by Bouchard and Elie [18] and follows the ideas of Section 2.1. The formulation (3.3) can be treated similarly, see Elie [35] for more details.

First, we define the Euler scheme X^π of X as usual. The Euler scheme $(Y^\pi, Z^\pi, \Gamma^\pi)$ for (Y, Z, Γ) is then constructed as above by the Backward induction

$$\begin{cases} \bar{Z}^\pi_{t_i} &= \frac{1}{\Delta t_i}\,\mathbb{E}\left[Y^\pi_{t_{i+1}}(W_{t_{i+1}} - W_{t_i}) \mid \mathcal{F}_{t_i}\right] \\ \bar{\Gamma}^\pi_{t_i} &= \frac{1}{\Delta t_i}\,\mathbb{E}\left[Y^\pi_{t_{i+1}}\int_E \rho(e)\bar{\mu}(de,(t_i,t_{i+1}]) \mid \mathcal{F}_{t_i}\right] \\ Y^\pi_{t_i} &= \mathbb{E}\left[Y^\pi_{t_{i+1}} \mid \mathcal{F}_{t_i}\right] + h\left(X^\pi_{t_i}, Y^\pi_{t_i}, \bar{Z}^\pi_{t_i}, \bar{\Gamma}^\pi_{t_i}\right)\Delta t_i \end{cases}$$

with terminal condition $Y^\pi_1 := g(X^\pi_1)$.

Remark 3.1.3. In the case where E is a finite set $\{e_1, \ldots, e_\kappa\}$, one can always choose ρ of the form $\rho^i(e_j) = \mathbf{1}_{i=j}/\lambda(\{e_i\})$ so that $\Gamma = (U(e_i))_{i\leq\kappa}$. The above procedure then allows to approximate U directly. It can obviously be extended to the approximation of any integrated version of $e \mapsto U(e)$ with respect to λ. We only concentrate here on Γ as defined above because this is the term which appears in h and its approximation is therefore necessary to approximate Y.

As in the no-jump case, rather standard BSDE techniques allow to show that the approximation error

$$\begin{aligned} \mathrm{Err}(\pi)^2 \;&:=\; \max_{i\leq n-1}\mathbb{E}\left[\sup_{t\in[t_i,t_{i+1}]}|Y_t - Y^\pi_{t_i}|^2\right] + \mathbb{E}\left[\sum_{i\leq n-1}\int_{t_i}^{t_{i+1}}|Z_t - \bar{Z}^\pi_{t_i}|^2 dt\right] \\ &\quad+\mathbb{E}\left[\sum_{i\leq n-1}\int_{t_i}^{t_{i+1}}|\Gamma_t - \bar{\Gamma}^\pi_{t_i}|^2 dt\right]\,, \end{aligned}$$

can be controlled in terms of $|\pi|$, $\mathcal{R}_Y^2(\pi)$, $\mathcal{R}_Z^2(\pi)$ and an additional term that takes jumps into account

$$\mathcal{R}_\Gamma^2(\pi) := \mathbb{E}\left[\sum_{i\le n-1}\int_{t_i}^{t_{i+1}} |\Gamma_t - \bar{\Gamma}_{t_i}|^2 dt\right] \text{ with } \bar{\Gamma}_{t_i} := \frac{1}{\Delta t_i}\mathbb{E}\left[\int_{t_i}^{t_{i+1}} \Gamma_s ds \mid \mathcal{F}_{t_i}\right].$$

Under our Lipschitz continuity assumptions, the $1/2$-Hölder continuity in time and Lipschitz continuity in space of the deterministic map v is rather standard. In view of the identities $Y_t = v(t, X_t)$ and $U_t(e) = v(t, X_{t-} + \beta(X_{t-}, e)) - v(t, X_{t-})$, see [18] for details, this readily implies that $\mathcal{R}_Y^2(\pi) + \mathcal{R}_\Gamma^2(\pi) \le C|\pi|$ for some $C > 0$.

It thus remains to study the squared modulus of continuity $\mathcal{R}_Z^2(\pi)$ of Z. Following the approach of Zhang [74], it was shown by Bouchard and Elie [18] that $\mathcal{R}_Z^2(\pi)$ can be controlled by $|\pi|$ whenever one of the following assumptions holds:

$Ha)$ For all $e \in E$, the map $x \in \mathbb{R}^d \mapsto \beta(x, e)$ admits a Jacobian matrix $D\beta(x, e)$ such that

$$(x, \xi) \in \mathbb{R}^d \times \mathbb{R}^d \mapsto a(x, \xi; e) := \xi'(D\beta(x, e) + I_d)\xi$$

satisfies, uniformly in $(x, \xi) \in \mathbb{R}^d \times \mathbb{R}^d$

$$a(x, \xi; e) \ \ge\ |\xi|^2 K^{-1} \quad \text{or} \quad a(x, \xi; e) \ \le\ -|\xi|^2 K^{-1},$$

for some $K > 0$.

$Hb)$ b, σ, $\beta(\cdot, e)$, h and g are C_b^1 with Lipschitz first derivatives, uniformly in $e \in E$.

Remark 3.1.4. The condition $Ha)$ ensures that the tangent process ∇X is invertible, with inverse satisfying suitable integrability conditions, which is necessary to reproduce the argument presented in Section 2.3, recall the definition of the term η. Under $Hb)$, it can be shown, see [35], that Dv is $1/2$-Hölder-continuous in time and Lipschitz-continuous in space. This allows to show that $\mathcal{R}_Z^2(\pi)$ can be controlled by $|\pi|$ by simply using the $1/2$-Hölder regularity of X in L^2.

This allows to provide a $|\pi|^{\frac{1}{2}}$ convergence rate.

Theorem 3.1.5 ([18]). *Let $Ha)$ or $Hb)$ hold, assume that λ is bounded and that the coefficients are Lipschitz, uniformly in the e-variable. Then,* $\mathrm{Err}(\pi) \le \mathcal{O}(|\pi|^{\frac{1}{2}})$.

Remark 3.1.6. So far, it seems that only the case of jump measures with finite activity has been treated. In [18], the compensatory measure is assumed to be bounded, but it can be generalised without much difficulties under suitable integrability conditions. Moreover, pure jump Lévy processes can be approximated by compound point measures by truncating the small jumps. It would therefore be natural to approximate a BSDE driven by a Lévy process with infinite activity by a sequence of BSDEs driven by compound point measures to which the above analysis would apply. This would introduce an additional error related to the approximation of the original random measure. We refer to [70] and the references therein for the approximation of Lévy processes.

3.2 Reflected BSDEs

Let us now turn to the approximation of the solution (Y, Z, A) of a reflected BSDE of the form

$$\begin{aligned} Y_t &= g(X_1) + \int_t^1 h\left(X_s, Y_s, Z_s\right) ds - \int_t^1 Z_s' dW_s + A_1 - A_t \\ Y_t &\geq f(X_t) \ , \ 0 \leq t \leq 1 \ , \end{aligned} \tag{3.4}$$

where A is a cadlag adapted non-decreasing process satisfying

$$\int_0^1 (Y_t - f(X_t))\, dA_t = 0 \ .$$

In the case where $h = 0$ and $f = g$, Y is the Snell envelope of $g(X)$, which, in finance, corresponds to the super-hedging price of the American option with payoff g. More generally, the solution (Y, Z, A) of (3.4) is related to semilinear PDEs with free boundaries of the form

$$\min\left\{-\mathcal{L}v - h(\cdot, v, \sigma' Dv) \ , \ v - f\right\} = 0 \ , \ v(1-, \cdot) = g \tag{3.5}$$

see [47].

For such BSDEs, it is natural to modify the Backward Euler scheme (2.4) as follows so as to take into account the reflection:

$$\begin{aligned} \bar{Z}_{t_i}^{\pi} &= (t_{i+1} - t_i)^{-1}\, \mathbb{E}\left[Y_{t_{i+1}}^{\pi} \Delta W_{t_i} \mid \mathcal{F}_{t_i}\right] \\ \tilde{Y}_{t_i}^{\pi} &= \mathbb{E}\left[Y_{t_{i+1}}^{\pi} \mid \mathcal{F}_{t_i}\right] + h(X_{t_i}^{\pi}, \tilde{Y}_{t_i}^{\pi}, \bar{Z}_{t_i}^{\pi})\Delta t_i \\ Y_{t_i}^{\pi} &= \mathcal{R}\left(t_i \ , \ X_{t_i}^{\pi} \ , \ \tilde{Y}_{t_i}^{\pi}\right) \ , \ i \leq n-1 \ , \end{aligned} \tag{3.6}$$

with the terminal condition $Y_1^{\pi} = g(X_1^{\pi})$ and where

$$\mathcal{R}(t, x, y) \ := \ y + [f(x) - y]^+ \mathbf{1}_{\{t \in \Re \setminus \{0, T\}\}} \ , \ (t, x, y) \in [0, 1] \times \mathbb{R}^{d+1}$$

for a partition $\Re := \{0 = r_0 < r_1 < \ldots < r_\kappa = 1\} \supset \pi$.

Numerical procedures for such Backward Euler schemes have been first studied in Bally and Pagès [3] in the case where $\pi = \Re$, h does not depend on Z and $g = f$. The fact that h is independent on Z allows them to use the formulation of Y in terms of optimal stopping problem to study its discrete time approximation error. They retrieve a $|\pi|^{-\frac{1}{2}}$ control for the convergence rate when the coefficients are Lipschitz continuous and improve it to $|\pi|$ when g is semiconvex and X can be perfectly simulated on the time-grid. The error on Z is not discussed.

A more general analysis has then been performed by Ma and Zhang [60], who obtained a bound for the convergence rate of order $|\pi|^{-\frac{1}{4}}$ when f depends on Z whenever it is C_b^2 and σ is uniformly elliptic. Note that the arguments leading to (2.10) cannot be used here, since this would involve taking the Malliavin derivative of A, which has no reason to be well-defined. Instead, they rely on a representation of Z of the form ($d = 1$ to avoid notational complexities):

$$Z_t = \mathbb{E}\left[g(X_1)N_1^t + \int_t^1 h\left(X_r, Y_r, Z_r\right) N_r^t dr + \int_t^1 N_r^t dA_r \mid \mathcal{F}_t\right] \sigma(X_t) \qquad (3.7)$$

where

$$N_s^t := \frac{1}{s-t} \int_t^s \sigma(X_r)^{-1} \nabla X_r dW_r \left(\nabla X_t\right)^{-1}.$$

Their argument can be formally explained in three steps. First, they derive a representation of the Z-component associated to a discretely reflected BSDE of the form

$$Y_t^b = g(X_1) + \int_t^1 h(\Theta_r^b) dr - \int_t^1 (Z_r^b)' dW_r + A_1^b - A_t^b \qquad (3.8)$$

where $\Theta^b := (X, Y^b, Z^b)$ and $A_t^b := \sum_{j=1}^{\kappa-1} \left[h(X_{r_j}) - Y_{r_j}^b\right]^+ \mathbf{1}_{r_j \le t}$. The first representation is obtained by considering the Malliavin derivatives of (Y, Z) as in Section 2.3, see Step 1, on each interval $[r_j, r_{j+1})$

$$Z_t^b = D_t Y_t^b = D_t Y_{r_{j+1}-}^b + \int_t^{r_{j+1}} D_t h(\Theta_r^b) dr - \int_t^{r_{j+1}} D_t Z_r^b dW_r \,, \quad t \in [r_j, r_{j+1})$$

where $D_t Y_{r_{j+1}-}^b = D_t Y_{r_{j+1}}^b + D_t \left[h(X_{r_{j+1}}) - Y_{r_{j+1}}^b\right]^+$ since, by construction, $Y_{r_{j+1}}^b = Y_{r_{j+1}}^b + \left[h(X_{r_{j+1}}) - Y_{r_{j+1}}^b\right]^+$. This implies that, for $t < r_{j+1}$,

$$Z_t^b = D_t g(X_1) + \int_t^1 D_t h(\Theta_r^b) dr - \int_t^1 D_t Z_r^b dW_r + \sum_{k=j+1}^{\kappa} D_t \left[h(X_{r_k}) - Y_{r_k}^b\right]^+ .$$

The fact that the discretely reflected BSDE is differentiable in the Malliavin sense can be easily proved by using a backward induction argument and the fact that $z \mapsto z^+$ is a Lipschitz function. Second, they perform an integration by parts, in the Malliavin sense, in order to get rid of the Malliavin derivatives, and obtain a formulation of the form

$$Z_t^b = \mathbb{E}\left[g(X_1)N_1^t + \int_t^1 h(\Theta_r^b) N_r^t dr + \sum_{k=0}^{\kappa} N_{r_k}^t \Delta A_{r_k}^b \mathbf{1}_{t < r_k} \mid \mathcal{F}_t\right] \sigma(X_t)$$

for $t \in [r_j, r_{j+1})$, where $\Delta A_{r_k}^b := \left[h(X_{r_k}) - Y_{r_k}^b\right]^+$. Finally, they pass to the limit on the grid of reflection times so as to recover (3.7) for the continuously reflected BSDE.

This formulation is in itself very interesting and can be viewed as a generalisation of the representation obtained in [38] for the linear non reflected case. However, it requires a uniform ellipticity condition on σ. Moreover, the exploding behaviour of N_r^t as $r \to t$ is a major drawback in the regularity analysis of Z.

Another regularity analysis was carried out in Bouchard and Chassagneux [16]. The main difference is that they do not perform the integration by parts but rather provide a representation of Z^b in terms of the next reflection time:

$$Z_t^b = \mathbb{E}\Big[Dg(X_1)\eta_1^b \mathbf{1}_{\tau_j=1} + Df(X_{\tau_j})\eta_{\tau_j}^b \mathbf{1}_{\tau_j<1} + \int_t^{\tau_j} D_x h(\Theta_r^b)\eta_r^b dr \mid \mathcal{F}_t\Big](\eta_t^b)^{-1}\sigma(X_t) \,, \tag{3.9}$$

for $t \in [r_j, r_{j+1})$, where $\tau_j := \inf\{r \in \Re \mid r \geq r_{j+1},\ f(X_r) > Y_r^b\} \wedge 1$ and η^b is defined as in (2.9) above with (X, Y^b, Z^b) in place of (X, Y, Z). This is the analogue of (2.10) up to the stopping time τ_j.

This formulation does not require any uniform ellipticity assumption on σ and turn out to be more flexible.

Let us expose the argument in the simpler case where $d = 1, g \equiv f, h \equiv 0, \nabla X \equiv I_d$ and σ is constant. Set

$$V_t^j := \mathbb{E}\left[Df(X_{\tau_j}) \mid \mathcal{F}_t\right]\sigma \quad \text{so that } Z_t^b = V_t^j \text{ if } t \in [r_j, r_{j+1}) \,,$$

and note that the martingale property of V^j implies that

$$\mathbb{E}\left[|V_t^j - V_{t_i}^j|^2\right] \;\leq\; \mathbb{E}\left[|V_{t_{i+1}}^j|^2 - |V_{t_i}^j|^2\right] \text{ for } t \in [t_i, t_{i+1}] \,.$$

Letting i_j be defined by $r_j = t_{i_j}$, recall that $\pi \subset \Re$, it follows that

$$|\pi|^{-1}\sum_{i=0}^{n-1}\mathbb{E}\left[\int_{t_i}^{t_{i+1}} |Z_t^b - Z_{t_i}^b|^2 dt\right] \leq \Sigma := \sum_{j=0}^{\kappa-1}\sum_{k=i_j}^{i_{j+1}-1}\mathbb{E}\left[|V_{t_{k+1}}^j|^2 - |V_{t_k}^j|^2\right]$$

where, by direct computations,

$$\Sigma \;=\; \sum_{j=0}^{\kappa-1}\mathbb{E}\left[|V_{r_{j+1}}^j|^2 - |V_{r_j}^j|^2\right] \leq C + \sum_{j=1}^{\kappa-1}\mathbb{E}\left[|V_{r_j}^{j-1}|^2 - |V_{r_j}^j|^2\right]$$

and $C > 0$ denotes a generic constant independent on π and $\Re$. On the other hand, it follows from the Cauchy–Schwarz inequality that, if Df is bounded,

$$|V_{r_j}^{j-1}|^2 - |V_{r_j}^j|^2 \;\leq\; |V_{r_j}^j - V_{r_j}^{j-1}|\,|V_{r_j}^j + V_{r_j}^{j-1}| \leq C\,\mathbb{E}\left[\beta \mid \mathcal{F}_{r_j}\right]\,|V_{r_j}^j - V_{r_j}^{j-1}|$$

where $\beta := 1 + \sup_{t\leq 1}|X_t|$. Moreover, if $Df \in C_b^2$, then

$$V_{r_j}^j - V_{r_j}^{j-1} \;=\; \mathbb{E}\left[(Df(X_{\tau_j}) - Df(X_{\tau_{j-1}})) \mid \mathcal{F}_{r_j}\right]\sigma \leq C\,\mathbb{E}\left[(\tau_j - \tau_{j-1}) \mid \mathcal{F}_{r_j}\right]\sigma \,.$$

Combining the above inequalities, this implies

$$\sum_{i=0}^{n-1}\mathbb{E}\left[\int_{t_i}^{t_{i+1}} |Z_t^b - Z_{t_i}^b|^2 dt\right] \;\leq\; C|\pi|\left(1 + \sum_{j=1}^{\kappa-1}\mathbb{E}\left[\mathbb{E}\left[\beta \mid \mathcal{F}_{r_j}\right]\mathbb{E}\left[\tau_j - \tau_{j-1} \mid \mathcal{F}_{r_j}\right]\right]\right)$$

which is of order of $|\pi|$ since the expectation on the right-hand side simply equals $\mathbb{E}\left[\beta(\tau_j - \tau_{j-1})\right]$.

The general case is treated along the same main argument: play with the difference of the next reflection times $\tau_j - \tau_{j-1}$. It allows to control the modulus of regularity $\mathcal{R}^2_{Z^b}(\pi)$ of Z^b, defined as $\mathcal{R}^2_Z(\pi)$ for Z^b is place of Z, as follows:

Theorem 3.2.1 ([16]). *Assume that all the coefficients are Lipschitz continuous. Then,*

$$\mathcal{R}^2_{Z^b}(\pi) \leq \sum_{i=0}^{n-1} \int_{t_i}^{t_{i+1}} \mathbb{E}\left[|Z^b_t - Z^b_{t_i}|^2\right] dt \quad \leq \quad C\,|\pi| \,\left(1 + \beta|\pi|^{-\frac{1}{2}} + \alpha(\Re)\right) ,$$

where $\alpha(\Re) = |\Re|^{-\frac{1}{2}}$ and $\beta = 1$ under the assumption
H_1) : f is C^1_b with Lipschitz first derivative
and, $\alpha(\Re) = 1$ and $\beta = 0$ under the assumption
H_2) : f is C^2_b with Lipschitz second derivative and σ is C^1_b with Lipschitz first derivative.

This regularity property of Z^b then allows to provide a convergence rate for the Euler scheme $(Y^\pi, \bar{Z}^\pi)$ to (Y^b, Z^b).

Theorem 3.2.2 ([16]). *Assume that the coefficients are Lipschitz. Then,*

$$\max_{i<n} \mathbb{E}\left[\sup_{t\in(t_i,t_{i+1}]} |Y^\pi_{t_{i+1}} - Y^b_t|^2\right]^{\frac{1}{2}} \quad \leq \quad O\left(\alpha_Y(\Re)\,|\pi|^{\frac{1}{2}} + \lambda\,|\pi|^{\frac{1}{4}}\right)$$

$$\sum_{i=0}^{n-1} \mathbb{E}\left[\int_{t_i}^{t_{i+1}} |\bar{Z}^\pi_{t_i} - Z^b_t|^2 dt\right] \quad \leq \quad O\left(\alpha_Z(\Re)\,|\pi| + \lambda\,|\pi|^{\frac{1}{2}}\right)$$

with $(\alpha_Y(\Re), \alpha_Z(\Re)) = (|\Re|^{-\frac{1}{4}}, |\Re|^{-1})$ and $\lambda = 1$ under H_1), and, $(\alpha_Y(\Re), \alpha_Z(\Re)) = (1, |\Re|^{-1})$ and $\lambda = 0$ under H_2).

They also show that the above result holds true with $\alpha_Z(\kappa) = |\Re|^{-\frac{1}{2}}$ under H_1), and $\alpha_Z(\kappa) = 1$ under H_2), when $(X^\pi_{t_i})_{i\leq n}$ is replaced by $(X_{t_i})_{i\leq n}$ in the Backward Euler scheme (3.6).

To obtain a convergence rate of the approximation error of Y, it then essentially suffices to consider the case $\Re = \pi$ and show that Y^b approximates Y at least at a rate $|\Re|^{\frac{1}{2}}$. The approximation of the Z-component is more involved and requires the introduction of another approximation scheme.

Theorem 3.2.3 ([16]). *Assume that the coefficients are Lipschitz. Then, there exists $C > 0$, independent on π, such that*

$$\mathbb{E}\left[\max_{i\le n}|Y^\pi_{t_i}-Y_{t_i}|^2\right]^{\frac{1}{2}}+\max_{i<n}\mathbb{E}\left[\sup_{t\in(t_i,t_{i+1}]}|Y^\pi_{t_{i+1}}-Y_t|^2\right]^{\frac{1}{2}}=\mathcal{O}(\alpha(\pi))$$

$$\mathcal{R}^2_Z(\pi)\le\sum_{i=0}^{n-1}\mathbb{E}\left[\int_{t_i}^{t_{i+1}}|\bar{Z}^\pi_{t_i}-Z_t|^2dt\right]=\mathcal{O}(|\pi|^{\frac{1}{2}})\,,$$

where $\alpha(\pi)=|\pi|^{\frac{1}{4}}$ under H_1) and $\alpha(\pi)=|\pi|^{\frac{1}{2}}$ under H_2).

When H_2) holds and $(X^\pi_{t_i})_{i\le n}$ is replaced by $(X_{t_i})_{i\le n}$ in the Backward Euler scheme (3.6), the (squared) error on Z and the (squared) modulus of regularity $\mathcal{R}^2_Z(\pi)$ are shown to be controlled by $|\pi|$, which corresponds to a convergence rate of order $|\pi|^{\frac{1}{2}}$ as in the non-reflected case.

Remark 3.2.4. The representation (3.9) of Z^b is of own interest as it provides a natural Monte-Carlo estimator for the so-called "delta" of Bermudean options. Such a result was already known in finance for American options, under a uniform ellipticity condition, see [41].

Remark 3.2.5. Extensions to doubly reflected and multivariate reflected BSDEs have been considered in Chassagneux [23] and [24].

3.3 BSDEs with random time horizon

By BSDE with random time horizon, we mean the solution (Y, Z) of

$$Y_t \;=\; g(X_\tau)+\int_t^1 \mathbf{1}_{s<\tau}h(X_s,Y_s,Z_s)ds-\int_t^1 Z'_s dW_s\;,\;t\le 1\,, \tag{3.10}$$

where τ is the first exit time of $(t, X_t)_{t\le T}$ from a cylindrical domain $D = [0, 1) \times \mathcal{O}$.

The first step toward the definition of a Backward Euler scheme, consists in choosing a suitable approximation of the random time horizon τ. In Bouchard and Menozzi [19], the authors consider the first exit time (on the grid) $\bar{\tau}$ of the Euler scheme X^π

$$\bar{\tau} := \inf\{t\in\pi\ :\ (t,\bar{X}_t)\notin D\}\,.$$

They then reproduce the Backward scheme (2.4)

$$\begin{cases} \bar{Z}^\pi_{t_i} &= \frac{1}{\Delta t_i}\,\mathbb{E}\left[Y^\pi_{t_{i+1}}\Delta W_{t_i}\mid\mathcal{F}_{t_i}\right]\\ Y^\pi_{t_i} &= \mathbb{E}\left[Y^\pi_{t_{i+1}}\mid\mathcal{F}_{t_i}\right]+h\left(X^\pi_{t_i},Y^\pi_{t_i},\bar{Z}^\pi_{t_i}\right)\Delta t_i\end{cases}$$

with the terminal condition

$$Y^\pi_1 = g(X^\pi_{\bar{\tau}})\,.$$

Following the approach of [20] and [74], they show that the discrete-time approximation error can be controlled as:

$$\mathrm{Err}(\pi)^2 \;\le\; C\left(|\pi| + \mathcal{R}_Y^2(\pi) + \mathcal{R}_Z^2(\pi) + \mathbb{E}\left[\xi|\tau - \bar{\tau}| + \mathbf{1}_{\bar{\tau}<\tau}\int_{\bar{\tau}}^{\tau}|Z_s|^2 ds\right]\right) \quad (3.11)$$

where $C > 0$ is independent of π and ξ is a positive random variable satisfying $\mathbb{E}\left[\xi^p\right] \le C'_p$ for all $p \ge 2$, for some C'_p independent of π.

Not surprisingly, the error now depends on an additional term which is related to the approximation error of τ by $\bar{\tau}$ which has been widely studied under the assumption that the coefficients are smooth enough and σ is uniformly elliptic. In particular, Gobet [39], and, Gobet and Menozzi [45] have derived an expansion of the form $\mathbb{E}\left[\tau - \bar{\tau}\right] = C|\pi|^{\frac{1}{2}} + o(|\pi|^{\frac{1}{2}})$.

A similar result can be obtained without uniform ellipticity, whenever $\mathcal{O}$ is a C^2 domain with a compact boundary satisfying a uniform non-characteristic condition.

Theorem 3.3.1 ([19]). *Assume that the coefficients of X are Lipschitz continuous and that $\mathcal{O}$ is a C^2 domain with a compact boundary satisfying a uniform non-characteristic condition. Then, for each $\varepsilon \in (0,1)$ and each positive random variable ξ satisfying $\mathbb{E}\left[\xi^p\right] < \infty$ for all $p \ge 1$,*

$$\mathbb{E}\left[\mathbb{E}\left[\xi\,|\tau - \bar{\tau}|\;|\;\mathcal{F}_{\tau_+\wedge\bar{\tau}}\right]^2\right] \;\le\; O_\varepsilon^\xi\left(|\pi|^{1-\varepsilon}\right) ,$$

where $\tau_+ := \inf\{t \in \pi \;:\; \tau \le t\}$.

This shows in particular that, for each $\varepsilon \in (0, 1/2)$,

$$\mathbb{E}\left[|\tau - \bar{\tau}|\right] \;\le\; O_\varepsilon\left(|\pi|^{\frac{1}{2}-\varepsilon}\right)$$

which, up to the ε term, is consistent with the result of Gobet and Menozzi [45]. Moreover, it is easily extended to the case where

$D1)$ $\mathcal{O}$ is an intersection of C^2 domains with compact boundary, and it satisfies a uniform non-characteristic condition outside the corners and a uniform ellipticity condition on a neighbourhood of the corners.

The next step toward a convergence rate consists in studying the modulus of regularity $\mathcal{R}_Y^2(\pi)$ and $\mathcal{R}_Z^2(\pi)$, and the last term $\mathbb{E}\left[\mathbf{1}_{\bar{\tau}<\tau}\int_{\bar{\tau}}^{\tau}|Z_s|^2 ds\right]$.

In contrast with the case where $\mathcal{O} = \mathbb{R}^d$, the uniform Lipschitz continuity of the (viscosity) solution v of the associated semilinear Cauchy Dirichlet problem

$$-\mathcal{L}v - h(\cdot, v, \sigma' Dv) = 0 \text{ on } D\,,\; v = g \text{ on } \partial_p D := ([0,1) \times \partial\mathcal{O}) \cup \left(\{1\} \times \bar{\mathcal{O}}\right)\,, \quad (3.12)$$

is by no means obvious, and usually requires additional assumptions than the simple Lipschitz continuity of the coefficients. It is shown in [19], by adapting standard barrier techniques, when g is smooth enough:

$Hg)$ $g \in C_b^{1,2}([0,1] \times \mathbb{R}^d)$,

and the domain satisfies a uniform exterior sphere condition as well as a uniform truncated interior cone condition:

$D2)$ For all $x \in \partial\mathcal{O}$, there is $y(x) \in \mathcal{O}^c$, $r(x) \in [L^{-1}, L]$, $L > 0$, and $\delta(x) \in B(0,1)$ such that $\bar{B}(y(x), r(x)) \cap \bar{\mathcal{O}} = \{x\}$ and $\{x' \in B(x, L^{-1}) \ : \ \langle x' - x, \delta(x)\rangle \geq (1 - L^{-1})|x' - x|\} \subset \bar{\mathcal{O}}$.

Here $B(z,r)$ denotes the open ball of centre z and radius r, $\bar{B}(z,r)$ is its closure.

Theorem 3.3.2 ([19]). *Assume that all the coefficients are Lipschitz continuous and that the conditions $Hg)$, $D1)$ and $D2)$ hold. Then, v is uniformly $1/2$-Hölder continuous in time and Lipschitz continuous in space.*

In particular, this allows to show that $\mathcal{R}_Y^2(\pi)$ is controlled by $|\pi|$ and the last term $\mathbb{E}\left[\mathbf{1}_{\bar{\tau}<\tau} \int_{\bar{\tau}}^{\tau} |Z_s|^2 ds\right]$ is controlled by $\mathbb{E}\left[|\tau - \bar{\tau}|\right] \leq C^{\varepsilon}|\pi|^{\frac{1}{2}-\varepsilon}$, for $\varepsilon > 0$.

It thus remains to study the modulus of continuity of Z. As in the reflected case, it is not possible to apply the techniques leading to (2.10) since it would involve taking the Malliavin derivative of $g(X_\tau)$, which has no reason to be well-defined. Instead, we can remark that, if v is smooth enough, $Dv(\cdot, X)\eta + \int_0^{\cdot} D_x h\left(\Theta_s\right)\eta_s ds$ is a martingale on $[0, \tau]$, recall the definition of η in (2.9). This implies that ($d = 1$ for notational simplicity)

$$Z_t = Dv(t, X_t)\sigma(X_t) = \mathbb{E}\left[Dv(\tau, X_\tau)\eta_\tau + \int_t^\tau D_x h\left(\Theta_s\right)\eta_s ds \mid \mathcal{F}_t\right]\eta_t^{-1}\sigma(X_t) .$$

This formula is very close to (2.10) except that Dg is replaced by Dv. But, since Dv is bounded, one can actually repeat the martingale-type argument of Step 3 of Section 2.3. The smoothness of v can be obtained for smooth coefficients and domains, under a uniform ellipticity condition of σ, which can then be relaxed by standard approximation arguments.

Theorem 3.3.3 ([19]). *Assume that all the coefficients are Lipschitz continuous and that the conditions $Hg)$, $D1)$ and $D2)$ hold. Then,*

$$\mathcal{R}_Y^2(\pi) + \mathcal{R}_Z^2(\pi) \quad \leq \quad \mathcal{O}(|\pi|) .$$

This provides the following $|\pi|^{\frac{1}{4}-\varepsilon}$ convergence rate.

Corollary 3.3.4 ([19]). *Assume that all the coefficients are Lipschitz continuous and that the conditions $Hg)$, $D1)$ and $D2)$ hold. Then, for all $\varepsilon \in (0,1)$,*

$$\mathrm{Err}(\pi) \quad \leq \quad O_\varepsilon\left(|\pi|^{\frac{1}{2}-\varepsilon}\right) .$$

As a matter of facts, the global error is driven by the approximation error of the exit time which propagates backward thanks to the Lipschitz continuity of v.

Remark 3.3.5. It is shown in [19] that one can actually achieve a convergence rate of order $|\pi|^{\frac{1}{2}-\varepsilon}$ if one computes the approximation error only up to $\bar{\tau} \wedge \tau$.

4 Extension to fully nonlinear PDEs

The discretisation of backward stochastic differential equations can be viewed as a probabilistic approach for the numerical resolution of semilinear partial differential equations in parabolic or elliptic form, and possibly with a nonlocal integral term. This restriction on the nonlinearity of the corresponding PDEs excludes the important general class of Hamilton–Jacobi–Bellman equations which appear naturally as the infinitesimal version of the dynamic programming principle for stochastic control problems.

An extension of the theory of backward SDEs to the second order case was initiated by [25] and is still in progress. An alternative approach for such an extension in the context of a specific form of the nonlinearity was also introduced by Peng in various papers, see e.g. [69], and is based on the new concept of G-Brownian motion.

We shall not elaborate more on this aspect. Instead, our objective is to concentrate on the probabilistic numerical implications as discussed in Section 5 of [25]. Following [37], we provide a natural presentation of the discrete-time approximation without appealing to the theory of backward stochastic differential equations. This presentation shows in particular the important connection between the discretisation schemes of this paper and the finite differences approximations of solutions of PDEs. Finally, we were not able to extend the methods of proofs outlined in the previous sections. Instead, the convergence results and the rates of convergence are derived by means of the PDE approach for monotonic schemes.

4.1 Discretisation

Consider the fully nonlinear parabolic PDE:

$$-\mathcal{L}v - F\left(\cdot, v, Dv, D^2v\right) = 0, \quad \text{on } [0,1) \times \mathbb{R}^d, \tag{4.1}$$

$$v(1,\cdot) = g, \quad \text{on } \in \mathbb{R}^d, \tag{4.2}$$

where, given bounded and continuous maps b and σ from $\mathbb{R}_+ \times \mathbb{R}^d$ to $\mathbb{M}^d$ and $\mathbb{R}^d$, the linear second order operator $\mathcal{L}$ is defined by

$$\mathcal{L}v := \frac{\partial v}{\partial t} + \langle b, Dv\rangle + \frac{1}{2}\mathrm{Tr}\left[\sigma\sigma' D^2 v\right],$$

and the nonlinearity is isolated in the map $F : (t,x,r,p,\gamma) \in \mathbb{R}_+ \times \mathbb{R}^d \times \mathbb{R} \times \mathbb{R}^d \times \mathcal{S}_d \longmapsto F(t,x,r,p,\gamma) \in \mathbb{R}$.

For a positive integer n, set $h := 1/n$, $t_i := ih$, $i = 1, \ldots, n$, and define:

$$\hat{X}_h^{t,x} := x + b(t,x)h + \sigma(t,x)(W_{t+h} - W_t), \tag{4.3}$$

which is the one-step Euler discretisation of the diffusion corresponding to the operator $\mathcal{L}$.

Assuming that the PDE (4.1) has a classical solution, it follows from Itô's formula that

$$\mathbb{E}_{t_i,x}\left[v\left(t_{i+1}, X_{t_{i+1}}\right)\right] = v\left(t_i, x\right) + \mathbb{E}_{t_i,x}\left[\int_{t_i}^{t_{i+1}} \mathcal{L}v(t, X_t)dt\right]$$

where we ignored the difficulties related to local martingale part, and $\mathbb{E}_{t_i,x} := \mathbb{E}[\cdot|X_{t_i} = x]$ denotes the expectation operator conditional on $\{X_{t_i} = x\}$. Since v solves the PDE (4.1), this provides

$$\mathbb{E}_{t_i,x}\left[v\left(t_{i+1}, X_{t_{i+1}}\right)\right] = v(t_i, x) - \mathbb{E}_{t_i,x}\left[\int_{t_i}^{t_{i+1}} F(\cdot, v, Dv, D^2v)(t, X_t)dt\right].$$

By approximating the Riemann integral, and replacing the process X by its Euler discretisation, this suggests the following approximation of the value function v

$$v^h(1,.) = g \quad \text{and} \quad v^h(t_i, x) = T_h[v^h](t_i, x) \tag{4.4}$$

where, for a function $\psi : \mathbb{R}_+ \times \mathbb{R}^d \longrightarrow \mathbb{R}$ with exponential growth, we denote:

$$\begin{aligned} T_h[\psi](t_i, x) &:= \mathbb{E}\left[\psi(t_{i+1}, \hat{X}_h^{t_i,x})\right] + hF\left(\cdot, \mathcal{D}_h\psi\right)(t_i, x), \\ \mathcal{D}_h\psi &:= \left(\mathcal{D}_h^0\psi, \mathcal{D}_h^1\psi, \mathcal{D}_h^2\psi\right), \end{aligned}$$

and for $k = 0, 1, 2$, it follows from an easy integration by parts argument that:

$$\mathcal{D}_h^k\psi(t_i, x) := \mathbb{E}[D^k\psi(t_{i+1}, \hat{X}_h^{t_i,x})] = \mathbb{E}[\psi(\hat{X}_h^{t_i,x})H_k^h(t_i, x)], \tag{4.5}$$

where

$$H_0^h = 1, \quad H_1^h = (\sigma')^{-1}\frac{W_h}{h}, \quad H_2^h = (\sigma')^{-1}\frac{W_hW_h' - h\mathbf{1}_d}{h^2}\sigma^{-1}. \tag{4.6}$$

Remark 4.1.1. In the semilinear case, i.e. F is independent of the argument γ, the numerical scheme coincides with the discretisation of BSDEs as studied in the previous sections of this paper.

Remark 4.1.2. Other choices can be made for the above integration by parts. For instance, we also have:

$$\mathcal{D}_2^h\varphi(t, x) = \mathbb{E}\left[\varphi(\hat{X}_h^{t,x})(\sigma')^{-1}\frac{W_{h/2}}{(h/2)}\frac{W_h' - W_{h/2}'}{(h/2)}\sigma^{-1}\right],$$

which shows that the backward scheme (4.4) is very similar to the probabilistic numerical algorithm suggested in [25].

Remark 4.1.3. So far, the choice of the drift and the diffusion coefficients b and σ in the nonlinear PDE (4.1) is arbitrary, and was only used to fix the underlying diffusion X. Our convergence result will however place some restrictions on this choice.

Remark 4.1.4. The integration by parts in (4.5) is intimately related to finite differences for numerical schemes on deterministic grids. We formally justify this connection for $d = 1$, $b \equiv 0$, and $\sigma = 1$ for simplicity:

- Let $\{w_j, j \geq 1\}$ be independent r.v. distributed as $\frac{1}{2}\left(\delta_{\sqrt{h}} + \delta_{-\sqrt{h}}\right)$. Then, the binomial random walk approximation of the Brownian motion $\hat{W}_{t_k} := \sum_{j=1}^{k} w_j$, $t_k := kh$, $k \geq 1$, suggests the following approximation:

$$\mathcal{D}_h^1 \psi(t,x) := \mathbb{E}\left[\psi(t+h, X_h^{t,x}) H_1^h\right] \quad \approx \quad \frac{\psi(t, x+\sqrt{h}) - \psi(t, x-\sqrt{h})}{2\sqrt{h}},$$

which is the centred finite differences approximation of the gradient.

- Let $\{w_j, j \geq 1\}$ be independent r.v. distributed as $\frac{1}{6}\left(\delta_{\{\sqrt{3h}\}} + 4\delta_{\{0\}} + \delta_{\{-\sqrt{3h}\}}\right)$. Then, the trinomial random walk approximation of the Brownian motion $\hat{W}_{t_k} := \sum_{j=1}^{k} w_j$, $t_k := kh$, $k \geq 1$, suggests the following approximation:

$$\mathcal{D}_h^2 \psi(t,x) := \mathbb{E}\left[\psi(t+h, X_h^{t,x}) H_2^h\right] \approx \frac{\psi(t, x+\sqrt{3h}) - 2\psi(t,x) + \psi(t, x-\sqrt{3h})}{3h},$$

which is the centred finite differences approximation of the Hessian.

In view of the above interpretation, the numerical scheme studied in this paper can be viewed as a mixed Monte Carlo–Finite Differences algorithm. The Monte Carlo component of the scheme consists in the choice of an underlying diffusion process X. The finite differences component of the scheme consists in approximating the remaining nonlinearity by means of the integration-by-parts (4.5).

4.2 Convergence of the discretisation

Definition 4.2.1. We say that (4.1) has strong comparison of bounded solutions if for any bounded upper semicontinuous viscosity supersolution $\overline{v}$ and any bounded lower semicontinuous subsolution $\underline{v}$ on $[0,1) \times \mathbb{R}^d$, satisfying

- either $\overline{v}(1,\cdot) \geq \underline{v}(1,\cdot)$,
- or the viscosity property of $\overline{v}$ and $\underline{v}$ holds true on $[0,1] \times \mathbb{R}^d$,

we have $\overline{v} \geq \underline{v}$.

The strong comparison principle is an important notion in the theory of viscosity solutions which allows to handle situations where the boundary condition g is not compatible with the equation.

In the sequel, we denote by F_r, F_p and F_γ the partial gradients of F with respect to r, p and γ, respectively. We recall that any Lipschitz function is differentiable a.e.

Assumption F. *The nonlinearity F is Lipschitz-continuous with respect to (r,p,γ) uniformly in (t,x) and $\sup_{(t,x)\in[0,T]\times\mathbb{R}^d} |F(t,x,0,0,0)| < \infty$. Moreover, F is uniformly elliptic and dominated by the diffusion of the linear operator $\mathcal{L}$, i.e.*

$$\varepsilon I_d \leq F_\gamma \leq a \text{ on } \mathbb{R}^d \times \mathbb{R} \times \mathbb{R}^d \times \mathcal{S}_d \quad \text{for some} \quad \varepsilon > 0. \tag{4.7}$$

Theorem 4.2.2. *Let Assumption F hold true, and assume that the fully nonlinear PDE* (4.1) *has strong comparison of bounded solutions. Then for every bounded function g, there exists a bounded function v so that*

$$v^h \longrightarrow v \qquad \textit{locally uniformly.}$$

In addition, v is the unique bounded viscosity solution of the relaxed boundary problem (4.1)–(4.2).

Remark 4.2.3. The restriction to bounded terminal data g in the above Theorem 4.2.2 can be relaxed by an immediate change of variable, thanks to the boundedness conditions on the coefficients b and σ, see [37].

Remark 4.2.4. Theorem 4.2.2 states that the right hand-side inequality of (4.7) (i.e. diffusion must dominate the nonlinearity in γ) is sufficient for the convergence of the Monte Carlo–Finite Differences scheme. We do not know whether this condition is necessary. However, in [37], it is shown that this condition is not sharp in the simple linear case, while the numerical experiments reveal that the method may have a poor performance in the absence of this condition.

The rest of this section is dedicated to the proof of Theorem 4.2.2 which is based on the convergence of monotonic schemes of Barles and Souganidis [9]. Since the probabilistic community is not familiar with this result we report a complete proof.

Properties of the discretisation scheme The approach of [9] is based on three main properties of the scheme.

1 The first ingredient is the so-called consistency property, which states that for every smooth function φ with bounded derivatives and $(t, x) \in [0, T] \times \mathbb{R}^d$:

$$\lim_{\substack{(t',x') \to (t,x) \\ (h,c) \to (0,0) \\ t'+h \leq 1}} \frac{(c+\varphi - T_h[c+\varphi])(t',x')}{h} = -\left(\mathcal{L}^X \varphi + F(\cdot, \varphi, D\varphi, D^2\varphi)\right)(t,x). \tag{4.8}$$

The verification of this property follows from an immediate application of Itô's formula.

2 The second ingredient is the so-called stability property:

$$\text{the family } (v^h)_h \text{ is } \mathbb{L}^\infty - \text{bounded, uniformly in } h. \tag{4.9}$$

This property is implied by the boundedness of g and $F(t, x, 0, 0, 0)$ (Assumption F), together with the stronger property that for every $\mathbb{L}^\infty$-bounded functions $\varphi, \psi : [0, 1] \times \mathbb{R}^d \longrightarrow \mathbb{R}$:

$$|T_h[\varphi] - T_h[\psi]|_\infty \leq |\varphi - \psi|_\infty (1 + Ch) \quad \text{for some} \quad C > 0, \tag{4.10}$$

For ease of presentation, we report the proof of (4.10) in the one-dimensional case; the general multi-dimensional case follows the same line of argument. Set $f := \varphi - \psi$. By the mean value theorem there exists some $\theta = (t, x, \bar{r}, \bar{p}, \bar{\gamma})$ such that:

$$\begin{aligned}
(T_h[\varphi] - T_h[\psi])\,(t,x) &= \mathbb{E}\left[f(\hat{X}_h^{t,x})\left(1 + hF_r(\theta) + \frac{W_h}{\sigma}F_p(\theta) + \frac{W_h^2 - h}{\sigma^2 h}F_\gamma(\theta)\right)\right] \\
&= \mathbb{E}\left[f(\hat{X}_h^{t,x})\left(1 - \frac{F_\gamma(\theta)}{\sigma^2} + hF_r - h\frac{F_p(\theta)^2}{4F_\gamma(\theta)} + A_h^2\right)\right] \quad (4.11)
\end{aligned}$$

where

$$A_h \;:=\; \frac{W_h}{\sigma\sqrt{h}}\sqrt{F_\gamma(\theta)} + \frac{F_p(\theta)\sqrt{h}}{2\sqrt{F_\gamma(\theta)}}. \qquad (4.12)$$

Since $1 - \sigma^{-2}F_\gamma \geq 0$ and $|F_\gamma^{-1}|_\infty < \infty$ by (4.7) of Assumption F, it follows from the Lipschitz property of F that

$$\begin{aligned}
|T_h[\varphi] - T_h[\psi]|_\infty &\leq |f|_\infty \left(1 - \sigma^{-2}F_\gamma(\theta) + \mathbb{E}[|A_h|^2] + Ch\right) \\
&= |f|_\infty \left(1 + h(4F_\gamma)^{-1}(\theta)(F_p)^2(\theta) + Ch\right) \;\leq\; |f|_\infty(1 + Ch).
\end{aligned}$$

3 The final ingredient – and the most important – is the monotonicity. In the context of our scheme (4.4), we have for all functions $\varphi, \psi : [0,1]\times\mathbb{R}^d \longrightarrow \mathbb{R}$ with exponential growth:

$$\varphi \leq \psi \;\Longrightarrow\; T_h[\varphi](t,x) \leq T_h[\psi](t,x) + Ch\,\mathbb{E}[(\psi - \varphi)(t+h, \hat{X}_h^{t,x})] \quad (4.13)$$

for some constant $C > 0$. To prove (4.13), we proceed as in the proof of the stability result to arrive at (4.11). Since $f := \psi - \varphi \geq 0$ in the present context, and $F_\gamma \leq \sigma^2$ by (4.7) of Assumption F, we deduce that:

$$(T_h[\psi] - T_h[\varphi])\,(t,x) \;\geq\; \mathbb{E}\left[f(\hat{X}_h^{t,x})\left(hF_r(\theta) - \frac{h}{4}\frac{F_p(\theta)^2}{F_\gamma(\theta)}\right)\right], \qquad (4.14)$$

and the required result follows from the Lipschitz property of F and the fact that $|F_\gamma^{-1}|_\infty < \infty$ by (4.7).

The Barles–Souganidis monotone scheme convergence argument Given the above consistency, stability and monotonicity properties, the convergence of the family $(v^h)_h$ towards some function v which is the unique viscosity solution of the PDE (4.1) follows from [9]. We report the full argument for completeness.

From the stability property the semi-relaxed limits

$$V_*(t,x) := \liminf_{(t',x',h)\to(t,x,0)} v^h(t',x') \quad\text{and}\quad V^*(t,x) := \limsup_{(t',x',h)\to(t,x,0)} v^h(t',x')$$

are finite lower-semicontinuous and upper-semicontinuous functions, respectively. Also, they obviously inherit the boundedness of the family $(v^h)_h$. The key-point is to show that V_* and V^* are respectively viscosity supersolution and subsolution of the PDE (4.1)–(4.2), which implies by the strong comparison assumption that $V^* \leq V_*$. Since the converse inequality is trivial, this shows that $V_* = V^*$ is the unique bounded solution of (4.1), thus completing the proof of convergence.

In the rest of this section, we show that V_* is a viscosity supersolution of the PDE (4.1). A symmetric argument applies to prove the corresponding property for V^*. Let $(t_0, x_0) \in [0,1) \times \mathbb{R}^d$ and $\varphi \in C^2\left([0,1] \times \mathbb{R}^d\right)$ be such that

$$0 = (V_* - \varphi)(t_0, x_0) = \min(V_* - \varphi).$$

Without loss of generality, we may assume that (t_0, x_0) is a strict minimiser of $(V_* - \varphi)$.

Let (h_n, t_n, x_n) be a sequence in $(0,1] \times B_1(t_0, x_0)$ with

$$(h_n, t_n, x_n) \longrightarrow (0, t_0, x_0) \quad \text{and} \quad v^{h_n}(t_n, x_n) \longrightarrow V_*(t_0, x_0),$$

and define (t'_n, x'_n) by

$$c_n := (v_*^{h_n} - \varphi)(t'_n, x'_n) = \min_{B_0}(v^{h_n} - \varphi),$$

where $B_0 \subset [0,1] \times \mathbb{R}^d$ is a closed ball containing (t_0, x_0) in its interior. By definition, we have $c_n + \varphi(t'_n, x'_n) = v^{h_n}(t'_n, x'_n) = T_{h_n}[v^{h_n}](t'_n, x'_n)$. Since $v^{h_n} \geq c_n + \varphi$, it follows from the monotonicity property (4.13) that:

$$c_n + \varphi(t'_n, x'_n) \geq T_{h_n}[c_n + v^{h_n}](t'_n, x'_n). \tag{4.15}$$

Since the sequence $(t'_n, x'_n)_n$ is bounded, it converges to some (t_1, x_1) after possibly passing to a subsequence. Observe that

$$(V_* - \varphi)(t_0, x_0) = \lim_{n\to\infty} (v^{h_n} - \varphi)(t_n, x_n) \geq \liminf_{n\to\infty} (v^{h_n} - \varphi)(t'_n, x'_n) \geq (V_* - \varphi)(t_1, x_1).$$

Since (t_0, x_0) is a strict minimiser of the difference $V_* - \varphi$, this shows that $(t_1, x_1) = (t_0, x_0)$, and $c_n = (v_*^{h_n} - \varphi)(t'_n, x'_n) \longrightarrow 0$. We now go back to (4.15), normalise by h_n and send n to infinity. By the consistency property (4.8), this provides the required result:

$$-\mathcal{L}\varphi(t_0, x_0) - F\left(., \varphi, D\varphi, D^2\varphi\right)(t_0, x_0) \geq 0.$$

Remark 4.2.5. In the context of nonlinear PDEs of the HJB type, see Assumption HJB below, Bonnans and Zidani [14] introduced finite differences approximations which obey to the monotonicity condition. The choice of the discretisation turns out to be pretty involved and is specific to each choice of control. Our probabilistic scheme has the nice property to be automatically monotonic.

4.3 Rate of convergence of the discretisation of HJB equations

We next provide an error estimate for our probabilistic numerical scheme. For ease of presentation, we assume that the nonlinearity F satisfies the condition

$$F_r - \frac{1}{4} F'_p F_\gamma^{-1} F_p \geq 0, \tag{4.16}$$

which implies that the monotonicity property (4.13) is strengthened to:

$$\varphi \leq \psi \implies T_h[\varphi] \leq T_h[\psi], \tag{4.17}$$

see (4.14). In [37], it is shown that our main results hold without Condition (4.16).

The subsequent argument for the derivation of the error estimate is crucially based on the following comparison principle satisfied by the scheme.

Lemma 4.3.1. *Let Assumption F holds true, and consider two arbitrary bounded functions φ and ψ satisfying:*

$$h^{-1}(\varphi - T_h[\varphi]) \leq g_1 \quad \text{and} \quad h^{-1}(\psi - T_h[\psi]) \geq g_2 \tag{4.18}$$

for some bounded functions g_1 and g_2. Then, for every $i = 0, \ldots, n$:

$$(\varphi - \psi)(t_i, x) \leq e^{\beta(T-t_i)}|(\varphi - \psi)^+(1, \cdot)|_\infty + (1-h)e^{\beta(1-t_i)}|(g_1 - g_2)^+|_\infty \tag{4.19}$$

for some parameter $\beta > |F_r|_\infty$.

Before stating precise results, let us explain the key idea as introduced by Krylov [52, 53, 54]. We also refer to the lecture notes by Bonnans [13] which provide a clear summary in the context of infinite horizon stochastic control problems.

Key-idea for the lower bound Given the solution v of the nonlinear PDE (4.1), suppose that there is a function $\underline{u}^\varepsilon$ satisfying

$$\underline{u}^\varepsilon \quad \text{is a classical subsolution of (4.1) and} \quad v - \underline{C}(\varepsilon) \leq \underline{u}^\varepsilon \leq v \tag{4.20}$$

for some function $\underline{C}(\varepsilon)$. Let

$$\underline{R}_h[\underline{u}^\varepsilon] \;:=\; \frac{\underline{u}^\varepsilon - T_h[\underline{u}^\varepsilon]}{h} + \mathcal{L}\underline{u}^\varepsilon + F\left(., \underline{u}^\varepsilon, D\underline{u}^\varepsilon, D^2\underline{u}^\varepsilon\right),$$

and let $\underline{R}(h, \varepsilon)$ be a bound on $\underline{R}_h[\underline{u}^\varepsilon]$:

$$|\underline{R}_h[\underline{u}^\varepsilon](t, x)| \leq \underline{R}(h, \varepsilon) \quad \text{for every} \quad (t, x) \in [0, T] \times \mathbb{R}^n. \tag{4.21}$$

Since $\underline{u}^\varepsilon$ is a subsolution of the nonlinear PDE (4.1), it follows that

$$\underline{u}^\varepsilon(t, x) \;\leq\; \mathbb{E}\left[\underline{u}^\varepsilon(t + h, \hat{X}_h^{t,x})\right] + h\left[F(., \mathcal{D}\underline{u}^\varepsilon)(t, x) + \underline{R}(h, \varepsilon)\right].$$

We next introduce the function $\underline{U}^\varepsilon$ defined by $\underline{U}^\varepsilon(1, x) = \underline{u}^\varepsilon(1, x)$ and

$$\underline{U}^\varepsilon(t_i, x) \;=\; \mathbb{E}\left[\underline{U}^\varepsilon(t_i + h, \hat{X}_h^{t_i,x})\right] + h\left[F(., \mathcal{D}\underline{U}^\varepsilon)(t_i, x) + \underline{R}(h, \varepsilon)\right].$$

Then, it follows from the comparison property of Lemma 4.3.1 that

$$\underline{u}^\varepsilon \;\leq\; \underline{U}^\varepsilon. \tag{4.22}$$

Moreover, arguing as in the proof of (4.10), we see that

$$\left|\underline{U}^\varepsilon - v^h\right| \;\le\; (1+Ch)\mathbb{E}_{t,x}\left[\left|\underline{U}^\varepsilon - v^h\right|(t+h,\hat{X}_h^{t,x})\right] + h\overline{R}(h,\varepsilon),$$

which provides the estimate:

$$\left|\underline{U}^\varepsilon - v^h\right| \;\le\; \underline{R}(h,\varepsilon). \tag{4.23}$$

We then deduce from (4.20), (4.22) and (4.23) that

$$v - v^h = v - \underline{u}^\varepsilon + \underline{u}^\varepsilon - v \le v - \underline{u}^\varepsilon + \underline{U}^\varepsilon - v^h \le \underline{C}(\varepsilon) + \underline{R}(h,\varepsilon),$$

and therefore

$$v - v^h \;\le\; \inf_{\varepsilon>0}\left(\underline{C}(\varepsilon) + \underline{R}(h,\varepsilon)\right). \tag{4.24}$$

The approximating subsolution The following construction of the function $\underline{u}^\varepsilon$ satisfying (4.20) requires the nonlinearity $F(t,x,r,p,\gamma)$ to be concave in (r,p,γ) with some additional conditions in order to ensure some regularity of the solution. Notice that the case where $F(t,x,r,p,\gamma)$ is convex in (r,p,γ) can also be dealt with by a symmetric argument (inverting the roles of supersolutions and subsolutions). For a bounded function $\psi(t,x)$ Lipschitz in x and $1/2$–Hölder continuous in t, we denote

$$|\psi|_1 \;:=\; |\psi|_\infty + \sup_{([0,1]\times\mathbb{R}^d)^2} \frac{\psi(t,x)-\psi(t',x')}{|x-x'|+|t-t'|^{1/2}}$$

Assumption HJB *The nonlinearity F is of the Hamilton–Jacobi–Bellman type:*

$$\begin{aligned} F(t,x,r,p,\gamma) \;&=\; \inf_{\alpha\in\mathcal{A}}\{\mathcal{L}^\alpha(t,x,r,p,\gamma)\} \\ \mathcal{L}^\alpha(t,x,r,p,\gamma) \;&:=\; \frac{1}{2}Tr[\sigma^\alpha\sigma^{\alpha\mathrm{T}}(t,x)\gamma] + b^\alpha(t,x)p + c^\alpha(t,x)r + f^\alpha(t,x) \end{aligned}$$

where b, σ, σ^α, b^α, c^α and f^α satisfy:

$$|b|_1 + |\sigma|_1 + \sup_{\alpha\in\mathcal{A}}\left(|\sigma^\alpha|_1 + |b^\alpha|_1 + |c^\alpha|_1 + |f^\alpha|_1\right) \;<\; \infty.$$

The above Assumption implies in particular that our nonlinear PDE satisfies a strong comparison result for bounded functions. Let w^ε be the unique viscosity solution of the nonlinear PDE with shaked coefficients (in the terminology of Krylov):

$$\begin{aligned} &-\tfrac{\partial w^\varepsilon}{\partial t} - \inf_{|e|\le 1}\bar{F}\left(t+\varepsilon e, x+\varepsilon e, w^\varepsilon(t,x), Dw^\varepsilon(t,x), D^2 w^\varepsilon(t,x)\right) = 0, \\ &w^\varepsilon(1,.) = g, \end{aligned} \tag{4.25}$$

where:

$$\bar{F}(t,x,r,p,\gamma) \;:=\; \langle b(t,x),p\rangle + \frac{1}{2}\mathrm{Tr}[\sigma\sigma'\gamma] + F(t,x,r,p,\gamma).$$

The existence of w^ε follows from a direct identification of this PDE as the HJB equation of a corresponding stochastic control problem. By the definition of w^ε together with a strong comparison argument, we then have that

$$w^\varepsilon \text{ is a subsolution of (4.1) and } |w^\varepsilon - v| \leq C\varepsilon, \tag{4.26}$$

for some constant $C > 0$. It also follows from classical estimates that for a Lipschitz–continuous final condition g:

$$w^\varepsilon \text{ is bounded, Lipschitz in } x, \text{ and } 1/2\text{–Hölder continuous in } t. \tag{4.27}$$

Let $\rho(t,x)$ be a C^∞ positive function supported in $\{(t,x) : t \in [0,1], |x| \leq 1\}$ with unit mass, and define

$$\underline{u}^\varepsilon(t,x) := w^\varepsilon * \rho^\varepsilon \quad \text{where} \quad \rho^\varepsilon(t,x) := \frac{1}{\varepsilon^{d+2}}\rho\left(\frac{t}{\varepsilon^2}, \frac{x}{\varepsilon}\right) \tag{4.28}$$

so that it follows from (4.27) that

$$\underline{u}^\varepsilon \quad \text{is } C^\infty, \text{ and} \quad \left|\partial_t^{\beta_0} D^\beta \underline{u}^\varepsilon\right| \leq C\varepsilon^{1-2\beta_0-|\beta|_1} \tag{4.29}$$

for any $(\beta_0, \beta) \in \mathbb{N} \times \mathbb{N}^d \setminus \{0\}$, where $|\beta|_1 := \sum_{i=1}^d \beta_i$, and $C > 0$ is some constant. Then, it follows from two successive applications of Itô's formula that

$$\underline{R}_h[\underline{u}^\varepsilon] \quad \leq \quad R(h,\varepsilon) \; := \; Ch\varepsilon^{-3} \tag{4.30}$$

for some constant C. Finally, by the concavity of F, the function $\underline{u}^\varepsilon$ inherits (4.26):

$$\underline{u}^\varepsilon \quad \text{is a viscosity subsolution of and} \quad |\underline{u}^\varepsilon - v| \leq C(\varepsilon) := C\varepsilon. \tag{4.31}$$

Since $\underline{u}^\varepsilon$ satisfies all the requirements of the previous section, we deduce that

$$v - v^h \quad \leq \quad \inf_{\varepsilon>0} C\left(\varepsilon + h\varepsilon^{-3}\right) \; \sim \; Ch^{1/4}. \tag{4.32}$$

The upper bound To obtain an upper bound on the error, we rely on the switching system method of Barles and Jakobsen [6] who derive an approximate supersolution in the context Hamilton-Jacobi-Bellman equations under the stronger condition:

Assumption HJB+ *The nonlinearity F satisfies* HJB, *and for any $\delta > 0$, there exists a finite set $\{\alpha_i\}_{i=1}^{M_\delta}$ such that for any $\alpha \in \mathcal{A}$:*

$$\inf_{1\leq i\leq M_\delta} |\sigma^\alpha - \sigma^{\alpha_i}|_\infty + |b^\alpha - b^{\alpha_i}|_\infty + |c^\alpha - c^{\alpha_i}|_\infty + |f^\alpha - f^{\alpha_i}|_\infty \;\leq\; \delta.$$

However, in the present context, the difference from the approximating supersolution to the solution is of $\mathbb{L}^\infty$–order $C\varepsilon^{1/3}$. Then, we obtain a lower bound on the error:

$$v - v^h \quad \geq \quad -C\inf_{\varepsilon}\left(\varepsilon^{1/3} + R(h,\varepsilon)\right) \;=\; -Ch^{1/10}.$$

The main result Summing up the above results, we have

Theorem 4.3.2. *Let the nonlinearity F be as in Assumption* HJB+. *Then, for any bounded Lipschitz final condition g:*

$$-Ch^{1/10} \leq v - v^h \leq Ch^{1/4}.$$

Remark 4.3.3. In the PDE Finite Differences literature, the rate of convergence is usually stated in terms of the discretisation in the space variable $|\Delta x|$. In our context of stochastic differential equation, notice that $|\Delta x|$ is or the order of $h^{1/2}$. Therefore, the above upper bound on the rate of convergence corresponds to the classical rate $|\Delta x|^{1/2}$.

Bibliography

[1] F. Antonelli and A. Kohatsu-Higa, *Filtration stability of backward SDE's*, Stochastic Analysis and Its Applications **18** (2000), pp. 11–37.

[2] V. Bally, *An approximation scheme for BSDEs and applications to control and nonlinear PDE's*, Pitman Research Notes in Mathematics Series, Longman **364** (1997).

[3] V. Bally and G. Pagès, *A quantization algorithm for solving discrete time multidimensional optimal stopping problems*, Bernoulli **9** (2002), pp. 1003–1049.

[4] V. Bally, G. Pagès, and J. Printems, *A quantization method for pricing and hedging multi-dimensional American style options*, Mathematical Finance **15** (2005), pp. 119–168.

[5] G. Barles, R. Buckdahn, and E. Pardoux, *Backward stochastic differential equations and integral-partial differential equations*, Stochastics and Stochastics Reports **60** (1997), pp. 57–83.

[6] G. Barles and E. R. Jakobsen, *On the convergence rate of approximation schemes for Hamilton-Jacobi-Bellman equations*, Mathematical Modelling and Numerical Analysis, ESAIM, M2AM **36** (2002), pp. 33–54.

[7] G. Barles and E. R. Jakobsen, *Error bounds For monotone approximation schemes for Hamilton-Jacobi-Bellman equations*, SIAM J. Numer. Anal. **43** (2005), pp. 540–558.

[8] G. Barles and E. R. Jakobsen, *Error bounds For monotone approximation schemes for parabolic Hamilton-Jacobi-Bellman equations*, Math. Comp. **76** (2007), pp. 1861–1893.

[9] G. Barles and P. E. Souganidis, *Convergence of approximation schemes for fully non-linear second order equation*, Asymptotic Analysis **4** (1991), pp. 271–283.

[10] D. Becherer, *Bounded solutions to Backward SDE's with Jumps for Utility Optimization and Indifference Hedging*, Annals of Applied Probability **16** (2006), pp. 2027-2054.

[11] C. Bender and J. Zhang, *Time discretization and markovian iteration for coupled FBSDEs*, Ann. Appl. Probab. **18** (2008), pp. 143–177 .

[12] J. M. Bismut, *Conjugate convex functions in optimal stochastic control*, J. Math. Anal. Appl. **44** (1973), pp. 384–404.

[13] F. Bonnans, *Quelques aspects numériques de la commande optimale stochastique*, Lecture Notes, 2009.

[14] F. Bonnans and H. Zidani, *Consistency of generalized finite difference schemes for the stochastic HJB equation*, SIAM J. Numerical Analysis **41** (2003), pp. 1008–1021.

[15] B. Bouchard, *A stochastic target formulation for optimal switching problems in finite horizon*, Stochastics **81** (2009) pp. 171-197.

[16] B. Bouchard and J.-F. Chassagneux, *Discrete time approximation for continuously and discretely reflected BSDEs*, Stochastic Processes and their Applications **118** (2008), pp. 2269–2293.

[17] B. Bouchard, I. Ekeland, and N. Touzi, *On the Malliavin approach to Monte Carlo approximation of conditional expectations*, Finance and Stochastics **8** (2004), pp. 45–71.

[18] B. Bouchard and R. Elie, *Discrete time approximation of decoupled forward-backward SDE with jumps*, Stochastic Processes And Their Applications **118** (2008), pp. 53–75

[19] B. Bouchard and S. Menozzi, *Strong approximation of BSDEs in a domain*, to appear in Bernoulli.

[20] B. Bouchard and N. Touzi, *Discrete-time approximation and Monte-Carlo simulation of backward stochastic differential equations*, Stochastic Processes And Their Applications **111** (2004), pp. 175–206.

[21] P. Briand, B. Delyon, and J. Mémin, *Donsker-type theorem for BSDE's*, Electronic Communications In Probability **6** (2001), pp. 1–14.

[22] J. F. Carrière, *Valuation of the early-exercise price for options using simulations and nonparametric regression*, Insurance : Mathematics And Economics **19** (1996), pp. 19–30.

[23] J.-F. Chassagneux, *Discrete-time approximation of doubly reflected BSDEs*, Advances in Applied Probability **41** (2009).

[24] J.-F. Chassagneux, *Processus réfléchis en finance et probabilités numérique*, PhD Thesis, University Paris-Diderot, 2008.

[25] P. Cheridito, H. M. Soner, N. Touzi, and N. Victoir, *Second order backward stochastic differential equations and fully non-linear parabolic PDEs*, Communications On Pure And Applied Mathematics **60** (2007), pp. 1081–1110.

[26] D. Chevance, *Numerical methods for backward stochastic differential equations*, in Numerical methods in finance 1997, Edt L.C.G. Rogers and D. Talay, Cambridge University Press, pp. 232–244.

[27] E. Clément, D. Lamberton, and P. Protter, *An analysis of a least squares regression method for American option pricing*, Finance And Stochastics **6** (2002), pp. 449–472.

[28] F. Coquet, J. Mémin, and L. Slominski, *On weak convergence of filtrations*, Séminaire De Probabilités, XXXV, pp. 306–328. Lecture Notes in Math. 1755 (2001) Springer, Berlin.

[29] F. Coquet, V. Mackevičius, and J. Mémin, *Stability in D of martingales and backward equations under discretization of filtration*, Stochastic Processes And Their Applications **75** (1998), pp. 235–248.

[30] R. W. R. Darling and E. Pardoux, *BSDE with random terminal time*, Annals Of Probability **25** (1997), pp. 1135–1159.

[31] F. Delarue and S. Menozzi, *A forward backward algorithm for quasi-linear PDEs*, Annals Of Applied Probability **16** (2006), pp. 140–184.

[32] F. Delarue and S. Menozzi, *An interpolated stochastic algorithm for quasi-linear PDEs*, Mathematics Of Computation **261** (2008), pp. 125–158.

[33] H. Dong, N. V. Krylov, *On the rate of convergence of finite-difference approximations for bellman's equations with constant coefficients*, St. Petersburg Math. J. **17** (2005), pp. 108–132.

[34] J. Jr. Douglas, J. Ma, and P. Protter, *Numerical methods for forward- backward stochastic differential equations*, Annals Of Applied Probability **6** (1996), pp. 940–968.

[35] R. Elie, *Contrôle stochastique et méthodes numériques en finance mathématique*, PHD thesis, University Paris-Dauphine, 2006.

[36] A. Eyraud-Loisel, *Backward stochastic differential equations with enlarged filtration. option hedging of an insider trader in a financial market with jumps*, Stochastic Processes And Their Applications **115** (2005), pp. 1745–1763,

[37] A. Fahim, N. Touzi, and X. Warin, *A probabilistic numerical scheme for fully nonlinear PDEs*, preprint, 2009.

[38] E. Fournié, J.-M. Lasry, J. Lebuchoux, P.-L. Lions, and N. Touzi, *Applications of Malliavin calculus to Monte Carlo methods in finance*, Finance And Stochastics **3** (1999), pp. 391–412.

[39] E. Gobet, *Schéma d'Euler pour diffusions tuées. Application aux options barrière*, Phd Thesis, University Paris-Diderot, 1998.

[40] E. Gobet, *Weak approximation of killed diffusion using Euler schemes*, Stochastic Processes And Their Applications **87** (2000), pp. 167–197.

[41] E. Gobet, *Revisiting the Greeks for European and American options*, Proceedings of the "International Symposium on Stochastic Processes and Mathematical Finance" at Ritsumeikan University, Kusatsu, Japan, March 2003. Edited by J. Akahori, S. Ogawa, S. Watanabe. World Scientific, pp. 53–71, 2004.

[42] E. Gobet and C. Labart, *Error expansion for the discretization of backward stochastic differential equations*, Stochastic Processes And Applications **117** (2007), pp. 803–829.

[43] E. Gobet, J. P. Lemor, and X. Warin, *Rate of convergence of empirical regression method for solving generalized BSDE*, Bernoulli **12** (2006), pp. 889–916.

[44] E. Gobet and A. Makhlouf, *L^2-time regularity of BSDEs with irregular terminal functions*, preprint 2008.

[45] E. Gobet and S. Menozzi, *Stopped diffusion processes: overshoots and boundary correction*, preprint PMA, University Paris-Diderot, 2007.

[46] Y. Hu, P. Imkeller, and M. Müller, *Utility maximization in incomplete markets*, Ann. Appl. Probab. **15** (2005), pp. 1691–1712.

[47] N. El Karoui, C. Kapoudjan, E. Pardoux, S. Peng, and M.C. Quenez, *Reflected solutions of backward stochastic differential equations and related obstacle problems for PDE's*, Annals of Probability **25** (1997), pp. 702–737.

[48] N. El Karoui, S. Pend and M.-C. Quenez, *Backward stochastic differential equations in finance*, Mathematical Finance **7** (1997), pp. 1–71.

[49] N. El Karoui, S. Peng, and M.-C. Quenez, *A dynamic maximum principle for the optimization of recursive utilities under constraints*, The Annals Of Applied Probability **11** (2001), pp. 664–693

[50] N. El Karoui and R. Rouge, *Pricing via utility maximization and entropy*, Mathematical Finance **10** (2000), pp. 259–276.

[51] I. Kharroubi, H. Pham, J. Ma, and J. Zhang, *Backward SDEs with constrained jumps and quasi-variational inequalities*, preprint 2008.

[52] N. V. Krylov, *On the rate of convergence of finite-difference approximations for bellman's equations*, St. Petersburg Math. J. **9** (1997), pp. 245–256.

[53] N. V. Krylov, *On the rate of convergence of finite-difference approximations for bellman's equations with variable coefficients*, Probability Theory And Related Fields **117** (2000), pp. 1–16.

[54] N. V. Krylov, *On the rate of convergence of finite-difference approximations for Bellman's equations with Lipschitz coefficients*, Applied Math. And Optimization **52** (2005), pp. 365–399.

[55] P.-L. Lions and H. Regnier, *Calcul du prix et des sensibilités d'une option américaine par une méthode de Monte Carlo*, preprint 2001.

[56] F. A. Longstaff and R. S. Schwartz, *Valuing american options by simulation : a simple least-square approach*, Review Of Financial Studies **14** (2001), pp. 113–147.

[57] J. Ma, P. Protter, J. San Martin, and S. Torres, *Numerical method for backward stochastic differential equations*, Annals Of Applied Probability **12** (2002), pp. 302–316.

[58] J. Ma, P. Protter, and J. Yong, *Solving forward-backward stochastic differential equations explicitly - a four step scheme*, Probability Theory And Related Fields **98** (1994), pp. 339–359.

[59] J. Ma and J. Zhang, *Path regularity of solutions to backward stochastic differential equations*, Probability Theory And Related Fields **122** (2002), pp. 163–190.

[60] J. Ma and J. Zhang, *Representations and regularities for solutions to BSDEs with reflections*, Stochastic Processes And Their Applications **115** (2005), pp. 539–569.

[61] D. Nualart, *The Malliavin Calculus and Related Topics*, Springer, Berlin, 1995.

[62] E. Pardoux, *Backward stochastic differential equations and viscosity solutions of semilinear parabolic and elliptic PDE's of second order*, In Stochastic Analysis and Related Topics: The Geilo Workshop 1996. L. Decreusefond, J. Gjerd, B. Oksendal, and A.S. Üstünel (eds.), Birkhäuser, pp. 79–127, 1998.

[63] E. Pardoux and S. Peng, *Adapted solution of a backward stochastic differential equation*, Systems and control letters **14** (1990), pp. 55–61.

[64] E. Pardoux and S. Peng, *Backward stochastic differential equations and quasilinear parabolic partial differential equations*, Lecture Notes In Control And Inform. Sci. **176** (1992), pp. 200–217.

[65] E. Pardoux, F. Pradeilles, and Z. Rao, *Probabilistic interpretation for a system of semilinear parabolic partial differential equations*, Ann. Inst. H. Poincare **33** (1997), pp. 467–490.

[66] S. Peng, *Probabilistic interpretation for systems of quasilinear parabolic partial differential equations*, Stochastics And Stochastics Reports **37** (1991), pp. 61–74.

[67] S. Peng, *Backward stochastic differential equations and applications to optimal control*, Appl Math Optim **27** (1993), pp. 125–144.

[68] S. Peng, *Backward stochastic differential equations, nonlinear expectations, nonlinear evaluations and risk measures*, Lecture Notes in Chinese Summer School in Mathematics Weihai, July 19–August 14, 2004.

[69] S. Peng, *G-Brownian motion and dynamic risk measure under volatility uncertainty*, arXiv:0711.2834v1, 2007.

[70] S. Rubenthaler, *Numerical simulation of the solution of a stochastic differential equation driven by a Levy process*, Stochastic Processes And Their Applications **103** (2003), pp. 311–349.

[71] A. B. Sow and E. Pardoux, *Probabilistic interpretation of a system of quasilinear parabolic PDEs*, Stochastics And Stochastics Reports **76** (2004), pp. 429–477.

[72] S. Tang and X. Li, *Necessary conditions for optimal control of stochastic systems with random jumps*, SIAM J. Control Optim. **32** (1994), pp. 1447–1475.

[73] J. Zhang, *Some fine properties of backward stochastic differential equations*, PhD thesis, Purdue University, 2001.

[74] J. Zhang, *A numerical scheme for BSDEs*, Annals of Applied Probability **14** (2004), pp. 459–488.

Author information

Bruno Bouchard, CEREMADE, Université Paris Dauphine and CREST-ENSAE, France.
Email: `bouchard@ceremade.dauphine.fr`

Romuald Elie, CEREMADE, Université Paris Dauphine and CREST-ENSAE, France.
Email: `elie@ceremade.dauphine.fr`

Nizar Touzi, CMAP, Ecole Polytechnique Paris, France.
Email: `nizar.touzi@polytechnique.edu`

Radon Series Comp. Appl. Math **8**, 125–164

Affine diffusion processes: theory and applications

Damir Filipović and Eberhard Mayerhofer

Abstract. We revisit affine diffusion processes on general and on the canonical state space in particular. A detailed study of theoretic and applied aspects of this class of Markov processes is given. In particular, we derive admissibility conditions and provide a full proof of existence and uniqueness through stochastic invariance of the canonical state space. Existence of exponential moments and the full range of validity of the affine transform formula are established. This is applied to the pricing of bond and stock options, which is illustrated for the Vasiček, Cox–Ingersoll–Ross and Heston models.

Key words. Affine processes, diffusions, bond option pricing, stochastic volatility, Riccati differential equations, differential inequalities, exponential moments.

AMS classification. 60J60, 91B70

1 Introduction

Affine Markov models have been employed in finance since decades, and they have found growing interest due to their computational tractability as well as their capability to capture empirical evidence from financial time series. Their main applications lie in the theory of term structure of interest rates, stochastic volatility option pricing and the modelling of credit risk (see [12] and the references therein). There is a vast literature on affine models. We mention here explicitly just the few articles [2, 4, 8, 10, 13, 14, 16, 20, 26, 27, 29] and [12] for a broader overview.

In this paper, we revisit the class of affine diffusion processes on subsets of $\mathbb{R}^d$ and on the canonical state space $\mathbb{R}_+^m \times \mathbb{R}^n$, in particular. In Section 2, we first provide necessary and sufficient conditions on the parameters of a diffusion process X to satisfy the affine transform formula

$$\mathbb{E}\left[e^{u^\top X(T)} \mid \mathcal{F}_t\right] = e^{\phi(T-t,\, u)+\psi(T-t,\, u)^\top X(t)}.$$

The functions ϕ and ψ in turn are given as solutions of a system of coupled Riccati equations. Arguing by stochastic invariance, in Section 3, we can further restrict the choice of admissible diffusion parameters.

Glasserman and Kim [16] showed recently that the affine transform formula holds whenever either side is well defined under the assumption of strict mean reversion. This is an extension of the findings in [12], where only sufficient conditions are given

Support from WWTF (Vienna Science and Technology Fund) gratefully acknowledged. We thank Paul Glasserman for helpful comments

in terms of analyticity of the right hand side. The strict mean reversion assumption, however, excludes the Heston stochastic volatility model. In our paper, we show that strict mean reversion is not needed (Theorem 3.3). As a by product, we obtain some non-trivial convexity results for Riccati equations. Having the full range of validity of the above transform formula under control, in Section 4, we can then proceed to pricing bond and stock options in affine models. Particular examples are the Vasiček and Cox–Ingersoll–Ross (CIR) short rate models in Section 5, and Heston's stochastic volatility model in Section 5.1.

The representation of affine short rate models bears some ambiguity with respect to linear transformations of the state process. This motivates the question whether there exists a classification method ensuring that affine short rate models with the same observable implications have a unique canonical representation. This topic has been addressed in [10, 9, 24, 8]. In Section 6, we recap this issue and show that the diffusion matrix of X can always be brought into block-diagonal form by a regular linear transform leaving the canonical state space invariant.

The existence and uniqueness question of the relevant stochastic differential equation is completely solved through stochastic invariance and the block-diagonal transformation in Section 7. The presented proof builds on the seminal result by Yamada and Watanabe [35] and provides strong solutions of the respective SDEs. Therefore we approach the existence issue differently from [12], which uses infinite divisibility on the canonical state space and the Markov semigroup theory and derives weak solutions only.

In the appendix, we provide some self contained proofs of existence and comparison statements for relevant systems of Riccati equations (Section B). Moreover, some moment lemmas from [12] in a more elaborated fashion can be found in Section A.

2 Definition and characterisation of affine processes

Fix a dimension $d \geq 1$ and a closed state space $\mathcal{X} \subset \mathbb{R}^d$ with non-empty interior. We let $b : \mathcal{X} \to \mathbb{R}^d$ be continuous, and $\rho : \mathcal{X} \to \mathbb{R}^{d\times d}$ be measurable and such that the diffusion matrix

$$a(x) = \rho(x)\rho(x)^\top$$

is continuous in $x \in \mathcal{X}$. Let W denote a d-dimensional Brownian motion defined on a filtered probability space $(\Omega, \mathcal{F}, (\mathcal{F}_t), \mathbb{P})$. Throughout, we assume that for every $x \in \mathcal{X}$ there exists a unique solution [1] $X = X^x$ of the stochastic differential equation

$$dX(t) = b(X(t))\,dt + \rho(X(t))\,dW(t), \quad X(0) = x. \tag{2.1}$$

Definition 2.1. We call X *affine* if the $\mathcal{F}_t$-conditional characteristic function of $X(T)$ is exponential affine in $X(t)$, for all $t \leq T$. That is, there exist $\mathbb{C}$- and $\mathbb{C}^d$-valued functions $\phi(t, u)$ and $\psi(t, u)$, respectively, with jointly continuous t-derivatives such

[1] If not otherwise stated, a solution of (2.1) is understood as a strong solution.

that $X = X^x$ satisfies

$$\mathbb{E}\left[e^{u^\top X(T)} \mid \mathcal{F}_t\right] = e^{\phi(T-t,\,u)+\psi(T-t,\,u)^\top X(t)} \tag{2.2}$$

for all $u \in \mathrm{i}\mathbb{R}^d$, $t \le T$ and $x \in \mathcal{X}$.

Since the conditional characteristic function is bounded by one, the real part of the exponent $\phi(T-t,u) + \psi(T-t,u)^\top X(t)$ in (2.2) has to be negative. Note that $\phi(t,u)$ and $\psi(t,u)$ for $t \ge 0$ and $u \in \mathrm{i}\mathbb{R}^d$ are uniquely[2] determined by (2.2), and satisfy the initial conditions $\phi(0,u) = 0$ and $\psi(0,u) = u$, in particular.

We first derive necessary and sufficient conditions for X to be affine.

Theorem 2.2. *Suppose X is affine. Then the diffusion matrix $a(x)$ and drift $b(x)$ are affine in x. That is,*

$$\begin{aligned} a(x) &= a + \sum_{i=1}^d x_i \alpha_i \\ b(x) &= b + \sum_{i=1}^d x_i \beta_i = b + \mathcal{B}x \end{aligned} \tag{2.3}$$

for some $d \times d$-matrices a and α_i, and d-vectors b and β_i, where we denote by

$$\mathcal{B} = (\beta_1, \ldots, \beta_d)$$

the $d \times d$-matrix with ith column vector β_i, $1 \le i \le d$. Moreover, ϕ and $\psi = (\psi_1, \ldots, \psi_d)^\top$ solve the system of Riccati equations

$$\begin{aligned} \partial_t \phi(t,u) &= \frac{1}{2}\psi(t,u)^\top a\ \psi(t,u) + b^\top \psi(t,u) \\ \phi(0,u) &= 0 \\ \partial_t \psi_i(t,u) &= \frac{1}{2}\psi(t,u)^\top \alpha_i\, \psi(t,u) + \beta_i^\top \psi(t,u), \quad 1 \le i \le d, \\ \psi(0,u) &= u. \end{aligned} \tag{2.4}$$

In particular, ϕ is determined by ψ via simple integration:

$$\phi(t,u) = \int_0^t \left(\frac{1}{2}\psi(s,u)^\top a\ \psi(s,u) + b^\top \psi(s,u)\right) ds.$$

Conversely, suppose the diffusion matrix $a(x)$ and drift $b(x)$ are affine of the form (2.3) *and suppose there exists a solution (ϕ, ψ) of the Riccati equations* (2.4) *such that $\phi(t,u) + \psi(t,u)^\top x$ has negative real part for all $t \ge 0$, $u \in \mathrm{i}\mathbb{R}^d$ and $x \in \mathcal{X}$. Then X is affine with conditional characteristic function* (2.2).

[2] In fact, $\phi(t,u)$ may be altered by multiples of $2\pi\mathrm{i}$. We uniquely fix the continuous function $\phi(t,u)$ by the initial condition $\phi(0,u) = 0$.

Proof. Suppose X is affine. For $T > 0$ and $u \in \mathrm{i}\mathbb{R}^d$ define the complex-valued Itô process

$$M(t) = \mathrm{e}^{\phi(T-t,u)+\psi(T-t,u)^\top X(t)}.$$

We can apply Itô's formula, separately to real and imaginary part of M, and obtain

$$dM(t) = I(t)\,dt + \psi(T-t,u)^\top \rho(X(t))\,dW(t), \quad t \le T,$$

with

$$\begin{aligned} I(t) &= -\partial_T \phi(T-t,u) - \partial_T \psi(T-t,u)^\top X(t) \\ &\quad + \psi(T-t,u)^\top b(X(t)) + \frac{1}{2}\psi(T-t,u)^\top a(X(t))\,\psi(T-t,u). \end{aligned}$$

Since M is a martingale, we have $I(t) = 0$ for all $t \le T$ a.s. Letting $t \to 0$, by continuity of the parameters, we thus obtain

$$\partial_T \phi(T,u) + \partial_T \psi(T,u)^\top x = \psi(T,u)^\top b(x) + \frac{1}{2}\psi(T,u)^\top a(x)\,\psi(T,u)$$

for all $x \in \mathcal{X}$, $T \ge 0$, $u \in \mathrm{i}\mathbb{R}^d$. Since $\psi(0,u) = u$, this implies that a and b are affine of the form (2.3). Plugging this back into the above equation and separating first order terms in x yields (2.4).

Conversely, suppose a and b are of the form (2.3). Let (ϕ, ψ) be a solution of the Riccati equations (2.4) such that $\phi(t,u)+\psi(t,u)^\top x$ has negative real part for all $t \ge 0$, $u \in \mathrm{i}\mathbb{R}^d$ and $x \in \mathcal{X}$. Then M, defined as above, is a uniformly bounded local martingale, and hence a martingale, with $M(T) = \mathrm{e}^{u^\top X(T)}$. Therefore $\mathbb{E}[M(T) \mid \mathcal{F}_t] = M(t)$, for all $t \le T$, which is (2.2), and the theorem is proved. □

We now recall an important global existence, uniqueness and regularity result for the above Riccati equations. We let K be a placeholder for either $\mathbb{R}$ or $\mathbb{C}$.

Lemma 2.3. *Let a and α_i be real $d \times d$-matrices, and b and β_i be real d-vectors, $1 \le i \le d$.*

1. *For every $u \in K^d$, there exists some $t_+(u) \in (0,\infty]$ such that there exists a unique solution $(\phi(\cdot,u), \psi(\cdot,u)) : [0, t_+(u)) \to K \times K^d$ of the Riccati equations* (2.4). *In particular, $t_+(0) = \infty$.*
2. *The domain*
$$\mathcal{D}_K = \{(t,u) \in \mathbb{R}_+ \times K^d \mid t < t_+(u)\}$$
is open in $\mathbb{R}_+ \times K^d$ and maximal in the sense that for all $u \in K^d$ either $t_+(u) = \infty$ or $\lim_{t \uparrow t_+(u)} \|\psi(t,u)\| = \infty$, respectively, .
3. *For every $t \in \mathbb{R}_+$, the t-section*
$$\mathcal{D}_K(t) = \{u \in K^d \mid (t,u) \in \mathcal{D}_K\}$$
is an open neighbourhood of 0 in K^d. Moreover, $\mathcal{D}_K(0) = K^d$ and $\mathcal{D}_K(t_1) \supseteq \mathcal{D}_K(t_2)$ for $0 \le t_1 \le t_2$.

4. *ϕ and ψ are analytic functions on $\mathcal{D}_K$.*
5. $\mathcal{D}_{\mathbb{R}} = \mathcal{D}_{\mathbb{C}} \cap (\mathbb{R}_+ \times \mathbb{R}^d)$.

Henceforth, we shall call $\mathcal{D}_K$ the maximal domain for equation (2.4).

Proof. Since the right-hand side of (2.4) is formed by analytic functions in ψ on K^d, part 1 follows from the basic theorems for ordinary differential equations, e.g. [1, Theorem 7.4]. In particular, $t_+(0) = \infty$ since $(\phi(\cdot, 0), \psi(\cdot, 0)) \equiv 0$ is the unique solution of (2.4) for $u = 0$. It is proved in [1, Theorems 7.6 and 8.3] that $\mathcal{D}_K$ is maximal and open, which is part 2. This also implies that all t-sections $\mathcal{D}_K(t)$ are open in K^d. The inclusion $\mathcal{D}_K(t_1) \supseteq \mathcal{D}_K(t_2)$ is a consequence of the maximality property from part 2. Whence part 3 follows. For a proof of part 4 see [11, Theorem 10.8.2]. Part 5 is obvious. □

We will provide in Section B below some substantial improvements of the properties stated in Lemma 2.3 for the canonical state space $\mathcal{X}$ introduced in the following section.

3 Canonical state space

There is an implicit trade off between the parameters a, α_i, b, β_i in (2.3) and the state space $\mathcal{X}$:

- a, α_i, b, β_i must be such that X does not leave the set $\mathcal{X}$, and
- a, α_i must be such that $a + \sum_{i=1}^d x_i \alpha_i$ is symmetric and positive semi-definite for all $x \in \mathcal{X}$.

To gain further explicit insight into this interplay, we now and henceforth assume that the state space is of the following canonical form

$$\mathcal{X} = \mathbb{R}_+^m \times \mathbb{R}^n$$

for some integers $m, n \geq 0$ with $m + n = d$.

Remark 3.1. This canonical state space covers most applications appearing in the finance literature. However, other choices for the state space of an affine process are possible:

1. For instance, the following example for $d = 1$ admits as state space any closed interval $\mathcal{X} \subset \mathbb{R}$ containing 0:

$$dX = -X\,dt, \quad X(0) = x \in \mathcal{X}.$$

   This degenerate diffusion process is affine, since $\mathrm{e}^{uX(T)} = \mathrm{e}^{u\mathrm{e}^{-(T-t)}X(t)}$ for all $t \leq T$ ([12], Section 12). In general, affine diffusion processes on compact state spaces have to be degenerate.
2. Matrix state-spaces S_d^+ ($d \geq 2$), the cone of symmetric positive definite matrices (see [5, 6, 15, 17, 19]).

3. Parabolic state-spaces, cf. [18], which are in turn, related to quadratic processes on the canonical state-space ([7], see also their Example 5.3, Section 5)

For the above canonical state space, we can give necessary and sufficient admissibility conditions on the parameters. The following terminology will be useful in the sequel. We define the index sets

$$I = \{1, \dots, m\} \quad \text{and} \quad J = \{m+1, \dots, m+n\}.$$

For any vector μ and matrix ν, and index sets M, N, we denote by

$$\mu_M = (\mu_i)_{i \in M}, \qquad \nu_{MN} = (\nu_{ij})_{i \in M,\, j \in N}$$

the respective sub-vector and -matrix.

Theorem 3.2. *The process X on the canonical state space $\mathbb{R}_+^m \times \mathbb{R}^n$ is affine if and only if $a(x)$ and $b(x)$ are affine of the form* (2.3) *for parameters a, α_i, b, β_i which are admissible in the following sense:*

$$\begin{aligned}
a,\ \alpha_i \ &\text{are symmetric positive semi-definite,}\\
a_{II} &= 0 \quad (\text{and thus } a_{IJ} = a_{JI}^\top = 0),\\
\alpha_j &= 0 \quad \text{for all} \quad j \in J\\
\alpha_{i,kl} = \alpha_{i,lk} &= 0 \quad \text{for } k \in I \setminus \{i\}, \text{for all } 1 \le i, l \le d,\\
b &\in \mathbb{R}_+^m \times \mathbb{R}^n,\\
\mathcal{B}_{IJ} &= 0,\\
\mathcal{B}_{II} \ &\text{has positive off-diagonal elements.}
\end{aligned} \tag{3.1}$$

In this case, the corresponding system of Riccati equations (2.4) *simplifies to*

$$\begin{aligned}
\partial_t \phi(t,u) &= \frac{1}{2}\psi_J(t,u)^\top a_{JJ}\, \psi_J(t,u) + b^\top \psi(t,u)\\
\phi(0,u) &= 0\\
\partial_t \psi_i(t,u) &= \frac{1}{2}\psi(t,u)^\top \alpha_i\, \psi(t,u) + \beta_i^\top \psi(t,u), \quad i \in I,\\
\partial_t \psi_J(t,u) &= \mathcal{B}_{JJ}^\top \psi_J(t,u),\\
\psi(0,u) &= u,
\end{aligned} \tag{3.2}$$

and there exists a unique global solution $(\phi(\cdot,u), \psi(\cdot,u)) : \mathbb{R}_+ \to \mathbb{C}_- \times \mathbb{C}_-^m \times \mathrm{i}\mathbb{R}^n$ for all initial values $u \in \mathbb{C}_-^m \times \mathrm{i}\mathbb{R}^n$. In particular, the equation for ψ_J forms an autonomous linear system with unique global solution $\psi_J(t,u) = \mathrm{e}^{\mathcal{B}_{JJ}^\top t}\, u_J$ for all $u_J \in \mathbb{C}^n$.

Before we prove the theorem, let us illustrate the admissibility conditions (3.1) for the diffusion matrix $\alpha(x)$ for dimension $d = 3$ and the corresponding cases $m = 0, 1, 2, 3$. Note that $\alpha(x) = a + \sum_{i=1}^m x_i \alpha_i$, hence in the case $m = 0$ we have

$$\alpha(x) \equiv a$$

for an arbitrary positive semi-definite symmetric 3×3-matrix a. For $m = 1$, we have

$$a = \begin{pmatrix} 0 & 0 & 0 \\ & + & * \\ & & + \end{pmatrix}, \quad \alpha_1 = \begin{pmatrix} + & * & * \\ & + & * \\ & & + \end{pmatrix},$$

for $m = 2$,

$$a = \begin{pmatrix} 0 & 0 & 0 \\ & 0 & 0 \\ & & + \end{pmatrix}, \quad \alpha_1 = \begin{pmatrix} + & 0 & * \\ & 0 & 0 \\ & & + \end{pmatrix}, \quad \alpha_2 = \begin{pmatrix} 0 & 0 & 0 \\ & + & * \\ & & + \end{pmatrix},$$

and for $m = 3$,

$$a = 0, \quad \alpha_1 = \begin{pmatrix} + & 0 & 0 \\ & 0 & 0 \\ & & 0 \end{pmatrix}, \quad \alpha_2 = \begin{pmatrix} 0 & 0 & 0 \\ & + & 0 \\ & & 0 \end{pmatrix}, \quad \alpha_3 = \begin{pmatrix} 0 & 0 & 0 \\ & 0 & 0 \\ & & + \end{pmatrix},$$

where we leave the lower triangle of symmetric matrices blank, $+$ denotes a non-negative real number and $*$ any real number such that positive semi-definiteness holds.

We continue with the proof of Theorem 3.2:

Proof. Suppose X is affine. That $a(x)$ and $b(x)$ are of the form (2.3) follows from Theorem 2.2. Obviously, $a(x)$ is symmetric positive semi-definite for all $x \in \mathbb{R}^m_+ \times \mathbb{R}^n$ if and only if $\alpha_j = 0$ for all $j \in J$, and a and α_i are symmetric positive semi-definite for all $i \in I$.

We extend the diffusion matrix and drift continuously to $\mathbb{R}^d$ by setting

$$a(x) = a + \sum_{i \in I} x_i^+ \alpha_i \quad \text{and} \quad b(x) = b + \sum_{i \in I} x_i^+ \beta_i + \sum_{j \in J} x_j \beta_j.$$

Now let x be a boundary point of $\mathbb{R}^m_+ \times \mathbb{R}^n$. That is, $x_k = 0$ for some $k \in I$. The stochastic invariance Lemma B.1 below implies that the diffusion must be "parallel to the boundary",

$$e_k^\top \Big(a + \sum_{i \in I \setminus \{k\}} x_i \alpha_i \Big) e_k = 0,$$

and the drift must be "inward pointing",

$$e_k^\top \Big(b + \sum_{i \in I \setminus \{k\}} x_i \beta_i + \sum_{j \in J} x_j \beta_j \Big) \geq 0.$$

Since this has to hold for all $x_i \geq 0$, $i \in I \setminus \{k\}$, and $x_j \in \mathbb{R}$, $j \in J$, we obtain the following set of admissibility conditions

$$
\begin{aligned}
a,\ \alpha_i &\ \text{are symmetric positive semi-definite,} \\
a\, e_k &= 0 \quad \text{for all } k \in I, \\
\alpha_i\, e_k &= 0 \quad \text{for all } i \in I \setminus \{k\}, \text{ for all } k \in I, \\
\alpha_j &= 0 \quad \text{for all } j \in J, \\
b &\in \mathbb{R}^m_+ \times \mathbb{R}^n, \\
\beta_i^\top e_k &\geq 0 \quad \text{for all } i \in I \setminus \{k\}, \text{ for all } k \in I, \\
\beta_j^\top e_k &= 0 \quad \text{for all } j \in J, \text{ for all } k \in I,
\end{aligned}
$$

which is equivalent to (3.1). The form of the system (3.2) follows by inspection.

Now suppose a, α_i, b, β_i satisfy the admissibility conditions (3.1). We show below that there exists a unique global solution $(\phi(\cdot, u), \psi(\cdot, u)) : \mathbb{R}_+ \to \mathbb{C}_- \times \mathbb{C}^m_- \times \mathrm{i}\mathbb{R}^n$ of (3.2), for all $u \in \mathbb{C}^m_- \times \mathrm{i}\mathbb{R}^n$. In particular, $\phi(t, u) + \psi(t, u)^\top x$ has negative real part for all $t \geq 0$, $u \in \mathrm{i}\mathbb{R}^d$ and $x \in \mathbb{R}^m_+ \times \mathbb{R}^n$. Thus the first part of the theorem follows from Theorem 2.2.

As for the global existence and uniqueness statement, in view of Lemma 2.3, it remains to show that $\psi(t, u)$ is $\mathbb{C}^m_- \times \mathrm{i}\mathbb{R}^n$-valued and $t_+(u) = \infty$ for all $u \in \mathbb{C}^m_- \times \mathrm{i}\mathbb{R}^n$. For $i \in I$, denote the right-hand side of the equation for ψ_i by

$$R_i(u) = \frac{1}{2} u^\top \alpha_i\, u + \beta_i^\top u,$$

and observe that

$$\operatorname{Re} R_i(u) = \frac{1}{2} \operatorname{Re} u^\top \alpha_i \operatorname{Re} u - \frac{1}{2} \operatorname{Im} u^\top \alpha_i \operatorname{Im} u + \beta_i^\top \operatorname{Re} u.$$

Let us denote $x_I^+ = (x_1^+, \dots, x_m^+)^\top$. Since $\operatorname{Re} \psi_J(t, u) = 0$, it follows from the admissibility conditions (3.1) and Corollary B.2 below, setting $f(t) = -\operatorname{Re} \psi(t, u)$,

$$b_i(t, x) = -\frac{1}{2} \alpha_{i,ii} \left(x_i^+\right)^2 + \frac{1}{2} \operatorname{Im} \psi(t, u)^\top \alpha_i \operatorname{Im} \psi(t, u) + \beta_{i,I}^\top x_I^+, \quad i \in I,$$

and $b_j(t, x) = 0$ for $j \in J$, that the solution $\psi(t, u)$ of (3.2) has to take values in $\mathbb{C}^m_- \times \mathrm{i}\mathbb{R}^n$ for all initial points $u \in \mathbb{C}^m_- \times \mathrm{i}\mathbb{R}^n$.

Further, for $i \in I$ and $u \in \mathbb{C}^d$, one verifies that

$$
\begin{aligned}
\operatorname{Re}\left(\overline{u_i} R_i(u)\right) &= \frac{1}{2} \alpha_{i,ii} |u_i|^2 \operatorname{Re} u_i + \operatorname{Re}\left(\overline{u_i}\, u_i \alpha_{i,iJ}\, u_J\right) \\
&\quad + \frac{1}{2} \operatorname{Re}\left(\overline{u_i}\, u_J^\top \alpha_{i,JJ}\, u_J\right) + \operatorname{Re}\left(\overline{u_i}\, \beta_i^\top u\right) \\
&\leq \frac{K}{2} \left(1 + \|(\operatorname{Re} u_I)^+\| + \|u_J\|^2\right) \left(1 + \|u_I\|^2\right)
\end{aligned}
$$

for some finite constant K which does not depend on u. We thus obtain

$$
\begin{aligned}
\partial_t \|\psi_I(t, u)\|^2 &= 2 \operatorname{Re}\left(\overline{\psi_I(t, u)}^\top R_I\left(\psi_I(t, u), \mathrm{e}^{\mathcal{B}_{JJ}^\top t} u_J\right)\right) \\
&\leq K g(t) \left(1 + \|\psi_I(t, u)\|^2\right)
\end{aligned}
$$

for

$$g(t) = \left(1 + \|(\operatorname{Re}\psi_I(t,u))^+\| + \|e^{\mathcal{B}_{JJ}^\top t} u_J\|^2\right).$$

Gronwall's inequality ([11, (10.5.1.3)]), applied to $(1 + \|\psi_I(t,u)\|^2)$, yields

$$\|\psi_I(t,u)\|^2 \le \|u_I\|^2 + K\left(1 + \|u_I\|^2\right) \int_0^t g(s) e^{K \int_s^t g(\xi)\, d\xi}\, ds. \tag{3.3}$$

From above, for all initial points $u \in \mathbb{C}_-^m \times i\mathbb{R}^n$, we know that $(\operatorname{Re}\psi_I(t,u))^+ = 0$ and therefore $t_+(u) = \infty$ by (3.3). Hence the theorem is proved. □

Now suppose X is affine with characteristics (2.3) satisfying the admissibility conditions (3.1). In what follows we show that not only can the functions $\phi(t,u)$ and $\psi(t,u)$ be extended beyond $u \in i\mathbb{R}^d$, but also the validity of the affine transform formula (2.2) carries over. This asserts exponential moments of $X(t)$ in particular and will prove most useful for deriving pricing formulas in affine factor models.

For any set $U \subset \mathbb{R}^k$ ($k \in \mathbb{N}$), we define the strip

$$\mathcal{S}(U) = \left\{ z \in \mathbb{C}^k \mid \operatorname{Re} z \in U \right\}$$

in $\mathbb{C}^k$. The proof of the following theorem builds on results that are derived in Sections A and B below.

Theorem 3.3. *Suppose X is affine with admissible parameters as given in* (3.1). *Let $\tau > 0$. Then*

1. $\mathcal{S}(\mathcal{D}_\mathbb{R}(\tau)) \subset \mathcal{D}_\mathbb{C}(\tau)$
2. $\mathcal{D}_\mathbb{R}(\tau) = M(\tau)$ *where*

$$M(\tau) = \left\{ u \in \mathbb{R}^d \mid \mathbb{E}\left[e^{u^\top X^x(\tau)}\right] < \infty \text{ for all } x \in \mathbb{R}_+^m \times \mathbb{R}^n \right\}.$$

3. $\mathcal{D}_\mathbb{R}(\tau)$ *and* $\mathcal{D}_\mathbb{R}$ *are convex sets.*

   Moreover, for all $0 \le t \le T$ *and* $x \in \mathbb{R}_+^m \times \mathbb{R}^n$,
4. (2.2) *holds for all* $u \in \mathcal{S}(\mathcal{D}_\mathbb{R}(T-t))$
5. (2.2) *holds for all* $u \in \mathbb{C}_-^m \times i\mathbb{R}^n$
6. $M(t) \supseteq M(T)$.

Proof. We first claim that, for every $u \in \mathbb{C}^d$ with $t_+(u) < \infty$, there exists some $i \in I$ and some sequence $t_n \uparrow t_+(u)$ such that

$$\lim_n (\operatorname{Re}\psi_i(t_n,u))^+ = \infty. \tag{3.4}$$

Indeed, otherwise we would have $\sup_{t\in[0,t_+(u))} \|(\operatorname{Re}\psi_I(t,u))^+\| < \infty$. But then (3.3) would imply $\sup_{t\in[0,t_+(u))} \|\psi_I(t,u)\| < \infty$, which is absurd. Whence (3.4) is proved.

In the following, we write

$$G(u,t,x) = \mathbb{E}\left[e^{u^\top X^x(t)}\right], \quad V(t,x) = \left\{ u \in \mathbb{R}^d \mid G(u,t,x) < \infty \right\}.$$

Since X is affine, by definition we have $\mathbb{R}_+ \times \mathrm{i}\mathbb{R}^d \subset \mathcal{D}_\mathbb{C}$ and (2.2) implies

$$G(u,t,x) = \mathrm{e}^{\phi(t,u)+\psi(t,u)^\top x} \tag{3.5}$$

for all $u \in \mathrm{i}\mathbb{R}^d$, $t \in \mathbb{R}_+$ and $x \in \mathbb{R}_+^m \times \mathbb{R}^n$. Moreover, by Lemma B.5, $\mathcal{D}_\mathbb{R}(t) = \mathcal{D}_\mathbb{C}(t) \cap \mathbb{R}^d$ is open and star-shaped around 0 in $\mathbb{R}^d$. Hence Lemma A.3 implies that $\mathcal{D}_\mathbb{R}(t) \subset V(t,x)$ and (3.5) holds for all $u \in \mathcal{D}_\mathbb{C}(t) \cap \mathcal{S}(\mathcal{D}_\mathbb{R}(t))$, for all $x \in \mathbb{R}_+^m \times \mathbb{R}^n$ and $t \in [0,\tau]$.

Now let $u \in \mathcal{D}_\mathbb{R}(\tau)$ and $v \in \mathbb{R}^d$, and define

$$\theta^* = \inf\{\theta \in \mathbb{R}_+ \mid u + \mathrm{i}\theta v \notin \mathcal{D}_\mathbb{C}(\tau)\}.$$

We claim that $\theta^* = \infty$. Arguing by contradiction, assume that $\theta^* < \infty$. Since $\mathcal{D}_\mathbb{C}(\tau)$ is open, this implies $u + \mathrm{i}\theta^* v \notin \mathcal{D}_\mathbb{C}(\tau)$, and thus

$$t_+(u + \mathrm{i}\theta^* v) \le \tau. \tag{3.6}$$

On the other hand, since $\mathcal{D}_\mathbb{R}(\tau)$ is open, $(1+\varepsilon)u \in \mathcal{D}_\mathbb{R}(\tau)$ for some $\varepsilon > 0$. Hence (3.5) holds and $G(t,(1+\varepsilon)u,x)$ is uniformly bounded in $t \in [0,\tau]$, by continuity of $\phi(t,(1+\varepsilon)u)$ and $\psi(t,(1+\varepsilon)u)$ in t. We infer that the class of random variables $\{\mathrm{e}^{(u+\mathrm{i}\theta^* v)^\top X(t)} \mid t \in [0,\tau]\}$ is uniformly integrable, see [34, 13.3]. Since $X(t)$ is continuous in t, we conclude by Lebesgue's convergence theorem that $G(t, u+\mathrm{i}\theta^* v, x)$ is continuous in $t \in [0,\tau]$, for all $x \in \mathbb{R}_+^m \times \mathbb{R}^n$. But for all $t < t_+(u + \mathrm{i}\theta^* v)$ we have $(t, u+\mathrm{i}\theta^* v) \in \mathcal{D}_\mathbb{C}(t) \cap \mathcal{S}(\mathcal{D}_\mathbb{R}(t))$, and thus (3.5) holds for all $x \in \mathbb{R}_+^m \times \mathbb{R}^n$. In view of (3.4), this contradicts (3.6). Whence $\theta^* = \infty$ and thus $u + \mathrm{i}v \in \mathcal{D}_\mathbb{C}(\tau)$. This proves 1[1].

Applying the above arguments to[2] $\mathbb{E}\big[\mathrm{e}^{u^\top X(T)} \mid \mathcal{F}_t\big] = G(T-t, u, X(t))$ with $T = t + \tau$ yields 4. Part 5 follows, since, by Theorem 3.2, $\mathbb{C}_-^m \times \mathrm{i}\mathbb{R}^n \subset \mathcal{S}(\mathcal{D}_\mathbb{R}(t))$ for all $t \in \mathbb{R}_+$.

As for 2, we first let $u \in \mathcal{D}_\mathbb{R}(\tau)$. From part 4 it follows that $u \in M(\tau)$. Conversely, let $u \in M(\tau)$, and define $\theta^* = \sup\{\theta \ge 0 \mid \theta u \in \mathcal{D}_\mathbb{R}(\tau)\}$. We have to show that $\theta^* > 1$. Assume, by contradiction, that $\theta^* \le 1$. From Lemma B.5, we know that there exists some $x^* \in \mathbb{R}_+^m \times \mathbb{R}^n$ such that

$$\lim_{\theta \uparrow \theta^*} \phi(\tau,\theta u) + \psi(\tau,\theta u)^\top x^* = \infty. \tag{3.7}$$

On the other hand, from part 4 and Jensen's inequality, we obtain

$$\mathrm{e}^{\phi(\tau,\theta u)+\psi(\tau,\theta u)^\top x^*} = G(\tau,\theta u,x^*) \le G(\tau,u,x^*)^\theta \le G(\tau,u,x^*) < \infty$$

for all $\theta < \theta^*$. But this contradicts (3.7), hence $u \in \mathcal{D}_\mathbb{R}(\tau)$, and part 2 is proved. Since $M(\tau)$ is convex, this also implies 3. Finally, part 6 follows from part 2 and Lemma 2.3. Whence the theorem is proved. □

Remark 3.4. Glasserman and Kim [16] proved the equality in Theorem 3.3 2, and the validity of the transform formula (2.2) for all u in an open neighbourhood of $\mathcal{D}_\mathbb{R}(T-t)$

[1] For an alternative proof of the above, see remark B.7

[2] Here we use the Markov property of X, see [25, Theorem 5.4.20].

in $\mathbb{C}^d$, under the additional assumption that $\mathcal{B}$ has strictly negative eigenvalues. That assumption, however, excludes the simple Heston stochastic volatility model in Section 5.1 below.

Remark 3.5. In Keller-Ressel [27, Theorem 3.18 and Lemma 3.19] it is shown that

$$M(\tau+\varepsilon) \subseteq \mathcal{D}_{\mathbb{R}}(\tau)$$

for all $\varepsilon > 0$, for a more general class of affine Markov processes X^x. Obviously, in our framework, this is implied by parts 2 and 6 of Theorem 3.3.

Remark 3.6. The convexity property of the maximal domain stated in Theorem 3.3 3 represents a non-trivial result for ordinary differential equations. Only in the mid 1990s have corresponding convexity results been derived in the analysis literature, see Lakshmikantham et al. [28].

4 Discounting and pricing in affine models

We let X be affine on the canonical state space $\mathbb{R}^m_+ \times \mathbb{R}^n$ with admissible parameters a, α_i, b, β_i as given in (3.1). Since we are interested in pricing, and to avoid a change of measure, we interpret $\mathbb{P} = \mathbb{Q}$ as risk-neutral measure in what follows.

A short rate model of the form

$$r(t) = c + \gamma^\top X(t), \tag{4.1}$$

for some constant parameters $c \in \mathbb{R}$ and $\gamma \in \mathbb{R}^d$, is called an affine short rate model. Special cases, for dimension $d = 1$, are the Vasiček and Cox–Ingersoll–Ross short rate models. We recall that an affine term structure model always induces an affine short rate model.

Now consider a T-claim with payoff $f(X(T))$. Here $f : \mathbb{R}^m_+ \times \mathbb{R}^n \to \mathbb{R}$ denotes a measurable payoff function, such that $f(X(T))$ meets the required integrability conditions

$$\mathbb{E}\big[\mathrm{e}^{-\int_0^T r(s)\,ds}\,|f(X(T))|\big] < \infty.$$

Its arbitrage price at time $t \le T$ is then given by

$$\pi(t) = \mathbb{E}\big[\mathrm{e}^{-\int_t^T r(s)\,ds} f(X(T)) \mid \mathcal{F}_t\big]. \tag{4.2}$$

A particular example is the T-bond with $f \equiv 1$. Our aim is to derive an analytic, or at least numerically tractable, pricing formula for (4.2). To this end we shall make use of a change of numeraire technique to price, e.g., Bond options and caplets. Denote the risk free bank account by $B(t) := \mathrm{e}^{\int_0^t r(s)ds}$. For fixed $T > 0$ it is easily observed that

$$\frac{1}{P(0,T)B(T)} > 0 \quad \text{and} \quad \mathbb{E}\left[\frac{1}{P(0,T)B(T)}\right] = 1,$$

hence we may introduce an equivalent probability measure $\mathbb{Q}^T \sim \mathbb{Q}$ on $\mathcal{F}_T$ by its Radon–Nikodym derivative

$$\frac{d\mathbb{Q}^T}{d\mathbb{Q}} = \frac{1}{P(0,T)B(T)}.$$

$\mathbb{Q}^T$ is called the T-forward measure. Note that for $t \leq T$,

$$\left.\frac{d\mathbb{Q}^T}{d\mathbb{Q}}\right|_{\mathcal{F}_t} = \mathbb{E}\Big[\frac{1}{P(0,T)B(T)} |\mathcal{F}_t\Big] = \frac{P(t,T)}{P(0,T)B(t)}. \tag{4.3}$$

As a first step towards establishing useful pricing formulas, we derive a formula for the $\mathcal{F}_t$-conditional characteristic function of $X(T)$ under $\mathbb{Q}^T$, which up to normalisation with $\mathbb{E}\big[\mathrm{e}^{-\int_t^T r(s)\,ds} \mid \mathcal{F}_t\big]$ equals,

$$\mathbb{E}\big[\mathrm{e}^{-\int_t^T r(s)\,ds}\,\mathrm{e}^{u^\top X(T)} \mid \mathcal{F}_t\big], \quad u \in \mathrm{i}\mathbb{R}^d \tag{4.4}$$

(use equation (4.3)).

Note that the following integrability condition 1 is satisfied in particular if r is uniformly bounded from below, that is, if $\gamma \in \mathbb{R}^m_+ \times \{0\}$.

Theorem 4.1. *Let $\tau > 0$. The following statements are equivalent:*

1. $\mathbb{E}\big[\mathrm{e}^{-\int_0^\tau r(s)\,ds}\big] < \infty$ *for all* $x \in \mathbb{R}^m_+ \times \mathbb{R}^n$.
2. *There exists a unique solution* $(\Phi(\cdot,u), \Psi(\cdot,u)) : [0,\tau] \to \mathbb{C} \times \mathbb{C}^d$ *of*

$$\begin{aligned}
\partial_t\Phi(t,u) &= \frac{1}{2}\Psi_J(t,u)^\top a_{JJ}\,\Psi_J(t,u) + b^\top\Psi(t,u) - c,\\
\Phi(0,u) &= 0,\\
\partial_t\Psi_i(t,u) &= \frac{1}{2}\Psi(t,u)^\top\alpha_i\,\Psi(t,u) + \beta_i^\top\Psi(t,u) - \gamma_i, \quad i \in I,\\
\partial_t\psi_J(t,u) &= \mathcal{B}_{JJ}^\top\Psi_J(t,u) - \gamma_J,\\
\Psi(0,u) &= u
\end{aligned} \tag{4.5}$$

for $u = 0$.

In either case, there exists an open convex neighbourhood U of 0 in $\mathbb{R}^d$ such that the system of Riccati equations (4.5) *admits a unique solution* $(\Phi(\cdot,u), \Psi(\cdot,u)) : [0,\tau] \to \mathbb{C} \times \mathbb{C}^d$ *for all $u \in \mathcal{S}(U)$, and* (4.4) *allows the following affine representation*

$$\mathbb{E}\big[\mathrm{e}^{-\int_t^T r(s)\,ds}\,\mathrm{e}^{u^\top X(T)} \mid \mathcal{F}_t\big] = \mathrm{e}^{\Phi(T-t,u)+\Psi(T-t,u)^\top X(t)} \tag{4.6}$$

for all $u \in \mathcal{S}(U)$, $t \leq T \leq t+\tau$ and $x \in \mathbb{R}^m_+ \times \mathbb{R}^n$.

Proof. We first enlarge the state space and consider the real-valued process

$$Y(t) = y + \int_0^t \big(c + \gamma^\top X(s)\big)\,ds, \quad y \in \mathbb{R}.$$

A moment's reflection reveals that $X' = (X, Y)$ is an $\mathbb{R}^m_+ \times \mathbb{R}^{n+1}$-valued diffusion process with diffusion matrix $a' + \sum_{i \in I} x_i \alpha'_i$ and drift $b' + \mathcal{B}' x'$ where

$$a' = \begin{pmatrix} a & 0 \\ 0 & 0 \end{pmatrix}, \quad \alpha'_i = \begin{pmatrix} \alpha_i & 0 \\ 0 & 0 \end{pmatrix}, \quad b' = \begin{pmatrix} b \\ c \end{pmatrix}, \quad \mathcal{B}' = \begin{pmatrix} \mathcal{B} & 0 \\ \gamma^T & 0 \end{pmatrix}$$

form admissible parameters. We claim that X' is an affine process.

Indeed, the candidate system of Riccati equations reads

$$\begin{aligned} \partial_t \phi'(t,u,v) &= \frac{1}{2} \psi'_J(t,u,v)^\top a_{JJ} \, \psi'_J(t,u,v) + b^\top \psi'_{\{1,\dots,d\}}(t,u,v) + \boxed{cv}, \\ \phi'(0,u,v) &= 0, \\ \partial_t \psi'_i(t,u,v) &= \frac{1}{2} \psi'(t,u,v)^\top \alpha_i \, \psi'(t,u,v) + \beta_i^\top \psi'(t,u,v) + \boxed{\gamma_i v}, \quad i \in I, \\ \partial_t \psi'_J(t,u,v) &= \mathcal{B}_{JJ}^\top \psi'_J(t,u,v) + \boxed{\gamma_J v}, \\ \partial_t \psi'_{d+1}(t,u,v) &= 0, \\ \psi'(0,u,v) &= \begin{pmatrix} u \\ v \end{pmatrix}. \end{aligned} \tag{4.7}$$

Here we replaced the constant solution $\psi'_{d+1}(\cdot, u, v) \equiv v$ by v in the boxes. Theorem 3.2 carries over and asserts a unique global $\mathbb{C}_- \times \mathbb{C}^m_- \times \mathrm{i}\mathbb{R}^{n+1}$-valued solution $(\phi'(\cdot,u,v), \psi'(\cdot,u,v))$ of (4.5) for all $(u,v) \in \mathbb{C}^m_- \times \mathrm{i}\mathbb{R}^n \times \mathrm{i}\mathbb{R}$. The second part of Theorem 2.2 thus asserts that X' is affine with conditional characteristic function

$$\mathbb{E}\left[\mathrm{e}^{u^\top X(T) + vY(T)} \mid \mathcal{F}_t\right] = \mathrm{e}^{\phi'(T-t,u,v) + \psi'(T-t,u,v)^\top X(t) + vY(t)}$$

for all $(u,v) \in \mathbb{C}^m_- \times \mathrm{i}\mathbb{R}^n \times \mathrm{i}\mathbb{R}$ and $t \le T$.

The theorem now follows from Theorem 3.3 once we set $\Phi(t,u) = \phi'(t,u,-1)$ and $\Psi(t,u) = \psi'_{\{1,\dots,d\}}(t,u,-1)$. □

Suppose, for the rest of this section, that either condition 1 or 2 of Theorem 4.1 is met. As immediate consequence of Theorem 4.1, we obtain the following explicit price formulas for T-bonds in terms of Φ and Ψ.

Corollary 4.2. *For any maturity $T \le \tau$, the T-bond price at $t \le T$ is given as*

$$P(t,T) = \mathrm{e}^{-A(T-t) - B(T-t)^\top X(t)}$$

where we denote

$$A(t) = -\Phi(t,0), \quad B(t) = -\Psi(t,0).$$

Moreover, for $t \le T \le S \le \tau$, the $\mathcal{F}_t$-conditional characteristic function of $X(T)$ under the S-forward measure $\mathbb{Q}^S$ is given by

$$\mathbb{E}_{\mathbb{Q}^S}\left[\mathrm{e}^{u^\top X(T)} \mid \mathcal{F}_t\right] = \frac{\mathrm{e}^{-A(S-T) + \Phi(T-t, u - B(S-T)) + \Psi(T-t, u - B(S-T))^\top X(t)}}{P(t,S)} \tag{4.8}$$

for all $u \in \mathcal{S}(U + B(S-T))$*, where* U *is the neighbourhood of* 0 *in* $\mathbb{R}^d$ *from Theorem 4.1.*

Proof. The bond price formula follows from (4.6) with $u = 0$.

Now let $t \le T \le S \le \tau$ and $u \in \mathcal{S}(U + B(S-T))$. We obtain from (4.6) by nested conditional expectation

$$\begin{aligned}\mathbb{E}\left[\mathrm{e}^{-\int_t^S r(s)\,ds}\mathrm{e}^{u^\top X(T)} \mid \mathcal{F}_t\right] &= \mathbb{E}\left[\mathrm{e}^{-\int_t^T r(s)\,ds}\,\mathbb{E}\left[\mathrm{e}^{-\int_T^S r(s)\,ds} \mid \mathcal{F}_T\right]\mathrm{e}^{u^\top X(T)} \mid \mathcal{F}_t\right]\\ &= \frac{\mathbb{E}\left[\mathrm{e}^{-\int_t^T r(s)\,ds}\mathrm{e}^{(u-B(S-T))^\top X(T)} \mid \mathcal{F}_t\right]}{\mathrm{e}^{A(S-T)}}\\ &= \frac{\mathrm{e}^{\Phi(T-t,u-B(S-T))+\Psi(T-t,u-B(S-T))^\top X(t)}}{\mathrm{e}^{A(S-T)}}.\end{aligned}$$

Normalising by $P(t,S)$ yields (4.8). □

For more general payoff functions f, we can proceed as follows.

- Either we recognise the $\mathcal{F}_t$-conditional distribution, say $q(t,T,dx)$, of $X(T)$ under the T-forward measure from its characteristic function (4.8). Or we derive $q(t,T,dx)$ via numerical inversion of the characteristic function (4.8), using e.g. fast Fourier transform (FFT). Then compute the price (4.2) by integration of f,

$$\pi(t) = P(t,T)\int_{\mathbb{R}_+^m\times\mathbb{R}^n} f(x)\,q(t,T,dx). \tag{4.9}$$

  Examples are given in Section 5 below.

- Or suppose f can be expressed by

$$f(x) = \int_{\mathbb{R}^d} \mathrm{e}^{(u+\mathrm{i}y)^\top x}\,\widetilde{f}(y)\,dy \tag{4.10}$$

  for some integrable function $\widetilde{f} : \mathbb{R}^d \to \mathbb{C}$ and some constant $u \in U$. Then we may apply Fubini's theorem to change the order of integration, which gives

$$\begin{aligned}\pi(t) &= \mathbb{E}\left[\mathrm{e}^{-\int_t^T r(s)\,ds}\int_{\mathbb{R}^d}\mathrm{e}^{(u+\mathrm{i}y)^\top X(T)}\widetilde{f}(y)\,dy \mid \mathcal{F}_t\right]\\ &= \int_{\mathbb{R}^d}\mathbb{E}\left[\mathrm{e}^{-\int_t^T r(s)\,ds}\,\mathrm{e}^{(u+\mathrm{i}y)^\top X(T)} \mid \mathcal{F}_t\right]\widetilde{f}(y)\,dy \\ &= \int_{\mathbb{R}^d}\mathrm{e}^{\Phi(T-t,u+\mathrm{i}y)+\Psi(T-t,u+\mathrm{i}y)^\top X(t)}\,\widetilde{f}(y)\,dy.\end{aligned} \tag{4.11}$$

  This integral can be numerically computed. An example is given in Section 5.1 below.

The function $\widetilde{f}$ in (4.10) can be found by Fourier transformation, as the following classical result indicates.

Lemma 4.3. *Let $f : \mathbb{R}^d \to \mathbb{C}$ be a measurable function and $u \in \mathbb{R}^d$ be such that the function $h(x) = \mathrm{e}^{-u^\top x} f(x)$ and its Fourier transform*

$$\hat{h}(y) = \int_{\mathbb{R}^d} h(x)\, \mathrm{e}^{-\mathrm{i} y^\top x}\, dx$$

are integrable on $\mathbb{R}^d$. Then (4.10) *holds for almost all $x \in \mathbb{R}^d$ for*

$$\widetilde{f} = \frac{1}{(2\pi)^d} \hat{h}.$$

Moreover, the right hand side of (4.10) *is continuous in x. Hence, if f is continuous then* (4.10) *holds for all $x \in \mathbb{R}^d$.*

Proof. From Fourier analysis, see [33, Chapter I, Corollary 1.21], we know that

$$h(x) = \frac{1}{(2\pi)^d} \int_{\mathbb{R}^d} \mathrm{e}^{\mathrm{i} y^\top x}\, \hat{h}(y)\, dy$$

for almost all $x \in \mathbb{R}^d$. Multiplying both sides with $\mathrm{e}^{u^\top x}$ yields the first claim.

From the Riemann–Lebesgue theorem ([33, Chapter I, Theorem 1.2]) we know that the right hand side of (4.10) is continuous in x. □

An example is the continuous payoff function

$$f(x) = (\mathrm{e}^x - K)^+$$

of a European call option with strike price K on the underlying stock price e^L, where L may be any affine function of X. Fix a real constant $p > 1$. Then $h(x) = \mathrm{e}^{-px} f(x)$ is integrable on $\mathbb{R}$. An easy calculation shows that its Fourier transform

$$\hat{h}(y) = \int_{\mathbb{R}} \mathrm{e}^{-px} f(x)\, \mathrm{e}^{-\mathrm{i} yx}\, dx = \frac{K^{1-p-\mathrm{i}y}}{(p+\mathrm{i}y)(p+\mathrm{i}y-1)}$$

is also integrable on $\mathbb{R}$. In view of Lemma 4.3, we thus conclude that, for $p > 1$,

$$(\mathrm{e}^x - K)^+ = \frac{1}{2\pi} \int_{\mathbb{R}} \mathrm{e}^{(p+\mathrm{i}y)x} \frac{K^{1-p-\mathrm{i}y}}{(p+\mathrm{i}y)(p+\mathrm{i}y-1)}\, dy, \tag{4.12}$$

which is of the desired form (4.10). We will apply this for the Heston stochastic volatility model in Section 5.1 below.

A related example is the following

$$(\mathrm{e}^x - K)^+ - \mathrm{e}^x = \frac{1}{2\pi} \int_{\mathbb{R}} \mathrm{e}^{(p+\mathrm{i}y)x} \frac{K^{1-p-\mathrm{i}y}}{(p+\mathrm{i}y)(p+\mathrm{i}y-1)}\, dy, \tag{4.13}$$

which holds for all $0 < p < 1$.

More examples of payoff functions with integral representation, including the above, can be found in [21].

5 Bond option pricing in affine models

We can further simplify formula (4.9) for a European call option on a S-bond with expiry date $T < S$ and strike price K. The payoff function is

$$f(x) = \left(\mathrm{e}^{-A(S-T)-B(S-T)^\top x} - K\right)^+.$$

We can decompose (4.2),

$$\pi^C(t;T,S) = P(t,S)\mathbb{Q}^S[E \mid \mathcal{F}_t] - KP(t,T)\mathbb{Q}^T[E \mid \mathcal{F}_t] \tag{5.1}$$

for the event $E = \{B(S-T)^\top X(T) \leq -A(S-T) - \log K\}$. The pricing of this bond option boils down to the computation of the probability of the event E under the S- and T-forward measures.

Similarly, the value of a put equals

$$\pi^P(t;T,S) = KP(t,T)\mathbb{Q}^T[E^c \mid \mathcal{F}_t] - P(t,S)\mathbb{Q}^S[E^c \mid \mathcal{F}_t] \tag{5.2}$$

for the event $E^c = \Omega \setminus E = \{B(S-T)^\top X(T) > -A(S-T) - \log K\}$.

In the following two subsections, we illustrate this approach for the Vasiček and Cox–Ingersoll–Ross short rate models.

5.0.1 Example: Vasiček short rate model

Example 5.1 (Vasiček short rate model). The state space is $\mathbb{R}$, and we set $r = X$ for the Vasiček short rate model

$$dr = (b + \beta r)\,dt + \sigma\,dW.$$

The system (4.5) reads

$$\begin{aligned}
\Phi(t,u) &= \frac{1}{2}\sigma^2 \int_0^t \Psi^2(s,u)\,ds + b\int_0^t \Psi(s,u)\,ds \\
\partial_t \Psi(t,u) &= \beta\Psi(t,u) - 1, \\
\Psi(0,u) &= u
\end{aligned}$$

which admits a unique global solution with

$$\begin{aligned}
\Psi(t,u) &= \mathrm{e}^{\beta t}u - \frac{\mathrm{e}^{\beta t} - 1}{\beta} \\
\Phi(t,u) &= \frac{1}{2}\sigma^2\left(\frac{u^2}{2\beta}(\mathrm{e}^{2\beta t} - 1) + \frac{1}{2\beta^3}(\mathrm{e}^{2\beta t} - 4\mathrm{e}^{\beta t} + 2\beta t + 3)\right. \\
&\quad \left. - \frac{u}{\beta^2}(\mathrm{e}^{2\beta t} - 2\mathrm{e}^{\beta t} + 2\beta)\right) + b\left(\frac{\mathrm{e}^{\beta t} - 1}{\beta}u + \frac{\mathrm{e}^{\beta t} - 1 - \beta t}{\beta^2}\right)
\end{aligned}$$

for all $u \in \mathbb{C}$. Hence (4.6) holds for all $u \in \mathbb{C}$ and $t \leq T$. In particular, by Corollary 4.2, the bond prices $P(t,T)$ can be determined by A and B,

$$B(t) = -\Psi(t,0) = \frac{\mathrm{e}^{\beta t} - 1}{\beta},$$

$$A(t) = -\Phi(t,0) = -\frac{\sigma^2}{4\beta^3}(\mathrm{e}^{2\beta t} - 4\mathrm{e}^{\beta t} + 2\beta t + 3) + b\frac{\mathrm{e}^{\beta t} - 1 - \beta t}{\beta^2}.$$

Hence, under the S-forward measure, $r(T)$ is $\mathcal{F}_t$-conditionally Gaussian distributed with (cf. [3], Chapter 3.2.1)

$$\mathbb{E}_{\mathbb{Q}^S}[r(T) \mid \mathcal{F}_t] = r(t)e^{-\beta(T-s)} + M^S(t,T),$$

$$\mathrm{Var}_{\mathbb{Q}^S}(r(T) \mid \mathcal{F}_t) = \sigma^2\frac{\mathrm{e}^{2\beta(T-t)} - 1}{2\beta},$$

where M^S is defined by

$$M^S(t,T) = \left(\frac{b}{\beta} - \frac{\sigma^2}{2\beta^2}\right)\left(1 - \mathrm{e}^{-\beta(T-t)}\right) + \frac{\sigma^2}{2\beta^2}\left[\mathrm{e}^{-\beta(S-T)} - \mathrm{e}^{-\beta(S+T-2t)}\right].$$

The bond option price formula for the Vasiček short rate model can now be derived via (5.1) and (5.2).

Example 5.2 (Cox–Ingersoll–Ross short rate model). The state space is $\mathbb{R}_+$, and we set $r = X$ for the Cox–Ingersoll–Ross short rate model

$$dr = (b + \beta r)\,dt + \sigma\sqrt{r}\,dW.$$

The system (4.5) reads

$$\begin{aligned} \Phi(t,u) &= b\int_0^t \Psi(s,u)\,ds, \\ \partial_t\Psi(t,u) &= \frac{1}{2}\sigma^2\Psi^2(t,u) + \beta\Psi(t,u) - 1, \\ \Psi(0,u) &= u. \end{aligned} \tag{5.3}$$

By Lemma 5.4 below, there exists a unique solution $(\Phi(\cdot,u), \Psi(\cdot,u)) : \mathbb{R}_+ \to \mathbb{C}_- \times \mathbb{C}_-$, and thus (4.6) holds, for all $u \in \mathbb{C}_-$ and $t \leq T$. The solution is given explicitly as

$$\Phi(t,u) = \frac{2b}{\sigma^2}\log\left(\frac{L_5(t)}{L_3(t) - L_4(t)u}\right)$$

$$\Psi(t,u) = -\frac{L_1(t) - L_2(t)u}{L_3(t) - L_4(t)u}$$

where $\lambda = \sqrt{\beta^2 + 2\sigma^2}$ and

$$\begin{aligned}
L_1(t) &= 2\left(\mathrm{e}^{\lambda t} - 1\right)\\
L_2(t) &= \lambda\left(\mathrm{e}^{\lambda t} + 1\right) + \beta\left(\mathrm{e}^{\lambda t} - 1\right)\\
L_3(t) &= \lambda\left(\mathrm{e}^{\lambda t} + 1\right) - \beta\left(\mathrm{e}^{\lambda t} - 1\right)\\
L_4(t) &= \sigma^2\left(\mathrm{e}^{\lambda t} - 1\right)\\
L_5(t) &= 2\lambda \mathrm{e}^{\frac{(\lambda-\beta)t}{2}}.
\end{aligned}$$

Some tedious but elementary algebraic manipulations show that the $\mathcal{F}_t$-conditional characteristic function of $r(T)$ under the S-forward measure $\mathbb{Q}^S$ is given by

$$\mathbb{E}_{\mathbb{Q}^S}\left[\mathrm{e}^{ur(T)} \mid \mathcal{F}_t\right] = \frac{\mathrm{e}^{-C_2(t,T,S)r(t)+\frac{C_2(t,T,S)r(t)}{1-C_1(t,T,S)u}}}{(1-C_1(t,T,S)u)^{\frac{2b}{\sigma^2}}}$$

where

$$C_1(t,T,S) = \frac{L_3(S-T)L_4(T-t)}{2\lambda L_3(S-t)}, \quad C_2(t,T,S) = \frac{L_2(T-t)}{L_4(T-t)} - \frac{L_1(S-t)}{L_3(S-t)}.$$

Comparing this with Lemma 5.3 below, we conclude that, up to scaling by $1/C_1(t,T,S)$, the $\mathcal{F}_t$-conditional distribution of $r(T)$ under the S-forward measure $\mathbb{Q}^S$ is noncentral χ^2 with $4b/\sigma^2$ degrees of freedom and parameter of noncentrality $2C_2(t,T,S)r(t)$. The corresponding density is therefore given by

$$\frac{1}{C_1(t,T,S)} f_{\chi^2\left(\frac{4b}{\sigma^2}, 2C_2(t,T,S)r(t)\right)}\left(\frac{x}{C_1(t,T,S)}\right), \quad x \in \mathbb{R}_+.$$

Combining this with (5.1)–(5.2), we obtain explicit European bond option price formulas.

As an application, we now compute cap prices. Let us consider a cap with strike rate κ and tenor structure $1/4 = T_0 < T_1 < \cdots < T_n$, with $T_i - T_{i-1} = 1/4$. Here, as usual, T_i denote the settlement dates and T_{i-1} the reset dates for the ith caplet, $i = 1, \ldots, n$ and T_n is the maturity of the cap. It is well known that the cash flow of a ith caplet at time T_i equals the $(1 + \kappa/4)$ multiple of the cash-flow at T_{i-1} of a put option on the T_i-bond with strike price $1/(1 + \kappa/4)$. Hence the cap price equals

$$Cp = \sum_{i=1}^{n} Cpl(i) = (1+\kappa/4)\sum_{i=1}^{n} P(0,T_{i-1})\mathbb{E}_{\mathbb{Q}^{T_{i-1}}}\left[\left(\frac{1}{1+\kappa/4} - P(T_{i-1},T_i)\right)^+\right].$$

In practice, cap prices are often quoted in Black implied volatilities. By definition, the implied volatility $\sigma_B > 0$ is the number, which, plugged into Black's formula, yields the cap value $Cp = \sum_{i=1}^{n} Cpl(i)$, where the ith caplet price is given as

$$Cpl(i) = \frac{1}{4}P(0,T_i)(F(T_{i-1},T_i)\Phi(d_1(i)) - \kappa\Phi(d_2(i)))$$

with

$$d_{1,2}(i) = \frac{\log\left(\frac{F(T_{i-1},T_i)}{\kappa}\right) \pm \frac{\sigma_B^2}{2}(T_{i-1}-t)}{\sigma_B\sqrt{T_{i-1}-t}},$$

where $F(T_{i-1},T_i) = 4\left(\frac{P(0,T_{i-1})}{P(0,T_i)} - 1\right)$ denotes the corresponding simple forward rate.

As parameters for the CIR model we assume

$$\sigma^2 = 0.033, \quad b = 0.08, \quad \beta = -0.9, \quad r_0 = 0.08.$$

In Table 5.1 we summarise the ATM[1] cap prices and implied volatilities for various maturities.

Maturity Years	*strike rate*	*cap price*	*implied volatility*
1	0.0843	0.0073	0.4506
2	0.0855	0.0190	0.3720
3	0.0862	0.0302	0.3226
4	0.0866	0.0406	0.2890
5	0.0868	0.0501	0.2647
6	0.0870	0.0588	0.2462
7	0.0871	0.0668	0.2316
8	0.0872	0.0742	0.2198
9	0.0873	0.0809	0.2100
10	0.0873	0.0871	0.2017
15	0.0875	0.1110	0.1744
20	0.0876	0.1265	0.1594
25	0.0876	0.1365	0.1502
30	0.0876	0.1430	0.1442

Table 5.1. ATM cap prices for the CIR model

Lemma 5.3 (Noncentral χ^2-Distribution). *The noncentral χ^2-distribution with $\delta > 0$ degrees of freedom and noncentrality parameter $\zeta > 0$ has density function*

$$f_{\chi^2(\delta,\zeta)}(x) = \frac{1}{2}\mathrm{e}^{-\frac{x+\zeta}{2}}\left(\frac{x}{\zeta}\right)^{\frac{\delta}{4}-\frac{1}{2}} I_{\frac{\delta}{2}-1}(\sqrt{\zeta x}), \quad x \geq 0$$

and characteristic function

$$\int_{\mathbb{R}_+} \mathrm{e}^{ux} f_{\chi^2(\delta,\zeta)}(x)\,dx = \frac{\mathrm{e}^{\frac{\zeta u}{1-2u}}}{(1-2u)^{\frac{\delta}{2}}}, \quad u \in \mathbb{C}_-.$$

[1] The cap with maturity T_n is at-the-money (ATM) if its strike rate κ equals the prevailing forward swap rate $4(P(0,T_0) - P(0,T_n))/\sum_{i=1}^n P(0,T_i)$.

Here $I_\nu(x) = \sum_{j\geq 0} \frac{1}{j!\Gamma(j+\nu+1)} \left(\frac{x}{2}\right)^{2j+\nu}$ *denotes the modified Bessel function of the first kind of order* $\nu > -1$.

Proof. See e.g. [23]. □

Lemma 5.4. *Consider the Riccati differential equation*

$$\partial_t G = AG^2 + BG - C, \quad G(0,u) = u, \tag{5.4}$$

where $A, B, C \in \mathbb{C}$ *and* $u \in \mathbb{C}$, *with* $A \neq 0$ *and* $B^2 + 4AC \in \mathbb{C} \setminus \mathbb{R}_-$. *Let* $\sqrt{\cdot}$ *denote the analytic extension of the real square root to* $\mathbb{C} \setminus \mathbb{R}_-$, *and define* $\lambda = \sqrt{B^2 + 4AC}$.

1. *On its maximal interval of existence* $[0, t_+(u))$, *the function*

$$G(t,u) = -\frac{2C\left(e^{\lambda t} - 1\right) - \left(\lambda\left(e^{\lambda t} + 1\right) + B\left(e^{\lambda t} - 1\right)\right)u}{\lambda\left(e^{\lambda t} + 1\right) - B\left(e^{\lambda t} - 1\right) - 2A\left(e^{\lambda t} - 1\right)u} \tag{5.5}$$

is the unique solution of equation (5.4). *Moreover,*

$$\int_0^t G(s,u)\,ds = \frac{1}{A}\log\left(\frac{2\lambda e^{\frac{\lambda - B}{2}t}}{\lambda(e^{\lambda t}+1) - B(e^{\lambda t}-1) - 2A(e^{\lambda t}-1)u}\right). \tag{5.6}$$

2. *If, moreover,* $A > 0$, $B \in \mathbb{R}$, $\mathrm{Re}\,(C) \geq 0$ *and* $u \in \mathbb{C}_-$ *then* $t_+(u) = \infty$ *and* $G(t,u)$ *is* $\mathbb{C}_-$*-valued.*

Proof. 1: Recall that the square root $\sqrt{z} := e^{1/2\log(z)}$ is the well defined analytic extension of the real square root to $\mathbb{C} \setminus \mathbb{R}_-$, through the main branch of the logarithm which can be written in the form $\log(z) = \int_{[0,z]} \frac{dz}{z}$. Hence we may write (5.4) as

$$\dot{G} = A(G - \lambda_+)(G - \lambda_-), \quad G(0,u) = u,$$

where $\lambda_\pm = \frac{-B \pm \sqrt{B^2+4AC}}{2A}$, and it follows that

$$G(t,u) = \frac{\lambda_+(u - \lambda_-) - \lambda_-(u - \lambda_+)e^{\lambda t}}{(u - \lambda_-) - (u - \lambda_+)e^{\lambda t}},$$

which can be seen to be equivalent to (5.5). As $\lambda_+ \neq \lambda_-$, numerator and denominator cannot vanish at the same time t, and certainly not for t near zero. Hence, by the maximality of $t_+(u)$, (5.5) is the solution of (5.4) for $t \in [0, t_+(u))$. Finally, the integral (5.6) is checked by differentiation.

2: We show along the lines of the proof of Theorem 3.2, that for this choice of coefficients global solutions exist for initial data $u \in \mathbb{C}_-$ and stay in $\mathbb{C}_-$. To this end, write $R(G) = AG^2 + BG - C$, then

$$\mathrm{Re}\,(R(G)) = A(\mathrm{Re}\,(G))^2 - A(\mathrm{Im}\,(G))^2 + B\,\mathrm{Re}\,(G) - \mathrm{Re}\,(C) \leq A(\mathrm{Re}\,(G))^2 + B\,\mathrm{Re}\,(G)$$

and since $A, B \in \mathbb{R}$ we have that $\mathrm{Re}\,(G(t,u)) \leq 0$ for all times $t \in [0, t_+(u))$, see Corollary B.2 below. Furthermore, we see that $\mathrm{Re}\,(\overline{G}R(G)) \leq (1 + |G|^2)(|B| + |C|)$, hence $\partial_t |G(t,u)|^2 \leq 2(1 + |G(t,u)|^2)(|B| + |C|)$. This implies, by Gronwall's inequality ([11, (10.5.1.3)]), that $t_+(u) = \infty$. Hence the lemma is proved. □

5.1 Heston stochastic volatility model

This affine model, proposed by Heston [20], generalises the Black–Scholes model by assuming a stochastic volatility.

Interest rates are assumed to be constant $r(t) \equiv r \geq 0$, and there is one risky asset (stock) $S = \mathrm{e}^{X_2}$, where $X = (X_1, X_2)$ is the affine process with state space $\mathbb{R}_+ \times \mathbb{R}$ and dynamics

$$dX_1 = (k + \kappa X_1)\,dt + \sigma\sqrt{2X_1}\,dW_1$$
$$dX_2 = (r - X_1)\,dt + \sqrt{2X_1}\left(\rho\,dW_1 + \sqrt{1-\rho^2}dW_2\right)$$

for some constant parameters $k, \sigma \geq 0$, $\kappa \in \mathbb{R}$, and some $\rho \in [-1, 1]$. In view of Remark 3.4, we note that here

$$\mathcal{B} = \begin{pmatrix} \kappa & 0 \\ -1 & 0 \end{pmatrix}$$

is singular, and hence cannot have strictly negative eigenvalues.

The implied risk-neutral stock dynamics read

$$dS = Sr\,dt + S\sqrt{2X_1}\,d\mathcal{W}$$

for the Brownian motion $\mathcal{W} = \rho W_1 + \sqrt{1-\rho^2}\,W_2$. We see that $\sqrt{2X_1}$ is the stochastic volatility of the price process S. They have possibly non-zero covariation

$$d\langle S, X_1\rangle = 2\rho\sigma S X_1\,dt.$$

The corresponding system of Riccati equations (3.2) is equivalent to

$$\begin{aligned}
\phi(t,u) &= k\int_0^t \psi_1(s,u)\,ds + ru_2 t \\
\partial_t\psi_1(t,u) &= \sigma^2\psi_1^2(t,u) + (2\rho\sigma u_2 + \kappa)\psi_1(t,u) + u_2^2 - u_2 \\
\psi_1(0,u) &= u_1 \\
\psi_2(t,u) &= u_2,
\end{aligned} \tag{5.7}$$

which, in view of Lemma 5.4 2 admits an explicit global solution if $u_1 \in \mathbb{C}_-$ and $0 \leq \operatorname{Re} u_2 \leq 1$. In particular, for $u_1 = 0$ and by setting $\lambda = \sqrt{(2\rho\sigma u_2 + \kappa)^2 + 4\sigma^2(u_2 - u_2^2)}$, the solution can be given explicitly as

$$\begin{aligned}
\phi(t,u) &= \frac{k}{\sigma^2}\log\left(\frac{2\lambda e^{\frac{\lambda-(2\rho\sigma u_2+\kappa)}{2}t}}{\lambda(e^{\lambda t}+1) - (2\rho\sigma u_2+\kappa)(e^{\lambda t}-1)}\right) + ru_2 t \\
\psi_1(t,u) &= -\frac{2(u_2 - u_2^2)(e^{\lambda t}-1)}{\lambda(e^{\lambda t}+1) - (2\rho\sigma u_2+\kappa)(e^{\lambda t}-1)} \\
\psi_2(t,u) &= u_2.
\end{aligned} \tag{5.8}$$

Furthermore, for $u = (0, 1)$, we obtain

$$\phi(t, 0, 1) = rt, \quad \psi(t, 0, 1) = (0, 1)^\top.$$

Theorem 3.3 thus implies that $S(T)$ has finite first moment, for any $T \in \mathcal{R}_+$, and

$$\mathbb{E}[\mathrm{e}^{-rT} S(T) \mid \mathcal{F}_t] = \mathrm{e}^{-rT} \mathbb{E}[\mathrm{e}^{X_2(T)} \mid \mathcal{F}_t] = \mathrm{e}^{-rT} \mathrm{e}^{r(T-t)+X_2(t)} = \mathrm{e}^{-rt} S(t),$$

for $t \leq T$, which is just the martingale property of S.

We now want to compute the price

$$\pi(t) = \mathrm{e}^{-r(T-t)} \mathbb{E}\left[(S(T) - K)^+ \mid \mathcal{F}_t\right]$$

of a European call option on $S(T)$ with maturity T and strike price K. Fix some $p > 1$ small enough with $(0, p) \in \mathcal{D}_\mathcal{R}(T)$. Formula (4.12) combined with (4.11) then yields

$$\begin{aligned} \pi(t) &= \frac{1}{2\pi} \mathrm{e}^{-r(T-t)} \\ &\times \int_\mathbb{R} \mathrm{e}^{\phi(T-t,0,p+\mathrm{i}y)+\psi_1(T-t,0,p+\mathrm{i}y)X_1(t)+(p+\mathrm{i}y)X_2(t)} \frac{K^{1-p-\mathrm{i}y}}{(p+\mathrm{i}y)(p+\mathrm{i}y-1)} \, dy. \end{aligned} \tag{5.9}$$

Alternatively, we may fix any $0 < p < 1$ and then, combining (4.13) with (4.11),

$$\begin{aligned} \pi(t) &= S(t) + \frac{1}{2\pi} \mathrm{e}^{-r(T-t)} \\ &\times \int_\mathbb{R} \mathrm{e}^{\phi(T-t,0,p+\mathrm{i}y)+\psi_1(T-t,0,p+\mathrm{i}y)X_1(t)+(p+\mathrm{i}y)X_2(t)} \frac{K^{1-p-\mathrm{i}y}}{(p+\mathrm{i}y)(p+\mathrm{i}y-1)} \, dy. \end{aligned} \tag{5.10}$$

Since we have explicit expressions (5.8) for $\phi(T-t, 0, p+\mathrm{i}y)$ and $\psi_1(T-t, 0, p+\mathrm{i}y)$, we only need to compute the integral with respect to y in (5.9) or (5.10) numerically. We have carried out numeric experiments for European option prices using MATLAB. Fastest results were achieved for values $p \approx 0.5$ by using (5.10) whereas keeping a constant error level the runtime explodes at $p \to 0,\ 1$, which is due to the singularities of the integrand. Also, an evaluation of residua

$$\frac{\pi(t=0, p=1/2) - \pi(t=0, p=1/2+\varepsilon)}{\pi(t=0, p=1/2)}$$

for $\varepsilon \in [0, 1/2) \cup (1/2, 1]$ suggests that (5.10) is numerically more stable than (5.9).

Next, we present implied volatilities obtained by (5.10) setting $p = 1/2$. As initial data for X and model parameters, we chose

$$X_1(0) = 0.02,\ X_2(0) = 0.00,\ \sigma = 0.1,\ \kappa = -2.0,\ k = 0.02,\ r = 0.01,\ \rho = 0.5.$$

Table 5.2 shows implied volatilities from call option prices at $t = 0$ for various strikes K and maturities T, computed with (5.10) for $p = 0.5$. These values are in well accordance with MC simulations (mesh size $T/500$, number of sample paths $= 10000$). The corresponding implied volatility surface is shown in Figure 5.1.

T-K	0.8000	0.9000	1.0000	1.1000	1.2000
0.5000	0.1611	0.1682	0.1785	0.1892	0.1992
1.0000	0.1513	0.1579	0.1664	0.1751	0.1835
1.5000	0.1464	0.1524	0.1594	0.1665	0.1734
2.0000	0.1438	0.1492	0.1551	0.1611	0.1668
2.5000	0.1424	0.1473	0.1524	0.1574	0.1623
3.0000	0.1417	0.1460	0.1505	0.1549	0.1591

Table 5.2. Implied volatilities for the Heston model

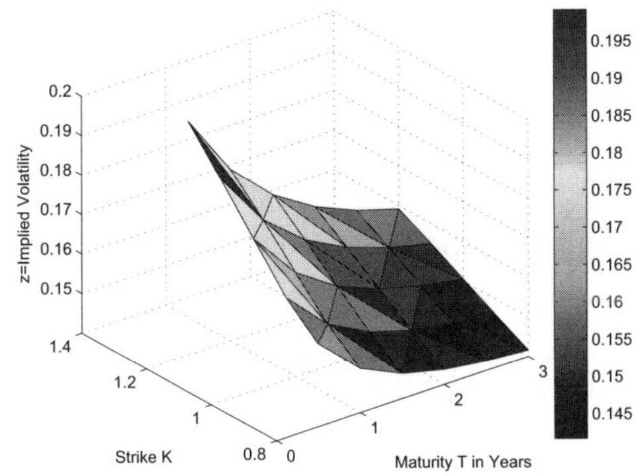

Figure 5.1. Implied volatility surface for the Heston model

Remark 5.5. We note that the Heston model is often written in the equivalent form

$$\begin{aligned} dv &= \bar{\kappa}(\eta - v)dt + \sigma\sqrt{v}\, dW_1 \\ dS &= rSdt + S\sqrt{v}\, d\mathcal{W}\,. \end{aligned}$$

To see the relation of the parameters of this form and the one used in this section, we simply set $v = 2X_1$, and then get

$$\begin{aligned} dX_1 &= (\bar{\kappa}\eta - \bar{\kappa}v)dt + \sigma\sqrt{2X_1}\, dW_1 \quad X_1(0) = X_{10} \\ \frac{dS}{S} &= rdt + \sqrt{2X_1}\, d\mathcal{W}, \quad S(0) = e^{X_2(0)} \end{aligned}$$

from which we read off

$$k = \bar{\kappa}\eta, \quad \kappa = -\bar{\kappa}, \quad X_{10} = v_0/2$$

and all other parameters coincide.

6 Affine transformations and canonical representation

As above, we let X be affine on the canonical state space $\mathbb{R}_+^m \times \mathbb{R}^n$ with admissible parameters a, α_i, b, β_i. Hence, in view of (2.1), for any $x \in \mathbb{R}_+^m \times \mathbb{R}^n$ the process $X = X^x$ satisfies

$$dX = (b + \mathcal{B}X)\, dt + \rho(X)\, dW, \quad X(0) = x, \tag{6.1}$$

and $\rho(x)\rho(x)^\top = a + \sum_{i \in I} x_i \alpha_i$.

It can easily be checked that for every invertible $d \times d$-matrix Λ, the linear transform $Y = \Lambda X$ satisfies

$$dY = \left(\Lambda b + \Lambda \mathcal{B} \Lambda^{-1} Y\right) dt + \Lambda \rho\left(\Lambda^{-1} Y\right) dW, \quad Y(0) = \Lambda x. \tag{6.2}$$

Hence, Y has again an affine drift and diffusion matrix

$$\Lambda b + \Lambda \mathcal{B} \Lambda^{-1} y \quad \text{and} \quad \Lambda \alpha(\Lambda^{-1} y) \Lambda^\top, \tag{6.3}$$

respectively.

On the other hand, the affine short rate model (4.1) can be expressed in terms of $Y(t)$ as

$$r(t) = c + \gamma^\top \Lambda^{-1} Y(t)\,. \tag{6.4}$$

This shows that Y and (6.4) specify an affine short rate model producing the same short rates, and thus bond prices, as X and (4.1). That is, an invertible linear transformation of the state process changes the particular form of the stochastic differential equation (6.1). But it leaves observable quantities, such as short rates and bond prices invariant.

This motivates the question whether there exists a classification method ensuring that affine short rate models with the same observable implications have a unique canonical representation. This topic has been addressed in [10, 9, 24, 8]. We now elaborate on this issue and show that the diffusion matrix $\alpha(x)$ can always be brought into block-diagonal form by a regular linear transform Λ with $\Lambda(\mathbb{R}_+^m \times \mathbb{R}^n) = \mathbb{R}_+^m \times \mathbb{R}^n$.

We denote by

$$\operatorname{diag}(z_1, \ldots, z_m)$$

the diagonal matrix with diagonal elements $z_1, \ldots, z_m$, and we write I_m for the $m \times m$-identity matrix.

Lemma 6.1. *There exists some invertible $d \times d$-matrix Λ with $\Lambda(\mathbb{R}_+^m \times \mathbb{R}^n) = \mathbb{R}_+^m \times \mathbb{R}^n$ such that $\Lambda \alpha(\Lambda^{-1} y) \Lambda^\top$ is block-diagonal of the form*

$$\Lambda \alpha(\Lambda^{-1} y) \Lambda^\top = \begin{pmatrix} \operatorname{diag}(y_1, \ldots, y_q, 0, \ldots, 0) & 0 \\ 0 & p + \sum_{i \in I} y_i \pi_i \end{pmatrix}$$

for some integer $0 \le q \le m$ and symmetric positive semi-definite $n \times n$ matrices $p, \pi_1, \ldots, \pi_m$. Moreover, Λb and $\Lambda \mathcal{B} \Lambda^{-1}$ meet the respective admissibility conditions (3.1) in lieu of b and $\mathcal{B}$.

Proof. From (2.3) we know that $\Lambda\alpha(x)\Lambda^\top$ is block-diagonal for all $x = \Lambda^{-1}y$ if and only if $\Lambda a\Lambda^\top$ and $\Lambda\alpha_i\Lambda^\top$ are block-diagonal for all $i \in I$. By permutation and scaling of the first m coordinate axes (this is a linear bijection from $\mathbb{R}^m_+ \times \mathbb{R}^n$ onto itself, which preserves the admissibility of the transformed b and $\mathcal{B}$), we may assume that there exists some integer $0 \le q \le m$ such that $\alpha_{1,11} = \cdots = \alpha_{q,qq} = 1$ and $\alpha_{i,ii} = 0$ for $q < i \le m$. Hence a and α_i for $q < i \le m$ are already block-diagonal of the special form

$$a = \begin{pmatrix} 0 & 0 \\ 0 & a_{JJ} \end{pmatrix}, \quad \alpha_i = \begin{pmatrix} 0 & 0 \\ 0 & \alpha_{i,JJ} \end{pmatrix}.$$

For $1 \le i \le q$, we may have non-zero off-diagonal elements in the ith row $\alpha_{i,iJ}$. We thus define the $n \times m$-matrix $D = (\delta_1, \ldots, \delta_m)$ with ith column $\delta_i = -\alpha_{i,iJ}$ and set

$$\Lambda = \begin{pmatrix} I_m & 0 \\ D & I_n \end{pmatrix}.$$

One checks by inspection that D is invertible and maps $\mathbb{R}^m_+ \times \mathbb{R}^n$ onto $\mathbb{R}^m_+ \times \mathbb{R}^n$. Moreover,

$$D\alpha_{i,II} = -\alpha_{i,JI}, \quad i \in I.$$

From here we easily verify that

$$\Lambda\alpha_i = \begin{pmatrix} \alpha_{i,II} & \alpha_{i,IJ} \\ 0 & D\alpha_{i,IJ} + \alpha_{i,JJ} \end{pmatrix},$$

and thus

$$\Lambda\alpha_i\Lambda^\top = \begin{pmatrix} \alpha_{i,II} & 0 \\ 0 & D\alpha_{i,IJ} + \alpha_{i,JJ} \end{pmatrix}.$$

Since $\Lambda a\, \Lambda^\top = a$, the first assertion is proved.

The admissibility conditions for Λb and $\Lambda\mathcal{B}\Lambda^{-1}$ can easily be checked as well. □

In view of (6.3), (6.4) and Lemma 6.1 we thus obtain the following result.

Theorem 6.2 (Canonical Representation). *Any affine short rate model* (4.1)*, after some modification of γ if necessary, admits an $\mathbb{R}^m_+ \times \mathbb{R}^n$-valued affine state process X with block-diagonal diffusion matrix of the form*

$$\alpha(x) = \begin{pmatrix} \operatorname{diag}(x_1, \ldots, x_q, 0, \ldots, 0) & 0 \\ 0 & a + \sum_{i\in I} x_i\alpha_{i,JJ} \end{pmatrix} \tag{6.5}$$

for some integer $0 \le q \le m$.

7 Existence and uniqueness of affine processes

All we said about the affine process X so far was under the premise that there exists a unique solution $X = X^x$ of the stochastic differential equation (2.1) on some appropriate state space $\mathcal{X} \subset \mathbb{R}^d$. However, if the diffusion matrix $\rho(x)\rho(x)^\top$ is affine then $\rho(x)$ cannot be Lipschitz continuous in x in general. This raises the question whether (2.1) admits a solution at all.

In this section, we show how X can always be realized as unique solution of the stochastic differential equation (2.1), which is (6.1), in the canonical affine framework $\mathcal{X} = \mathbb{R}_+^m \times \mathbb{R}^n$ and for particular choices of $\rho(x)$.

We recall from Theorem 2.2 that the affine property of X imposes explicit conditions on $\rho(x)\rho(x)^\top$, but not on $\rho(x)$ as such. Indeed, for any orthogonal $d \times d$-matrix D, the function $\rho(x)D$ yields the same diffusion matrix, $\rho(x)DD^\top\rho(x)^\top = \rho(x)\rho(x)^\top$, as $\rho(x)$.

On the other hand, from Theorem 3.2 we know that any admissible parameters a, α_i, b, β_i in (2.3) uniquely determine the functions $(\phi(\cdot, u), \psi(\cdot, u)) : \mathbb{R}_+ \to \mathbb{C}_- \times \mathbb{C}_-^m \times \mathrm{i}\mathbb{R}^n$ as solution of the Riccati equations (3.2), for all $u \in \mathbb{C}_-^m \times \mathrm{i}\mathbb{R}^n$. These in turn uniquely determine the law of the process X. Indeed, for any $0 \le t_1 < t_2$ and $u_1, u_2 \in \mathbb{C}_-^m \times \mathrm{i}\mathbb{R}^n$, we infer by iteration of (2.2)

$$
\begin{aligned}
\mathbb{E}\big[\mathrm{e}^{u_1^\top X(t_1)+u_2^\top X(t_2)}\big] &= \mathbb{E}\Big[\mathrm{e}^{u_1^\top X(t_1)}\mathbb{E}\big[\mathrm{e}^{u_2^\top X(t_2)} \mid \mathcal{F}_{t_1}\big]\Big] \\
&= \mathbb{E}\big[\mathrm{e}^{u_1^\top X(t_1)}\mathrm{e}^{\phi(t_2-t_1,u_2)+\psi(t_2-t_1,u_2)^\top X(t_1)}\big] \\
&= \mathrm{e}^{\phi(t_2-t_1,u_2)+\phi(t_1,u_1+\psi(t_2-t_1,u_2))+\psi(t_1,u_1+\psi(t_2-t_1,u_2))^\top x}.
\end{aligned}
$$

Hence the joint distribution of $(X(t_1), X(t_2))$ is uniquely determined by the functions ϕ and ψ. By further iteration of this argument, we conclude that every finite dimensional distribution, and thus the law, of X is uniquely determined by the parameters a, α_i, b, β_i.

We conclude that the law of an affine process X, while uniquely determined by its characteristics (2.3), can be realized by infinitely many variants of the stochastic differential equation (6.1) by replacing $\rho(x)$ by $\rho(x)D$, for any orthogonal $d \times d$-matrix D. We now propose a canonical choice of $\rho(x)$ as follows:

- In view of (6.2) and Lemma 6.1, every affine process X on $\mathbb{R}_+^m \times \mathbb{R}^n$ can be written as $X = \Lambda^{-1}Y$ for some invertible $d \times d$-matrix Λ and some affine process Y on $\mathbb{R}_+^m \times \mathbb{R}^n$ with block-diagonal diffusion matrix. It is thus enough to consider such $\rho(x)$ where $\rho(x)\rho(x)^\top$ is of the form (6.5). Obviously, $\rho(x) \equiv \rho(x_I)$ is a function of x_I only.
- Set $\rho_{IJ}(x) \equiv 0$, $\rho_{JI}(x) \equiv 0$, and

$$\rho_{II}(x_I) = \mathrm{diag}(\sqrt{x_1}, \dots, \sqrt{x_q}, 0, \dots, 0).$$

Chose for $\rho_{JJ}(x_I)$ any measurable $n \times n$-matrix-valued function satisfying

$$\rho_{JJ}(x_I)\rho_{JJ}(x_I)^\top = a + \sum_{i \in I} x_i \alpha_{i,JJ}. \tag{7.1}$$

In practice, one would determine $\rho_{JJ}(x_I)$ via Cholesky factorisation, see e.g. [31, Theorem 2.2.5]. If $a + \sum_{i\in I} x_i\alpha_{i,JJ}$ is strictly positive definite, then $\rho_{JJ}(x_I)$

turns out to be the unique lower triangular matrix with strictly positive diagonal elements and satisfying (7.1). If $a+\sum_{i\in I} x_i\alpha_{i,JJ}$ is merely positive semi-definite, then the algorithm becomes more involved. In any case, $\rho_{JJ}(x_I)$ will depend measurably on x_I.

- The stochastic differential equation (6.1) now reads

$$\begin{aligned} dX_I &= (b_I + \mathcal{B}_{II}X_I)\,dt + \rho_{II}(X_I)\,dW_I \\ dX_J &= (b_J + \mathcal{B}_{JI}X_I + \mathcal{B}_{JJ}X_J)\,dt + \rho_{JJ}(X_I)\,dW_J \\ X(0) &= x \end{aligned} \tag{7.2}$$

Lemma 7.2 below asserts the existence and uniqueness of an $\mathbb{R}^m_+ \times \mathbb{R}^n$-valued solution $X = X^x$, for any $x \in \mathbb{R}^m_+ \times \mathbb{R}^n$.

We thus have shown:

Theorem 7.1. *Let a, α_i, b, β_i be admissible parameters. Then there exists a measurable function $\rho : \mathbb{R}^m_+ \times \mathbb{R}^n \to \mathbb{R}^{d\times d}$ with $\rho(x)\rho(x)^\top = a + \sum_{i\in I} x_i\alpha_i$, and such that, for any $x \in \mathbb{R}^m_+ \times \mathbb{R}^n$, there exists a unique $\mathbb{R}^m_+ \times \mathbb{R}^n$-valued solution $X = X^x$ of (6.1).*

Moreover, the law of X is uniquely determined by a, α_i, b, β_i, and does not depend on the particular choice of ρ.

The proof of the following lemma uses the concept of a weak solution. The interested reader will find detailed background in e.g. [25, Section 5.3].

Lemma 7.2. *For any $x \in \mathbb{R}^m_+ \times \mathbb{R}^n$, there exists a unique $\mathbb{R}^m_+ \times \mathbb{R}^n$-valued solution $X = X^x$ of (7.2).*

Proof. First, we extend ρ continuously to $\mathbb{R}^d$ by setting $\rho(x) = \rho(x_1^+, \dots, x_m^+)$, where we denote $x_i^+ = \max(0, x_i)$.

Now observe that X_I solves the autonomous equation

$$dX_I = (b_I + \mathcal{B}_{II}X_I)\,dt + \rho_{II}(X_I)\,dW_I, \quad X_I(0) = x_I. \tag{7.3}$$

Obviously, there exists a finite constant K such that the linear growth condition

$$\|b_I + \mathcal{B}_{II}x_I\|^2 + \|\rho(x_I)\|^2 \le K(1 + \|x_I\|^2)$$

is satisfied for all $x \in \mathbb{R}^m$. By [22, Theorems 2.3 and 2.4] there exists a weak solution[1] of (7.3). On the other hand, (7.3) is exactly of the form as assumed in [35, Theorem 1], which implies that pathwise uniqueness[2] holds for (7.3). The Yamada–Watanabe

[1] A weak solution consists of a filtered probability space $(\Omega, \mathcal{F}, (\mathcal{F}_t), \mathbb{P})$ carrying a continuous adapted process X_I and a Brownian motion W_I such that (7.3) is satisfied. The crux of a weak solution is that X_I is not necessarily adapted to the filtration generated by the Brownian motion W_I. See [35, Definition 1] or [25, Definition 5.3.1].

[2] Pathwise uniqueness holds if, for any two weak solutions (X_I, W_I) and (X_I', W_I) of (7.3) defined on the the same probability space $(\Omega, \mathcal{F}, \mathbb{P})$ with common Brownian motion W_I and with common initial value $X_I(0) = X_I'(0)$, the two processes are indistinguishable: $\mathbb{P}[X_I(t) = X_I'(t)$ for all $t \ge 0] = 1$. See [35, Definition 2] or [25, Section 5.3].

Theorem, see [35, Corollary 3] or [25, Corollary 5.3.23], thus implies that there exists a unique solution $X_I = X_I^{x_I}$ of (7.3), for all $x_I \in \mathbb{R}^m$.

Given $X_I^{x_I}$, it is then easily seen that

$$X_J(t) = \mathrm{e}^{\mathcal{B}_{JJ}t}\Big(x_J + \int_0^t \mathrm{e}^{-\mathcal{B}_{JJ}s}(b_J + \mathcal{B}_{JI}X_I(s))\,ds + \int_0^t \mathrm{e}^{-\mathcal{B}_{JJ}s}\rho_{JJ}(X_I(s))\,dW_J(s)\Big)$$

is the unique solution to the second equation in (7.2).

Admissibility of the parameters b and β_i and the stochastic invariance Lemma B.1 eventually imply that $X_I = X_I^{x_I}$ is $\mathbb{R}_+^m$-valued for all $x_I \in \mathbb{R}_+^m$. Whence the lemma is proved. □

A On the regularity of characteristic functions

This auxiliary section provides some analytic regularity results for characteristic functions, which are of independent interest. These results enter the main text only via the proof of Theorem 3.3. This section may thus be skipped at the first reading.

Let ν be a bounded measure on $\mathbb{R}^d$, and denote by

$$G(z) = \int_{\mathbb{R}^d} \mathrm{e}^{z^\top x}\,\nu(dx)$$

its characteristic function[1] for $z \in \mathrm{i}\mathbb{R}^d$. Note that $G(z)$ is actually well defined for $z \in \mathcal{S}(V)$ where

$$V = \Big\{ y \in \mathbb{R}^d \Big| \int_{\mathbb{R}^d} \mathrm{e}^{y^\top x}\,\nu(dx) < \infty \Big\}.$$

We first investigate the interplay between the (marginal) moments of ν and the corresponding (partial) regularity of G.

Lemma A.1. *Denote $g(y) = G(iy)$ for $y \in \mathbb{R}^d$, and let $k \in \mathbb{N}$ and $1 \le i \le d$.*

If $\partial_{y_i}^{2k} g(0)$ exists then

$$\int_{\mathbb{R}^d} |x_i|^{2k}\,\nu(dx) < \infty.$$

On the other hand, if $\int_{\mathbb{R}^d} \|x\|^k\,\nu(dx) < \infty$ then $g \in C^k$ and

$$\partial_{y_{i_1}} \cdots \partial_{y_{i_l}} g(y) = \mathrm{i}^l \int_{\mathbb{R}^d} x_{i_1} \cdots x_{i_l}\,\mathrm{e}^{\mathrm{i}y^\top x}\,\nu(dx)$$

for all $y \in \mathbb{R}^d$, $1 \le i_1, \ldots, i_l \le d$ and $1 \le l \le k$.

[1] This is a slight abuse of terminology, since the characteristic function $g(y) = G(iy)$ of ν is usually defined on real arguments $y \in \mathbb{R}^d$. However, it facilitates the subsequent notation.

Proof. As usual, let e_i denote the ith standard basis vector in $\mathbb{R}^d$. Observe that $s \mapsto g(se_i)$ is the characteristic function of the image measure of ν on $\mathbb{R}$ by the mapping $x \mapsto x_i$. Since $\partial_s^{2k} g(se_i)|_{s=0} = \partial_{y_i}^{2k} g(0)$, the assertion follows from the one-dimensional case, see [30, Theorem 2.3.1].

The second part of the lemma follows by differentiating under the integral sign, which is allowed by dominated convergence. □

Lemma A.2. *The set V is convex. Moreover, if $U \subset V$ is an open set in $\mathbb{R}^d$, then G is analytic on the open strip $\mathcal{S}(U)$ in $\mathbb{C}^d$.*

Proof. Since $G : \mathbb{R}^d \to [0, \infty]$ is a convex function, its domain $V = \{y \in \mathbb{R}^d \mid G(y) < \infty\}$ is convex, and so is every level set $V_l = \{y \in \mathbb{R}^d \mid G(y) \le l\}$ for $l \ge 0$.

Now let $U \subset V$ be an open set in $\mathbb{R}^d$. Since any convex function on $\mathbb{R}^d$ is continuous on the open interior of its domain, see [32, Theorem 10.1], we infer that G is continuous on U. We may thus assume that $U_l = \{y \in \mathbb{R}^d \mid G(y) < l\} \cap U \subset V_l$ is open in $\mathbb{R}^d$ and non-empty for $l > 0$ large enough.

Let $z \in \mathcal{S}(U_l)$ and (z_n) be a sequence in $\mathcal{S}(U_l)$ with $z_n \to z$. For n large enough, there exists some $p > 1$ such that $pz_n \in \mathcal{S}(U_l)$. This implies $p\mathrm{Re}\, z_n \in V_l$ and hence

$$\int_{\mathbb{R}^d} \left|\mathrm{e}^{z_n^\top x}\right|^p \nu(dx) \le l.$$

Hence the class of functions $\{\mathrm{e}^{z_n^\top x} \mid n \in \mathbb{N}\}$ is uniformly integrable with respect to ν, see [34, 13.3]. Since $\mathrm{e}^{z_n^\top x} \to \mathrm{e}^{z^\top x}$ for all x, we conclude by Lebesgue's convergence theorem that

$$\left|G(z_n) - G(z)\right| \le \int_{\mathbb{R}^d} \left|\mathrm{e}^{z_n^\top x} - \mathrm{e}^{z^\top x}\right| \nu(dx) \to 0.$$

Hence G is continuous on $\mathcal{S}(U_l)$.

It thus follows from the Cauchy formula, see [11, Section IX.9], that G is analytic on $\mathcal{S}(U_l)$ if and only if, for every $z \in \mathcal{S}(U_l)$ and $1 \le i \le d$, the function $\zeta \mapsto G(z+\zeta e_i)$ is analytic on $\{\zeta \in \mathbb{C} \mid z + \zeta e_i \in \mathcal{S}(U_l)\}$. Here, as usual, we denote e_i the ith standard basis vector in $\mathbb{R}^d$.

We thus let $z \in \mathcal{S}(U_l)$ and $1 \le i \le d$. Then there exists some $\varepsilon_- < 0 < \varepsilon_+$ such that $z + \zeta e_i \in \mathcal{S}(U_l)$ for all $\zeta \in \mathcal{S}([\varepsilon_-, \varepsilon_+])$. In particular, $\left|\mathrm{e}^{(z+\varepsilon_- e_i)^\top x}\right| \nu(dx)$ and $\left|\mathrm{e}^{(z+\varepsilon_+ e_i)^\top x}\right| \nu(dx)$ are bounded measures on $\mathbb{R}^d$. By dominated convergence, it follows that the two summands

$$\begin{aligned} G(z+\zeta e_i) = &\int_{\{x_i<0\}} \mathrm{e}^{(\zeta-\varepsilon_-)x_i}\, \mathrm{e}^{(z+\varepsilon_- e_i)^\top x}\, \nu(dx) \\ &+ \int_{\{x_i \ge 0\}} \mathrm{e}^{(\zeta-\varepsilon_+)x_i}\, \mathrm{e}^{(z+\varepsilon_+ e_i)^\top x}\, \nu(dx), \end{aligned}$$

are complex differentiable, and thus G is analytic, in $\zeta \in \mathcal{S}((\varepsilon_-, \varepsilon_+))$. Whence G is analytic on $\mathcal{S}(U_l)$. Since $\mathcal{S}(U) = \cup_{l>0}\mathcal{S}(U_l)$, the lemma follows. □

In general, V does not have an open interior in $\mathbb{R}^d$. The next lemma provides sufficient conditions for the existence of an open set $U \subset V$ in $\mathbb{R}^d$.

Lemma A.3. *Let U' be an open neighbourhood of 0 in $\mathbb{C}^d$ and h an analytic function on U'. Suppose that $U = U' \cap \mathbb{R}^d$ is star-shaped around 0 and $G(z) = h(z)$ for all $z \in U' \cap \mathrm{i}\mathbb{R}^d$. Then $U \subset V$ and $G = h$ on $U' \cap \mathcal{S}(U)$.*

Proof. We first suppose that $U' = P_\rho$ for the open polydisc

$$P_\rho = \left\{ z \in \mathbb{C}^d \mid |z_i| < \rho_i,\ 1 \le i \le d \right\},$$

for some $\rho = (\rho_1, \dots, \rho_d) \in \mathbb{R}^d_{++}$. Note the symmetry $\mathrm{i}P_\rho = P_\rho$.

As in Lemma A.1, we denote $g(y) = G(iy)$ for $y \in \mathbb{R}^d$. By assumption, $g(y) = h(\mathrm{i}y)$ for all $y \in P_\rho \cap \mathbb{R}^d$. Hence g is analytic on $P_\rho \cap \mathbb{R}^d$, and the Cauchy formula, [11, Section IX.9], yields

$$g(y) = \sum_{i_1,\dots,i_d \in \mathbb{N}_0} c_{i_1,\dots,i_d} y_1^{i_1} \cdots y_d^{i_d} \quad \text{for } y \in P_\rho \cap \mathbb{R}^d$$

where $\sum_{i_1,\dots,i_d \in \mathbb{N}_0} c_{i_1,\dots,i_d} z_1^{i_1} \cdots z_d^{i_d} = h(\mathrm{i}z)$ for all $z \in P_\rho$. This power series is absolutely convergent on P_ρ, that is,

$$\sum_{i_1,\dots,i_d \in \mathbb{N}_0} \left| c_{i_1,\dots,i_d} \right| \left| z_1^{i_1} \cdots z_d^{i_d} \right| < \infty \quad \text{for all } z \in P_\rho.$$

From the first part of Lemma A.1, we infer that ν possesses all moments, that is, $\int_{\mathbb{R}^d} \|x\|^k \,\nu(dx) < \infty$ for all $k \in \mathbb{N}$. From the second part of Lemma A.1 thus

$$c_{i_1,\dots,i_d} = \frac{\mathrm{i}^{i_1+\dots+i_d}}{i_1! \cdots i_d!} \int_{\mathbb{R}^d} x_1^{i_1} \cdots x_d^{i_d} \,\nu(dx).$$

From the inequality $|x_i|^{2k-1} \le (x_i^{2k} + x_i^{2k-2})/2$, for $k \in \mathbb{N}$, and the above properties, we infer that for all $z \in P_\rho$,

$$\int_{\mathbb{R}^d} \mathrm{e}^{\sum_{i=1}^d |z_i|\,|x_i|} \,\nu(dx) = \sum_{i_1,\dots,i_d \in \mathbb{N}_0} \frac{\left| z_1^{i_1} \cdots z_d^{i_d} \right|}{i_1! \cdots i_d!} \int_{\mathbb{R}^d} \left| x_1^{i_1} \cdots x_d^{i_d} \right| \nu(dx) < \infty$$

Hence $P_\rho \cap \mathbb{R}^d \subset V$, and Lemma A.2 implies that G is analytic on $\mathcal{S}(P_\rho \cap \mathbb{R}^d)$. Since the power series for G and h coincide on $P_\rho \cap \mathrm{i}\mathbb{R}^d$, we conclude that $G = h$ on P_ρ, and the lemma is proved for $U' = P_\rho$.

Now let U' be an open neighbourhood of 0 in $\mathbb{C}^d$. Then there exists some open polydisc $P_\rho \subset U'$ with $\rho \in \mathbb{R}^d_{++}$. By the preceding case, we have $P_\rho \cap \mathbb{R}^d \subset V$ and $G = h$ on P_ρ. In view of Lemma A.2 it thus remains to show that $U = U' \cap \mathbb{R}^d \subset V$.

To this end, let $a \in U$. Since U is star-shaped around 0 in $\mathbb{R}^d$, there exists some $s_1 > 1$ such that $sa \in U$ for all $s \in [0, s_1]$ and $h(sa)$ is analytic in $s \in (0, s_1)$. On the other hand, there exists some $0 < s_0 < s_1$ such that $sa \in P_\rho \cap \mathbb{R}^d$ for all $s \in [0, s_0]$, and $G(sa) = h(sa)$ for $s \in (0, s_0)$. This implies

$$\int_{\{a^\top x \ge 0\}} \mathrm{e}^{sa^\top x} \,\nu(dx) = h(sa) - \int_{\{a^\top x < 0\}} \mathrm{e}^{sa^\top x} \,\nu(dx)$$

for $s \in (0, s_0)$. By Lemma A.2, the right hand side is an analytic function in $s \in (0, s_1)$. We conclude by Lemma A.4 below, for μ defined as the image measure of ν on $\mathbb{R}_+$ by the mapping $x \mapsto a^\top x$, that $a \in V$. Hence the lemma is proved. □

Lemma A.4. *Let μ be a bounded measure on $\mathbb{R}_+$, and h an analytic function on $(0, s_1)$, such that*

$$\int_{\mathbb{R}_+} \mathrm{e}^{sx}\,\mu(dx) = h(s) \tag{A.1}$$

for all $s \in (0, s_0)$, for some numbers $0 < s_0 < s_1$. Then (A.1) *also holds for $s \in (0, s_1)$.*

Proof. Denote $f(s) = \int_{\mathbb{R}_+} \mathrm{e}^{sx}\,\mu(dx)$ and define $s_\infty = \sup\{s > 0 \mid f(s) < \infty\} \geq s_0$, such that

$$f(s) = +\infty \quad \text{for } s > s_\infty. \tag{A.2}$$

We assume, by contradiction, that $s_\infty < s_1$. Then there exists some $s_* \in (0, s_\infty)$ and $\varepsilon > 0$ such that $s_* < s_\infty < s_* + \varepsilon$ and such that h can be developed in an absolutely convergent power series

$$h(s) = \sum_{k \geq 0} \frac{c_k}{k!}(s - s_*)^k \quad \text{for } s \in (s_* - \varepsilon, s_* + \varepsilon).$$

In view of Lemma A.2, f is analytic, and thus $f = h$, on $(0, s_\infty)$. Hence we obtain, by dominated convergence,

$$c_k = \frac{d^k}{ds^k} h(s)\Big|_{s=s_*} = \frac{d^k}{ds^k} f(s)\Big|_{s=s_*} = \int_{\mathbb{R}_+} x^k \mathrm{e}^{s_* x}\,\mu(dx) \geq 0.$$

By monotone convergence, we conclude that for all $s \in (s_*, s_* + \varepsilon)$,

$$\begin{aligned} h(s) &= \sum_{k \geq 0} \int_{\mathbb{R}_+} \frac{x^k}{k!}(s - s_*)^k \mathrm{e}^{s_* x}\,\mu(dx) \\ &= \int_{\mathbb{R}_+} \sum_{k \geq 0} \frac{x^k}{k!}(s - s_*)^k \mathrm{e}^{s_* x}\,\mu(dx) = \int_{\mathbb{R}_+} \mathrm{e}^{sx}\,\mu(dx). \end{aligned}$$

But this contradicts (A.2). Whence $s_\infty \geq s_1$, and the lemma is proved. □

B Invariance and comparison results for differential equations

In this section we deliver invariance and comparison results for stochastic and ordinary differential equations, which are used in the proofs of the main Theorems 3.2, 3.3 and 4.1 and Lemma 7.2 above.

We start with an invariance result for the stochastic differential equation (2.1).

Lemma B.1. *Suppose b and ρ in* (2.1) *admit a continuous and measurable extension to $\mathbb{R}^d$, respectively, and such that a is continuous on $\mathbb{R}^d$. Let $u \in \mathbb{R}^d \setminus \{0\}$ and define the half space*

$$H = \{x \in \mathbb{R}^d \mid u^\top x \geq 0\},$$

its interior $H^0 = \{x \in \mathbb{R}^d \mid u^\top x > 0\}$, and its boundary $\partial H = \{x \in H \mid u^\top x = 0\}$.

1. *Fix* $x \in \partial H$ *and let* $X = X^x$ *be a solution of* (2.1). *If* $X(t) \in H$ *for all* $t \geq 0$, *then necessarily*

$$u^\top a(x)\, u = 0 \tag{B.1}$$

$$u^\top b(x) \geq 0. \tag{B.2}$$

2. *Conversely, if* (B.1) *and* (B.2) *hold for all* $x \in \mathbb{R}^d \setminus H^0$, *then any solution* X *of* (2.1) *with* $X(0) \in H$ *satisfies* $X(t) \in H$ *for all* $t \geq 0$.

Intuitively speaking, (B.1) means that the diffusion must be "parallel to the boundary", and (B.2) says that the drift must be "inward pointing" at the boundary of H.

Proof. Fix $x \in \partial H$ and let $X = X^x$ be a solution of (2.1). Hence

$$u^\top X(t) = \int_0^t u^\top b(X(s))\, ds + \int_0^t u^\top \rho(X(s))\, dW(s).$$

Since a and b are continuous, there exists a stopping time $\tau_1 > 0$ and a finite constant K such that

$$\left| u^\top b(X(t \wedge \tau_1)) \right| \leq K$$

and

$$\left\| u^\top \rho(X(t \wedge \tau_1)) \right\|^2 = u^\top a(X(t \wedge \tau_1))\, u \leq K$$

for all $t \geq 0$. In particular, the stochastic integral part of $u^\top X(t \wedge \tau_1)$ is a martingale. Hence

$$\mathbb{E}\big[u^\top X(t \wedge \tau_1)\big] = \mathbb{E}\left[\int_0^{t \wedge \tau_1} u^\top b(X(s))\, ds\right], \quad t \geq 0.$$

We now argue by contradiction, and assume first that $u^\top b(x) < 0$. By continuity of b and $X(t)$, there exists some $\varepsilon > 0$ and a stopping time $\tau_2 > 0$ such that $u^\top b(X(t)) \leq -\varepsilon$ for all $t \leq \tau_2$. In view of the above this implies

$$\mathbb{E}\big[u^\top X(\tau_2 \wedge \tau_1)\big] < 0.$$

This contradicts $X(t) \in H$ for all $t \geq 0$, whence (B.2) holds.

As for (B.1), let $C > 0$ be a finite constant and define the stochastic exponential $Z_t = \mathcal{E}(-C \int_0^t u^\top \rho(X)\, dW)$. Then Z is a strictly positive local martingale. Integration by parts yields

$$u^\top X(t) Z(t) = \int_0^t Z(s) \left(u^\top b(X(s)) - C\, u^\top a(X(s))\, u \right) ds + M(t)$$

where M is a local martingale. Hence there exists a stopping time $\tau_3 > 0$ such that for all $t \geq 0$,

$$\mathbb{E}\big[u^\top X(t \wedge \tau_3) Z(t \wedge \tau_3)\big] = \mathbb{E}\left[\int_0^{t \wedge \tau_3} Z(s) \left(u^\top b(X(s)) - C\, u^\top a(X(s))\, u \right) ds\right].$$

Now assume that $u^\top a(x)\,u > 0$. By continuity of a and $X(t)$, there exists some $\varepsilon > 0$ and a stopping time $\tau_4 > 0$ such that $u^\top a(X(t))\,u \geq \varepsilon$ for all $t \leq \tau_4$. For $C > K/\varepsilon$, this implies

$$\mathbb{E}\big[u^\top X(\tau_4 \wedge \tau_3 \wedge \tau_1) Z(\tau_4 \wedge \tau_3 \wedge \tau_1)\big] < 0.$$

This contradicts $X(t) \in H$ for all $t \geq 0$. Hence (B.1) holds, and part 1 is proved.

As for part 2, suppose (B.1) and (B.2) hold for all $x \in \mathbb{R}^d \setminus H^0$, and let X be a solution of (2.1) with $X(0) \in H$. For $\delta, \varepsilon > 0$ define the stopping time

$$\tau_{\delta,\varepsilon} = \inf\left\{t \mid u^\top X(t) \leq -\varepsilon \text{ and } u^\top X(s) < 0 \text{ for all } s \in [t-\delta, t]\right\}.$$

Then on $\{\tau_{\delta,\varepsilon} < \infty\}$ we have $u^\top \rho(X(s)) = 0$ for $\tau_{\delta,\varepsilon} - \delta \leq s \leq \tau_{\delta,\varepsilon}$ and thus

$$0 > u^\top X(\tau_{\delta,\varepsilon}) - u^\top X(\tau_{\delta,\varepsilon} - \delta) = \int_{\tau_{\delta,\varepsilon}-\delta}^{\tau_{\delta,\varepsilon}} u^\top b(X(s))\,ds \geq 0,$$

a contradiction. Hence $\tau_{\delta,\varepsilon} = \infty$. Since $\delta, \varepsilon > 0$ were arbitrary, we conclude that $u^\top X(t) \geq 0$ for all $t \geq 0$, as desired. Whence the lemma is proved. □

It is straightforward to extend Lemma B.1 towards a polyhedral convex set $\cap_{i=1}^k H_i$ with half-spaces $H_i = \{x \in \mathbb{R}^d \mid u_i^\top x \geq 0\}$, for some elements $u_1, \dots, u_k \in \mathbb{R}^d \setminus \{0\}$ and some $k \in \mathbb{N}$. This holds in particular for the canonical state space $\mathbb{R}^m_+ \times \mathbb{R}^n$. Moreover, Lemma B.1 includes time-inhomogeneous[1] ordinary differential equations as special case. The proofs of the following two corollaries are left to the reader.

Corollary B.2. *Let $H_i = \{x \in \mathbb{R}^d \mid x_i \geq 0\}$ denote the ith canonical half space in $\mathbb{R}^d$, for $i = 1, \dots, m$. Let $b : \mathbb{R}_+ \times \mathbb{R}^d \to \mathbb{R}^d$ be a continuous map satisfying, for all $t \geq 0$,*

$$b(t,x) = b(t, x_1^+, \dots, x_m^+, x_{m+1}, \dots, x_d) \quad \text{for all } x \in \mathbb{R}^d, \text{ and}$$
$$b_i(t,x) \geq 0 \quad \text{for all } x \in \partial H_i,\ i = 1, \dots, m.$$

Then any solution f of

$$\partial_t f(t) = b(t, f(t))$$

with $f(0) \in \mathbb{R}^m_+ \times \mathbb{R}^n$ satisfies $f(t) \in \mathbb{R}^m_+ \times \mathbb{R}^n$ for all $t \geq 0$.

Corollary B.3. *Let $B(t)$ and $C(t)$ be continuous $\mathbb{R}^{m\times m}$- and $\mathbb{R}^m_+$-valued parameters, respectively, such that $B_{ij}(t) \geq 0$ whenever $i \neq j$. Then the solution f of the linear differential equation in $\mathbb{R}^m$*

$$\partial_t f(t) = B(t)\,f(t) + C(t)$$

with $f(0) \in \mathbb{R}^m_+$ satisfies $f(t) \in \mathbb{R}^m_+$ for all $t \geq 0$.

Here and subsequently, we let $\succeq$ denote the partial order on $\mathbb{R}^m$ induced by the cone $\mathbb{R}^m_+$. That is, $x \succeq y$ if $x - y \in \mathbb{R}^m_+$. Then Corollary B.3 may be rephrased, for $C(t) \equiv 0$, by saying that the operator $e^{\int_0^t B(s)\,ds}$ is $\succeq$-order preserving, i.e. $e^{\int_0^t B(s)\,ds}\mathbb{R}^m_+ \subseteq \mathbb{R}^m_+$.

[1] Time-inhomogeneous differential equations can be made homogeneous by enlarging the state space.

Next, we consider time-inhomogeneous Riccati equations in $\mathbb{R}^m$ of the special form

$$\partial_t f_i(t) = A_i f_i(t)^2 + B_i^\top f(t) + C_i(t), \quad i = 1, \dots, m, \tag{B.3}$$

for some parameters $A, B, C(t)$ satisfying the following admissibility conditions

$$\begin{aligned} A &= (A_1, \dots, A_m) \in \mathbb{R}^m, \\ B_{i,j} &\geq 0 \quad \text{for } 1 \leq i \neq j \leq m, \\ C(t) &= (C_1(t), \dots, C_m(t)) \text{ continuous } \mathbb{R}^m\text{-valued.} \end{aligned} \tag{B.4}$$

The following lemma provides a comparison result for (B.3). It shows, in particular, that the solution of (B.3) is uniformly bounded from below on compacts with respect to $\succeq$ if $A \succeq 0$.

Lemma B.4. *Let $A^{(k)}, B, C^{(k)}$, $k = 1, 2$, be parameters satisfying the admissibility conditions* (B.4)*, and*

$$A^{(1)} \preceq A^{(2)}, \quad C^{(1)}(t) \preceq C^{(2)}(t). \tag{B.5}$$

Let $\tau > 0$ and $f^{(k)} : [0, \tau) \to \mathbb{R}^m$ be solutions of (B.4) *with A and C replaced by $A^{(k)}$ and $C^{(k)}$, respectively, $k = 1, 2$. If $f^{(1)}(0) \preceq f^{(2)}(0)$ then $f^{(1)}(t) \preceq f^{(2)}(t)$ for all $t \in [0, \tau)$. If, moreover, $A^{(1)} = 0$ then*

$$e^{Bt}\left(f^{(1)}(0) + \int_0^t e^{-Bs} C^{(1)}(s)\, ds\right) \preceq f^{(2)}(t)$$

for all $t \in [0, \tau)$.

Proof. The function $f = f^{(2)} - f^{(1)}$ solves

$$\begin{aligned} \partial_t f_i(t) &= A_i^{(2)} \left(f_i^{(2)}(t)\right)^2 - A_i^{(1)} \left(f_i^{(1)}(t)\right)^2 + B_i^\top f + C_i^{(2)}(t) - C_i^{(1)}(t) \\ &= \left(A_i^{(2)} - A_i^{(1)}\right)\left(f_i^{(2)}(t)\right)^2 + A_i^{(1)}\left(f_i^{(2)}(t) + f_i^{(1)}(t)\right) f_i(t) \\ &\quad + B_i^\top f(t) + C_i^{(2)}(t) - C_i^{(1)}(t) \\ &= \widetilde{B}_i(t)^\top f(t) + \widetilde{C}_i(t), \end{aligned}$$

where we write

$$\begin{aligned} \widetilde{B}_i(t) &= B_i + A_i^{(1)}\left(f_i^{(2)}(t) + f_i^{(1)}(t)\right) e_i, \\ \widetilde{C}_i(t) &= \left(A_i^{(2)} - A_i^{(1)}\right)\left(f_i^{(2)}(t)\right)^2 + C_i^{(2)}(t) - C_i^{(1)}(t). \end{aligned}$$

Note that $\widetilde{B} = (\widetilde{B}_{i,j})$ and $\widetilde{C}$ satisfy the assumptions of Corollary B.3 in lieu of B and C, and $f(0) \in \mathbb{R}_+^m$. Hence Corollary B.3 implies $f(t) \in \mathbb{R}_+^m$ for all $t \in [0, \tau)$, as desired. The last statement of the lemma follows by the variation of constants formula for $f^{(1)}(t)$. □

After these preliminary comparison results for the Riccati equation (B.3), we now can state and prove an important result for the system of Riccati equations (3.2). The following is an essential ingredient of the proof of Theorem 3.3. It is inspired by the line of arguments in Glasserman and Kim [16].

Lemma B.5. *Let $\mathcal{D}_{\mathbb{R}}$ denote the maximal domain for the system* (3.2) *of Riccati equations. Let $(\tau, u) \in \mathcal{D}_{\mathbb{R}}$. Then*

1. *$\mathcal{D}_{\mathbb{R}}(\tau)$ is star-shaped around zero.*
2. *$\theta^* = \sup\{\theta \geq 0 \mid \theta u \in \mathcal{D}_{\mathbb{R}}(\tau)\}$ satisfies either $\theta^* = \infty$ or $\lim_{\theta \uparrow \theta^*} \|\psi_I(t, \theta u)\| = \infty$. In the latter case, there exists some $x^* \in \mathbb{R}^m_+ \times \mathbb{R}^n$ such that $\lim_{\theta\uparrow\theta^*} \phi(\tau, \theta u) + \psi(\tau, \theta u)^\top x^* = \infty$.*

Proof. We first assume that the matrices α_i are block-diagonal, such that $\alpha_{i,iJ} = 0$, for all $i = 1, \ldots, m$.

Fix $\theta \in (0, 1]$. We claim that $\theta u \in \mathcal{D}_{\mathbb{R}}(\tau)$. It follows by inspection that $f^{(\theta)}(t) = \frac{\psi_I(t,\theta u)}{\theta}$ solves (B.3) with

$$A_i^{(\theta)} = \frac{1}{2}\theta\alpha_{i,ii}, \quad B = \mathcal{B}_{II}^\top, \quad C_i^{(\theta)}(t) = \beta_{i,J}^\top \psi_J(t, u) + \frac{1}{2}\psi_J(t, u)^\top \theta\alpha_{i,JJ}\psi_J(t, u),$$

and $f(0) = u$. Lemma B.4 thus implies that $f^{(\theta)}(t)$ is nice behaved, as

$$\mathrm{e}^{\mathcal{B}_{II}^\top t}\left(u + \int_0^t \mathrm{e}^{-\mathcal{B}_{II}^\top s} C^{(0)}(s)\, ds\right) \preceq f^{(\theta)}(t) \preceq \psi_I(t, u), \tag{B.6}$$

for all $t \in [0, t_+(\theta u)) \cap [0, \tau]$. By the maximality of $\mathcal{D}_{\mathbb{R}}$ we conclude that $\tau < t_+(\theta u)$, which implies $\theta u \in \mathcal{D}_{\mathbb{R}}(\tau)$, as desired. Hence $\mathcal{D}_{\mathbb{R}}(\tau)$ is star-shaped around zero, which is part 1.

Next suppose that $\theta^* < \infty$. Since $\mathcal{D}_{\mathbb{R}}(\tau)$ is open, this implies $\theta^* u \notin \mathcal{D}_{\mathbb{R}}(\tau)$ and thus $t_+(\theta^* u) \leq \tau$. From part 1 we know that $(t, \theta u) \in \mathcal{D}_{\mathbb{R}}$ for all $t < t_+(\theta^* u)$ and $0 \leq \theta \leq \theta^*$. On the other hand, there exists a sequence $t_n \uparrow t_+(\theta^* u)$ such that $\|\psi_I(t_n, \theta^* u)\| > n$ for all $n \in \mathcal{N}$. By continuity of ψ on $\mathcal{D}_{\mathbb{R}}$, we conclude that there exists some sequence $\theta_n \uparrow \theta^*$ with $\|\psi_I(t_n, \theta_n u) - \psi_I(t_n, \theta^* u)\| \leq 1/n$ and hence

$$\lim_n \|\psi_I(t_n, \theta_n u)\| = \infty. \tag{B.7}$$

Applying Lemma B.4 as above, where initial time $t = 0$ is shifted to t_n, yields

$$g_n := \mathrm{e}^{\mathcal{B}_{II}^\top(\tau - t_n)}\left(f^{(\theta_n)}(t_n) + \int_{t_n}^\tau \mathrm{e}^{\mathcal{B}_{II}^\top(t_n - s)} C^{(0)}(s)\, ds\right) \preceq f^{(\theta_n)}(\tau).$$

Corollary B.3 implies that $\mathrm{e}^{\mathcal{B}_{II}^\top(\tau - t_n)}$ is $\succeq$-order preserving. That is, $\mathrm{e}^{\mathcal{B}_{II}^\top(\tau-t_n)}\mathbb{R}^m_+ \subseteq \mathbb{R}^m_+$. Hence, in view of (B.6) for $f^{(\theta_n)}(t_n)$,

$$\begin{aligned} g_n &\succeq \mathrm{e}^{\mathcal{B}_{II}^\top(\tau - t_n)}\left(\mathrm{e}^{\mathcal{B}_{II}^\top t_n}\left(u + \int_0^{t_n} \mathrm{e}^{-\mathcal{B}_{II}^\top s} C^{(0)}(s)\, ds\right) + \int_{t_n}^\tau \mathrm{e}^{\mathcal{B}_{II}^\top(t_n - s)} C^{(0)}(s)\, ds\right) \\ &= \mathrm{e}^{\mathcal{B}_{II}^\top \tau}\left(u + \int_0^\tau \mathrm{e}^{-\mathcal{B}_{II}^\top s} C^{(0)}(s)\, ds\right). \end{aligned}$$

On the other hand, elementary operator norm inequalities yield

$$\|g_n\| \geq \mathrm{e}^{-\|\mathcal{B}_{II}\|\tau} \|f^{(\theta_n)}(t_n)\| - \mathrm{e}^{\|\mathcal{B}_{II}\|\tau} \tau \sup_{s\in[0,\tau]} \|C^{(0)}(s)\|.$$

Together with (B.7), this implies $\|g_n\| \to \infty$. From Lemma B.6 below we conclude that $\lim_n f^{(\theta_n)}(\tau)^\top y^* = \infty$ for some $y^* \in \mathbb{R}^m_+$. Moreover, in view of Lemma B.4, we know that $f^{(\theta)}(\tau)^\top y^*$ is increasing θ. Therefore $\lim_{\theta\uparrow\theta^*} f^{(\theta)}(\tau)^\top y^* = \infty$. Applying (B.6) and Lemma B.6 below again, this also implies that $\lim_{\theta\uparrow\theta^*} \|f^{(\theta)}(\tau)\| = \infty$. It remains to set $x^* = (y^*, 0)$ and observe that $b_I \in \mathbb{R}^m_+$ and thus

$$\phi(\tau, \theta u) = \int_0^\tau \left(\frac{1}{2} \psi_J(t, \theta u)^\top a_{JJ}\, \psi_J(t, \theta u) + b_I^\top \psi_I(t, \theta u) + b_J^\top \psi_J(t, \theta u) \right) dt$$

is uniformly bounded from below for all $\theta \in [0, \theta^*)$. Thus the lemma is proved under the premise that the matrices α_i are block-diagonal for all $i = 1, \dots, m$.

The general case of admissible parameters a, α_i, b, β_i is reduced to the preceding block-diagonal case by a linear transformation along the lines of Lemma 6.1. Indeed, define the invertible $d \times d$-matrix Λ

$$\Lambda = \begin{pmatrix} I_m & 0 \\ D & I_n \end{pmatrix}$$

where the $n \times m$-matrix $D = (\delta_1, \dots, \delta_m)$ has ith column vector

$$\delta_i = \begin{cases} -\frac{\alpha_{i,iJ}}{\alpha_{i,ii}}, & \text{if } \alpha_{i,ii} > 0 \\ 0, & \text{else.} \end{cases}$$

It is then not hard to see that $\Lambda(\mathbb{R}^m_+ \times \mathbb{R}^n) = \mathbb{R}^m_+ \times \mathbb{R}^n$, and

$$\widetilde{\phi}(t, u) = \phi(t, \Lambda^\top u), \quad \widetilde{\psi}(t, u) = \left(\Lambda^\top\right)^{-1} \psi(t, \Lambda^\top u)$$

satisfy the system of Riccati equations (3.2) with a, α_i, b, and $\mathcal{B} = (\beta_1, \dots, \beta_d)$ replaced by the admissible parameters

$$\widetilde{a} = \Lambda a \Lambda^\top, \quad \widetilde{\alpha_i} = \Lambda \alpha_i \Lambda^\top, \quad \widetilde{b} = \Lambda b, \quad \widetilde{\mathcal{B}} = \Lambda \mathcal{B} \Lambda^{-1}.$$

Moreover, $\widetilde{\alpha_i}$ are block-diagonal, for all $i = 1, \dots, m$. By the first part of the proof, the corresponding maximal domain $\widetilde{\mathcal{D}_{\mathbb{R}}}(\tau)$, and hence also $\mathcal{D}_{\mathbb{R}}(\tau) = \Lambda^\top \widetilde{\mathcal{D}_{\mathbb{R}}}(\tau)$, is star-shaped around zero. Moreover, if $\theta^* < \infty$, then

$$\lim_{\theta\uparrow\theta^*} \|\psi_I(\tau, \theta u)\| = \lim_{\theta\uparrow\theta^*} \left\| \widetilde{\psi}_I \left(\tau, \theta \left(\Lambda^\top\right)^{-1} u\right) \right\| = \infty,$$

and there exists some $x^* \in \mathbb{R}^m_+ \times \mathbb{R}^n$ such that

$$\begin{aligned} &\lim_{\theta\uparrow\theta^*} \phi(\tau, \theta u) + \psi(\tau, \theta u)^\top x^* \\ &\quad = \lim_{\theta\uparrow\theta^*} \widetilde{\phi}\left(\tau, \theta \left(\Lambda^\top\right)^{-1} u\right) + \widetilde{\psi}\left(\tau, \theta \left(\Lambda^\top\right)^{-1} u\right)^\top \Lambda x^* = \infty. \end{aligned}$$

Hence the lemma is proved. □

Lemma B.6. *Let $c \in \mathbb{R}^m$, and (c_n) and (d_n) be sequences in $\mathbb{R}^m$ such that*

$$c \preceq c_n \preceq d_n$$

for all $n \in \mathcal{N}$. Then the following are equivalent

1. $\|c_n\| \to \infty$
2. $c_n^\top y^* \to \infty$ *for some* $y^* \in \mathbb{R}^m_+ \setminus \{0\}$.

In either case, $\|d_n\| \to \infty$ *and* $d_n^\top y^* \to \infty$.

Proof. $1 \Rightarrow 2$: since $\|c_n\|^2 = \sum_{i=1}^m (c_n^\top e_i)^2$ and $c_n^\top e_i \geq c^\top e_i$, we conclude that $c_n^\top e_i \to \infty$ for some $i = 1, \dots, m$.

$2 \Rightarrow 1$: this follows from $\|c_n^\top y^*\| \leq \|c_n\| \|y^*\|$.

The last statement now follows since $d_n^\top y^* \geq c_n^\top y^*$. □

Finally, we sketch an alternative proof of Theorem 3.3 part 1 which avoids probabilistic arguments.

Remark B.7. We may without loss of generality assume block-diagonal form of α_i, $i = 1, \dots, d$ (cf. the final part of the proof of Lemma B.5). Assume, by contradiction, that for some $v \in \mathbb{R}^d$, $t_+(u + iv) < t_+(u)$. Then, as in the first proof, we may deduce the existence of $t_n \uparrow t_+(u + iv)$ such that

$$\lim_n (\operatorname{Re} \psi_i(t_n, u + iv))^+ = \infty. \tag{B.8}$$

holds for some $i \in \{1, \dots, m\}$. Set $g(t, u+iv) := \operatorname{Re}(\psi_t, u+iv)$, $h := \operatorname{Im}(\psi(t, u+iv)$. Then for $i = 1, \dots, m$ the following differential inequality holds,

$$\begin{aligned} \dot{g}_i(t, u + iv) &= \frac{1}{2}\alpha_{i,ii}(g_i^2 - h_i^2) + g_J^\top \alpha_{i,JJ} g_J - h_J^\top \alpha_{i,JJ} h_J + \beta_i^\top g \\ &\leq \frac{1}{2}\alpha_{i,ii} g_i^2 + g_J^\top \alpha_{i,JJ} g_J + \beta_i^\top g \end{aligned} \tag{B.9}$$

and $g(t = 0, u + iv)) = \psi(t = 0, u) = u$. Hence noting $g_J(t, u + iv) = \psi_J(t, u)$ we obtain by Lemma B.4 for all $t \in (0, t_+(u + iv))$

$$\operatorname{Re} \psi(t, u + iv) = g(t, u + iv) \preceq \psi(t, u).$$

On the other hand, $\psi_I(t, u) \preceq M$ for some positive constant $M \in \mathbb{R}^m_+$, for all $t \in [0, t_+(u + iv)]$, hence $\operatorname{Re} \psi_i(t, u + iv) \leq M_i$, which contradicts (B.8).

Bibliography

[1] H. Amann, *Ordinary differential equations*, de Gruyter Studies in Mathematics, vol. 13, Walter de Gruyter & Co., Berlin, 1990, An introduction to nonlinear analysis, Translated from the German by Gerhard Metzen. MR MR1071170 (91e:34001)

[2] L. B. G. Andersen and V. V. Piterbarg, *Moment explosions in stochastic volatility models*, Finance and Stochastics 11 (2007), pp. 29–50. MR MR2284011 (2008a:65016)

[3] D. Brigo and F. Mercurio, *Interest rate models—theory and practice*, second. ed., Springer Finance, Springer-Verlag, Berlin, 2006, With smile, inflation and credit. MR MR2255741 (2007d:91002)

[4] R. H. Brown, Schaefer S. M, Rogers, L. C. G., S. Mehta, and J. Pezier, *Interest Rate Volatility and the Shape of the Term Structure [and Discussion]*, Philosophical Transactions of the Royal Society of London, Series A 347 (1994), pp. 563–576.

[5] M.-F. Bru, *Wishart Processes*, Journal of Theoretical Probability 4 (1991), pp. 725–751.

[6] B. Buraschi, P. Porchia, and F. Trojani, *Correlation risk and optimal portfolio choice*, Working paper, University St.Gallen, 2006.

[7] Li Chen, Damir Filipović, and H. Vincent Poor, *Quadratic term structure models for risk-free and defaultable rates*, Math. Finance 14 (2004), pp. 515–536. MR MR2092921 (2005f:91066)

[8] P. Cheridito, D. Filipović, and R. L. Kimmel, *A Note on the Dai–Singleton canonical representation of affine term structure models*, Forthcoming in Mathematical Finance, 2008.

[9] P. Collin-Dufresne, R. S. Goldstein, and C. S. Jones, *Identification of Maximal Affine Term Structure Models*, J. of Finance 63 (2008), pp. 743–795.

[10] Q. Dai and K. J. Singleton, *Specification Analysis of Affine Term Structure Models*, J. of Finance 55 (2000), pp. 1943–1978.

[11] J. Dieudonné, *Foundations of modern analysis*, Pure and Applied Mathematics, Vol. X, Academic Press, New York, 1960. MR MR0120319 (22 #11074)

[12] D. Duffie, D. Filipović, and W. Schachermayer, *Affine processes and applications in finance*, Ann. Appl. Probab. 13 (2003), pp. 984–1053. MR MR1994043 (2004g:60107)

[13] D. Duffie and R. Kan, *A Yield-Factor Model of Interest Rates*, Mathematical Finance 6 (1996), pp. 379–406.

[14] Darrell Duffie, Jun Pan, and Kenneth Singleton, *Transform analysis and asset pricing for affine jump-diffusions*, Econometrica. Journal of the Econometric Society 68 (2000), pp. 1343–1376. MR MR1793362 (2001m:91081)

[15] J. Fonseca, M. Grasseli, and C. Tebaldi, *A Multi-Factor volatility Heston model*, forthcoming in Quantitative Finance, 2009.

[16] P. Glasserman and K-K. Kim, *Moment Explosions and Stationary Distributions in Affine Diffusion Models*, To appear in Mathematical Finance, 2008/2009.

[17] C. Gourieroux and R. Sufana, *Wishart Quadratic Term structure models*, Working paper, CREF HRC Montreal, 2003.

[18] C. Gourieroux and R. Sufana, *A Classification of Two Factor Affine Diffusion Term Structure Models*, J. of Financial Econometrics 4 (2006), pp. 31–52.

[19] M. Grasseli and C. Tebaldi, *Solvable Affine Term structure models*, Mathematical Finance 18 (2008), pp. 135–153.

[20] S. Heston, *A closed-form solution for options with stochastic volatility with appliactions to bond and currency options*, Rev. of Financial Studies.

[21] F. Hubalek, J. Kallsen, and L. Krawczyk, *Variance-optimal hedging for processes with stationary independent increments*, Ann. Appl. Probab. 16 (2006), pp. 853–885. MR MR2244435 (2007k:60205)

[22] N. Ikeda and S. Watanabe, *Stochastic differential equations and diffusion processes*, North-Holland Mathematical Library, vol. 24, North-Holland Publishing Co., Amsterdam, 1981. MR MR637061 (84b:60080)

[23] Norman L. Johnson and Samuel Kotz, *Distributions in statistics. Continuous univariate distributions. 2.*, Houghton Mifflin Co., Boston, Mass., 1970. MR MR0270476 (42 #5364)

[24] S. Joslin, *Can Unspanned Stochastic Volatility Models Explain the Cross Section of Bond Volatilities?*, Working Paper, Stanford University, 2006.

[25] I. Karatzas and S. E. Shreve, *Brownian motion and stochastic calculus*, second. ed., Graduate Texts in Mathematics, vol. 113, Springer-Verlag, New York, 1991. MR MR1121940 (92h:60127)

[26] M. Keller-Ressel, *Moment Explosions and Long-Term Behavior of Affine Stochastic Volatility Models*, to appear in Mathematical Finance.

[27] ———, *Affine Processes- Theory and Applications in Finance*, PhD thesis Vienna University of Technology (January, 2009).

[28] V. Lakshmikantham, N. Shahzad, and W. Walter, *Convex dependence of solutions of differential equations in a Banach space relative to initial data*, Nonlinear Anal. 27 (1996), pp. 1351–1354. MR MR1408875 (97e:34109)

[29] Roger W. Lee, *The moment formula for implied volatility at extreme strikes*, Mathematical Finance. An International Journal of Mathematics, Statistics and Financial Economics 14 (2004), pp. 469–480. MR MR2070174 (2005b:91122)

[30] E. Lukacs, *Characteristic functions*, Hafner Publishing Co., New York, 1970, Second edition, revised and enlarged. MR MR0346874 (49 #11595)

[31] Arnold Neumaier, *Introduction to numerical analysis*, Cambridge University Press, Cambridge, 2001. MR MR1854534 (2002g:65002)

[32] R. T. Rockafellar, *Convex analysis*, Princeton Landmarks in Mathematics, Princeton University Press, Princeton, NJ, 1997, Reprint of the 1970 original, Princeton Paperbacks. MR MR1451876 (97m:49001)

[33] E. M. Stein and G. Weiss, *Introduction to Fourier analysis on Euclidean spaces*, Princeton University Press, Princeton, N.J., 1971, Princeton Mathematical Series, No. 32. MR MR0304972 (46 #4102)

[34] D. Williams, *Probability with martingales*, Cambridge Mathematical Textbooks, Cambridge University Press, 1991.

[35] T. Yamada and S. Watanabe, *On the uniqueness of solutions of stochastic differential equations*, J. Math. Kyoto Univ. 11 (1971), pp. 155–167. MR MR0278420 (43 #4150)

Author information

Damir Filipović, Vienna Institute of Finance, University of Vienna and Vienna University of Economics and Business Administration, Heiligenstädter Straße 46-48, A-1190 Wien, Austria.
Email: damir.filipovic@vif.ac.at

Eberhard Mayerhofer, Vienna Institute of Finance, University of Vienna and Vienna University of Economics and Business Administration, Heiligenstädter Straße 46-48, A-1190 Wien, Austria.
Email: eberhard.mayerhofer@vif.ac.at

Radon Series Comp. Appl. Math **8**, 165–181

Multilevel quasi-Monte Carlo path simulation

Michael B. Giles and Benjamin J. Waterhouse

Abstract. This paper reviews the multilevel Monte Carlo path simulation method for estimating option prices in computational finance, and extends it by combining it with quasi-Monte Carlo integration using a randomised rank-1 lattice rule. Using the Milstein discretisation of the stochastic differential equation, it is demonstrated that the combination has much lower computational cost than either one on its own for evaluating European, Asian, lookback, barrier and digital options.

Key words. Multilevel, Monte Carlo, quasi-Monte Carlo, computational finance.

AMS classification. 11K45, 60H10, 60H35, 65C05, 65C30, 68U20

1 Introduction

Giles [4, 5] has recently introduced a multilevel Monte Carlo path simulation method for the pricing of financial options. This improves the computational efficiency of Monte Carlo path simulation by combining results using different numbers of timesteps. This can be viewed as a generalisation of the two-level method of Kebaier [9] and is also similar in approach to Heinrich's multilevel method for parametric integration [7]. The first paper [5] (which was the second to appear in print due to a publication backlog) introduced the multilevel Monte Carlo method and proved that it can lower the computational complexity of path-dependent Monte Carlo evaluations. It also presented numerical results using the simplest Euler-Maruyama discretisation. The second paper [4] demonstrated that the computational cost can be further reduced by using the Milstein discretisation. This has the same weak order of convergence but an improved first order strong convergence, and it is the strong order of convergence which is central to the efficiency of the multilevel method.

In this paper we review the key ideas and introduce a new ingredient, the use of quasi-Monte Carlo (QMC) integration based on a randomised rank-1 lattice rule which further reduces the computational cost. To set the scene, we consider a scalar SDE with general drift and volatility terms,

$$\mathrm{d}S(t) = a(S,t)\,\mathrm{d}t + b(S,t)\,\mathrm{d}W(t), \quad 0 < t < T, \tag{1.1}$$

with given initial data S_0. In the case of European and digital options, we are interested in the expected value of a function of the terminal state, $f(S(T))$, but in

First author: supported by Microsoft Corporation, the UK Engineering and Physical Sciences Research Council and the Oxford-Man Institute of Quantitative Finance.
Second author: supported by the Australian Research Council through a Linkage project between the University of New South Wales and Macquarie Bank.

the case of Asian, lookback and barrier options the valuation depends on the entire path $S(t), 0<t<T$.

Using a simple Monte Carlo method with a numerical discretisation with first order weak convergence, to achieve a r. m. s. error of ϵ would require $O(\epsilon^{-2})$ independent paths, each with $O(\epsilon^{-1})$ timesteps, giving a computational complexity which is $O(\epsilon^{-3})$. With the Euler–Maruyama discretisation the multilevel method reduces the cost to $O(\epsilon^{-2}(\log \epsilon)^2)$ for a European option with a payoff with a uniform Lipschitz bound [5], while the use of the Milstein discretisation further reduces the cost to $O(\epsilon^{-2})$ for a larger class of options, including Asian, lookback, barrier and digital options [4]. The paper begins by reviewing the multilevel approach, first with the Euler path discretisation and then with the superior Milstein discretisation. QMC methods based on rank-1 lattice rules are then introduced, with particular attention to Brownian Bridge construction and the use of randomisation to obtain confidence intervals. The combined multilevel QMC algorithm is presented and the following section provides numerical results for a range of options.

2 Multilevel Monte Carlo method

Consider Monte Carlo path simulations with different timesteps $h_l = 2^{-l}T$, $l = 0, 1, \ldots, L$. Thus on the coarsest level, $l=0$, the simulations use just 1 timestep, while on the finest level, $l=L$, the simulations use 2^L timesteps. For a given Brownian path $W(t)$, let P denote the payoff, and let $\widehat{P}_l$ denote its approximation using a numerical discretisation with timestep h_l. Because of the linearity of the expectation operator, it is clearly true that

$$\mathbb{E}[\widehat{P}_L] = \mathbb{E}[\widehat{P}_0] + \sum_{l=1}^{L} \mathbb{E}[\widehat{P}_l - \widehat{P}_{l-1}]. \tag{2.1}$$

This expresses the expectation on the finest level as being equal to the expectation on the coarsest level plus a sum of corrections which give the difference in expectation between simulations using different numbers of timesteps. The idea behind the multilevel method is to independently estimate each of the expectations on the right-hand side in a way which minimises the overall variance for a given computational cost.

Let $\widehat{Y}_0$ be an estimator for $\mathbb{E}[\widehat{P}_0]$ using N_0 samples, and let $\widehat{Y}_l$ for $l>0$ be an estimator for $\mathbb{E}[\widehat{P}_l - \widehat{P}_{l-1}]$ using N_l paths. The simplest estimator is a mean of N_l independent samples, which for $l>0$ is

$$\widehat{Y}_l = N_l^{-1} \sum_{i=1}^{N_l} \left(\widehat{P}_l^{(i)} - \widehat{P}_{l-1}^{(i)} \right). \tag{2.2}$$

The key point here is that the quantity $\widehat{P}_l^{(i)} - \widehat{P}_{l-1}^{(i)}$ comes from two discrete approximations with different timesteps but the same Brownian path. The variance of this simple estimator is $\mathbb{V}[\widehat{Y}_l] = N_l^{-1} V_l$ where V_l is the variance of a single sample. Combining this with independent estimators for each of the other levels, the variance of the combined estimator $\sum_{l=0}^{L} \widehat{Y}_l$ is $\sum_{l=0}^{L} N_l^{-1} V_l$, while its computational cost is proportional to $\sum_{l=0}^{L} N_l\, h_l^{-1}$. Treating the N_l as continuous variables, the variance is minimised for a fixed computational cost by choosing N_l to be proportional to $\sqrt{V_l\, h_l}$.

In the particular case of an Euler discretisation, provided $a(S,t)$ and $b(S,t)$ satisfy certain conditions [2, 10, 21] there is $O(h^{1/2})$ strong convergence. From this it follows that $\mathbb{V}[\widehat{P}_l - P] = O(h_l)$ for a European option with a Lipschitz continuous payoff. Hence for the simple estimator (2.2), the single sample variance V_l is $O(h_l)$, and the optimal choice for N_l is asymptotically proportional to h_l. Setting $N_l = O(\epsilon^{-2} L\, h_l)$, the variance of the combined estimator $\widehat{Y}$ is $O(\epsilon^2)$. If L is chosen such that $L = \log \epsilon^{-1}/\log 2 + O(1)$, as $\epsilon \to 0$, then $h_L = 2^{-L} = O(\epsilon)$, and so the bias error $\mathbb{E}[\widehat{P}_L - P]$ is $O(\epsilon)$ due to standard results on weak convergence. Consequently, we obtain a mean square error which is $O(\epsilon^2)$, with a computational complexity which is $O(\epsilon^{-2} L^2) = O(\epsilon^{-2} (\log \epsilon)^2)$.

This analysis is generalised in the following theorem [5]:

Theorem 2.1. *Let P denote a functional of the solution of stochastic differential equation (1.1) for a given Brownian path $W(t)$, and let $\widehat{P}_l$ denote the corresponding approximation using a numerical discretisation with timestep $h_l = M^{-l}\, T$.*

If there exist independent estimators $\widehat{Y}_l$ based on N_l Monte Carlo samples, and positive constants $\alpha \geq \frac{1}{2}, \beta, c_1, c_2, c_3$ such that

i) $\left|\mathbb{E}[\widehat{P}_l - P]\right| \leq c_1\, h_l^\alpha$

ii) $\mathbb{E}[\widehat{Y}_l] = \begin{cases} \mathbb{E}[\widehat{P}_0], & l = 0 \\ \mathbb{E}[\widehat{P}_l - \widehat{P}_{l-1}], & l > 0 \end{cases}$

iii) $\mathbb{V}[\widehat{Y}_l] \leq c_2\, N_l^{-1} h_l^\beta$

iv) C_l, the computational complexity of $\widehat{Y}_l$, is bounded by

$$C_l \leq c_3\, N_l\, h_l^{-1},$$

then there exists a positive constant c_4 such that for any $\epsilon < e^{-1}$ there are values L and N_l for which the multilevel estimator

$$\widehat{Y} = \sum_{l=0}^{L} \widehat{Y}_l,$$

has a mean-square-error with bound

$$MSE \equiv \mathbb{E}\left[\left(\widehat{Y} - \mathbb{E}[P]\right)^2\right] < \epsilon^2$$

with a computational complexity C with bound

$$C \leq \begin{cases} c_4\, \epsilon^{-2}, & \beta > 1, \\ c_4\, \epsilon^{-2} (\log \epsilon)^2, & \beta = 1, \\ c_4\, \epsilon^{-2-(1-\beta)/\alpha}, & 0 < \beta < 1. \end{cases}$$

3 Milstein discretisation

The theorem proves that the best order of complexity is achieved using discretisations with $\beta > 1$. To achieve this for a scalar SDE, we use the Milstein discretisation of equation (1.1) which is

$$\widehat{S}_{n+1} = \widehat{S}_n + a_n\, h + b_n\, \Delta W_n + \frac{1}{2}\frac{\partial b_n}{\partial S}\, b_n \left((\Delta W_n)^2 - h \right). \tag{3.1}$$

In the above equation, the subscript n is used to denote the timestep index, and a_n, b_n and $\partial b_n/\partial S$ are evaluated at $\widehat{S}_n, t_n$.

All of the numerical results to be presented are for the case of geometric Brownian motion for which the SDE is

$$\mathrm{d}S(t) = r\, S\, \mathrm{d}t + \sigma\, S\, \mathrm{d}W(t), \quad 0 < t < T.$$

By switching to the new variable $X = \log S$, it is possible to construct numerical approximations which are exact, but here we directly simulate the geometric Brownian motion using the Milstein method as an indication of the behaviour with more complicated models, for example those with a local volatility function $\sigma(S,t)$.

The Milstein discretisation defines the numerical approximation at the discrete times t_n. Within the time interval $[t_n, t_{n+1}]$ we use a constant coefficient Brownian interpolation conditional on the two end values,

$$\widehat{S}(t) = \widehat{S}_n + \lambda\, (\widehat{S}_{n+1} - \widehat{S}_n) + b_n \Big(W(t) - W_n - \lambda\, (W_{n+1} - W_n) \Big), \tag{3.2}$$

where

$$\lambda = \frac{t - t_n}{t_{n+1} - t_n}.$$

For the fine path, standard results on i) the expected average value, ii) the distribution of the minimum, and iii) the probability of crossing a certain value, will be used to obtain the value $\widehat{P}_l$ for Asian, lookback and barrier options, respectively.

Exactly the same approach could also be used on the coarse path with half as many timesteps to obtain $\widehat{P}_{l-1}$. However, this would not give an estimator $\widehat{Y}_l$ with variance convergence rate $\beta > 1$. To achieve the better convergence rate, we first use the value of the underlying Brownian motion $W(t)$ at the midpoint (which has already been sampled and used for the fine path calculation) to define an interpolated midpoint

$$\widehat{S}_{n+\frac{1}{2}} = \frac{1}{2}(\widehat{S}_{n+1} + \widehat{S}_n) + b_n \left(W_{n+\frac{1}{2}} - \frac{1}{2}(W_{n+1} + W_n) \right). \tag{3.3}$$

We can then use the Brownian interpolation (with volatility b_n) on each of the half-intervals $[t_n, t_{n+\frac{1}{2}}]$ and $[t_{n+\frac{1}{2}}, t_{n+1}]$ which each correspond to one of the timesteps on the fine path. A key point in this construction is that we have not altered the expected value for $\widehat{P}_{l-1}$, averaged over all underlying Brownian paths $W(t)$, compared to its evaluation on level $l-1$ on which it corresponds to the finer path; see [4] for further discussion of this important point.

4 Quasi-Monte Carlo method

QMC methods approximate an integral on a high-dimensional hypercube with an N-point equal-weight quadrature rule of the form

$$\int_{[0,1]^d} f(x)\,\mathrm{d}x \approx \frac{1}{N}\sum_{i=0}^{N-1} f(x_i).$$

This is the same form which is used in the Monte Carlo method. However, rather than choosing the d-dimensional points x_i uniformly from the unit cube, as is the case with the Monte Carlo method, QMC methods choose the points in some deterministic manner.

Sobol′ sequences [20] and digital nets [15] are two popular choices of QMC points, which have been previously used for financial applications [6, 12]. In this paper we use a rank-1 lattice rule [19] in which the points have the particularly simple construction

$$x_i = \left\{\frac{i}{N} z\right\},$$

where z is a d-dimensional vector with integer components and the notation $\{\cdot\}$ denotes taking the fractional part of each component of the argument and disregarding the integer part so that x_i lies within the half-open unit cube.

For Monte Carlo integration it is well known that the error is $O(N^{-1/2})$. In one dimension, the lattice rule is equivalent to a rectangle rule and can achieve $O(N^{-1})$ convergence of the error, for a sufficiently smooth integrand. For larger dimensions, it may be shown that for integrands with sufficient smoothness and dimensions which become progressively less important, there exist lattice rules for which the error decays at $O(N^{-1+\varepsilon})$ for all $\varepsilon > 0$, see [11]. Unfortunately, many integrands in mathematical finance applications do not have the required smoothness and so we may not apply the theory to claim the $O(N^{-1+\varepsilon})$ convergence. However, experimentation suggests that this rate can in fact be achieved for many finance problems [3].

Two key aspects of the implementation of QMC methods are randomisation and the factorisation of the covariance matrix. If we neglect for the moment the discretisation errors which arise from finite timesteps, the standard Monte Carlo method has the attractive feature that it provides both an unbiased estimate of the desired value and a confidence interval for that estimate. The QMC method lacks this feature but it can be regained by re-defining the ith point to be

$$x_i = \left\{\frac{i}{N} z + \boldsymbol{\Delta}\right\}.$$

For a given offset vector $\boldsymbol{\Delta} \in [0,1)^d$, this defines a set of N points, for which one can compute the average

$$\widehat{Y} = \frac{1}{N}\sum_{i=0}^{N-1} f(x_i).$$

If we now treat $\boldsymbol{\Delta}$ as a random variable then the expected value of $\widehat{Y}$ is equal to the desired integral, and therefore $\widehat{Y}$ is an unbiased estimator. By choosing a number of different random offsets $\boldsymbol{\Delta}_1, \ldots, \boldsymbol{\Delta}_q$ ($q = 32$ is used in this paper) and computing a separate $\widehat{Y}_j$ for each, one can construct a confidence interval in the usual way.

For a scalar SDE with n_T timesteps, the dimensionality of the problem is $d = n_T$, and the factorisation of the covariance matrix concerns the question of how best to map the different dimensions of the hypercube to the n_T Wiener increments in the Milstein discretisation. The expected value of a financial product whose value is determined by an asset whose dynamics are described by (1.1), discretised at times $t_n = nh$, is given by the integral

$$\int_{\mathbb{R}^d} p(x)\, \frac{\exp\left(-\frac{1}{2}\, x^T \Sigma^{-1} x\right)}{(2\pi)^{d/2}\, \sqrt{\det \Sigma}}\, \mathrm{d}x.$$

Here $p(x)$ is the payoff function and the d-dimensional matrix $\Sigma_{i,j} = \min(t_i, t_j)$ is the covariance matrix for the elements of x which are the underlying Wiener path values W_n. Taking a matrix A such that $A\,A^T = \Sigma$, and making the substitutions $x = A\,y$ and $y = \Phi^{-1}(z)$ where Φ^{-1} is the inverse of the cumulative Normal distribution function taken componentwise, this can be reformulated as an integral over the unit cube

$$\int_{\mathbb{R}^d} p(A\,y)\, \frac{\exp\left(-\frac{1}{2} y^T y\right)}{(2\pi)^{d/2}}\, \mathrm{d}y = \int_{[0,1]^d} p(A\,\Phi^{-1}(z))\, \mathrm{d}z.$$

For Monte Carlo integration the choice of the matrix A makes no difference, but for QMC integration it is very important [18, 6, 12]. While any choice of A such that $A\,A^T = \Sigma$ is suitable, there are three established ways in which the matrix A may be chosen. Firstly, A may be chosen to be the Cholesky factor of Σ. This is the simplest method and corresponds to taking the nth component of x_i to define ΔW_n through

$$\Delta W_n = \sqrt{h}\, \Phi^{-1}(x_{i,n}).$$

This would correctly map a uniform $[0, 1]$ distribution for x_n into a Normal distribution for ΔW_n with zero mean and variance h. This method is often referred to as the standard construction and is usually used for Monte Carlo integration due to the simplicity of its construction.

A second way in which A may be chosen is to use a Brownian Bridge construction [18, 6]. Under this method, the first component of x is used to define $W(T)$, the second component defines $W(T/2)$ (conditional on the first), the third and fourth components define $W(T/4)$ and $W(3T/4)$ (conditional on the first two), and so on. Note that in the standard and Brownian Bridge constructions, the matrix A is not explicitly used, but rather implicitly used in the recursive construction.

The final way is known as the “Principal Components Analysis” (PCA) method. In this method A is chosen to be the matrix with n^{th} column equal to $\sqrt{\lambda_n} v_n$ where λ_n is the nth largest eigenvalue of Σ and v_n is the corresponding eigenvector [6].

Several authors [18, 12, 6] have found the Brownian Bridge and PCA constructions to be much better for some problems, although it is known that there are problems from mathematical finance for which the standard construction performs much better than the Brownian Bridge, see [17]. In our numerical experiments we use the Brownian

Bridge construction, since for our applications it consistently outperforms the standard construction.

The final implementation issue is the choice of the generating vector z. We use a vector using the construction algorithm of Dick *et al.* [8]. This particular type of lattice rule is said to be *embedded* since it can be used as a sequence with differing values of N. The construction algorithm is particularly efficient due to the fast FFT implementation technique of Nuyens and Cools [16].

5 Multilevel QMC algorithm

At level l in the multilevel formulation, N_l is defined to be the number of QMC points, and $\widehat{Y}_l$ is the computed average of $\widehat{P}_l$ (for $l=0$) or $\widehat{P}_l - \widehat{P}_{l-1}$ (for $l>0$) over the 32 sets of N_l QMC lattice points, each set having a different random offset. An unbiased estimate of its variance V_l is computed in the usual way from the differing values for the 32 averages.

On the assumption that there is first order weak convergence, the remaining bias at the finest level $\mathbb{E}[P-\widehat{P}_L]$ is approximately equal to $\widehat{Y}_L$. Being more cautious (to allow for the possibility that $\widehat{Y}_l$ changes sign as l increases before settling into its first order asymptotic convergence) we estimate the magnitude of the bias using

$$\max\left\{\frac{1}{2}|\widehat{Y}_{L-1}|, |\widehat{Y}_L|\right\}.$$

The mean square error is the sum of the combined variance $\sum_{l=0}^{L} V_l$ (where V_l is now the variance of $\widehat{V}_l$) plus the square of the bias $\mathbb{E}[P-\widehat{P}_L]$. We choose to make each of these smaller than $\epsilon^2/2$, so that overall we achieve a user-specified RMS accuracy of ϵ. The variance is reduced by increasing the number of lattice points on each level, while the bias is reduced by increasing the level of path refinement (i.e. increasing L).

Given this outline strategy, the multilevel QMC algorithm proceeds as follows:

1. start with $L=0$
2. get an initial estimate for V_L using 32 random offsets and $N_L=1$
3. while $\displaystyle\sum_{l=0}^{L} V_l > \epsilon^2/2$, double N_l on the level with largest $V_l \,/\, (2^l N_l)$
4. if $L<2$ or the bias estimate is greater than $\epsilon/\sqrt{2}$, set $L := L+1$ and go to step 2

Step 3 is based on the fact that doubling N_l will eliminate most of the variance V_l at a cost proportional to the product of the number of timesteps 2^l and the number of lattice points N_l. The choice of level l aims to maximise the reduction in variance per unit cost.

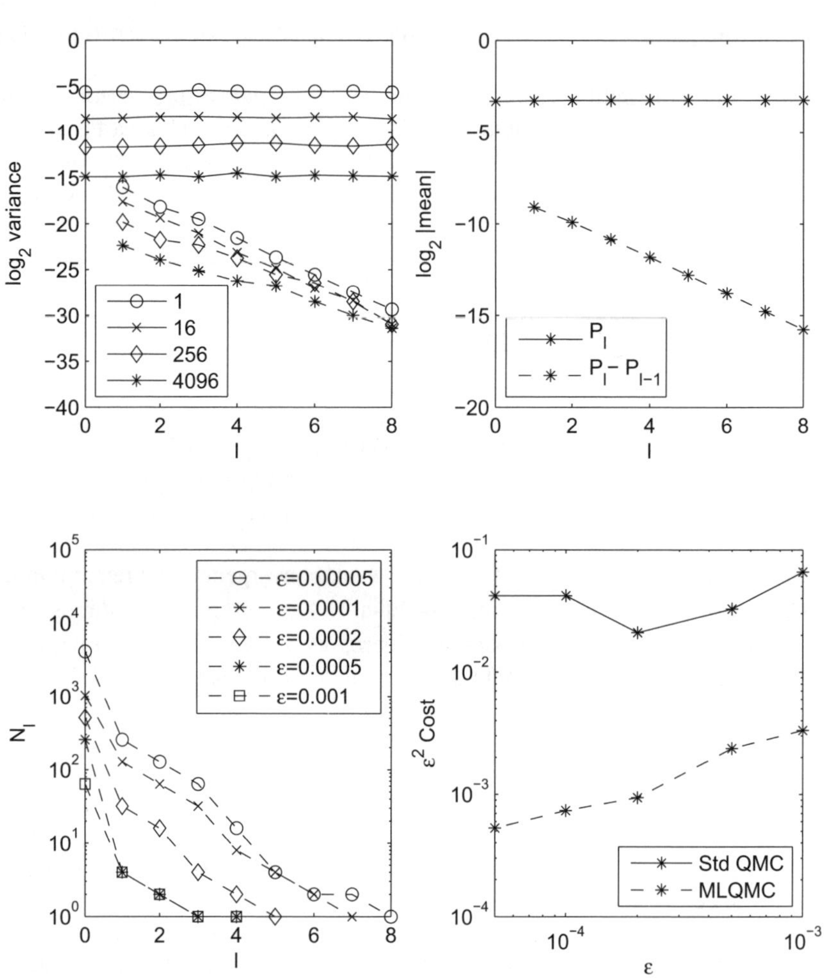

Figure 6.1. European call option

6 Numerical results

6.1 European call option

The European call option we consider has the discounted payoff

$$P = \exp(-rT)\ (S(T) - K)^{+},$$

where the notation $(x)^{+}$ denotes $\max(0, x)$. Figure 6.1 shows the numerical results for parameters $S(0)=1$, $K=1$, $T=1$, $r=0.05$, $\sigma=0.2$.

The solid lines in the top left plot show the behaviour of the variance $\widehat{P}_l$, while the dashed lines show the variance of $\widehat{P}_l - \widehat{P}_{l-1}$. The four sets of calculations use different numbers of lattice points. The calculations with just one lattice point correspond to standard Monte Carlo. The calculations with 16, 256 and 4094 lattice points show the variance of the average over the set of lattice points multiplied by the number of lattice points; for standard Monte Carlo this quantity would be independent of the number of points, and therefore this is a fair basis of comparison which accounts for the cost of 4096 points being 4096 times greater than a single point. The solid line results show that the QMC method on its own is very effective in reducing the variance compared to the standard Monte Carlo method. The dashed line results show that in conjunction with the multilevel approach the QMC is effective at reducing the variance on the coarsest levels, but the benefits diminish on the finer levels. This is probably because the multilevel approach itself extracts much of the low-dimensional content in the integrand, so that on the finer levels the correction is predominantly high-dimensional and so the QMC approach is less effective. However, most of the computational cost of the multilevel method is on the coarsest levels, and so we will see that the combination does reduce the overall cost significantly.

The top right plot shows that $\mathbb{E}[\widehat{P}_l - \widehat{P}_{l-1}]$ is approximately $O(h_l)$, corresponding to the expected first order weak convergence. Each line in the bottom left plot shows the values for $N_l, l = 0, \ldots, L$, with the values decreasing with level l as expected. It can also be seen that the value for L, the maximum level of timestep refinement, increases as the value for ϵ decreases, requiring a lower bias error.

The bottom right plot shows the variation with ϵ of $\epsilon^2 C$ where the computational complexity C is defined as

$$C = 32 \sum_l 2^l N_l,$$

which is the total number of fine grid timesteps on all levels. One line shows the results for the multilevel QMC method and the other shows the corresponding cost of a standard QMC simulation of the same accuracy, i.e. the same bias error corresponding to the same value for L, and the same variance. It can be seen that $\epsilon^2 C$ is roughly constant for the standard QMC method, and this is at a level which is comparable to that achieved previously using the multilevel method on its own. However, combining the multilevel method with QMC gives additional savings of factor 20–100, with the computational cost being approximately proportional to ϵ^{-1}. This is the best one could hope for using QMC since in the best cases its error is inversely proportional to the number of points, and hence, at best, inversely proportional to the computational cost.

6.2 Asian option

The Asian option we consider has the discounted payoff

$$P = \exp(-rT)\ \max\left(0, \overline{S} - K\right),$$

where

$$\overline{S} = T^{-1} \int_0^T S(t)\, \mathrm{d}t.$$

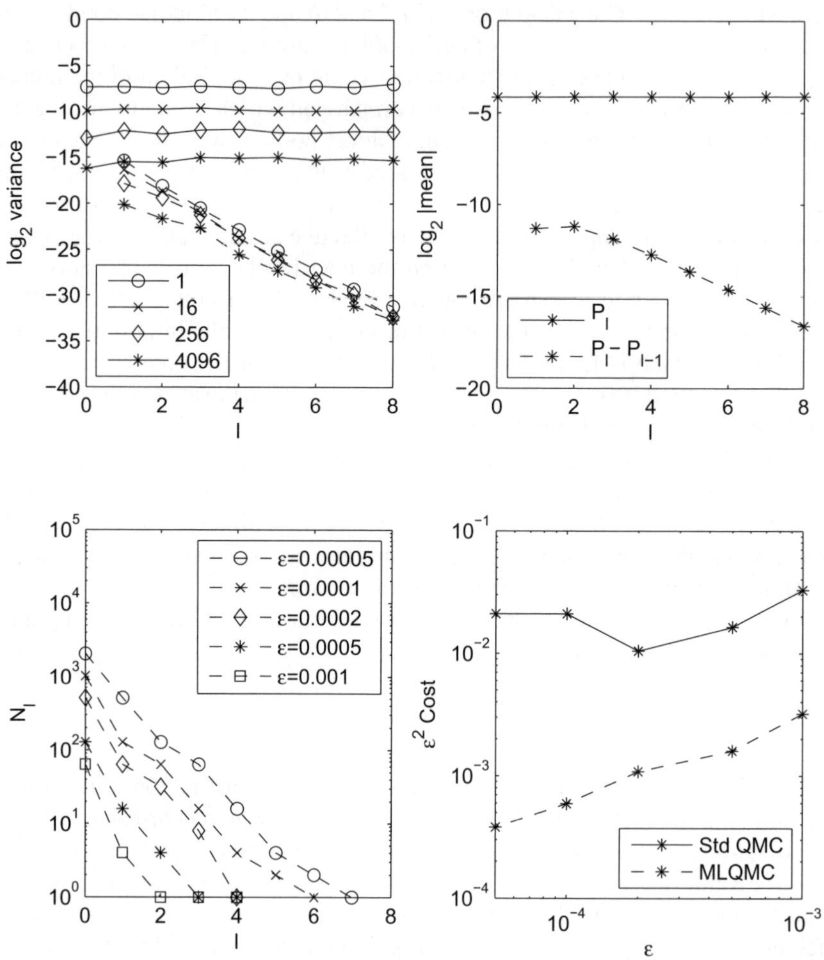

Figure 6.2. Asian option

On the fine path, integrating (3.2) and using standard Brownian Bridge results (see section 3.1 in [6]) gives

$$\overline{\widehat{S}} = T^{-1} \sum_{0}^{n_T - 1} \left(\frac{1}{2}\, h\, (\widehat{S}_n + \widehat{S}_{n+1}) + b_n \Delta I_n \right),$$

where

$$\Delta I_n = \int_{t_n}^{t_{n+1}} (W(t) - W(t_n))\, \mathrm{d}t \; - \; \frac{1}{2}\, h \Delta W$$

is a $N(0, h^3/12)$ Normal random variable, independent of ΔW. The coarse path approximation is similar except that the values for ΔI_n are derived from the fine path values, noting that

$$\begin{aligned}
\int_{t_n}^{t_n+2h} (W(t) - W(t_n))\,\mathrm{d}t \;-\; h(W(t_n+2h) - W(t_n)) \\
= \int_{t_n}^{t_n+h} (W(t) - W(t_n))\,\mathrm{d}t \;-\; \frac{1}{2}h\,(W(t_n+h) - W(t_n)) \\
+ \int_{t_n+h}^{t_n+2h} (W(t) - W(t_n+h))\,\mathrm{d}t \;-\; \frac{1}{2}h\,(W(t_n+2h) - W(t_n+h)) \\
+\frac{1}{2}h\,(W(t_n+h) - W(t_n)) - \frac{1}{2}h\,(W(t_n+2h) - W(t_n+h)),
\end{aligned}$$

and hence

$$\Delta I^c = \Delta I^{f1} + \Delta I^{f2} + \frac{1}{2}h(\Delta W^{f1} - \Delta W^{f2}),$$

where ΔI^c is the value for the coarse timestep, and ΔI^{f1} and ΔW^{f1} are the values for the first fine timestep, and ΔI^{f2} and ΔW^{f2} are the values for the second fine timestep.

Figure 6.2 shows the numerical results for parameters $S(0) = 1$, $K = 1$, $T = 1$, $r = 0.05$, $\sigma = 0.2$. The top left plot shows the behaviour of the variance of both $\widehat{P}_l$ and $\widehat{P}_l - \widehat{P}_{l-1}$. The standard QMC method is effective at reducing the variance on all levels, but with the multilevel estimator its effectiveness diminishes at the finer levels. The bottom two plots again have results from five multilevel calculations for different values of ϵ. It can be seen that $\epsilon^2 C$ is very roughly constant for the standard QMC method (again at a level comparable to that achieved previously by the multilevel method on its own [4]), while $\epsilon^2 C$ decreases significantly with decreasing ϵ for the combined multilevel QMC method.

6.3 Lookback option

The lookback option we consider has the discounted payoff

$$P = \exp(-rT)\big(S(T) - \min_{0<t<T} S(t)\big).$$

For the fine path calculation on the time interval $[t_n, t_{n+1}]$, a standard Brownian interpolation result (see section 6.4 in [6]) gives the minimum value as

$$\widehat{S}^f_{n,min} = \frac{1}{2}\left(\widehat{S}^f_n + \widehat{S}^f_{n+1} - \sqrt{\left(\widehat{S}^f_{n+1} - \widehat{S}^f_n\right)^2 - 2\,b_n^2\,h\log U_n}\,\right), \tag{6.1}$$

where U_n is a uniform random variable on $[0, 1]$. Taking the minimum over all timesteps gives an approximation to $\min_{0<t<T} S(t)$ from which $\widehat{P}_l$ is calculated.

For the coarse path calculation, $\widehat{P}_{l-1}$ is defined similarly, except that for each timestep the mid-point value is first constructed using (3.3), and then the minimum over the

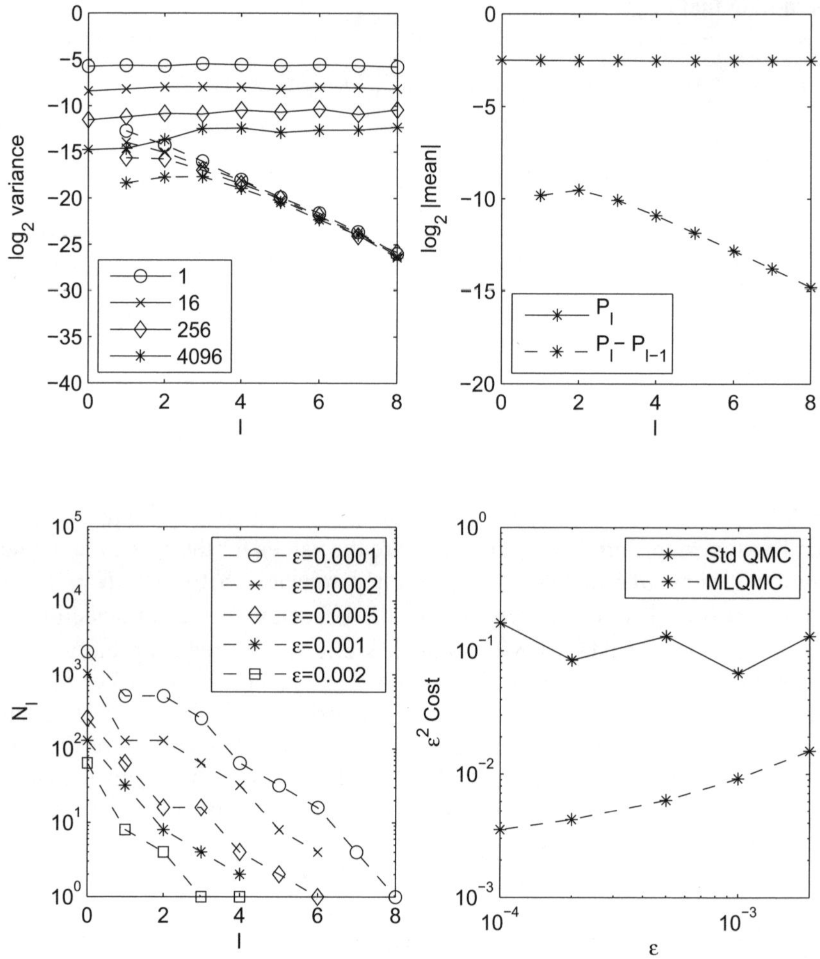

Figure 6.3. Lookback option

timestep is given by

$$\widehat{S}^c_{m,min} = \min\left\{\frac{1}{2}\left(\widehat{S}^c_m + \widehat{S}^c_{m+\frac{1}{2}} - \sqrt{\left(\widehat{S}^c_{m+\frac{1}{2}} - \widehat{S}^c_n\right)^2 - b_m^2\, h \log U_{2m-1}}\right),\right.$$
$$\left.\frac{1}{2}\left(\widehat{S}^c_{m+\frac{1}{2}} + \widehat{S}^c_{m+1} - \sqrt{\left(\widehat{S}^c_{m+1} - \widehat{S}^c_{m+\frac{1}{2}}\right)^2 - b_m^2\, h \log U_{2m}}\right)\right\}. \tag{6.2}$$

Note the re-use of the uniform random variables U_{2m-1} and U_{2m} from the two fine timesteps corresponding to this coarse timestep; it is this which ensures that the minimum from the coarse path is very close to the minimum from the fine path, resulting in a low variance for $\widehat{P}_l - \widehat{P}_{l-1}$.

Figure 6.3 shows the results for parameters $S(0)=1$, $T=1$, $r=0.05$, $\sigma=0.2$. The results are qualitatively similar to the previous two cases. There is almost no improvement from using QMC on the finer levels, but nevertheless there is a big reduction in the overall cost compared to the multilevel method without QMC [4].

6.4 Barrier option

The barrier option which is considered is a down-and-out call for which the discounted payoff is

$$P = \exp(-rT)\ (S(T) - K)^+\ \mathbf{1}_{\tau>T},$$

where $\mathbf{1}_{\tau>T}$ is an indicator function taking value 1 if the argument is true, and zero otherwise, and the barrier crossing time τ is $\tau = \inf_{t>0}\{S(t) < B\}$.

For the fine path simulation, following a standard approach for continuously monitored barrier crossings (see section 6.4 in [6]), the conditional expectation of the payoff can be expressed as

$$\exp(-rT)\ (\widehat{S}^f_{n_T} - K)^+ \prod_{n=0}^{n_T-1} \widehat{p}_n,$$

where $\widehat{p}_n$, the probability the interpolated path did not cross the barrier during the nth timestep, is equal to

$$\widehat{p}_n = 1 - \exp\left(\frac{-2\,(\widehat{S}^f_n - B)^+ (\widehat{S}^f_{n+1} - B)^+}{b_n^2\, h}\right). \tag{6.3}$$

For the coarse path calculation, we again use equation (3.3) to construct a midpoint value $\widehat{S}^c_{m+1/2}$ for each timestep. Given this value, the probability that the Brownian interpolation path does not cross the barrier during the mth coarse timestep is

$$\begin{aligned}\widehat{p}^c_m \;=\;& \left\{1 - \exp\left(\frac{-2\,(\widehat{S}^c_m - B)^+ (\widehat{S}^c_{m+1/2} - B)^+}{b_m^2\, h}\right)\right\}\\ &\times \left\{1 - \exp\left(\frac{-2\,(\widehat{S}^c_{m+1/2} - B)^+ (\widehat{S}^c_{m+1} - B)^+}{b_m^2\, h}\right)\right\}.\end{aligned} \tag{6.4}$$

Figure 6.4 has the results for parameters $S(0)=1$, $K=1$, $B=0.85$, $T=1$, $r=0.05$, $\sigma = 0.2$. The main features are similar, but the variance V_l decreases with level at a slightly lower rate in this case [4] and consequently $\epsilon^2 C$ for the combined multilevel QMC method does not decrease quite as much as ϵ is reduced compared to the previous examples.

6.5 Digital option

The digital option which is considered has the discounted payoff

$$P = \exp(-rT)\ \mathbf{1}_{S(T)>K}.$$

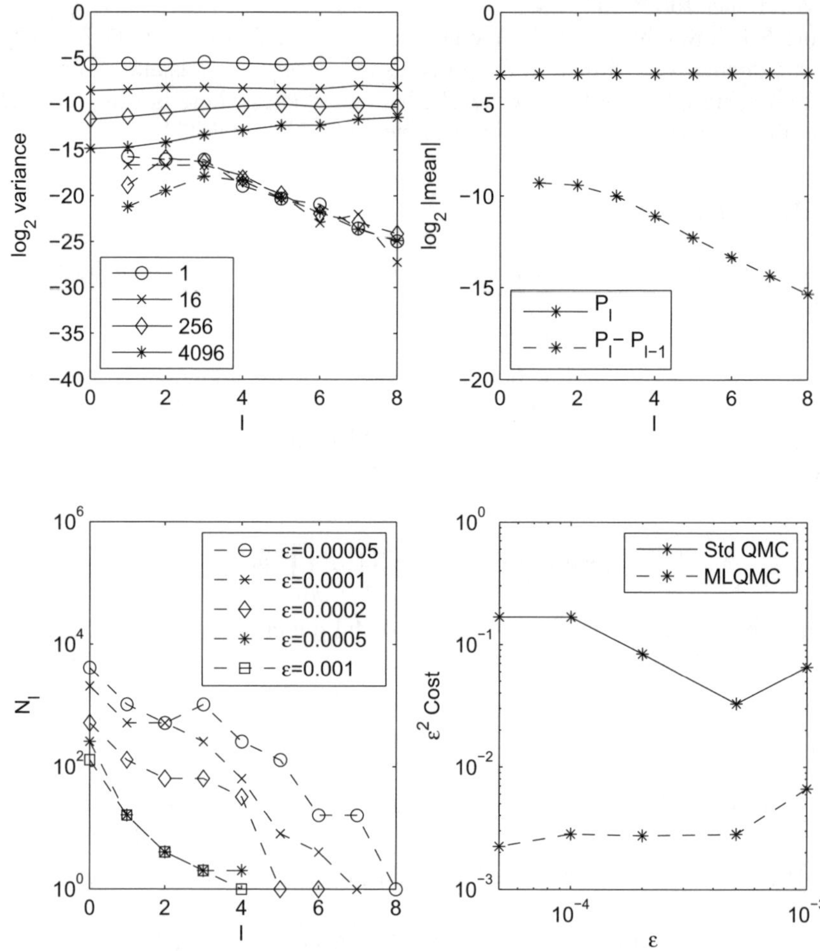

Figure 6.4. Barrier option

To achieve a good multilevel variance convergence rate, we follow the same procedure used previously [4], smoothing the payoff using the technique of conditional expectation (see section 7.2.3 in [6]) in which we terminate the path calculations one timestep before reaching the terminal time T. If $\widehat{S}^f_{n_T-1}$ denotes the fine path value at this time, then if we approximate the motion thereafter as a simple Brownian motion with constant drift a_{n_T-1} and volatility b_{n_T-1}, the probability that $\widehat{S}^f_{n_T} > K$ after one further timestep is

$$\widehat{p}^f = \Phi\left(\frac{\widehat{S}^f_{n_T-1} + a_{n_T-1}h - K}{b_{n_T-1}\sqrt{h}}\right), \tag{6.5}$$

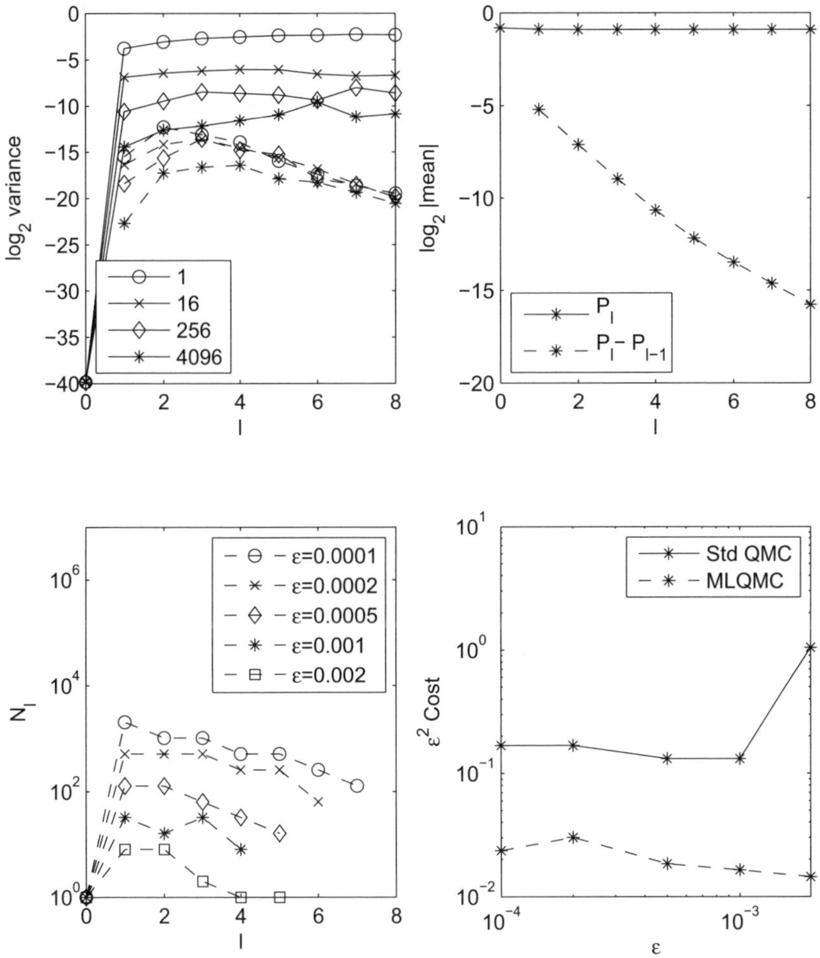

Figure 6.5. Digital option

where Φ is the cumulative Normal distribution. For the fine-path payoff $\widehat{P}_l^f$ we therefore use $\widehat{P}_l^f = \exp(-rT)\,\widehat{p}^f$.

For the coarse-path payoff, we note that given the Brownian increment ΔW for the first half of the last timestep, which is already known because it corresponds to the last of the computed timesteps in the fine path calculation, then the probability that $\widehat{S}^c_{n_T/2} > K$ is

$$\widehat{p}^c = \Phi\left(\frac{\widehat{S}^c_{n_T/2-1}+a_{n_T-1}h+b_{n_T-1}\Delta W - K}{b_{n_T-1}\sqrt{h/2}}\right), \tag{6.6}$$

where $a_{n_T/2-1}$ and $b_{n_T/2-1}$ are the drift and volatility based on $\widehat{S}^c_{n_T/2-1}$.

Figure 6.5 has the results for parameters $S(0)=1$, $K=1$, $T=1$, $r=0.05$, $\sigma=0.2$. One strikingly different feature is that the variance of the level 0 estimator, V_0, is zero. This is because at level $l=0$ there would usually be only one timestep, and so here it is not simulated at all; one simply uses equation (6.5) to evaluate the payoff. This essentially eliminates the cost of the level 0 calculation, which is where the QMC method is usually most effective. Consequently, the cost of the combined multilevel QMC method remains approximately proportional to ϵ^{-2}, and is only slightly lower than the results obtained previously for the multilevel method without QMC [4]. However, we still get a factor 5–10 computational savings compared to the standard QMC on its own.

7 Conclusions and future work

In this paper we have demonstrated the benefits of combining rank-1 lattice rule quasi-Monte Carlo integration with multilevel Monte Carlo path simulation. Together, the computational cost is lower than using either one on its own.

There are two major directions for future research. The first is the extension of the algorithms to multi-dimensional SDEs, for which the Milstein discretisation usually requires the simulation of Lévy areas [6, 10]. Current investigations indicate that this can be avoided for European options with a Lipschitz payoff through the use of antithetic variables. However, the extension to more difficult payoffs, such as the Asian, lookback, barrier and digital options considered in this paper, looks more challenging and the direct simulation of the Lévy areas may be necessary.

The second direction for future research is the numerical analysis of multilevel methods. Müller-Gronbach and Ritter [14], Giles, Higham and Mao [13] and Avikainen [1] have obtained bounds on the convergence of the multilevel method using the Euler discretisation for different classes of output functional, but additional research is required for the Milstein discretisation.

Bibliography

[1] R. Avikainen, *Convergence rates for approximations of functionals of SDEs*, Finance and Stochastics (to appear) (2009).

[2] V. Bally and D. Talay, *The law of the Euler scheme for stochastic differential equations, I: convergence rate of the distribution function*, Probability Theory and Related Fields 104 (1995), pp. 43–60.

[3] G.W. Wasilkowski, F.Y. Kuo and B.J. Waterhouse, *Randomly shifted lattice rules for unbounded integrands*, J. Complexity 22 (2006), pp. 630–651.

[4] M.B. Giles, *Monte Carlo and Quasi-Monte Carlo Methods 2006*, ch. Improved multilevel Monte Carlo convergence using the Milstein scheme, pp. 343–358, Springer-Verlag, 2007.

[5] ———, *Multilevel Monte Carlo path simulation*, Operations Research 56 (2008), pp. 981–986.

[6] P. Glasserman, *Monte Carlo Methods in Financial Engineering*, Springer-Verlag, New York.

[7] S. Heinrich, *Lecture Notes in Computer Science*, vol. 2179, ch. Multilevel Monte Carlo Methods, pp. 58–67, Springer-Verlag, 1.

[8] F. Pillichshammer, J. Dick and B.J. Waterhouse, *The construction of good extensible rank-1 lattices*, Math. Comp. 77 (2007), pp. 2345–2373.

[9] A. Kebaier, *Statistical Romberg extrapolation: a new variance reduction method and applications to options pricing*, Annals of Applied Probability 14 (2005), pp. 2681–2705.

[10] P.E. Kloeden and E. Platen, *Numerical Solution of Stochastic Differential Equations*, Springer-Verlag, Berlin.

[11] F.Y. Kuo and I.H. Sloan, *Lifting the curse of dimensionality*, Notices of the AMS 52 (2005), pp. 1320–1328.

[12] P. L'Ecuyer, *Proceedings of the 2004 Winter Simulation Conference*, ch. Quasi-Monte Carlo methods in finance, pp. 1645–1655, IEEE Press, 2004.

[13] D. Higham, M.B. Giles and X. Mao, *Analysing multilevel Monte Carlo for options with non-globally Lipschitz payoff*, Finance and Stochastics (to appear) (2009).

[14] T. Müller-Gronbach and K. Ritter, *Monte Carlo and Quasi-Monte Carlo Methods 2006*, ch. Minimal errors for strong and weak approximation of stochastic differential equations, pp. 53–82, Springer-Verlag, 2007.

[15] H. Niederreiter, *Random Number Generation and Quasi-Monte Carlo Methods*, SIAM.

[16] D. Nuyens and R. Cools, *Fast algorithms for component-by-component construction of rank-1 lattice rules in shift-invariant reproducing kernel Hilbert spaces*, Math. Comp. 75 (2006), pp. 903–920, (electronic).

[17] A. Papageorgiou, *The Brownian bridge does not offer a consistent advantage in Quasi-Monte Carlo integration*, J. Complex. 18 (2002), pp. 171–186.

[18] W.J. Morokoff, R.E. Caflisch and A.B. Owen, *Valuation of Mortgage Backed Securities Using Brownian Bridges to Reduce Effective Dimension*, J. Comput. Finance 1 (1997), pp. 27–46.

[19] I.H. Sloan and S. Joe, *Lattice Methods for Multiple Integration*, Oxford University Press.

[20] I.M. Sobol′, *Distribution of points in a cube and approximate evaluation of integrals*, Akademija Nauk SSSR. Žurnal Vyčislitel′noĭ Matematiki i Matematičeskoĭ Fiziki 7 (1967), pp. 784–802.

[21] D. Talay and L. Tubaro, *Expansion of the global error for numerical schemes solving stochastic differential equations*, Stochastic Analysis and Applications 8 (1990), pp. 483–509.

Author information

Michael B. Giles, Mathematical Institute and Oxford-Man Institute of Quantitative Finance, Oxford University, Oxford, United Kingdom.
Email: mike.giles@maths.ox.ac.uk

Benjamin J. Waterhouse, School of Mathematics and Statistics, University of New South Wales, Sydney, Australia.
Email: ben.waterhouse@unsw.edu.au

Radon Series Comp. Appl. Math **8**, 183–203

Modelling default and prepayment using Lévy processes: an application to asset backed securities

Henrik Jönsson, Wim Schoutens, and Geert Van Damme

Abstract. The securitization of financial assets is a form of structured finance, developed by the U.S. banking world in the early 1980's (in Mortgage-Backed-Securities format) in order to reduce regulatory capital requirements by removing and transferring risk from the balance sheet to other parties. Today, virtually any form of debt obligations and receivables has been securitised, resulting in an approximately \$2.5 trillion ABS outstanding in the U.S. alone,* a market which is rapidly spreading to Europe, Latin-America and Southeast Asia.

Though no two ABS contracts are the same and therefore each deal requires its very own model, there are three important features which appear in virtually any securitization deal: default risk, Loss-Given-Default and prepayment risk. In this paper we will only be concerned with default and prepayment and discuss a number of traditional (continuous) and Lévy-based (pure jump) methods for modelling the latter risks. After briefly explaining the methods and their underlying intuition, the models are applied to a simple ABS deal in order to determine the rating of the notes. It turns out that the pure jump models produce lower (i.e. more conservative) ratings than the traditional methods (e.g. Vasicek), which are clearly incapable of capturing the shock-driven nature of losses and prepayments.

Key words. Lévy processes, default probability, prepayment probability, rating, asset-backed securities.

AMS classification. 60G35, 62P05, 91B28, 91B70

1 Introduction

Securitisation is the process whereby an institution packs and sells a number of financial assets to a special entity, created specifically for this purpose and therefore termed the Special Purpose Entity (SPE) or Special Purpose Vehicle (SPV), which funds this purchase by issuing notes secured by the revenues from the underlying pool of assets. In general, we can say that securitization is the transformation of illiquid assets (for instance, mortgages, auto loans, credit card receivables and home equity loans) into liquid assets (marketable securities that can be sold in securities markets).

This form of structured finance was initially developed by the U.S. banking world

*Source: SIFMA, Q2 2008.

First author: H. Jönsson is funded by the European Investment Bank's EIBURS programme "Quantitative Analysis and Analytical Methods to Price Securitization Deals". Part of this research has been done during the time while H. Jönsson was an EU-Marie Curie Intra-European Fellow with funding from the European Community's Sixth Framework Programme (MEIF-CT-2006-041115)

in the early 1980's (in Mortgage-Backed-Securities format) in order to reduce regulatory capital requirements by removing and transferring risk from the balance sheet to other parties. Over the years, however, the technique has spread to many other industries (also outside the U.S.) and the goal shifted from reducing capital requirements to funding and hedging. Today, virtually any form of debt obligations and receivables has been securitised, with companies showing a seemingly infinite creativity in allocating the revenues from the pool to the noteholders (respecting their seniority). This results in an approximately $2.5 trillion ABS market in the U.S. alone, which is rapidly spreading to Europe, Latin-America and Southeast Asia.

Unlike the nowadays very popular Credit Default Swap, ABS contracts are not yet standardised. This lack of uniformity implies that each deal requires a new model. However, there are certain features that emerge in virtually any ABS deal, the most important ones of which are default risk, amortisation of principal value (and thus prepayment risk) and Loss-Given-Default (LGD). Since defaults, losses and accelerated principal repayments can substantially alter the projected cashflows and therefore the planned investment horizon, it is of key importance to adequately describe and model these phenomena when pricing securitization deals.

In the current ABS practice, the probability of default is generally modeled by means of a sigmoid function, such as the Logistic function, or by Vasicek's one-factor model, whereas the prepayment rate and the LGD rate are assumed to be constant (or at least deterministic) over time and independent of default. However, it is intuitively clear that each of these events is coming unexpectedly and is generally driven by the overall economy, hence infecting many borrowers at the same time, causing jumps in the default and prepayment term structures. Therefore it is essential to model the latter by stochastic processes that include jumps. Furthermore, it is unrealistic to assume that prepayment rates and loss rates are time-independent and uncorrelated, neither with each other, nor with default rates. For instance, a huge economic downturn will most likely result in a large number of defaults and a significant increase of the interest rates, causing huge losses and a decrease in prepayments. Reality indeed shows a negative correlation between default and prepayment.

In this paper, we propose a number of alternative techniques that can be applied to stochastically model default, prepayment and Loss-Given-Default, introducing dependence between the latter as well. The models we propose are based on Lévy processes, a well know family of jump-diffusion processes that have already proven their modelling abilities in other settings like equity and fixed income (cf. Schoutens [12]). The text is organised as follows. In the following section we present four models for the default term structure. In Section 3 we discuss three models for the prepayment term structure. Numerical results are presented in Section 4, where the default and prepayment models are built into a cashflow model in order to determine the cumulative expected loss rate, the Weighted Average Life (WAL) and the corresponding rating of two subordinated notes of a simple ABS deal. Section 5 concludes the paper.

2 Default models

In this section we will briefly discuss four models for the default term structure, respectively based on

1. the generalised Logistic function;
2. a strictly increasing Lévy process;
3. Vasicek's Normal one-factor model;
4. the generic one-factor Lévy model [1], with an underlying shifted Gamma process.

We will focus on the time interval between the issue ($t = 0$) of the ABS notes and the weighted average time to maturity ($t = T$) of the underlying assets. In the sequel we will use the term default *curve* to refer to the default term structure. By default *distribution*, we mean the distribution of the cumulative default rate at time T. Hence, the endpoint of the default curve is a random draw from the default distribution.

2.1 Generalised logistic default model

Traditional methods typically use a sigmoid (S-shaped) function to model the term structure of defaults. One famous example of such sigmoid functions is the (generalised) Logistic function (Richards [10]), defined as

$$F(t) = \frac{a}{1 + b\mathrm{e}^{-c(t-t_0)}}, \tag{2.1}$$

where $F(t)$ satisfies the following ODE

$$\frac{\mathrm{d}F(t)}{\mathrm{d}t} = c\left(1 - \frac{F(t)}{a}\right) F(t), \tag{2.2}$$

with $a, b, c, t_0 > 0$ being constants and $t \in [0, T]$.

In the context of default curve modelling, $\mathrm{P_d}(t) := F(t)$ is the cumulative default rate at time t. Note that when $b = 1$, t_0 corresponds to the inflection point in the loss buildup, i.e. $\mathrm{P_d}$ grows at an increasing rate before time t_0 and at a decreasing rate afterwards. Furthermore, $\lim_{t\to+\infty} F(t) = a$, thus a controls the right endpoint of the default curve. For sufficiently large T we can therefore approximate the cumulative default rate at maturity by a, i.e. $\mathrm{P_d}(T) \approx a$. Hence, a is a random draw from a predetermined default distribution (e.g. the Log-Normal distribution) and each different value for a will give rise to a new default curve. This makes the Logistic function suitable for scenario analysis. Finally, the parameter c controls the spread of the Logistic curve around t_0. In fact, c determines the growth rate of the Logistic curve, i.e. the proportional increase in one unit of time, as can be seen from equation (2.2). Values of c between 0.10 and 0.20 produce realistic default curves.

The left panel of Figure 2.1 shows five default curves, generated by the Logistic function with parameters $b = 1, c = 0.1, t_0 = 55$, $T = 120$ and decreasing values of

a, drawn from a Log-Normal distribution with mean 0.20 and standard deviation 0.10. Notice the apparent inflection in the default curve at $t = 55$. The probability density function (p.d.f.) of the cumulative default rate at time T is shown on the right.

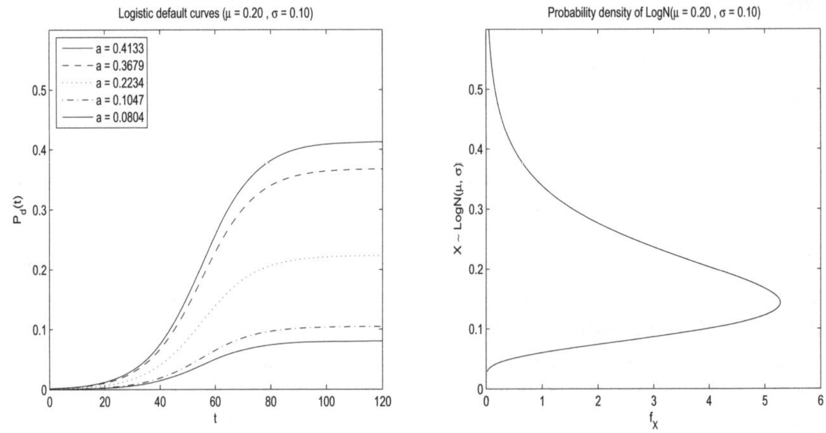

Figure 2.1. Logistic default curve (left) and Lognormal default distribution (right).

It has to be mentioned that the Logistic function (2.1) has several drawbacks when it comes to modelling a default curve. First of all, assuming real values for the parameters, the Logistic function does not start at 0, i.e. $\mathrm{P_d}(0) > 0$. Moreover, a is only an approximation of the cumulative default rate at maturity, but in general we have that $\mathrm{P_d}(T) < a$. Hence $\mathrm{P_d}$ has to be rescaled, in order to guarantee that a is indeed the cumulative default rate in the interval $[0, T]$. Secondly, the Logistic function is a deterministic function of time (the only source of randomness is in the choice of the endpoint), whereas defaults generally come as a surprise. And finally, the Logistic function is continuous and hence unable to deal with the shock-driven behaviour of defaults.

In the next sections we will describe three default models that (partly) solve the above mentioned problems. The first two problems will be solved by using a stochastic (instead of deterministic) process that starts at 0, whereas the shocks will be captured by introducing jumps in the model.

2.2 Lévy portfolio default model

In order to tackle the shortcomings of the Logistic model, we propose to model the default term structure by the process[1]

$$\mathrm{P_d} = \left\{ \mathrm{P_d}(t) = 1 - \mathrm{e}^{-\lambda_t^d}, \quad t \geq 0 \right\}, \tag{2.3}$$

[1] This can be linked to the world of intensity-based default modelling. See Lando [6] and Schönbucher [11] for a more detailed exposition. Cariboni and Schoutens [3] incorporate jump dynamics into intensity models.

with $\lambda^d = \{\lambda_t^d : t \geq 0\}$ a strictly increasing Lévy process. The latter introduces both jump dynamics and stochasticity, i.e. $\mathrm{P_d}(t)$ is a random variable, for all $t > 0$. Therefore, in order to simulate a default curve, we must first draw a realization of the process λ^d. Moreover, $\mathrm{P_d}(0) = 0$, since by the properties of a Lévy process $\lambda_0^d = 0$.

In this paper we assume that λ^d is a Gamma process $G = \{G_t : t \geq 0\}$ with shape parameter a and scale parameter b, hence $\lambda_t^d \sim \mathrm{Gamma}\,(at, b)$, for $t > 0$. Hence, the cumulative default rate at maturity follows the law $1 - \mathrm{e}^{-\lambda_T^d}$, where $\lambda_T^d \sim \mathrm{Gamma}\,(aT, b)$. Using this result, the parameters a and b can be found as the solution to the following system of equations

$$\begin{cases} \mathrm{E}\left[1 - \mathrm{e}^{-\lambda_T^d}\right] & = & \mu_d; \\ \mathrm{Var}\left[1 - \mathrm{e}^{-\lambda_T^d}\right] & = & \sigma_d^2, \end{cases} \tag{2.4}$$

for predetermined values of the mean μ_d and standard deviation σ_d of the default distribution. Explicit expressions for the left hand sides of (2.4) can be found, by noting that the expected value and the variance can be written in terms of the characteristic function of the Gamma distribution.

The left panel of Figure 2.2 shows five default curves, generated by the process (2.3) with parameters $a \approx 0.024914, b \approx 12.904475$ and $T = 120$, such that the mean and standard deviation of the default distribution are 0.20 and 0.10. Note that all curves start at zero, include jumps and are fully stochastic functions of time, in the sense that in order to construct a new default curve, one has to rebuild the whole intensity process over $[0, T]$, instead of just changing its endpoint. The corresponding default p.d.f. is again shown on the right. Recall, in this case, that $\mathrm{P_d}(T)$ follows the law $1 - \mathrm{e}^{-\lambda_T^d}$, with $\lambda_T^d \sim \mathrm{Gamma}\,(aT, b)$.

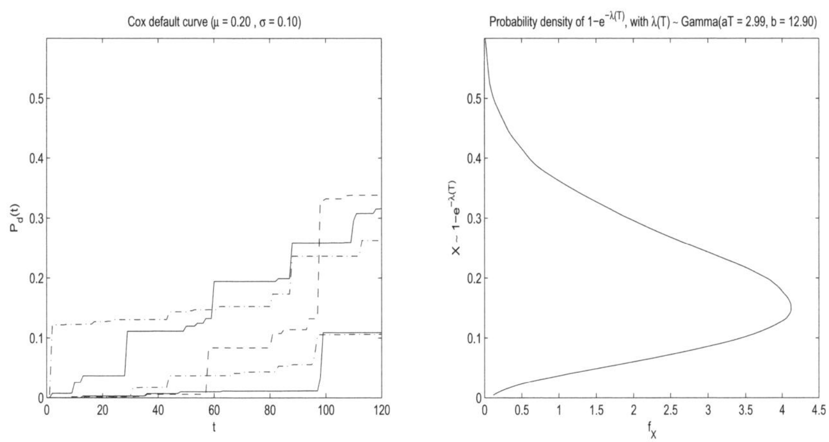

Figure 2.2. Lévy portfolio default curves (left) and corresponding default distribution (right).

2.3 Normal one-factor default model

The Normal one-factor (structural) model (Vasicek [14], Li [7]) models the cash position $V^{(i)}$ of a borrower, where $V^{(i)}$ is described by a geometric Brownian motion,

$$\begin{aligned} V_T^{(i)} &= V_0^{(i)} \exp\left[a\left(\mu_T^{(i)}, \sigma_T^{(i)}\right) + b\left(\mu_T^{(i)}, \sigma_T^{(i)}\right) W_T^{(i)}\right] \\ &\stackrel{d}{=} V_0^{(i)} \exp\left[a\left(\mu_T^{(i)}, \sigma_T^{(i)}\right) + b\left(\mu_T^{(i)}, \sigma_T^{(i)}\right) Z_i\right], \end{aligned} \tag{2.5}$$

for $i = 1, 2, \ldots, N$, with N the number of loans in the asset pool. Here $\stackrel{d}{=}$ denotes equality in distribution and $Z_i \sim \mathrm{N}(0,1)$. Furthermore, Z_i satisfies

$$Z_i = \sqrt{\rho} X + \sqrt{1-\rho} X_i, \tag{2.6}$$

with $X, X_1, X_2, \ldots, X_N \stackrel{\text{i.i.d.}}{\sim} \mathrm{N}(0,1)$. It is easy to verify that $\rho = \mathrm{Corr}(Z_i, Z_j)$, for all $i \neq j$. The latter parameter is calibrated to match a predetermined value for the standard deviation σ of the default distribution.

A borrower is said to default at time t, if his financial situation has deteriorated so dramatically that $V_T^{(i)}$ hits a predetermined lower bound B_t^d, which (as can be seen from (2.5)) is equivalent to saying that Z_i hits some barrier H_t^d. The latter barrier is chosen such that the expected probability of default before time t matches the default probabilities observed in the market, where it is assumed that the latter follow a homogeneous Poisson process with intensity λ, i.e. H_t^d satisfies

$$\Pr\left[Z_i \leq H_t^d\right] = \Phi\left[H_t^d\right] = \Pr\left[N_t > 0\right] = 1 - \mathrm{e}^{-\lambda t}, \tag{2.7}$$

where λ is set such that $\Pr\left[Z_i \leq H_T^d\right] = \mu_d$, with μ_d the predetermined value for the mean of the default distribution. From (2.7) it then follows that

$$\lambda = -\log\left(\left[1-\mu_d\right]^{\frac{1}{T}}\right) \tag{2.8}$$

and hence

$$H_t^d = \Phi^{-1}\left[1 - (1-\mu_d)^{\frac{t}{T}}\right], \tag{2.9}$$

with Φ the standard Normal cumulative distribution function.

Given a sample of (correlated) standard Normal random variables $Z = (Z_1, Z_2, \ldots, Z_N)$, the default curve is then given by

$$\mathrm{P_d}(t; Z) = \frac{\sharp\left\{Z_i \leq H_t^d; i = 1, 2, \ldots, N\right\}}{N}, \quad t \geq 0. \tag{2.10}$$

In order to simulate default curves, one must thus first generate a sample of standard Normal random variables Z_i satisfying (2.6), and then, at each (discrete) time t, count the number of Z_i's that are less than or equal to the value of the default barrier H_t^d at that time.

The left panel of Figure 2.3 shows five default curves, generated by the Normal one-factor model (2.6) with $\rho \approx 0.121353$, such that the mean and standard deviation of the default distribution are 0.20 and 0.10. All curves start at zero and are fully stochastic, but unlike the Lévy portfolio model the Normal one-factor default model does not include any jump dynamics. Therefore, as will be seen later, this model is unable to deal with the shock-driven nature of defaults and as such generates ratings that are too optimistic (high). The corresponding default p.d.f. is again shown in the right panel.

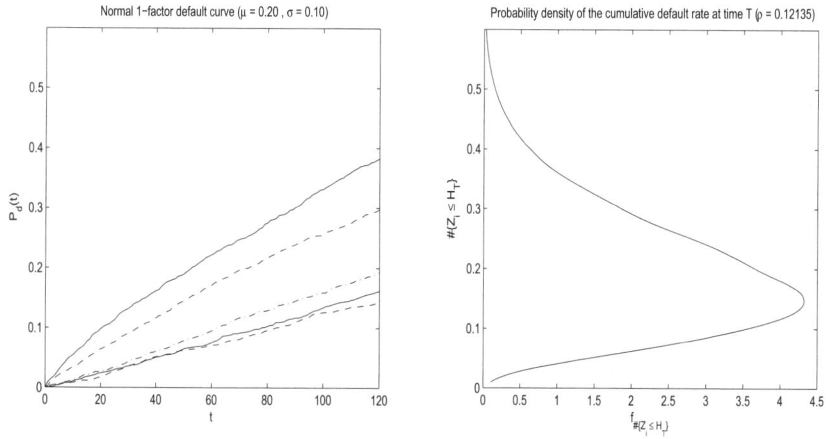

Figure 2.3. Normal one-factor default curves (left) and corresponding default distribution (right).

2.4 Generic one-factor Lévy default model

The generic one-factor Lévy model [1] is comparable to and in fact is a generalisation of the Normal one-factor model. Instead of describing a borrower's cash position by a geometric Brownian motion, $V^{(i)}$ is now modeled with a geometric Lévy model, i.e.

$$V_T^{(i)} = V_0^{(i)} \exp\left[A_T^{(i)}\right], \tag{2.11}$$

for $i = 1, 2, \ldots, N$. The process $A^{(i)} = \left\{A_t^{(i)} : t \geq 0\right\}$ is a Lévy process and satisfies

$$A_T^{(i)} = Y_\rho + Y_{1-\rho}^{(i)}, \tag{2.12}$$

with $Y, Y^{(1)}, Y^{(2)}, \ldots, Y^{(N)}$ i.i.d. Lévy processes, based on the same mother infinitely divisible distribution L, such that $\mathrm{E}\left[Y_1\right] = 0$ and $\mathrm{Var}\left[Y_1\right] = 1$, which implies that $\mathrm{Var}\left[Y_t\right] = t$. From this it is clear that $\mathrm{E}\left[A_T^{(i)}\right] = 0$ and $\mathrm{Var}\left[A_T^{(i)}\right] = 1$, such that $\mathrm{Corr}\left(A_T^{(i)}, A_T^{(j)}\right) = \rho$, for all $i \neq j$. As with the Normal one-factor model, the cross-correlation ρ will be calibrated to match a predetermined standard deviation for the default distribution.

As for the Normal one-factor model, we again say that a borrower defaults at time t, if $A_T^{(i)}$ hits a predetermined barrier H_t^d at that time, where H_t^d satisfies

$$\Pr\left[A_T^{(i)} \leq H_t^d\right] = 1 - \mathrm{e}^{-\lambda t}, \tag{2.13}$$

with λ given by (2.8).

In this paper we assume that $Y, Y^{(1)}, Y^{(2)}, \ldots, Y^{(N)}$ are i.i.d. shifted Gamma processes, i.e. $Y = \{Y_t = t\widetilde{\mu} - G_t : t \geq 0\}$, where G is a Gamma process, with shape parameter a and scale parameter b. From (2.12) and the fact that a Gamma distribution is infinitely divisible it then follows that

$$A_T^{(i)} \stackrel{d}{=} \widetilde{\mu} - \widetilde{X} \stackrel{d}{=} \widetilde{\mu} - [X + X_i], \tag{2.14}$$

with $X \sim \text{Gamma}(a\rho, b)$ and $X_i \sim \text{Gamma}(a(1-\rho), b)$ mutually independent and $X_i \sim \text{Gamma}(a, b)$. If we take $\widetilde{\mu} = \frac{a}{b}$ and $b = \sqrt{a}$, we ensure that $\mathrm{E}\big[A_T^{(i)}\big] = 0$, $\mathrm{Var}\big[A_T^{(i)}\big] = 1$ and $\mathrm{Corr}\big(A_T^{(i)}, A_T^{(j)}\big) = \rho$, for all $i \neq j$.

Furthermore, from (2.13), (2.14) and the expression for λ it follows that

$$H_t^d = \widetilde{\mu} - \Gamma_{a,b}^{-1}\left[(1-\mu_d)^{\frac{t}{T}}\right], \tag{2.15}$$

where $\Gamma_{a,b}$ denotes the cumulative distribution function of a $\text{Gamma}(a, b)$ distribution.

In order to simulate default curves, we first have to generate a sample of random variables $A_T = \big(A_T^{(1)}, A_T^{(2)}, \ldots, A_T^{(N)}\big)$ satisfying (2.12), with $Y, Y^{(1)}, Y^{(2)}, \ldots, Y^{(N)}$ i.i.d. Shifted-Gamma processes and then, at each (discrete) time t, count the number of $A_T^{(i)}$'s that are less than or equal to the value of the default barrier H_t^d at that time. Hence, the default curve is given by

$$\mathrm{P_d}(t; A_T) = \frac{\sharp\{A_T^{(i)} \leq H_t^d; i = 1, 2, \ldots, N\}}{N}, \quad t \geq 0. \tag{2.16}$$

The left panel of Figure 2.4 shows five default curves, generated by the Gamma one-factor model (2.12) with $(\widetilde{\mu}, a, b) = (1, 1, 1)$, and $\rho \approx 0.095408$, such that the mean and standard deviation of the default distribution are 0.20 and 0.10. Again, all curves start at zero and are fully stochastic. Furthermore, when comparing the curves of the one-factor shifted Gamma-Lévy model (hereafter termed the shifted Gamma-Lévy model or Gamma one-factor model) to the ones generated by the Lévy portfolio default model, one might be tempted to conclude that the former model does not include jumps. However, it does, but the jumps are embedded in the underlying dynamics of the asset return A_T. The corresponding default p.d.f. is shown in the right panel. Compared to the previous three default models, the default p.d.f. generated by the shifted Gamma-Lévy model seems to be squeezed around μ_d and has a significantly larger kurtosis.

It should also be mentioned that the latter default distribution has a rather heavy right tail, with a substantial probability mass at the 100 % default rate. This can be explained by looking at the right-hand side of equation (2.14). Since both terms between brackets are strictly positive and hence cannot compensate each other (unlike the Normal one-factor model), $A_T^{(i)}$ is bounded from above by $\widetilde{\mu}$. Hence, starting with a large systematic risk factor X, things can only get worse, i.e. the term between brackets can only increase and therefore $A_{i,T}$ can only decrease, when adding the idiosyncratic risk factor X_i. This implies that when we have a substantially large common factor (close to $\Gamma_{a,b}^{-1}[1-\mu_d]$, cf. (2.15)), it is very likely that all borrowers will default, i.e. that $A_T^{(i)} \leq H_T^d$ for all $i = 1, 2, \ldots, N$.

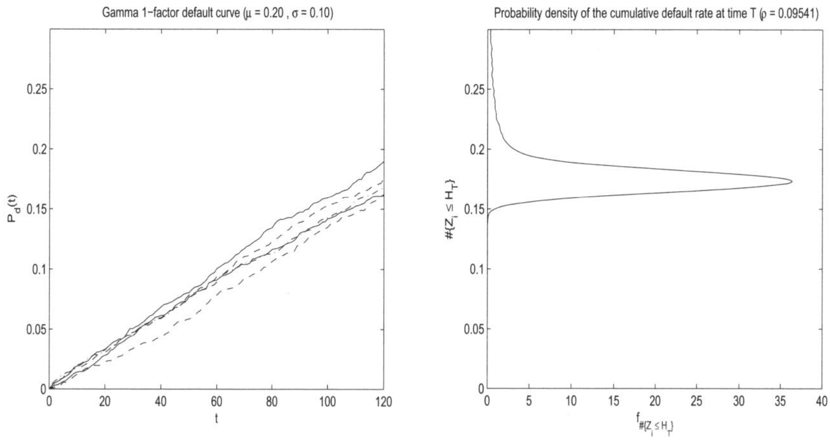

Figure 2.4. Gamma 1-factor default curves (left) and corresponding default distribution (right).

3 Prepayment models

In this section we will briefly discuss three models for the prepayment term structure, respectively based on

1. constant prepayment;
2. a strictly increasing Lévy process;
3. Vasicek's Normal one-factor model.

As before, we will use the terms prepayment *curve* and prepayment *distribution* to refer to the prepayment term structure and the distribution of the cumulative prepayment rate at maturity T.

3.1 Constant prepayment model

The idea of constant prepayment stems from the former Public Securities Association[1] (PSA). The basic assumption is that the (monthly) amount of prepayment begins at 0

[1] In 1997 the PSA changed its name to The Bond Market Association (TBMA), which merged with the Securities Industry Association on November 1, 2006, to form the Securities Industry and Financial Markets Association (SIFMA).

and rises at a constant rate of increase α until reaching its characteristic steady state rate at time t_{00}, after which the prepayment rate remains constant until maturity T. Note that t_{00} is generally not the same as the inflection point t_0 of the default curve.

The corresponding marginal (e.g. monthly) and cumulative prepayment curves are given by

$$\mathrm{cpr}(t) = \begin{cases} \alpha t & ; \quad 0 \leq t \leq t_{00} \\ \alpha t_{00} & ; \quad t_{00} \leq t \leq T \end{cases} \tag{3.1}$$

and

$$\mathrm{CPR}(t) = \begin{cases} \frac{\alpha t^2}{2} & ; \quad 0 \leq t \leq t_{00} \\ -\frac{\alpha t_{00}^2}{2} + \alpha t_{00} t & ; \quad t_{00} \leq t \leq T\,. \end{cases} \tag{3.2}$$

From (3.1) it is obvious that the marginal prepayment rate increases at a speed of α per period before time t_{00} and remains constant afterwards. Consequently, the cumulative prepayment curve (3.2) increases quadratically on the interval $[0, t_{00}]$ and linearly on $[t_{00}, T]$. Given t_{00} and $\mathrm{CPR}(T)$, i.e. the cumulative prepayment rate at maturity, the constant rate of increase α equals

$$\alpha = \frac{\mathrm{CPR}(T)}{T t_{00} - \frac{t_{00}^2}{2}}. \tag{3.3}$$

Hence, once t_{00} and $\mathrm{CPR}(T)$ are fixed, the marginal and cumulative prepayment curves are completely deterministic. Moreover, the CPR model does not include jumps. Due to these features, the CPR model is an unrealistic representation of real-life prepayments, which are shock-driven and typically show some random effects. In the next sections we will describe two models that (partially) solve these problems.

Figure 3.1 shows the marginal and cumulative prepayment curve, in case the steady state t_{00} is reached after 48 months and the cumulative prepayment rate at maturity equals $\mathrm{CPR}(T) = 0.20$. The corresponding constant rate of increase is $\alpha = 0.434$bps.

3.2 Lévy portfolio prepayment model

The Lévy portfolio prepayment model is completely analogous to the Lévy portfolio default model described in Section 2.2, with λ_t^d replaced λ_t^p. Although there is empirical evidence that defaults and prepayments are negatively correlated, in the simulation study in Section 4 we assumed the above mentioned processes to be mutually independent.

Evidently, also the Lévy portfolio prepayment curves start at zero, are fully stochastic and include jumps, solving the above mentioned problems of the CPR model.

3.3 Normal one-factor prepayment model

The Normal one-factor prepayment model starts from the same underlying philosophy as its default equivalent of Section 2.3. We again model the cash position $V^{(i)}$ of a borrower. Just as a borrower is said to default if his financial situation has deteriorated so

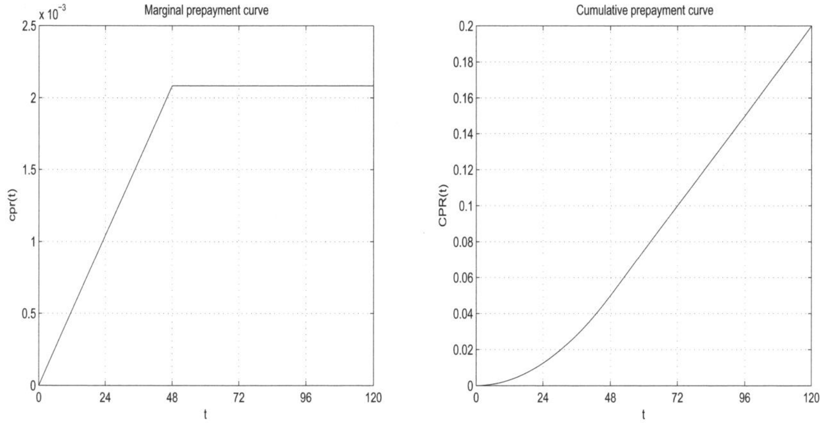

Figure 3.1. Marginal (left) and cumulative (right) constant prepayment curve.

dramatically that $V^{(i)}$ hits some predetermined lower bound B_t^d, we state that a borrower will decide to prepay if his financial health has improved sufficiently, so that $V^{(i)}$ (or equivalently Z_i) hits a prespecified upper bound B_t^p (H_t^p).

The barrier H_t^p is chosen such that the expected probability of prepayment before time t equals the (observed) cumulative prepayment rate $\mathrm{CPR}(t)$, given by (3.2), i.e.

$$\Pr\left[Z_i \geq H_t^p\right] = 1 - \Phi\left[H_t^p\right] = \mathrm{CPR}(t), \tag{3.4}$$

which implies,

$$H_t^p = \Phi^{-1}\left[1 - \mathrm{CPR}(t)\right], \tag{3.5}$$

with Φ the standard Normal cumulative distribution function.

In order to simulate prepayment curves, we must thus draw a sample of standard Normal random variables $Z = (Z_1, Z_2, \ldots, Z_N)$ satisfying (2.6), and then, at each (discrete) time t, count the number of Z_i's that are greater than or equal to the value of the prepayment barrier H_t^p at that time. The prepayment curve is then given by

$$\mathrm{P_p}(t; Z) = \frac{\sharp\{Z_i \geq H_t^p : i = 1, 2, \ldots, N\}}{N}, \quad t \geq 0. \tag{3.6}$$

The left panel of Figure 3.2 shows five prepayment curves, generated by the Normal one-factor model (2.6) with $\rho \approx 0.121353$, such that the mean and standard deviation of the prepayment distribution are 0.20 and 0.10 (as for the default model). The fact that the cross-correlation coefficient ρ is the same as the one of the default model is a direct consequence of the symmetry of the Normal distribution. The curves start at zero and are fully stochastic, but the model lacks jump dynamics. As will be seen later on, ignoring prepayment shocks results in an overestimation of the weighted average life of an ABS, which in turn produces higher ratings. The corresponding prepayment p.d.f. is shown in the right panel.

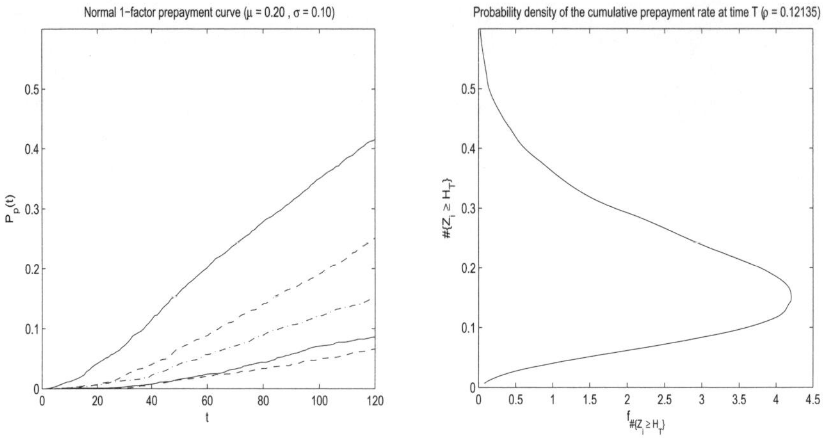

Figure 3.2. Normal one-factor prepayment curves (left) and corresponding prepayment distribution (right).

4 Numerical results

4.1 Introduction

One can now build these default and prepayment models into any scenario generator for rating and analysing asset-backed securities. Any combination of the above described default and prepayment models is meaningful, except for the combination of the shifted Gamma(-Lévy) default model with the Normal one-factor prepayment model. In that case the borrower's cash position would be modeled by two different processes: one to obtain his default probability and another one for his prepayment probability, which is neither consistent nor realistic.

Hence, all together we can construct 11 different scenario generators. Table 4.1 summarises the possible combinations of default and prepayment models.

		Prepayment models		
		CPR	Lévy portfolio	Normal one-factor
Default models	Logistic	ok	ok	ok
	Lévy portfolio	ok	ok	ok
	Normal one-factor	ok	ok	ok
	Gamma one-factor	ok	ok	nok

Table 4.1. Possible combinations of default and prepayment models.

We will now apply each of the above mentioned 11 default-prepayment combinations to derive the expected loss, the WAL and the corresponding rating of two (subordinated) notes backed by a homogeneous pool of commercial loans. Table 4.2 lists the specifications of the ABS deal under consideration (cf. Raynes & Rutledge [9]).

ASSETS		
Initial balance of the asset pool	V_0	\$30,000,000
Number of loans in the asset pool	N_0	2,000
Weighted Average Maturity of the assets	WAM	10 years
Weighted Average Coupon of the assets	WAC	12% p.a.
Payment frequency		monthly
Reserve target		5%
Eligible reinvestment rate		3.92% p.a.
Loss-Given-Default	LGD	50%
Lag		5 months
LIABILITIES		
Initial balance of the senior note	A_0	\$24,000,000
Premium of the senior note	r_A	7% p.a.
Initial balance of the subordinated note	B_0	\$6,000,000
Premium of the subordinated note	r_B	9% p.a.
Servicing fee	r_{sf}	1% p.a.
Servicing fee shortfall rate	r_{sf-sh}	20% p.a.
Payment method		Pro-rata Sequential

Table 4.2. Specifications of the ABS deal.

Note that the cash collected (from the pool) and distributed (to the note holders) by the SPV, in a particular period, contains both principal and interest. Each period, principal (scheduled, prepaid and recoveries from default) and interest collections are combined into a pool, which is then used to pay the interest and principal (in this order) due to the investors. Whatever cash is left after fulfilling the interest obligations is used to pay the principal due (scheduled principal + prepaid principal + defaulted face value) on the notes, according to the priority rules. From this it is evident that default and prepayment will have a significant effect on the amortisation of the notes and (consequently) on the interest received by the note holders.

Furthermore, as can be seen from Table 4.2, the ABS deal under consideration benefits from credit enhancement under the form of a reserve account, required to be equal to 5% of the balance of the asset pool at the end of each payment period. The funds available in this account are reinvested at the 10-year US Treasury rate (of May 22, 2008) and will be used to fulfil the payment obligations, in case the collections in a specific period are insufficient to cover the expenses. In order to achieve the targeted reserve amount of 5% of the asset pool's balance at the end of each payment period, before being transferred to the owners of the SPV, any excess cash is first used to replenish the reserve account. Hence it is possible that the owners of the SPV are not compensated in certain periods, or in the worst case not at all. On the other hand, there may also be periods in which the SPV owners receive a substantial amount of cash. This especially happens in periods with a high number of defaults and/or prepayments, where the outstanding balance of the asset pool suddenly decreases very fast, requiring the reserve account to be reduced in order to match the targeted 5% of the asset pool at the end of the payment period.

Furthermore, unless explicitly stated otherwise, the parameter values mentioned in Table 4.3 will be used.

Mean of the default distribution	μ_{d}	20%
Standard deviation of the default distribution	σ_{d}	10%
Mean of the prepayment distribution	μ_{p}	20%
Standard deviation of the prepayment distribution	σ_{p}	10%
Parameters of the Logistic curve	b	1
	c	0.1
	t_0	55 months
Steady state of the prepayment curve	t_{00}	45 months

Table 4.3. Default parameter values for the default and prepayment models.

Finally, before moving on to the actual sensitivity analysis, we introduce two important concepts, i.e. the DIRR and the WAL of an ABS. By DIRR we mean the difference between the promised and the realised internal rate of return. The WAL is defined as

$$\mathrm{WAL} = \frac{1}{P}\left(\sum_{t=1}^{T} t \cdot P_t + T\Big[P - \sum_{t=1}^{T} P_t\Big] \right), \tag{4.1}$$

where P_t is the total principal paid at time t and P is the initial balance of the note. The term between the square brackets accounts for principal shortfall, in the sense that if the note is not fully amortised after its legal maturity, we assume that the non-amortised amount is redeemed at the legal maturity date[1]. Clearly, both the DIRR and the WAL are non-negative. Furthermore, by inspecting the rating table mentioned [4] and [5], it is obvious that there is some interplay between the DIRR and the WAL: of two notes with the same DIRR, the one with the highest WAL will have the highest rating. For instance, consider two notes A_1 and A_2 with a DIRR of 0.03%, but with respective WALs of 4 and 5 years. Then note A_1 will get a Aa3 rating, whereas the A_2 note gets a Aa2 rating. Obviously, of two notes with the same WAL, the one with the highest DIRR will get the lowest rating.

[1] This method is proposed in Mazataud and Yomtov [8]. Moreover, in Moody's ABSROM™ application (v 1.0) the WAL of a note is calculated as $\frac{\sum_{t=0}^{T-1} F_t}{F_0}$, with F_t the note's outstanding balance at time t. Hence $F_0 = P$. It is left as an exercise to the reader to verify that this formula is equivalent to formula (4.1).

4.2 Sensitivity analysis

Tables 5.1–5.3 contain ratings – based on the Moody's Idealised Cumulative Expected Loss Rates[2]– and DIRRs and WALs of the two ABS notes, obtained with each of the 11 default-prepayment combinations, for several choices of μ_d and μ_p. The figures mentioned in these tables are averages based on a Monte Carlo simulation with 1,000,000 scenarios.

More specifically, in Table 5.1 we investigate what happens to the ratings if μ_d is changed, while holding μ_p and σ_p constant[3], whereas Table 5.1 provides insight in the impact of a change in μ_p, while keeping μ_d and σ_d fixed.

Unless stated otherwise, the (principal) collections from the asset pool are distributed across the note holders according to a pro-rata payment method, i.e. proportionally with the note's outstanding balances. However, Table 5.3 presents the ratings using both pro rata and sequential payment method, where the subordinated B note starts amortising only after the outstanding balance of the senior A note is fully redeemed, in both cases assuming that there exists a reserve account. The effect of having no reserve account in the pro rata case is also shown in Table 5.3.

4.2.1 Influence of μ_d

From Table 5.1 we may conclude that when increasing the average cumulative default rate the credit rating of the notes stays the same or is lowered for all combinations of default and prepayment models.

For the model dependence we first analyse the rating columns for the A note. For $\mu_d = 10\%$ we can see that all but the two pairs with the Gamma one-factor default model give Aaa ratings, indicating that the rating is not so model-dependent for a relatively low cumulative default rate.

Increasing μ_d to 20%, the rating using the Normal one-factor default model stays at Aaa regardless of prepayment models. For the Logistic default model the rating is changed to Aa1 for all combination of prepayment models and for the Gamma one-factor model the rating is Aa3. It is only for the Lévy portfolio default model that we can see a small difference between the CPR model and the two other prepayment models.

Finally, assuming that $\mu_d = 40\%$, the Lévy portfolio prepayment model in combination with either the Logistic or the Normal one-factor default model gives lower ratings than the other two prepayment models. For the other default models no dependence on the prepayment model can be traced.

Analysing the influence of the prepayment model, it is worth noticing that the Lévy portfolio model always gives the lowest WAL and the highest DIRR for any default model, compared to the other two prepayment models. This can be explained by looking at the typical path of a Lévy portfolio process (cf. Figure 2.2). Note that such a path does not increase continuously, but moves up with jumps, between which the curve remains rather flat. Translated to the prepayment phenomena, this means that

[2] See Cifuentes and O'Connor [4] and Cifuentes and Wilcox [5] for further details.

[3] In order to keep μ_p and σ_p fixed, also the cross-correlation ρ must remain fixed, since there is a unique parameter ρ for each pair $(\mu_\text{p}, \sigma_\text{p})$ (or equivalently $(\mu_\text{d}, \sigma_\text{d})$). This explains why also σ_d changes if μ_d changes.

there will be times when a large number of borrowers decide to prepay, followed by a period where there are virtually no prepayments, until the next time where a substantial amount of the remaining debtors prepays. Obviously, this results in a very irregular cash inflow, which will cause difficulties when trying to honour the payment obligations. Indeed, as previously explained, in payment periods with a jump in the prepayment rate, the outstanding balance of the asset pool and consequently the reserve account will be significantly decreased, which in turn increases the probability of future interest and principal shortfalls, leading to higher DIRRs. Moreover, since a shock-driven prepayment model increases the probability that a substantial number of borrowers will choosc to prepay very early in the life of the loan, it is not surprising that the Lévy portfolio prepayment model produces lower WALs than the other two models. Finally, as explained before, higher DIRRs and lower WALs lead to lower ratings.

The Gamma one-factor model always gives the lowest rating, and a look at the DIRR and WAL columns gives the explanation for this, namely, the DIRRs for the Gamma one-factor model is always much higher than for any of the other default models but the WALs is almost the same leading to a lower rating. The Normal one-factor default model gives in general the highest rating, which can be explained by the fact that it produces the lowest DIRRs.

For the B note the general tendency is that the rating is lowered when the mean cumulative default rate is increased. It is worth mentioning that the Normal one-factor model gives the highest rating among the default models and that the Gamma one-factor model gives the lowest rating for $\mu_d = 10\%$ and the Lévy portfolio model gives the lowest for $\mu_d = 40\%$. Thus, the jump-driven default models produce the lowest ratings. The Lévy portfolio prepayment model combined with the Lévy portfolio, Normal one-factor or Gamma one-factor default model gives generally the lowest rating compared to the other prepayment models, for reasons explained in the previous paragraph.

4.2.2 Influence of μ_{p}

The influence of changing the mean cumulative prepayment rate is given in Table 5.2. A comparison to Table 5.1 learns that the ratings are less sensitive to changes in the mean prepayment rate than they are to changes in the expected default rate, as the rating transitions caused by the former are significantly smaller.

Furthermore, any of the above made observations concerning specific prepayment or default models remains valid also here. Especially it still holds that the Normal one-factor default model gives the same or higher rating of both notes than the other default models and that the jump-driven models give the lowest ratings, for each of the prepayment models.

4.2.3 Influence of the reserve account

Table 5.3 provides insight into the effect of incorporating a reserve account (credit enhancement) into the cash flow waterfall of an ABS deal. The results in this table show no surprises: since assuming there is no reserve account implies that there are less funds available for reimbursing the note investors (on the contrary, any excess cash is fully transferred to the SPV owners) it is evident that removing the reserve account will lead to higher DIRRs and WALs and lower ratings. This is indeed what we see, when comparing the above mentioned two tables. Notice that the effect is greater for the B note. This is of course due to its subordinated status.

4.2.4 Influence of the payment method

Table 5.3 shows the impact of choosing either the pro-rata or the sequential payment method, for allocating the (principal) collections to the different notes. What is clear from the definition of the two payment methods is that sequential payment will shorten the WAL of the A note and increase the WAL of the B note. Consulting Moody's Idealised Cumulative Expected Loss Rate table one can see that an increase in WAL, keeping the DIRR fixed, will result in a higher rating. The expected decrease and increase in WAL for the A note and B note, respectively, are evident. In fact, the WAL increases on average with a factor 1.72 (or 3.8 years) for the B note, going from pro rata to sequential payment. The decrease of the WAL for the A note is on average with a factor 0.82 (or 0.95 years). Thus the change in WAL is much more dramatic for the B note than for the A note. So based only on the change of the WALs, without taking the change in DIRR into account, we can directly assume that the rating would improve for the B note and for the A note we would expect the rating to stay the same or be lowered. However, taking the change in DIRR into account, we see that the the actual rating of both the A note and the B note stays the same or improves going from pro rata to sequential payment. The improvement of the A note rating is due to the fact that the DIRR is smaller for the sequential case than for the pro rata case, compensating for the decrease in WAL. For the B note the changes of the DIRRs are not enough to influence the rating improvements due to the increases in WALs.

5 Conclusion

Traditional models for the rating and the analysis of ABSs are typically based on Normal distribution assumptions and Brownian motion driven dynamics. The Normal distribution belongs to the class of the so-called light tailed distributions. This means that extreme events, shock, jumps, crashes, etc. are not incorporated in the Normal distribution based models. However looking at empirical data and certainly in the light of the current financial crisis, it are these extreme events that can have a dramatical impact on the product. In order to do a better assessment, new models incorporating these features are needed. This paper has introduced a whole battery of new models based on more flexible distributions incorporating extreme events and jumps in the sample paths. We observe that the jump-driven models in general produce lower ratings than the traditional models.

	Note A								
	Rating			DIRR (bp)			WAL (year)		
Model pair	$\mu_d = 10\%$	$\mu_d = 20\%$	$\mu_d = 40\%$	$\mu_d = 10\%$	$\mu_d = 20\%$	$\mu_d = 40\%$	$\mu_d = 10\%$	$\mu_d = 20\%$	$\mu_d = 40\%$
Logistic – CPR	Aaa	Aa1	Aa3	0.026746	0.3466	5.3712	5.4867	5.2742	4.8642
Logistic – Lévy portfolio	Aaa	Aa1	A1	0.039664	0.48683	7.4258	5.343	5.1311	4.729
Logistic – Normal one-factor	Aaa	Aa1	Aa3	0.027104	0.3278	5.1148	5.4869	5.2745	4.8656
Lévy portfolio – CPR	Aaa	Aaa	A1	0.0017992	0.16105	9.0857	5.4799	5.2529	4.7895
Lévy portfolio – Lévy portfolio	Aaa	Aa1	A1	0.0067859	0.34616	12.044	5.3355	5.1101	4.6543
Lévy portfolio – Normal one-factor	Aaa	Aa1	A1	0.0032759	0.20977	9.0265	5.4795	5.2532	4.7912
Normal one-factor – CPR	Aaa	Aaa	Aa2	0.00036114	0.034631	2.9626	5.4775	5.2427	4.7309
Normal one-factor – Lévy portfolio	Aaa	Aaa	Aa3	0.00060627	0.055516	3.6883	5.3335	5.0986	4.5895
Normal one-factor – Normal one-factor	Aaa	Aaa	Aa2	0.00014211	0.017135	2.0175	5.4774	5.2427	4.7303
Gamma one-factor – CPR	Aa1	Aa3	A2	1.4443	4.6682	18.431	5.4828	5.2599	4.7939
Gamma one-factor – Lévy portfolio	Aa2	Aa3	A2	2.5931	4.9614	20.385	5.3427	5.1167	4.6503

	Note B								
	Rating			DIRR (bp)			WAL (year)		
Model pair	$\mu_d = 10\%$	$\mu_d = 20\%$	$\mu_d = 40\%$	$\mu_d = 10\%$	$\mu_d = 20\%$	$\mu_d = 40\%$	$\mu_d = 10\%$	$\mu_d = 20\%$	$\mu_d = 40\%$
Logistic – CPR	Aa1	A1	Baa3	0.93026	10.581	139.46	5.4901	5.3124	5.3358
Logistic – Lévy portfolio	Aa1	A1	Baa3	1.1996	13.624	164.07	5.3471	5.1771	5.2456
Logistic – Normal one-factor	Aa1	A1	Baa3	0.93764	10.906	140.55	5.4903	5.3135	5.3391
Lévy portfolio – CPR	Aa1	A2	Baa3	1.4051	17.801	175.75	5.4949	5.3525	5.4753
Lévy portfolio – Lévy portfolio	Aa2	A2	Ba1	1.9445	21.891	195.61	5.3526	5.2204	5.373
Lévy portfolio – Normal one-factor	Aa1	A2	Baa3	1.6019	18.35	175.49	5.4951	5.353	5.4738
Normal one-factor – CPR	Aaa	Aa1	Baa1	0.033692	1.5642	57.936	5.4777	5.2502	4.9709
Normal one-factor – Lévy portfolio	Aaa	Aa2	Baa2	0.041807	1.9829	65.669	5.3337	5.1071	4.8421
Normal one-factor – Normal one-factor	Aaa	Aa1	Baa1	0.023184	1.156	48.936	5.4776	5.2491	4.9498
Gamma one-factor – CPR	Aa3	A2	Baa2	6.288	20.736	85.662	5.4955	5.3022	4.9739
Gamma one-factor – Lévy portfolio	A1	A3	Baa3	15.293	28.406	120.76	5.3631	5.1588	4.8351

Table 5.1. Ratings, DIRR and WAL of the ABS notes, for different combinations of default and prepayment models and mean cumulative default rate $\mu_d = 0.10, 0.20, 0.40$ and mean cumulative prepayment rate $\mu_p = 0.20$.

	Note A								
	Rating			DIRR (bp)			WAL (year)		
Model pair	$\mu_p = 10\%$	$\mu_p = 20\%$	$\mu_p = 40\%$	$\mu_p = 10\%$	$\mu_p = 20\%$	$\mu_p = 40\%$	$\mu_p = 10\%$	$\mu_p = 20\%$	$\mu_p = 40\%$
Logistic – CPR	Aa1	Aa1	Aa1	0.31365	0.3466	0.27714	5.4309	5.2742	4.9611
Logistic – Lévy portfolio	Aa1	Aa1	Aa1	0.34552	0.48683	0.90665	5.365	5.1311	4.618
Logistic – Normal one-factor	Aa1	Aa1	Aa1	0.30488	0.3278	0.25706	5.431	5.2745	4.9633
Lévy portfolio – CPR	Aaa	Aaa	Aa1	0.10416	0.16105	0.42327	5.4093	5.2529	4.9404
Lévy portfolio – Lévy portfolio	Aaa	Aa1	Aa1	0.14828	0.34616	1.4266	5.3439	5.1101	4.5982
Lévy portfolio – Normal one-factor	Aaa	Aa1	Aa1	0.11976	0.20977	0.51304	5.4096	5.2532	4.9424
Normal one-factor – CPR	Aaa	Aaa	Aaa	0.023787	0.034631	0.046599	5.3995	5.2427	4.9292
Normal one-factor – Lévy portfolio	Aaa	Aaa	Aaa	0.03208	0.055516	0.1094	5.3335	5.0986	4.5842
Normal one-factor – Normal one-factor	Aaa	Aaa	Aaa	0.016874	0.017135	0.018373	5.3995	5.2427	4.9291
Gamma one-factor – CPR	Aa3	Aa3	Aa2	5.811	4.6682	2.8855	5.4149	5.2599	4.9497
Gamma one-factor – Lévy portfolio	Aa3	Aa3	Aa2	6.7492	4.9614	3.2188	5.3487	5.1167	4.6043

	Note B								
	Rating			DIRR (bp)			WAL (year)		
Model pair	$\mu_p = 10\%$	$\mu_p = 20\%$	$\mu_p = 40\%$	$\mu_p = 10\%$	$\mu_p = 20\%$	$\mu_p = 40\%$	$\mu_p = 10\%$	$\mu_p = 20\%$	$\mu_p = 40\%$
Logistic – CPR	A1	A1	A2	8.9089	10.581	14.756	5.4642	5.3124	5.0111
Logistic – Lévy portfolio	A1	A1	A3	9.9097	13.624	26.681	5.4011	5.1771	4.695
Logistic – Normal one-factor	A1	A1	A2	9.0211	10.906	14.436	5.4646	5.3135	5.0123
Lévy portfolio – CPR	A1	A2	A3	14.216	17.801	27.506	5.4994	5.3525	5.0628
Lévy portfolio – Lévy portfolio	A1	A2	Baa1	15.687	21.891	42.04	5.4384	5.2204	4.7511
Lévy portfolio – Normal one-factor	A1	A2	A3	14.318	18.35	28.531	5.4992	5.353	5.0644
Normal one-factor – CPR	Aa1	Aa1	Aa2	1.3334	1.5642	2.0323	5.4064	5.2502	4.9375
Normal one-factor – Lévy portfolio	Aa1	Aa2	Aa3	1.4404	1.9829	3.4481	5.3406	5.1071	4.5943
Normal one-factor – Normal one-factor	Aa1	Aa1	Aa1	1.1397	1.156	1.2153	5.4059	5.2491	4.9356
Gamma one-factor – CPR	Baa1	A2	A1	54.297	20.736	11.785	5.4614	5.3022	4.9848
Gamma one-factor – Lévy portfolio	A3	A3	A2	42.16	28.406	17.871	5.3945	5.1588	4.6418

Table 5.2. Ratings, DIRR and WAL of the ABS notes, for different combinations of default and prepayment models and mean cumulative default rate $\mu_d = 0.20$ and mean cumulative prepayment rate $\mu_p = 0.10, 0.20, 0.40$.

	Note A								
	Rating			DIRR (bp)			WAL (year)		
Model pair	Reserve (Sq)	Reserve (PR)	No Reserve (PR)	Reserve (Sq)	Reserve (PR)	No Reserve (PR)	Reserve (Sq)	Reserve (PR)	No Reserve (PR)
Logistic – CPR	Aaa	Aa1	Aa1	0.02813	0.3466	0.71815	4.3424	5.2742	5.2752
Logistic – Lévy portfolio	Aaa	Aa1	Aa1	0.036137	0.48683	1.0068	4.1747	5.1311	5.1327
Logistic – Normal one-factor	Aaa	Aa1	Aa1	0.032607	0.327	0.7184	4.3475	5.2745	5.2755
Lévy portfolio – CPR	Aaa	Aaa	Aa1	0.064217	0.16105	0.71116	4.3189	5.2529	5.28
Lévy portfolio – Lévy portfolio	Aaa	Aa1	Aa1	0.094711	0.34616	1.1772	4.1503	5.1101	5.1379
Lévy portfolio – Normal one-factor	Aaa	Aa1	Aa1	0.056669	0.0977	0.80489	4.323	5.2532	5.2802
Normal one-factor – CPR	Aaa	Aaa	Aaa	0.020445	0.034631	0.16448	4.2894	5.2427	5.2437
Normal one-factor – Lévy portfolio	Aaa	Aaa	Aa1	0.018883	0.055516	0.26287	4.1185	5.0986	5.0997
Normal one-factor – Normal one-factor	Aaa	Aaa	Aaa	0.013992	0.017135	0.051144	4.2858	5.2427	5.2437
Gamma one-factor – CPR	Aa3	Aa3	Aa3	4.1521	4.6682	5.9435	4.3125	5.2599	5.264
Gamma one-factor – Lévy portfolio	Aa3	Aa3	Aa3	4.2954	4.9614	6.5872	4.1417	5.1167	5.1207

	Note B								
	Rating			DIRR (bp)			WAL (year)		
Model pair	Reserve (Sq)	Reserve (PR)	No Reserve (PR)	Reserve (Sq)	Reserve (PR)	No Reserve (PR)	Reserve (Sq)	Reserve (PR)	No Reserve (PR)
Logistic – CPR	Aa3	A1	A3	10.792	10.581	38.957	9.0526	5.3124	5.4739
Logistic – Lévy portfolio	Aa3	A1	Baa1	13.567	13.624	46.316	9.0201	5.1771	5.3522
Logistic – Normal one-factor	Aa3	A1	A3	10.952	10.906	39.955	9.0348	5.3135	5.4763
Lévy portfolio – CPR	Aa3	A2	Baa1	17.089	17.801	67.004	9.1082	5.3525	5.6466
Lévy portfolio – Lévy portfolio	A1	A2	Baa2	20.412	21.891	75.017	9.0832	5.2204	5.5242
Lévy portfolio – Normal one-factor	Aa3	A2	Baa1	17.566	18.35	67.608	9.0939	5.353	5.6467
Normal one-factor – CPR	Aa1	Aa1	A1	1.5244	1.5642	8.5988	9.0657	5.2502	5.3062
Normal one-factor – Lévy portfolio	Aa1	Aa2	A1	1.9526	1.9829	10.739	9.0305	5.1071	5.171
Normal one-factor – Normal one-factor	Aa1	Aa1	Aa3	1.1428	1.156	5.4548	9.078	5.2491	5.2995
Gamma one-factor – CPR	A1	A2	A3	20.322	20.736	30.589	9.0956	5.3022	5.341
Gamma one-factor – Lévy portfolio	A1	A3	A3	28.065	28.406	37.646	9.063	5.1588	5.1986

Table 5.3. Ratings, DIRR and WAL of the ABS notes, for different combinations of default and prepayment models with and without reserve account for sequential (Sq) and pro rata (PR) payment. Mean cumulative default rate $\mu_d = 0.20$ and mean cumulative prepayment rate $\mu_p = 0.20$.

Bibliography

[1] Albrecher, H., Ladoucette, S. and Schoutens, W. (2006), A Generic One-Factor Lévy Model for Pricing Synthetic CDOs, Advances in Mathematical Finance, M. Fu, R. Jarrow, J. Yen, R.J. Elliott (eds.), pp. 259–278, Birkhäuser, Boston.

[2] Bielecki, T. (2008), Rating SME Transactions, Moody's Investors Service.

[3] Cariboni, J., and Schoutens, W. (2008), Jumps in Intensity Models: Investigating the performance of Ornstein-Uhlenbeck processes, Metrika, Vol. 69, No. 2-3, pp. 173–198.

[4] Cifuentes, A. and O'Connor, G. (1996), The Binomial Expansion Method Applied to CBO/CLO Analysis, Moody's Special Report.

[5] Cifuentes, A. and Wilcox, C. (1998), The Double Binomial Method and its Application to a Special Case of CBO Structures, Moody's Special Report.

[6] Lando, D. (1994), On Cox Processes and Credit Risky Securities, Review of Derivatives Research, Vol. 2, No. 2-3, (December 1998), pp. 99–120.

[7] Li, A. (1995), A One-Factor Lognormal Markovian Interest Rate Model: Theory and Implementation, Advances in Futures and Options Research, Vol. 8, pp. 229–239.

[8] Mazataud, P. and Yomtov, C. (2000), The Lognormal Method Applied to ABS Analysis, Moody's Special Report.

[9] Raynes, S. and Rutledge, A. (2003), The Analysis of Structured Securities: Precise Risk Measurement and Capital Allocation, Oxford University Press.

[10] Richards, F. J. (1959), A flexible growth function for empirical use, Journal of Experimental Botany, Vol. 10, No. 2, pp. 290–300.

[11] Schönbucher, P. J. (2003), Credit Derivatives Pricing Models, Wiley Finance.

[12] Schoutens, W. (2003), Lévy Processes in Finance: Pricing Financial Derivatives, John Wiley & Sons, Chichester.

[13] Uzun, H. and Webb, E. (2007), Securitization and risk: empirical evidence on US banks, The Journal of Risk Finance, Vol. 8, No. 1, pp. 11–23.

[14] Vasicek, O. (1987), Probability of Loss on Loan Portfolio, Technical Report, KMV Corporation, 1987.

Author information

Henrik Jönsson, EURANDOM, Eindhoven University of Technology, Eindhoven, The Netherlands.
Email: jonsson@eurandom.tue.nl

Wim Schoutens, Department of Mathematics, K.U. Leuven, Celestijnenlaan 200 B, B-3001 Leuven, Belgium.
Email: wim@schoutens.be

Geert Van Damme, Department of Mathematics, K.U. Leuven, Celestijnenlaan 200 B, B-3001 Leuven, Belgium.
Email: geert.vandamme@wis.kuleuven.be

Radon Series Comp. Appl. Math **8**, 205–222

Adaptive variance reduction techniques in finance

Benjamin Jourdain

Abstract. This paper gives an overview of adaptive variance reduction techniques recently developed for financial applications. More precisely, we explain how information available in the random drawings made to compute the expectation of interest may be used at the same time to optimise control variates, importance sampling or stratified sampling.

Key words. Variance reduction techniques, control variates, importance sampling, stratification, sample average optimisation.

AMS classification. 65C05, 90C15, 91-08

1 Introduction

In mathematical finance, the price of a European option is expressed as the expectation under the risk neutral probability measure of the discounted payoff of the option. Sensitivities of the price with respect to various parameters, the so-called greeks, and in particular the delta which is of paramount importance for hedging purposes, may also be expressed as expectations. The simplest and most natural numerical approach to compute these expectations, the Monte Carlo method, is widely used in banks. According to the central limit theorem, the precision of the empirical mean approximation of the expectation of a random variable is proportional to the standard deviation of this variable. Variance reduction techniques aim at improving this precision by computing the empirical mean of independent copies of a random variable with the same expectation as the original one but with a lower variance. These techniques may be classified into two categories :

- the ones which guarantee that the variance of the new variable will be lower than the variance of the original one : antithetic variables and conditioning. In general, the variance reduction ratio obtained with these techniques is not very large.

- the ones which may lead to a more significant variance reduction ratio but may also increase the variance depending on whether they are properly implemented : control variates and importance sampling.

Stratified sampling is at the boundary between these two classes : when the allocation of the random drawings into the strata is made proportionally to their probabilities,

This research benefited from the support of the French National Research Agency (ANR) under the program ANR-05-BLAN-0299 and of the "Chair Risques Financiers", Fondation du Risque

variance reduction is guaranteed. Nevertheless, to improve efficiency, one should try other allocation rules but then the variance may increase.

Adaptive methods have been developed to ensure a proper implementation of the second category of variance reduction techniques : information available in the random drawings made to compute the expectation of interest is used to optimise the variance reduction technique at the same time. In general, they save computation time in comparison to their more natural and earlier investigated alternative : optimise the variance reduction technique on a first pilot set of random drawings and then compute the empirical mean of the resulting random variable on a second set of independent drawings. Such two stages procedures lead to unbiased estimators whereas, in general, adaptive estimators are only asymptotically unbiased.

Sections 2, 3 and 4 are respectively devoted to adaptive control variates, adaptive importance sampling and adaptive stratified sampling. Since we are interested in financial applications, we will pay in what follows particular attention to the computation of $\mathbb{E}(f(G))$ where G is a standard d-dimensional normal random vector and $f : \mathbb{R}^d \to \mathbb{R}$. Indeed, the price and hedging ratios of European options written on underlying assets evolving according to a multi-dimensional Black Scholes model may be expressed in this way. When the underlyings evolve according to a more general stochastic differential equation, Euler discretisation of this equation leads to approximations of the price and hedging ratios by expectations of the previous form, for a possibly high dimensional normal vector G and a complicated function f. Notice that in the present volume, Giles and Waterhouse [11] present an interesting multilevel path simulation technique which enables to reduce the time-discretisation bias by computing the expectation corresponding to a refined time-grid. In order to reduce the computation time necessary to obtain a balanced statistical error, they suggest to combine results using different time-steps numbers. In the end, their method consists in computing $\mathbb{E}(f(G))$ for an even higher-dimensional and more complicated function f than the one derived from standard Euler discretisation.

2 Adaptive control variates

Let us first illustrate the basic ideas of adaptive variance reduction on the simple example of linearly parametrised control variates (see for instance [21], [24] or Section 4.1 in [12]) before dealing with general parametrisation.

2.1 Linearly parametrised control variates

Suppose that we want to compute the expectation $\mathbb{E}(Y)$ of a real random variable Y and that $Z = (Z^1, \dots, Z^d)^*$ is a related $\mathbb{R}^d$-valued centred random vector with Y and Z both square-integrable. We also assume, up to removing some coordinates of Z, that the covariance matrix $\mathrm{Cov}(Z)$ of Z is non-singular and we denote by $\mathrm{Cov}(Y, Z) = \mathbb{E}(YZ)$ the covariance between Y and Z . In finance, typically $Y = e^{-rT} f(X^1_T, \dots, X^d_T)$ where f is the payoff of a European option with maturity T written on d underlying assets $X^1, \ldots, X^d$ with respective initial prices $x^1, \ldots, x^d$ and since the discounted price of each asset is a martingale under the risk neutral measure, one may choose

$$Z = (X_T^1 - e^{rT}x^1, \ldots, X_T^d - e^{rT}x^d)^*.$$

For $\theta \in \mathbb{R}^d$, since $\mathbb{E}(Y - \theta.Z) = \mathbb{E}(Y)$, one may approximate the expectation of interest $\mathbb{E}(Y)$ by the empirical mean $M_n(\theta) \stackrel{\text{def}}{=} \frac{1}{n}\sum_{j=1}^n (Y_j - \theta.Z_j)$ where $((Y_j, Z_j))_{j\geq 1}$ are independent copies of (Y, Z). The classical estimator $\frac{1}{n}\sum_{j=1}^n Y_j$ corresponds to the choice $\theta = 0$. The variance of $M_n(\theta)$, equal to $\frac{v(\theta)}{n}$ where

$$v(\theta) \stackrel{\text{def}}{=} \mathrm{Var}(Y - \theta.Z) = \mathrm{Var}(Y) - 2\theta.\mathrm{Cov}(Y, Z) + \theta.\mathrm{Cov}(Z)\theta,$$

is minimal for $\theta_\star = \mathrm{Cov}(Z)^{-1}\mathrm{Cov}(Y, Z)$. Of course, when $\mathbb{E}(Y)$ is unknown, so is $\theta_\star$. But one may estimate the covariances $\mathrm{Cov}(Z)$ and $\mathrm{Cov}(Y, Z)$, respectively, by

$$C_n \stackrel{\text{def}}{=} \frac{1}{n}\sum_{j=1}^n Z_j Z_j^* - \left(\frac{1}{n}\sum_{j=1}^n Z_j\right)\left(\frac{1}{n}\sum_{j=1}^n Z_j^*\right)$$

$$\text{and } D_n \stackrel{\text{def}}{=} \frac{1}{n}\sum_{j=1}^n Y_j Z_j - \left(\frac{1}{n}\sum_{j=1}^n Y_j\right)\left(\frac{1}{n}\sum_{j=1}^n Z_j\right).$$

Let N be the smallest index n such that no strict affine subspace of $\mathbb{R}^d$ contains $\{Z_1, \ldots, Z_n\}$. Since $\mathrm{Cov}(Z)$ is non-singular, N is a.s. finite. Moreover C_n is non-singular if and only if $n \geq N$. For $n \geq N$, one may approximate $\theta_\star$ by the estimator $\theta_n \stackrel{\text{def}}{=} C_n^{-1}D_n$ which converges a.s. to $\theta_\star$ when $n \to \infty$. The derived adaptive control variate estimator $M_n(\theta_n) = \frac{1}{n}\sum_{j=1}^n (Y_j - \theta_n.Z_j)$ of $\mathbb{E}(Y)$ is biased in general (but not when (Y, Z) is a Gaussian vector or more generally when $\mathbb{E}(Y|Z) = \mathbb{E}(Y) + \theta_\star.Z$). Nevertheless, $M_n(\theta_n)$ is a.s. convergent to $\mathbb{E}(Y)$. Moreover, writing

$$\sqrt{n}(M_n(\theta_n) - \mathbb{E}(Y)) = \begin{pmatrix} 1 \\ \theta_n \end{pmatrix} . \frac{1}{\sqrt{n}}\sum_{j=1}^n \begin{pmatrix} Y_j - \mathbb{E}(Y) \\ Z_j \end{pmatrix},$$

one deduces from the central limit theorem governing the convergence in law of the second term in the product and Slutsky's theorem that $M_n(\theta_n)$ is asymptotically normal with optimal asymptotic variance $v(\theta_\star)$. To sum up,

Proposition 2.1. *The vector* $(\theta_n, M_n(\theta_n))$ *converges a.s. to* $(\theta_\star, \mathbb{E}(Y))$ *and*

$$\sqrt{n}(M_n(\theta_n) - \mathbb{E}(Y)) \xrightarrow{\mathcal{L}} \mathcal{N}_1(0, v(\theta_\star)).$$

Variance reduction is guaranteed in the limit since $v(\theta_\star) \leq v(0) = \mathrm{Var}(Y)$, the inequality being an equality only when Y and Z are uncorrelated. When $v(\theta_\star) = 0$ i.e. when $Y = \mathbb{E}(Y) + \theta_\star.Z$ then for all $n \geq N$, $\theta_n = \theta_\star$ and $M_n(\theta_n) = \mathbb{E}(Y)$ (see [19]). This situation is not likely to occur in financial applications but an example in the context of Markov chains is given in [14] which also discusses the asymptotic properties of other adaptive estimators of $\mathbb{E}(Y)$.

One could also approximate $\mathbb{E}(Y)$ by the unbiased estimator $M_n(\tilde{\theta}_m)$ with

$$\tilde{\theta}_m = \left(\sum_{k=1}^m \tilde{Z}_k\tilde{Z}_k^* - \frac{1}{m}\sum_{k=1}^m \tilde{Z}_k \sum_{k=1}^m \tilde{Z}_k^*\right)^{-1}\left(\sum_{k=1}^m \tilde{Y}_k\tilde{Z}_k - \frac{1}{m}\sum_{k=1}^m \tilde{Y}_k\sum_{k=1}^m \tilde{Z}_k\right)$$

where $((\tilde{Y}_k, \tilde{Z}_k))_{k\geq 1}$ are i.i.d. copies of (Y, Z) independent of $((Y_j, Z_j))_{j\geq 1}$. This is an example of the two stages procedure mentioned in the introduction. But it is a pity not to use the drawings $((\tilde{Y}_k, \tilde{Z}_k))_{1\leq k\leq m}$ made to compute $\tilde{\theta}_m$ also in the computation of the expectation of interest.
Let us finally mention that θ_n introduced above as a sample average approximation of the optimal parameter $\theta_\star$ also has another interpretation. The vector θ_n minimises the sample approximation $v_n(\theta) = \frac{1}{n}\sum_{j=1}^n (Y_j - \theta.Z_j)^2 - \left(\frac{1}{n}\sum_{j=1}^n (Y_j - \theta.Z_j)\right)^2$ of $v(\theta)$. For more complex variance reduction techniques involving a parameter, no explicit expression of the optimal parameter $\theta_\star$ is in general available. So defining θ_n as an estimator of $\theta_\star$ is no longer possible. But the alternative definition of θ_n as the parameter minimising the sample average approximation of the variance remains possible. We will see applications to generally parametrised control variates in the next paragraph and to importance sampling for normal random vectors in Section 3.

2.2 General parametrisation

General parametrisation of control variates for the computation of the expectation $\mathbb{E}(Y)$ of a square-integrable random variable Y is addressed by Kim and Henderson [19, 20]. Let $\Theta \subset U \subset \mathbb{R}^p$ with Θ compact and U bounded open, Z be a d-dimensional random vector related to Y, $h : U \times \mathbb{R}^d \mapsto \mathbb{R}$ be such that

$$\forall\theta \in U,\ \mathbb{E}(h^2(\theta, Z)) < +\infty \text{ and } \mathbb{E}(h(\theta, Z)) = 0,$$

and $((Y_j, Z_j))_{j\geq 1}$ be a sequence of independent copies of (Y, Z).
For any $\theta \in U$, $M_n(\theta) \stackrel{\text{def}}{=} \frac{1}{n}\sum_{j=1}^n (Y_j - h(\theta, Z_j))$ is an unbiased and a.s. convergent estimator of the expectation of interest $\mathbb{E}(Y)$. Moreover $\text{Var}(M_n(\theta)) = \frac{v(\theta)}{n}$ where $v(\theta) \stackrel{\text{def}}{=} \text{Var}(Y - h(\theta, Z))$.

Let $m \geq 2$. When for all $z \in \mathbb{R}^d$, $U \ni \theta \mapsto h(\theta, z)$ is C^1, the unbiased estimator $\frac{1}{m-1}\sum_{j=1}^m (Y_j - h(\theta, Z_j) - M_m(\theta))^2$ of $v(\theta)$ is differentiable on U with respect to θ with gradient equal to $\frac{2}{m-1}\sum_{j=1}^m (Y_j - h(\theta, Z_j) - M_m(\theta))\nabla_\theta\left[h(\theta, Z_j) - \frac{1}{m}\sum_{k=1}^m h(\theta, Z_k)\right]$.

Let $(\gamma_l)_{l\geq 0}$ be a sequence of positive steps such that $\sum_l \gamma_l = \infty$ and $\sum_l \gamma_l^2 < \infty$. Starting from $\theta_0 \in \Theta$, Kim and Henderson [19, 20] suggest to optimise $v(\theta)$ with respect to θ using the following gradient-based stochastic approximation procedure:

$$\begin{cases} A_{l+1} &= \frac{1}{m}\sum_{j=lm+1}^{(l+1)m} (Y_j - h(\theta_l, Z_j)) \\ \theta_{l+1} &= \Pi_\Theta\Big(\theta_l - \frac{2\gamma_l}{m-1}\sum_{j=lm+1}^{(l+1)m} (Y_j - h(\theta_l, Z_j) - A_{l+1}) \\ & \qquad \times\nabla_\theta\left[h(\theta, Z_j) - \frac{1}{m}\sum_{k=lm+1}^{(l+1)m} h(\theta_l, Z_k)\right]\Big|_{\theta=\theta_l}\Big) \end{cases}$$

where Π_Θ denotes a projection of points outside Θ back into Θ. Using the law of large numbers and the central limit theorem for martingales, they study the asymptotic behaviour of the associated estimator $\mu_k \stackrel{\text{def}}{=} \frac{1}{k}\sum_{l=1}^k A_l$ of $\mathbb{E}(Y)$.

Theorem 2.2. *Assume that for all $z \in \mathbb{R}^d$, $U \ni \theta \mapsto h(\theta, z)$ is C^1 and that*

$$\mathbb{E}\left(\sup_{\theta \in U} |\nabla_\theta(\theta, Z)| \left(1 + \sup_{\theta \in U} |Y - h(\theta, Z)|\right)\right) < +\infty.$$

Then μ_k converges a.s. to $\mathbb{E}(Y)$ as $k \to \infty$. If moreover θ_k converges a.s. to a random variable θ_∞, then $\sqrt{km}(\mu_k - \mathbb{E}(Y)) \xrightarrow{\mathcal{L}} \sqrt{v(\theta_\infty)} \times G$ where $G \sim \mathcal{N}_1(0,1)$ is independent from θ_∞ and $\frac{1}{k(m-1)}\sum_{l=1}^k \sum_{j=(l-1)m+1}^{lm} (Y_j - h(\theta_{l-1}, Z_j) - A_l)^2$ converges a.s. to $v(\theta_\infty)$.
Last, if Θ is a box i.e. $\Theta = \prod_{i=1}^p [a_i, b_i]$ and $\exists \theta_0 \in \Theta$ such that

$$\mathbb{E}\left(Y^4 + \sup_{\theta \in U} |\nabla_\theta(\theta, Z)|^4 + h^4(\theta_0, Z)\right) < +\infty,$$

then the distance of θ_k to the set S of first order critical points of v on Θ converges a.s. to 0 and, when S is discrete, θ_k converges a.s. to an S-valued random variable θ_∞.

Kim and Henderson also study in [19, 20] the estimator $M_n(\tilde{\theta}_m)$ obtained by a two stages procedures where $\tilde{\theta}_m$ is obtained as a first order critical point of the sample average estimator of the variance $\frac{1}{m-1}\sum_{k=1}^m (\tilde{Y}_k - h(\theta, \tilde{Z}_k) - \frac{1}{m}\sum_{j=1}^m (\tilde{Y}_j - h(\theta, \tilde{Z}_j)))^2$ computed on a sequence $((\tilde{Y}_k, \tilde{Z}_k))_{k\geq 1}$ of independent copies of (Y, Z) independent from $((Y_j, Z_j))_{j \geq 1}$.

In [20], the behaviour of both estimators is illustrated on the example of the pricing of barrier options.

3 Importance sampling for normal random vectors

Adaptive importance sampling techniques have been developed to approximate multidimensional integrals over the unit hypercube (see [25] and the reference therein) or in the context of Markov chains (see for instance [3], [8]). But research on this topic in view of financial applications was centred on normal random vectors due to the importance of this specific case for models given by stochastic differential equations. That is why the present section is devoted to the computation of $\mathbb{E}(f(G))$ where G is distributed according to the standard d-dimensional normal law $\mathcal{N}_d(0, I_d)$ and $f : \mathbb{R}^d \to \mathbb{R}$.

We assume that

$$\mathbb{P}(f(G) \neq 0) > 0 \text{ and } \forall \theta \in \mathbb{R}^d,\ \mathbb{E}(f^2(G)e^{-\theta.G}) < +\infty. \tag{3.1}$$

The second hypothesis is implied for instance by the existence of a finite moment of order $2 + \varepsilon$ with $\varepsilon > 0$ for $|f(G)|$. Let $(G_j)_{j\geq 1}$ be i.i.d. copies of G. For $\theta \in \mathbb{R}^d$, since

$$\mathbb{E}\left(f(G+\theta)e^{-\theta.G - \frac{|\theta|^2}{2}}\right) = \mathbb{E}(f(G)), \tag{3.2}$$

$M_n(\theta) \stackrel{\mathrm{def}}{=} \frac{1}{n}\sum_{j=1}^n f(G_j+\theta)e^{-\theta.G_j-\frac{|\theta|^2}{2}}$ is an a.s. convergent and asymptotically normal estimator of $\mathbb{E}(f(G))$ with variance $\mathrm{Var}(M_n(\theta)) = \frac{v(\theta)-\mathbb{E}^2(f(G))}{n}$, where

$$\begin{aligned} v(\theta) &\stackrel{\mathrm{def}}{=} \mathbb{E}\Big(f^2(G+\theta)e^{-2\theta.G-|\theta|^2}\Big) \quad (3.3)\\ &= \mathbb{E}\Big(f^2(G+\theta)e^{-\theta.(G+\theta)+\frac{|\theta|^2}{2}}e^{-\theta.G-\frac{|\theta|^2}{2}}\Big)\\ &= \mathbb{E}\Big(f^2(G)e^{-\theta.G+\frac{|\theta|^2}{2}}\Big). \end{aligned}$$

Notice that the translated normal variable $G+\theta$ has the density $p_\theta(x) = (2\pi)^{-\frac{d}{2}}e^{-\frac{|x-\theta|^2}{2}}$ and that the importance sampling ratio $\frac{p_0}{p_\theta}(G+\theta) = e^{-\theta.G-\frac{|\theta|^2}{2}}$ appears as a factor in the left-hand-side of (3.2). The interest of the class of importance sampling estimators $M_n(\theta)$ parametrised by the translation vector $\theta \in \mathbb{R}^d$ is that a very simple analytic mapping (addition of θ) permits to transform an i.i.d. sample of the standard normal law $\mathcal{N}_d(0, I_d)$ into an i.i.d. sample of $\mathcal{N}_d(\theta, I_d)$. This feature is particularly convenient to compute and study adaptive estimators in which the parameter evolves during the simulation.

Under (3.1) the function v is

1. C^∞ with derivatives obtained by differentiation under the expectation (3.4) :

$$\nabla_\theta v^f(\theta) = \mathbb{E}\Big((\theta-G)f^2(G)e^{-\theta.G+\frac{|\theta|^2}{2}}\Big)$$

$$\nabla^2_\theta v^f(\theta) = \mathbb{E}\Big((I_d+(\theta-G)(\theta-G)^*)f^2(G)e^{-\theta.G+\frac{|\theta|^2}{2}}\Big).$$

2. strongly convex.

Therefore

$$\exists!\theta_\star \in \mathbb{R}^d \;:\; v(\theta_\star) = \inf_{\theta\in\mathbb{R}^d} v(\theta).$$

This suggests to approximate $\mathbb{E}(f(G))$ by $M_n(\theta_\star)$ but $\theta_\star$ is unknown. Unlike in the analogous example of linear control variates developed in Section 2, no explicit expression is available for $\theta_\star$. Methods aimed at approximating $\theta_\star$ have been developed in the literature. These methods are based

- either on deterministic optimisation : in [13], the authors suggest to choose θ maximising $\mathbb{R}^d \ni x \mapsto \log|f(x)| - \frac{|x|^2}{2}$ and justify this approximation by a large deviations asymptotics.
- or on stochastic optimisation procedures analogous to the ones presented in Section 2.2 : gradient-based stochastic approximation ([27] [26]), adaptive Robbins–Monro procedures [2, 1, 16, 23], robust optimisation of the sample average approximation of v by Newton's algorithm [15].

Let us now describe those stochastic optimisation procedures more precisely.

3.1 Gradient based stochastic approximation and adaptive Robbins–Monro algorithms

In [27] and [26], the authors suggest to minimise $v(\theta)$ over a compact convex subset Θ of $\mathbb{R}^d$ by the following iterative procedure using an integer $m \in \mathbb{N}^*$, a sequence $(\tilde{G}_k)_{k\geq 1}$ of independent copies of G (possibly equal to $(G_j)_{j\geq 1}$) and a sequence of positive steps $(\gamma_l)_{l\geq 0}$ s.t. $\sum_l \gamma_l = \infty$ and $\sum_l \gamma_l^2 < \infty$:

- start with $\theta_0 \in \Theta$,
- at step $l \geq 0$ compute $g_l = \frac{1}{m}\sum_{lm+1}^{(l+1)m}(\theta_l - \tilde{G}_k) f^2(\tilde{G}_k) e^{-\theta_l.\tilde{G}_k + \frac{|\theta_l|^2}{2}}$ approximating $\nabla_\theta v(\theta_l)$, then define θ_{l+1} as the projection $\theta_l - \gamma_l g_l$ on Θ.

Proposition 3.1. *Under* (3.1)*, the sequence* $(\theta_l)_{l\geq 1}$ *converges a.s. to the unique* $\theta_\Theta \in \Theta$ *such that* $v(\theta_\Theta) = \inf_{\theta\in\Theta} v(\theta)$.

The papers [27, 26] do not deal with asymptotic properties of the estimators $M_n(\theta_l)$ as $n, l \to \infty$. These questions are addressed by Arouna [2, 1] who also gets rid of the compact Θ. More precisely, he obtains a sequence $(\theta_n)_{n\geq 1}$ adapted to the filtration $(\sigma(G_1,\ldots,G_n))_{n\geq 1}$ by stabilising the Robbins–Monro algorithm corresponding to the choice $m = 1$ and $(\tilde{G}_k)_{k\geq 1} = (G_j)_{j\geq 1}$ with Chen's projection technique [6, 5]. Let $\theta_0 \in \mathbb{R}^d$, $\sigma_0 = 0$ and $(s_n)_{n\geq 0}$ be an increasing sequence of positive numbers tending to infinity with n and s.t. $s_0 \geq |\theta_0|$. The sequence (θ_n, σ_n) is defined inductively by

$$\forall n \in \mathbb{N}, \begin{cases} \theta_{n+\frac{1}{2}} = \theta_n - \gamma_n(\theta_n - G_{n+1}) f^2(G_{n+1}) e^{-\theta_n.G_{n+1} + \frac{|\theta_n|^2}{2}} \\ \text{if } |\theta_{n+\frac{1}{2}}| \leq s_{\sigma_n} \text{ then } \theta_{n+1} = \theta_{n+\frac{1}{2}} \text{ and } \sigma_{n+1} = \sigma_n \\ \text{if } |\theta_{n+\frac{1}{2}}| > s_{\sigma_n} \text{ then } \theta_{n+1} = \theta_0 \text{ and } \sigma_{n+1} = \sigma_n + 1\,. \end{cases}$$

Here σ_n is the number of projections made during the n first iterations.

Theorem 3.2. *Under* (3.1)*, the total number of projections* $\lim_{n\to\infty}\sigma_n$ *is finite and* θ_n *converges a.s. to* $\theta_\star$ *as* $n \to \infty$.

If moreover $\mathbb{E}(f^{4+\varepsilon}(G)) < +\infty$*, then as* $n \to \infty$,

$$\begin{pmatrix} M_n \\ S_n \end{pmatrix} \stackrel{\text{def}}{=} \frac{1}{n}\sum_{j=1}^{n} \begin{pmatrix} f(G_j + \theta_{j-1}) e^{-\theta_{j-1}.G_j - \frac{|\theta_{j-1}|^2}{2}} \\ f^2(G_j + \theta_{j-1}) e^{-2\theta_{j-1}.G_j - |\theta_{j-1}|^2} \end{pmatrix} \xrightarrow{\text{a.s.}} \begin{pmatrix} \mathbb{E}(f(G)) \\ v(\theta_\star) \end{pmatrix},$$

and $\sqrt{n}(M_n - \mathbb{E}(f(G))) \xrightarrow{\mathcal{L}} \mathcal{N}_1(0, v(\theta_*) - \mathbb{E}^2(f(G)))$.

As a consequence, $\sqrt{\frac{n}{S_n - M_n^2}}(M_n - \mathbb{E}(f(G))) \xrightarrow{\mathcal{L}} \mathcal{N}_1(0,1)$ which enables to construct confidence intervals for the expectation of interest $\mathbb{E}(f(G))$. The first statement follows from the verifiable sufficient conditions given by Lelong [22] for the convergence of randomly truncated stochastic algorithms. Originally, Arouna [2] checked the a.s. convergence of θ_n to $\theta_\star$ only under some explicit restrictive growth assumption on the sequence $(s_n)_n$. In [1], remarking that

$$\mathbb{E}\Big(f(G_n + \theta_{n-1}) e^{-\theta_{n-1}.G_n - \frac{|\theta_{n-1}|^2}{2}} \Big| \sigma(G_1,\ldots,G_{n-1})\Big) = \mathbb{E}(f(G)),$$

he derived the second statement using the law of large numbers and the central limit theorem for martingales

The previous algorithm takes advantage of the characterisation of $\theta_\star$ as the unique root of the equation $\mathbb{E}((\theta - G)f^2(G)e^{-\theta.G+\frac{|\theta|^2}{2}}) = 0$. Remarking that for all $\theta \in \mathbb{R}^d$, $\mathbb{E}((\theta - G)f^2(G)e^{-\theta.G+\frac{|\theta|^2}{2}}) = e^{|\theta|^2}\mathbb{E}((2\theta - G)f^2(G - \theta))$, Lemaire and Pagès [23] characterise $\theta_\star$ as the unique root of $\mathbb{E}((2\theta - G)f^2(G - \theta)) = 0$. When

$$\exists c, \alpha > 0,\ \exists \beta \in [0, 2),\ \forall x \in \mathbb{R}^d, |f(x)| \leq ce^{\alpha|x|^\beta}$$

then the Robbins–Monro procedure

$$\forall n \in \mathbb{N},\ \theta_{n+1} = \theta_n - \gamma_n e^{-2^\beta \alpha|\theta_n|^\beta}(2\theta_n - G_{n+1})f^2(G_{n+1} - \theta_n)$$

is stable without projections and Theorem 3.2 still holds with this new definition for the sequence $(\theta_n)_{n\geq 1}$. In particular, when f is bounded, α may be chosen equal to 0 and the factor $e^{-2^\beta \alpha|\theta_n|^\beta}$ is then equal to 1.

In [16], Kawai combines importance sampling with control variates remarking that for $\theta, \lambda \in \mathbb{R}^d$, the expectation and variance of the random variable

$$[f(G+\theta) - \lambda.(G+\theta)]e^{-\theta.G-\frac{|\theta|^2}{2}}$$

are respectively equal to $\mathbb{E}(f(G))$ and $v(\theta, \lambda) - \mathbb{E}^2(f(G))$ where

$$v(\theta, \lambda) \stackrel{\text{def}}{=} \mathbb{E}\Big((f(G) - \lambda.G)^2 e^{-\theta.G+\frac{|\theta|^2}{2}}\Big).$$

The function v is strictly convex in θ for fixed λ and strictly convex in λ for fixed θ. Let $g(\theta)$ (resp. $h(\lambda)$) denote the unique vector in $\mathbb{R}^d$ s.t. $v(\theta, g(\theta)) = \inf_{\lambda\in\mathbb{R}^d} v(\theta, \lambda)$ (resp. $v(h(\lambda), \lambda) = \inf_{\theta\in\mathbb{R}^d} v(\theta, \lambda)$). According to Kawai [16], the functions $v(\theta, g(\theta))$ and $v(h(\lambda), \lambda)$ are still strictly convex (but the proof of this statement does not seem correct) and there exists a unique $\theta_\star \in \mathbb{R}^d$ (resp. $\lambda_\star \in \mathbb{R}^d$) s.t. $v(\theta_\star, g(\theta_\star)) = \inf_{\theta\in\mathbb{R}^d} v(\theta, g(\theta))$ (resp. $v(h(\lambda_\star), \lambda_\star) = \inf_{\lambda\in\mathbb{R}^d} v(h(\lambda), \lambda)$). He proposes for (θ_n, λ_n) a two-scale Robbins Monro procedure with Chen's projection technique and increment

$$\begin{pmatrix} -\gamma_n(\theta_n - G_{n+1})(f(G_{n+1}) - \lambda_n.G_{n+1})^2 e^{-\theta_n.G_{n+1}+\frac{|\theta_n|^2}{2}} \\ 2\tilde{\gamma}_n(f(G_{n+1}) - \lambda_n.G_{n+1})G_{n+1}e^{-\theta_n.G_{n+1}+\frac{|\theta_n|^2}{2}} \end{pmatrix},$$

where $\tilde{\gamma}_n$ is another sequence of positive steps s.t. $\sum_n \tilde{\gamma}_n = +\infty$ and $\sum_n \tilde{\gamma}_n^2 < +\infty$. The sequence (θ_n, λ_n) converges a.s. to $(\theta_\star, g(\theta_\star))$ or $(h(\lambda_\star), \lambda_\star)$ depending on whether $\lim_{n\to\infty} \frac{\gamma_n}{\tilde{\gamma}_n}$ is equal to 0 or $+\infty$. Moreover the analogue of Theorem 3.2 holds in this setting, the estimator of $\mathbb{E}(f(G))$ being defined as

$$M_n = \frac{1}{n}\sum_{j=1}^{n}[f(G+\theta_{j-1}) - \lambda_{j-1}.(G+\theta_{j-1})]e^{-\theta_{j-1}.G_j-\frac{|\theta_{j-1}|^2}{2}}.$$

In [17], Kawai adapts the previous algorithm when the Gaussian random vector G is replaced by an infinitely divisible random vector (stochastic approximation by Robbins–Monro procedures of the parameter θ only is treated in [18]). In finance, problems involving such vectors arise for instance when the Brownian motion driving continuous time models is replaced by a Lévy process. Kawai pays particular attention to the case of independent gamma distributed components. This particular distribution has the following nice property: after the exponential change of measure (also called Esscher transform) considered in the present section, the law of a gamma random variable is the same as the law of this random variable multiplied by a constant under the original probability measure. In comparison with the Gaussian case, addition is replaced by multiplication.

Let us finally mention that an adaptive simulated annealing procedure has been recently developed by del Baño Rollin and Lázaro-Camí [7] to optimise antithetic variates. More precisely, using appropriate coordinates on the orthogonal group, the authors propose a Robbins–Monro procedure with an additional noise to compute a sequence $(O_n)_{n\geq 1}$ of orthogonal matrices converging to $O_\star$ minimising $\mathbb{E}(f(G)f(OG))$ other all orthogonal matrices O. The additional noise, obtained from a sequence $(\tilde{G}_j)_{j\geq 1}$ of random vectors i.i.d. according to $\mathcal{N}(0, I_d)$ independent of $(G_j)_{j\geq 1}$, vanishes when n tends to infinity and avoids that the algorithm remains trapped in a critical point at which $\mathbb{E}(f(G)f(OG))$ is not minimal. The derived estimator

$$M_n = \frac{1}{4n}\sum_{j=1}^{n}\Big(f(G_j) + f(O_jG_j) + f(\tilde{G}_j) + f(O_j\tilde{G}_j)\Big)$$

of $\mathbb{E}(f(G))$ is then a.s. convergent and asymptotically normal with asymptotic variance $\frac{1}{4}\left(\mathrm{Var}(f(G)) + \mathrm{Cov}(f(G), f(OG))\right)$.

3.2 Robust sample average optimisation

In order to save computation time, we introduce in [15] a parameter reduction. Indeed, numerical simulations show that, for a model driven by a Brownian motion, it is not useful to use different parameters for the increments of a single Brownian component. Let $A \in \mathbb{R}^{d\times d'}$ be a matrix with rank $d' \leq d$. We define $\tau_\star$ as the unique minimiser of the strongly convex and continuous function $\mathbb{R}^{d'} \ni \tau \mapsto v(A\tau)$. The sample average approximation of $v(A\tau)$ is given by $v_n(A\tau)$, where the C^∞ function

$$v_n(\theta) = \frac{1}{n}\sum_{j=1}^{n} f^2(G_j)e^{-\theta.G_j + \frac{|\theta|^2}{2}}$$

is strongly convex as soon as $f(G_j) \neq 0$ for some $j \in \{1, \ldots, n\}$ which holds a.s. for n large enough by (3.1). The unique minimiser τ_n of $\tau \mapsto v_n(A\tau)$ is characterised by the equality $\nabla_\tau v_n(A\tau) = 0$, which also writes $\nabla_\tau u_n(\tau) = 0$, where

$$u_n(\tau) = \frac{|A\tau|^2}{2} + \log\left(\sum_{j=1}^n f^2(G_j)e^{-A\tau\cdot G_j}\right)$$

$$\nabla_\tau u_n(\tau) = A^*A\tau - \frac{\sum_{j=1}^n A^*G_j f^2(G_j)e^{-A\tau\cdot G_j}}{\sum_{j=1}^n f^2(G_j)e^{-A\tau\cdot G_j}}$$

$$\nabla_\tau^2 u_n(\tau) = A^*A + \frac{\sum_{j=1}^n A^*G_jG_j^*Af^2(G_j)e^{-A\tau\cdot G_j}}{\sum_{j=1}^n f^2(G_j)e^{-A\tau\cdot G_j}} - \frac{\left(\sum_{j=1}^n A^*G_jf^2(G_j)e^{-A\tau\cdot G_j}\right)\left(\sum_{j=1}^n A^*G_jf^2(G_j)e^{-A\tau\cdot G_j}\right)^*}{\left(\sum_{j=1}^n f^2(G_j)e^{-A\tau\cdot G_j}\right)^2}.$$

The lowest eigenvalue of the Hessian matrix $\nabla_\tau^2 u_n$ is always larger than the one of A^*A. Therefore τ_n can easily and precisely be computed by a few iterations of Newton's algorithm using the above explicit expressions of $\nabla_\tau u_n$ and $\nabla_\tau^2 u_n$. Notice that the computation of the gradient and the Hessian of u_n is not too time-consuming since the points G_i, at which the payoff function f is evaluated, remain constant during the optimisation procedure.

Convergence of τ_n to $\tau_\star$ is a consequence of classical results concerning M-estimators.

Proposition 3.3. *1. Under* (3.1), *τ_n and $v_n(A\tau_n)$ converge a.s. to $\tau_\star$ and $v(A\tau_\star)$.*

2. If moreover

$$\forall\theta \in \mathbb{R}^d,\ \mathbb{E}\left(f^4(G)e^{-\theta.G}\right) < +\infty,\ \textit{then}\ \sqrt{n}(\tau_n - \tau_\star) \xrightarrow{\mathcal{L}} \mathcal{N}_{d'}(0, B^{-1}CB^{-1})$$

$$\textit{where}\ B = A^*\nabla_\theta^2 v(A\tau_\star)A\ \textit{and}\ C = \mathrm{Cov}\big(A^*(A\tau_\star - G)f^2(G)e^{-A\tau_\star\cdot G + \frac{|A\tau_\star|^2}{2}}\big).$$

In [15], we obtain convergence of $M_n(A\tau_n)$ to the expectation $\mathbb{E}(f(G))$ assuming that f is continuous and satisfies some growth assumption (see Theorem 3.5 below). When $d' = 1$, continuity may be replaced by a monotonicity assumption introduced in the next definition.

Definition 3.4. We say that a function $h : \mathbb{R}^d \to \mathbb{R}$

- is A-nondecreasing (resp. A-nonincreasing) if

$$\forall x \in \mathbb{R}^d,\ \tau \in \mathbb{R} \mapsto h(x + A\tau) \text{ is nondecreasing (resp. nonincreasing)},$$

- is A-monotonic if it is either A-nondecreasing or A-nonincreasing,
- belongs to $\mathcal{V}_A$ if h may be decomposed as the sum of two A-monotonic functions h_1 and h_2 such that

$$\exists\lambda > 0,\ \exists\beta \in [0, 2),\ \forall x \in \mathbb{R},\ |h_i(x)| \le \lambda e^{|x|^\beta} \text{ for } i = 1, 2.$$

When $d = 1$, $\mathcal{V}_1$ simply consists of functions with finite variation satisfying the previous growth assumption. The asymptotic properties of $M_n(A\tau_n)$ stated in the next theorem are proved in [15].

Theorem 3.5. *Assume* (3.1) *and that f admits a decomposition* $f = f_1 + 1_{\{d'=1\}} f_2$ *with*

1. *f_1 a continuous function s.t.* $\forall M > 0, \mathbb{E}\Big(\sup_{|\theta|\le M} |f_1(G+\theta)|\Big) < +\infty$,
2. *$f_2 \in \mathcal{V}_A$ defined above.*

Then, for any deterministic integer-valued sequence $(\nu_n)_n$ going to ∞ with n, $M_n(A\tau_{\nu_n})$ converges a.s. to $\mathbb{E}(f(G))$.

Assume (3.1), $\forall\theta \in \mathbb{R}^d$, $\mathbb{E}\left(f^4(G)e^{-\theta.G}\right) < +\infty$ *and that f admits a decomposition* $f = f_1 + f_2 + 1_{\{d'=1\}} f_3$ *with*

1. *f_1 a C^1 function s.t.*
$$\forall M > 0,\ \mathbb{E}\Big(\sup_{|\theta|\le M} |f_1(G+\theta)| + \sup_{|\theta|\le M} |\nabla f_1(G+\theta)|\Big) < +\infty,$$
2. $\exists \alpha \in \left((\sqrt{d'^2+8d'} - d')/4, 1\right], \beta \in [0,2), \lambda > 0,$
$$\forall x, y \in \mathbb{R}^d, |f_2(x) - f_2(y)| \le \lambda e^{|x|^\beta \vee |y|^\beta} |x-y|^\alpha,$$
3. $f_3 \in \mathcal{V}_A$.

Then $\sqrt{n}(M_n(A\tau_n) - \mathbb{E}(f(G))) \xrightarrow{\mathcal{L}} \mathcal{N}_1\left(0, v(A\tau_\star) - \mathbb{E}^2(f(G))\right)$.

In contrast to the estimator M_n constructed using Robbins–Monro procedures in the previous section, there is no martingale structure for $M_n(A\tau_{\nu_n})$. This explains why we need some regularity assumptions on the function f. Except for $d' = 1$, asymptotic normality with optimal asymptotic variance $v(A\tau_\star) - \mathbb{E}^2(f(G))$ requires more regularity on f than a.s. convergence. Note that $\frac{\sqrt{d'^2+8d'}-d'}{4}$ is increasing with d', equals $\frac{1}{2}$ for $d' = 1$ and converges to 1 as $d' \to \infty$. Therefore the choice $\alpha = 1$ is always possible for f_2. So all the financial payoffs except the discontinuous ones (barrier or digital options) satisfy the assumption made on f_2 to ensure the asymptotic normality of the adaptive estimator $M_n(A\tau_n)$. If $\mathrm{Var}(f(G)) > 0$, then the previous results imply that
$$\sqrt{\frac{n}{v_n(A\tau_n) - M_n^2(A\tau_n)}}(M_n(A\tau_n) - \mathbb{E}(f(G))) \xrightarrow{\mathcal{L}} \mathcal{N}_1(0,1),$$
and one may easily derive confidence intervals for $\mathbb{E}(f(G))$.

The numerical experiments performed in [15] suggest that strong convergence and asymptotic normality of $M_n(A\tau_n)$ still hold under less restrictive assumptions on f than those stated in the previous theorem.

4 Stratified sampling

We are interested in the computation of $c = \mathbb{E}(f(X))$ where X is an $\mathbb{R}^d$-valued random vector and $f : \mathbb{R}^d \to \mathbb{R}$ a measurable function such that $\mathbb{E}(f^2(X)) < \infty$. We suppose that $(A_i)_{1\le i\le I}$ is a partition of $\mathbb{R}^d$ into I *strata* such that $p_i = \mathbb{P}[X \in A_i]$ is known explicitly for $i \in \{1, \dots, I\}$. Up to removing some strata, we assume from now on that p_i is positive for all $i \in \{1, \dots, I\}$. The stratified Monte-Carlo estimator of c (see [12, p.209–235] and the references therein for a detailed presentation) is based on the equality $\mathbb{E}(f(X)) = \sum_{i=1}^I p_i \mathbb{E}(f(X^i))$ where X^i denotes a random variable distributed according to the conditional law of X given $X \in A_i$. Indeed, when the variables X_i can be simulated, it is possible to estimate each expectation in the right-hand side using n_i i.i.d drawings of X^i. Let $n = \sum_{i=1}^I n_i$ be the total number of drawings (in all the strata) and $q_i = n_i/n$ denote the proportion of drawings made in stratum i.

Then $\widehat{c}$ is defined by

$$\widehat{c} = \sum_{i=1}^{I} \frac{p_i}{n_i} \sum_{j=1}^{n_i} f(X_j^i) = \frac{1}{n} \sum_{i=1}^{I} \frac{p_i}{q_i} \sum_{j=1}^{q_i n} f(X_j^i),$$

where for each i the X_j^i, $1 \le j \le n_i$, are distributed like X^i, and all the X_j^i, for $1 \le i \le I, 1 \le j \le n_i$ are drawn independently. This stratified sampling estimator can be implemented for instance when X is distributed according to the standard normal law on $\mathbb{R}^d$, $A_i = \{x \in \mathbb{R}^d : y_{i-1} \le \theta.x < y_i\}$ where $-\infty = y_0 < y_1 < \cdots < y_{I-1} < y_I = +\infty$ and $\theta \in \mathbb{R}^d$ is such that $|\theta| = 1$. Indeed, then one has $p_i = N(y_i) - N(y_{i-1})$ with $N(.)$ denoting the cumulative distribution function of the one-dimensional normal law and when U is uniformly distributed on $[0, 1]$ and independent from X, then

$$X + (N^{-1}[N(y_{i-1}) + U(N(y_i) - N(y_{i-1}))] - \theta.X)\theta$$

follows the conditional law of X given $y_{i-1} \le \theta.X < y_i$.

We have $\mathbb{E}(\widehat{c}) = c$ and

$$\mathrm{Var}(\widehat{c}) = \sum_{i=1}^{I} \frac{p_i^2\sigma_i^2}{n_i} = \frac{1}{n}\sum_{i=1}^{I} \frac{p_i^2\sigma_i^2}{q_i} = \frac{1}{n}\sum_{i=1}^{I}\left(\frac{p_i\sigma_i}{q_i}\right)^2 q_i \ge \frac{1}{n}\left(\sum_{i=1}^{I}\frac{p_i\sigma_i}{q_i}q_i\right)^2, \tag{4.1}$$

where $\sigma_i^2 = \mathrm{Var}(f(X^i)) = \mathrm{Var}(f(X)|X \in A_i)$ for all $1 \le i \le I$.

In the sequel, we assume $\sigma_{i_0} > 0$ for at least one index i_0.

Let $(X_j)_{j\ge 1}$ be i.i.d. drawings of X. The variance of the crude Monte Carlo estimator $\frac{1}{n}\sum_{j=1}^n f(X_j)$ of $\mathbb{E}(f(X))$ is

$$\frac{1}{n}\left(\sum_{i=1}^{I} p_i\left(\sigma_i^2 + \mathbb{E}^2(f(X^i))\right) - \left(\sum_{i=1}^{I} p_i\mathbb{E}(f(X^i))\right)^2\right) \ge \frac{1}{n}\sum_{i=1}^{I} p_i\sigma_i^2.$$

For given strata, the stratified estimator achieves variance reduction if the allocations n_i or equivalently the proportions q_i are properly chosen. For instance, for the so-called proportional allocation $q \equiv p$, the variance of the stratified estimator is equal to the previous lower bound of the variance of the crude Monte Carlo estimator. For the *optimal allocation* $q_i^\star \stackrel{\text{def}}{=} p_i\sigma_i / \sum_{j=1}^{I} p_j\sigma_j$, $1 \le i \le I$, the lower bound in (4.1) is attained. Then

$$\operatorname{Var}(\widehat{c}) = \frac{1}{n}\Big(\sum_{i=1}^{I} p_i\sigma_i\Big)^2 \stackrel{\text{def}}{=} \frac{\sigma_\star^2}{n}.$$

In general, when the conditional expectations $\mathbb{E}(f(X)|X \in A_i) = \mathbb{E}(f(X^i))$ are unknown, then so are the conditional variances σ_i^2. Therefore optimal allocation of the drawings is not feasible at once. One can of course estimate the conditional variances and the optimal proportions by a first Monte Carlo algorithm and run a second Monte Carlo procedure with drawings independent from the first one to compute the stratified estimator corresponding to these estimated proportions. But why not use the drawings made in the first Monte Carlo procedure also for the final computation of the conditional expectations?

Instead of running two successive Monte Carlo procedures, one can think to obtain a first estimation of the σ_i's, using the first drawings of the X^i's made to compute the stratified estimator. One could then estimate the optimal allocations before making further drawings allocated in the strata according to these estimated proportions. One can next obtain another estimation of the σ_i's, compute again the allocations and so on. This is the principle of the adaptive allocation procedure proposed in [10] and described in the next section. Then, we will present the adaptive algorithm proposed in [9] in order to optimise the strata themselves.

4.1 Adaptive optimal allocation

Let N^k (resp. N_i^k) denote the total number of random drawings X_j^i made in all the strata (resp. in stratum i) at the end of step k of the following algorithm :

1. At step 1, allocate the N^1 first drawings in the strata proportionally to the p_i and estimate $\mathbb{E}(f(X^i))$ and σ_i, $1 \le i \le I$,
2. At the beginning of step $k \ge 2$, compute the vector $(n_1, \dots, n_I) \in \mathbb{R}_+^I$ obtained by allocating the $N^k - N^{k-1}$ new drawings

   - either proportionally to the estimations $p_i\widehat{\sigma}_i^{k-1} / \sum_{l=1}^{I} p_l\widehat{\sigma}_l^{k-1}$ of the $q_i^\star$ available at the end of step $k-1$,
   - or in order to minimise the estimated variance $\sum_{i=1}^{I}(p_i\widehat{\sigma}_i^{k-1})^2/N_i^k$ of the stratified estimator after step k under the constraints $\sum_{i=1}^{I} N_i^k = N^k$ and $\forall i$, $N_i^k \ge N_i^{k-1}$. The explicit solution of this constrained optimisation problem is given in [10].

   Then convert $(n_1, \dots, n_I)$ to $\mathbb{N}^I$ by the following rounding procedure preserving the sum : $n_i^k = \lfloor \sum_{l=1}^{i} n_l \rfloor - \lfloor \sum_{l=1}^{i-1} n_l \rfloor$ and allocate n_i^k new drawings in stratum i. Refine the estimations $\widehat{c}_i^k$ and $\widehat{\sigma}_i^k$ of $\mathbb{E}(f(X^i))$ and σ_i using these new drawings.

In fact, one has to modify this algorithm in order to enforce at least one drawing in each stratum at each step. Indeed, if $\widehat{\sigma}_{i_0}^1 = 0$ whereas $\sigma_{i_0} > 0$, then no drawings are made after step $k=1$ in the stratum i_0 and $\frac{1}{N^k_{i_0}}\sum_{j=1}^{N^k_{i_0}} f(X^{i_0}_j) = \frac{1}{N^1_{i_0}}\sum_{j=1}^{N^1_{i_0}} f(X^{i_0}_j)$ does not converges to $\mathbb{E}(f(X^{i_0}))$ when $k\to+\infty$ which prevents the stratified estimator $\sum_{i=1}^I \frac{p_i}{N^k_i}\sum_{j=1}^{N^k_i} f(X^i_j)$ from converging to $\mathbb{E}(f(X))$. Choosing the sequence $(N^k)_{k\geq 1}$ so that $N^k \geq N^{k-1}+I$ for all $k\geq 2$, enforcing one drawing in each stratum at each step k, and allocating the remaining $N^k - N^{k-1} - I$ drawings according the previous procedure permits to overcome this difficulty. Then $\forall 1\leq i\leq I$, $\forall k\geq 1$, $N^k_i \geq k$ and the following result is proved in [10] by first checking that the proportions $\frac{N^k_i}{N^k}$ converge a.s. to the optimal ones $q^\star_i$ as $k\to\infty$ and then applying the central limit theorem for martingales :

Theorem 4.1.

$$\mathbb{P}\left(\sum_{i=1}^I \frac{p_i}{N^k_i}\sum_{j=1}^{N^k_i} f(X^i_j) \xrightarrow[k\to\infty]{} \mathbb{E}(f(X))\right) = 1.$$

If, moreover, $\sigma_{i_0}>0$ for some $i_0\in\{1,\dots,I\}$ and $\lim_{k\to+\infty}\frac{k}{N^k}=0$, then

$$\sqrt{N^k}\left(\sum_{i=1}^I \frac{p_i}{N^k_i}\sum_{j=1}^{N^k_i} f(X^i_j) - \mathbb{E}(f(X))\right) \xrightarrow[k\to\infty]{\mathcal{L}} \mathcal{N}_1\left(0,\sigma^2_\star\right)$$

with $\sigma^2_\star = \left(\sum_{i=1}^I p_i\sigma_i\right)^2$ the asymptotic variance for the optimal allocation.

As a consequence, $\frac{\sqrt{N^k}}{\sum_{i=1}^I p_i\widehat{\sigma}^k_i}\left(\sum_{i=1}^I \frac{p_i}{N^k_i}\sum_{j=1}^{N^k_i} f(X^i_j) - \mathbb{E}(f(X))\right) \xrightarrow[k\to\infty]{\mathcal{L}} \mathcal{N}_1(0,1)$ and one may easily construct confidence intervals for $\mathbb{E}(f(X))$. Numerical experiments performed in [10] on the pricing of arithmetic average Asian options in the Black–Scholes model show that adaptive allocation permits to divide the variance obtained with proportional allocation by a factor up to 50.

Another stratified sampling algorithm in which the optimal proportions and the conditional expectations are estimated using the same drawings has been proposed in [4] for quantile estimation. More precisely, for a total number of drawings equal to N, the authors suggest to allocate the N^γ with $0<\gamma<1$ first ones proportionally to the probabilities of the strata and then use the estimation of the optimal proportions obtained from these first drawings to allocate the $N-N^\gamma$ remaining ones. Their stratified estimator is also asymptotically normal with asymptotic variance equal to the optimal one. In practice, N is finite and it seems better to take advantage of all the drawings and not only the N^γ first ones to modify adaptively the allocation between the strata.

4.2 Adaptive optimisation of the strata for normal random vectors

Let us now consider the problem of optimally designing the strata when they are parametrised in the following way : for $1\leq i\leq I$, $A_i = \left\{x\in\mathbb{R}^d : \theta.x\in[y_{i-1},y_i)\right\}$ where

$$-\infty = y_0 < y_1 < \cdots < y_{I-1} < y_I = +\infty \text{ and } \theta\in\mathbb{R}^d \text{ is s.t. } |\theta| = 1.$$

In [9], we address a more general parametrisation where the strata are defined by hyperrectangles but the present section is devoted to the particular case of a single stratification direction.

Our aim is to approximate the parameters $(\theta, y_1, \ldots, y_{I-1})$ defining the strata which minimise the standard deviation $\sigma_\star = \sum_{i=1}^{I} p_i\sigma_i$ obtained either by optimal allocation or with the adaptive allocation algorithm described above. This standard deviation $\sigma_\star$ is equal to

$$\sum_{i=1}^{I} \sqrt{(\nu_\theta(1, y_i) - \nu_\theta(1, y_{i-1}))(\nu_\theta(f^2, y_i) - \nu_\theta(f^2, y_{i-1})) - (\nu_\theta(f, y_i) - \nu_\theta(f, y_{i-1}))^2}.$$

where $\nu_\theta(h, y) \stackrel{\text{def}}{=} \mathbb{E}(h(X)1_{\{\theta.X \le y\}})$ for $y \in \mathbb{R}$ and $h : \mathbb{R}^d \to \mathbb{R}$ such that $h(X)$ is integrable. According to the following lemma proved in [9] it is possible to express the gradient of $\nu_\theta(h, y)$ in terms of conditional expectations.

Lemma 4.2. *When $\theta.X$ admits a density p_θ w.r.t. the Lebesgue measure on the real line and under further technical regularity assumptions not precised here,*

$$\partial_y \nu_\theta(h, y) = p_\theta(y)\mathbb{E}(h(X)|\theta.X = y)$$
$$\nabla_\theta \nu_\theta(h, y) = -p_\theta(y)\mathbb{E}(Xh(X)|\theta.X = y).$$

We suppose from now on that $X \sim \mathcal{N}_d(0, I_d)$ is a standard normal random vector. Then $p_\theta(y) = \frac{e^{-y^2/2}}{\sqrt{2\pi}}$ and

$$\forall i \in \{1, \ldots, I\},\ \mathbb{E}(h(X)|\theta.X = y) = \mathbb{E}[h(X^i + (y - \theta.X^i)\theta)].$$

At each step k of the above optimal allocation algorithm, this enables us

1. to estimate the gradient of $\sigma_\star$ w.r.t. $(y_1, \ldots, y_{I-1})$ and θ using the orthogonal projections on the boundaries of the random drawings X_j^i made at this step in the strata,
2. to perform a gradient descent step to update the stratification direction and boundaries.

In practice, the differences $N^k - N^{k-1}$ should be large enough not to increase significantly the computation time needed to calculate the crude Monte Carlo estimator. As a consequence, the Monte Carlo estimator of the gradient is precise and the optimisation of the strata parameters is rather a noisy gradient descent than a stochastic algorithm.

According to our numerical experiments, optimising the direction θ works : the gradient procedure converges to some limit and this ensures effective variance reduction. On examples involving discontinuous payoffs such as barrier options, the optimal direction computed with our algorithm is significantly different and more efficient than the one derived analytically in [13] using some large deviations asymptotics. Numerical optimisation of the strata boundaries was far less convincing. In [9], we explain this numerical observation by the following asymptotic analysis performed in the limit $I \to \infty$. We parametrise the boundaries by a positive probability density g on $\mathbb{R}$ with c.d.f. $G(y) = \int_{-\infty}^{y} g(z)dz$ and set $y_i = G^{-1}(\frac{i}{I})$ for $i \in \{0, \dots, I\}$.

Theorem 4.3. • *Let* $d \geq 2$. *If for*

$$h \in \{p_\theta, p_\theta \times \mathbb{E}(f(X)|\theta.X = \cdot), p_\theta \times \mathbb{E}(f^2(X)|\theta.X = \cdot)\}, \quad \int_{\mathbb{R}} \frac{h^2}{g}(y)dy < +\infty,$$

then $\lim_{I\to\infty} \sigma_\star(I) = \mathbb{E}\left(\sqrt{\mathrm{Var}(f(X)|\theta.X)}\right)$.

- *When* $d = 1$, *and* f *is a locally bounded function on the real line with a locally integrable distribution derivative* f' *such that* $\mathrm{esssup}_{dy} \frac{p_\theta + |f'|}{g}(y) < +\infty$, *then* $\lim_{I\to\infty} I\sigma_\star(I) = \frac{1}{\sqrt{12}} \int_{\mathbb{R}} \frac{|f'|p_\theta}{g}(y)dy$.

The fact that, in the practical case $d \geq 2$, the limit does not depend on g means that under optimal or adaptive allocation, the choice of the boundaries of the strata is not important when the number of strata is large. So only the stratification direction θ should be optimised.

Note that the optimised direction θ computed by our algorithm can be used to design Latin hypercube or Quasi Monte Carlo (see [12]) estimators of $\mathbb{E}(f(X))$. When X is a standard normal random vector, for any orthogonal matrix $O \in \mathbb{R}^{d\times d}$, $\mathbb{E}(f(X)) = \mathbb{E}(f(OX))$, but the convergence properties of Latin hypercube or QMC estimators associated with the variable $f(OX)$ crucially depend on O. Unfortunately, it is very difficult to estimate these rates of convergence and adaptive optimisation of the matrix O seems unreachable. As Latin hypercube or QMC methods somehow consist in stratifying each canonical direction, choosing the first column of O equal to θ should be effective.

Bibliography

[1] Bouhari Arouna, *Adaptative Monte Carlo method, a variance reduction technique*, Monte Carlo Methods Appl. 10 (2004), pp. 1–24. MR MR2054568 (2004m:62159)

[2] ______, *Robbins Monro algorithms and variance reduction in finance*, J. of Comput. Finance 7 (Winter 2003/04), pp. 35–61.

[3] Keith Baggerly, Dennis Cox, and Rick Picard, *Exponential convergence of adaptive importance sampling for Markov chains*, J. Appl. Probab. 37 (2000), pp. 342–358. MR MR1780995 (2001e:65008)

[4] Claire Cannamela, Josselin Garnier, and Bertrand Looss, *Controlled stratification for quantile estimation*, Ann. Appl. Stat. 2 (2008), pp. 1554–1580.

[5] Han Fu Chen, Guo Lei, and Ai Jun Gao, *Convergence and robustness of the Robbins-Monro algorithm truncated at randomly varying bounds*, Stochastic Process. Appl. 27 (1988), pp. 217–231. MR MR931029 (89b:62180)

[6] Han Fu Chen and Yun Min Zhu, *Stochastic approximation procedures with randomly varying truncations*, Sci. Sinica Ser. A 29 (1986), pp. 914–926. MR MR869196 (88b:62158)

[7] Sebastian del Baño Rollin and Joan-Andreu Lázaro-Camí, *Antithetic variates in higher dimension*, Preprint ArXiv 0902.4211 (2009).

[8] Paul Dupuis and Hui Wang, *Dynamic importance sampling for uniformly recurrent Markov chains*, Ann. Appl. Probab. 15 (2005), pp. 1–38. MR MR2115034 (2006b:60042)

[9] Pierre Étoré, Gersende Fort, Benjamin Jourdain, and Éric Moulines, *On adaptive stratification*, Preprint ArXiv:0809.1135 (2008).

[10] Pierre Étoré and Benjamin Jourdain, *Adaptive optimal allocation in stratified sampling methods*, Methodol. Comput. Appl. Probab. (To appear).

[11] Michael B. Giles and Ben J. Waterhouse, *Multilevel quasi-Monte Carlo path simulation*, Radon Series Comp. Appl. Math. 8 (2009).

[12] Paul Glasserman, *Monte Carlo methods in financial engineering*, Applications of Mathematics (New York), vol. 53, Springer-Verlag, New York, 2004, Stochastic Modelling and Applied Probability. MR MR1999614 (2004g:65005)

[13] Paul Glasserman, Philip Heidelberger, and Perwez Shahabuddin, *Asymptotically optimal importance sampling and stratification for pricing path-dependent options*, Math. Finance 9 (1999), pp. 117–152. MR MR1849001 (2002m:91035)

[14] Shane G. Henderson and Burt Simon, *Adaptive simulation using perfect control variates*, J. Appl. Probab. 41 (2004), pp. 859–876. MR MR2074828 (2005h:65009)

[15] Benjamin Jourdain and Jérôme Lelong, *Robust adaptive importance sampling for normal random vectors*, Ann. Appl. Probab. (To appear).

[16] Reiichiro Kawai, *Adaptive Monte Carlo variance reduction with two-time-scale stochastic approximation*, Monte Carlo Methods Appl. 13 (2007), pp. 197–217. MR MR2349428 (2008h:62195)

[17] ———, *Adaptive Monte Carlo variance reduction for Lévy processes with two-time-scale stochastic approximation*, Methodol. Comput. Appl. Probab. 10 (2008), pp. 199–223. MR MR2399681

[18] ———, *Optimal importance sampling parameters search for Lévy processes via stochastic approximation*, SIAM J. Numer. Anal. 47 (2008), pp. 293–307.

[19] Sujin Kim and Shane G. Henderson, *Adaptive control variates*, Proceedings of the 2004 Winter Simulation Conference (2004), pp. 621–629.

[20] ———, *Adaptive control variates for finite-horizon simulation*, Math. Oper. Res. 32 (2007), pp. 508–527. MR MR2348231 (2008i:65005)

[21] Stephen Lavenberg, Thomas Moeller, and Peter Welch, *Statistical Results on Control Variables with Application to Queuing Network Simulation*, Oper. Res. 30 (1982), pp. 182–202.

[22] Jérôme Lelong, *Almost sure convergence of randomly truncated stochastic algorithms under verifiable conditions*, Stat. Probab. Letters 78 (2008), pp. 2632–2636.

[23] Vincent Lemaire and Gilles Pagès, *Unconstrained Recursive Importance Sampling*, Preprint ArXiv:0807.0762 (2008).

[24] Barry L. Nelson, *Control variate remedies*, Oper. Res. 38 (1990), pp. 974–992. MR MR1095954

[25] Teemu Pennanen and Matti Koivu, *An adaptive importance sampling technique*, Monte Carlo and quasi-Monte Carlo methods 2004, Springer, Berlin, 2006, pp. 443–455. MR MR2208724 (2006k:65065)

[26] Yi Su and Michael Fu, *Optimal importance sampling in securities pricing*, J. Comput. Finance 5 (2002), pp. 26–50.

[27] Felicia Vázquez-Abad and Daniel Dufresne, *Accelerated simulation for pricing Asian options*, Proceedings of the 1998 Winter Simulation Conference (1998), pp. 1493–1500.

Author information

Benjamin Jourdain, Université Paris-Est, CERMICS, Project team MathFi ENPC-INRIA-UMLV, 6 et 8 avenue Blaise Pascal, 77455 Marne La Vallée, Cedex 2, France.
Email: jourdain@cermics.enpc.fr

Radon Series Comp. Appl. Math **8**, 223–244

Regularisation of inverse problems and its application to the calibration of option price models

Stefan Kindermann and Hanna K. Pikkarainen

Abstract. We give an overview of the most important features of inverse and ill-posed problems and their solutions by regularisation. We point out links to the problem of model calibration in financial mathematics by a survey on the calibration of option price models using Tikhonov regularisation.

Key words. Inverse problems, Tikhonov regularisation, model calibration, option pricing.

AMS classification. 65J20, 91B28

1 Introduction

Inverse problems are a well-established field in mathematics, combining theory and application in a fascinating manner. The ill-posedness of many of interesting inverse problems is the most salient feature that make these problems difficult to solve. The classical way to cope with the ill-posedness, is to apply regularisation. The theory of regularisation is built on a sound mathematical basis and is one of the cornerstones in the research of inverse and ill-posed problems.

With the emergence of advanced models in financial mathematics, the need for a robust model calibration arose in financial applications. Apart from rather simple models these calibration problems turned out to be ill-posed and inverse problems in many instances. Thus, there was and is a strong requirement for well-founded regularisation methods in financial mathematics.

The purpose of this paper is to give an overview of the regularisation theory and related methods in inverse problems. This overview is motivated by the problem of model calibration for option pricing, which is one of the most prominent examples where regularisation "pays off". We would like to highlight aspects of (mainly) Tikhonov regularisation with this application (like the calibration of a local volatility in the Dupire model) in mind. Hopefully, this introduction can shed some light on the necessity and the importance of regularisation and can serve as a very basic user's guide for non-specialists and researchers in mathematical finance. Finally, we give a survey on recent results where regularisation has been successfully applied to option pricing and related calibration problems.

2 Inverse problems

A well-known definition [23] states that inverse problems are concerned with determining causes for a desired or an observed effect. This general statement does not answer the question why inverse problems require a special mathematical theory. It is rather the case that most inverse problems are ill-posed in the sense of Hadamard [31, 23], and therefore need an extra mathematical treatment. Hadamard calls a problem well-posed if for all data a solution exists, the solution is unique and the solution depends stably on the data. If at least one of these conditions is violated, a problem is called ill-posed.

The ill-posedness is the distinguished feature of most inverse problems such that the notations *inverse problem* and *ill-posed problems* is in many cases used synonymously. Ill-posedness also leads to the distinction between direct and inverse problems: These are problems that are inverse to each other but the direct problem is usually the well-posed one while the inverse problem might be an ill-posed one. Ill-posedness is furthermore mainly apparent in the instability of an inverse problem. A solution to an inverse problem does not depend stably (in an appropriate topology) on the input.

A well-known example that demonstrates this dichotomy is the relation between the volatility and the option price in standard option price models. Given, for instance, a standard Dupire model, a direct problem is stated as follows: If the (possibly nonconstant) volatility function is known, find the associated option price for a European call option. A calculation of the direct problem here amounts to solving a parabolic partial differential equation with known coefficients (the volatility). From standard PDE theory it follows that a solution to this problem exists and is unique (under not too strong assumptions) and the option price depends stably on the input parameter, the volatility.

All this is not the case for the corresponding inverse problem: In the same model, let us assume to have full or partial information on the option price. An inverse problem is stated as follows: Given the option price, find a/the volatility that generates this option price via the Black–Scholes PDE. Now this is an ill-posed problem, because even if a solution to this inverse problem might exist and might be unique, the volatility will not depend on the option price in a continuous way (using, e.g., Sobolev space norms, see, e.g., [20]).

The instability is the main obstacle in calculating the volatility from the option price in a reliable way. Hence, it has to be taken into account if we want to calibrate an option price model in a *robust* way. Note that in general for calibration the use of the inverse problem theory is not always needed: If we want to find any parameter that fits the data, and do not want to draw conclusions about this parameter, this can be done by standard algorithms. Instances where such an approach is relevant are Black-Box models, or in the control theory where one is satisfied with a good fit. Mathematically speaking, here only the convergence of the model to the data in the data space is required. However, if we expect that there is a “real” meaningful parameter behind the model and if we want to extract reliable information from the data about this parameter, we ask for a method which not only provides a good fit, but also approximates this exact parameter in a satisfying manner. We therefore want to have a method which shows convergence in the parameter/solution space. Solving unstable inverse problems in a naive way does not necessarily give convergence in the parameter space: Or as a rule of thumb: *for ill-posed problems a good fit to the data does not mean a good fit of the unknown parameter.*

We should be aware that the option prices are not exact. There exists a bid-ask spread which can be seen as some kind of noise in the data. If we cannot calculate a volatility from the data in a stable way, the result is useless because it does not give any information on the 'true' volatility. The calculated solution can depend on the type of the computer used, on the accuracy used and on many other parameters. For ill-posed problems with instability, computational experience shows that with naive algorithms, useless solutions far from the true one are not exceptional but rather common.

This, however, is not the end of the story. Fortunately, there is a way to calculate approximate solutions to inverse problems in a stable way by so-called regularisation methods. This theory is the centre point for solving inverse problems algorithmically.

The theory of inverse problems can efficiently be treated by formulating them as abstract operator equations. It is common to introduce the so-called input-to-output mapping (the forward mapping or the parameter-to-solution mapping) which is just the mapping of an unknown in an inverse problem to the data generated by this unknown. Let us denote the unknown by x and the data by y. Then the forward mapping can be written as the operator

$$F : D(F) \subset X \to Y,\ x \mapsto F(x)$$

acting between topological spaces X and Y (usually Hilbert or Banach spaces). In the option price example above, the forward operator would take a volatility σ and map it to the solution of the associated Black–Scholes equation. The theory of PDEs can be used to show that this mapping is well-defined and continuous if appropriate spaces are chosen for X and Y. Although this is a compact notation, the computation of the forward operator can be quite difficult and expensive; for instance, it involves solving one or several PDEs.

Given the operator F we can express the inverse problem as an operator equation: Given the data y, find the/a solution x such that

$$F(x) = y. \tag{2.1}$$

The inverse problem is ill-posed if the operator F does not possess a continuous inverse on Y.

Let us look closer at the problem of ill-posedness: As we have already pointed out, out of the three conditions for well-posedness, the violation of stable dependence on the data is the most severe one. Violation of the other conditions can partially be remedied by an appropriate generalisation of the definition of a solution. For instance, it is standard to use least squares solutions for the case when the data are not in the range of F: We call x a least squares solution if it minimises the error $\|F(x) - y\|_Y$:

$$x = \operatorname{argmin}\{\|F(z) - y\|_Y \mid z \in D(F)\}, \tag{2.2}$$

where Y is a norm space and $\|\cdot\|_Y$ is a norm in Y. Here it is clear that a least squares solution can exist even if y is not in the range of F. However, such a generalised solution does not necessarily solve the problem of existence completely because it might not exist as well. For linear problems in Hilbert spaces the set of data y for which a least squares solution exists is well understood: it is the domain of definition of the pseudoinverse (see [23]), and hence a dense subset of the data space Y. Thus, a least squares solution generalises the notion of solution to data out of a dense subset of Y. In the nonlinear case, the question of existence of a least squares solution is not easy to answer. To circumvent this question, it is quite common to use the assumption of attainability, i.e., y is in the range of F. Of course, this trivially implies the existence of a least squares solution. If we assume that our model is correct (i.e., F is an exact description of the parameter-to-data mapping) and there is no noise in the data, attainability is a reasonable assumption. For an analysis of nonlinear ill-posed problems without the attainability assumption we refer to [4].

Next we discuss the uniqueness: A least squares solution is usually not unique. In fact, if F has a null space, adding an element of the null space to a least squares solution yields another least squares solution. If we want to reduce that ambiguity, we can define the minimum norm least squares solution $x^\dagger$:

$$x^\dagger = \text{argmin}\{\|x\|_X \mid x \text{ is a least squares solution}\}, \tag{2.3}$$

where X is a norm space and $\|\cdot\|_X$ is a norm in X. Thus we simply select a least squares solution which has the minimal norm among all least squares solutions. It is often helpful to change this definition and include an a-priori guess x^* to get the x^*-minimum norm least squares solution:

$$x^\dagger = \text{argmin}\{\|x - x^*\|_X \mid x \text{ is a least squares solution}\}. \tag{2.4}$$

Without further assumptions, a minimum norm least squares solution does not have to exist nor has to be unique. It exists if a least squares solution exists, and if F is linear or F is injective it is unique. Showing that a forward operator is injective is usually a difficult task for parameter identification problems, requiring sophisticated methods (for an overview see for instance [42]). From a practical point of view it is useful to separate the question of the uniqueness (the injectivity of F) from the reconstruction procedure and simply postulate the existence of a minimum norm least squares solution.

With these definitions we can now specify more precisely what we understand by a solution to an inverse problem: Given some data y we want to find an (x^*)-minimum norm least squares solution for equation (2.1). The use of minimum norm least squares solutions is especially useful for the case where F is a linear operator. Then the mapping from y to the minimum norm least squares solution $x^\dagger$ (for those y for which it is defined) is called the pseudoinverse $F^\dagger$. For a linear problem we can completely characterise when a problem is ill-posed, namely exactly when the pseudoinverse is not continuous. Using well-known tools from functional analysis (e.g., the open mapping theorem) it can further be concluded that for a linear problem between Hilbert spaces the pseudoinverse is not continuous (in particular, the problem is ill-posed) when the range of F is not closed in Y, or equivalently if the domain of definition of $F^\dagger$ is not the whole data space Y (but only a dense subset). This is a quite useful characterisation of the instability. Note that in the linear case the domain of definition of $F^\dagger$ coincides with the set of y for which a least squares solution exists. This links the existence question in Hadamard's definition to the stability question: If for all y a least squares solution exists, the problem is well posed and vice versa.

Another important conclusion concerns the discrete case. If we have a discretised problem (i.e., X and Y are finite-dimensional spaces), for linear problems the range of F is always closed and a least squares solution exists for all y. This means that in finite-dimensions, there is no ill-posedness. This is illustrated by the quote *ill-posedness resides in infinite dimensions* [28]. Strictly speaking, this only holds true for the linear case, but as a rule of thumb — finite-dimensional problems are well-posed — it is quite useful in the nonlinear case, too. This explains to some extent why naive algorithms can work for some inverse problems: If we are only interested in finding a *finite* number of parameters of a solution, we are in the discrete case and the inverse problem is stable. This is in particular the case when we take a parametric approach: We suppose that the volatility is taken out of a specific set of functions parameterised by a finite number of coefficients (e.g., Gaussians where only the mean is unknown). The same argument applies if we assume a constant volatility (the Black–Scholes model): Then we only have to find one number and the problem is not ill-posed. In these cases no regularisation is needed because the problem is stable anyway. However, if the dimension of the solution space becomes larger and larger, the problem might still be stable in a strict sense but – if the underlying infinite-dimensional problem is ill-posed – it will become *ill-conditioned.* The modulus of continuity of the pseudoinverse will be very large and hence again a naive computation will fail for the same reasons as in the infinite-dimensional case. So even for finite-dimensional problems regularisation is necessary, if the dimension of the solution space becomes large. (There is not a general answer to the question when the dimension is too large; this depends on the problem (on the mapping properties of F)).

For nonlinear problems, there is no standard notion of a pseudoinverse and thus it is more involved to find conditions when a problem is ill-posed. There is an important class of problems which are ill-posed and for which ill-posedness can easily be shown: If the operator F is compact, it cannot have a continuous inverse. Thus one way to prove ill-posedness is to show that F is a compact mapping in appropriate spaces. This was done for the option price example in [20].

The last important definition concerns the noise: In the deterministic theory, the noise is treated as a rather arbitrary (but deterministic) function that is additively added to the exact data. Here we can distinguish between the exact data y and the noisy data y_δ where y is assumed to be in the range of F while y_δ is a version with additive noise:

$$y_\delta = y + n, \quad y = F(x^\dagger).$$

Of course, in practical applications only noisy data y_δ are available, neither y nor n are given. What might be known is an upper bound on the amount of noise, the so-called noise level $\delta \in \mathbb{R}^+$:

$$\delta := \|y_\delta - y\|_Y, \quad y = F(x^\dagger).$$

Besides this approach, there is a stochastic version of the theory of inverse problems, coming from statistics: Most of this is concerned with the case when the noise is not a fixed function but a random variable. Moreover, quite often it is assumed that the distribution of the noise n is fixed (such as a Gaussian noise): Here some analogue to the deterministic noise level is the variance of the noise

$$\sigma^2 = \mathbb{E}\|y_\delta - y\|_Y^2.$$

If the assumption of a Gaussian noise is dropped and the distribution of the noise is not a-priori fixed, another general theory for stochastic inverse problems uses general metrics for stochastic variables, such as the Ky Fan or the Prokhorov metric (see [24, 35]). Again a kind of noise level can be defined as the distance of y_δ to y in these metrics.

3 Regularisation

After we outlined the main problems in inverse problems, the immediate question is how to solve an inverse problem in a stable way. As we have already indicated this can be done by regularisation.

The basic idea of regularisation is as simple as ingenious: Instead of solving the original ill-posed problem we solve a neighboring well-posed one.

Let us at first describe the method for linear problems. In abstract operator notation, regularisation can be formulated as follows: Instead of computing a solution from noisy data by the pseudoinverse $F^\dagger y_\delta$, which is unstable and might not even exist for some data, we use an operator R_α that is stable and approximates $F^\dagger$ and compute a regularised solution

$$x_\alpha^\delta = R_\alpha y_\delta. \tag{3.1}$$

Since we changed the operator $F^\dagger$, this will not give the right solution $x^\dagger$, however, under appropriate conditions, the regularised solution will be close to $x^\dagger$.

The important properties that we require of R_α is that it is stable and approximates $F^\dagger$. If $F^\dagger$ is discontinuous, these properties are opposite to each other: In view of the Banach–Steinhaus theorem it is impossible to approximate a discontinuous operator pointwise by a continuous one. Thus one has to find a compromise between the approximation and the stability. This is achieved by defining a family of regularisation operators R_α, depending on a *regularisation parameter* $\alpha > 0$ which controls the compromise between the approximation and the stability. A regularisation operator should at least have the following properties:

- *Stability*: For any $\alpha > 0$ R_α is a stable operator.
- *Approximation*:
$$\lim_{\alpha \to 0} R_\alpha y = F^\dagger y \quad \forall\, y \in D(F^\dagger).$$

For nonlinear problems, the pseudoinverse is not necessarily defined, but the basic properties of a regularisation operator stay the same. It should be a stable (possibly nonlinear) operator that approximates a minimum norm solution for those y for which a minimum norm solution exists.

Before we discuss convergence issues, we have to emphasise the role of the regularisation parameter α. It is not possible to find a good parameter α independent of the data. Instead, the regularisation parameter has to be chosen depending on the data. So in general α has to be a function of the available noisy data y_δ and/or the noise level δ. We can distinguish three types of so-called parameter choices:

- a-priori parameter choice rules where $\alpha = \alpha(\delta)$, i.e., α depends only on the noise level possibly including information about the a-priori smoothness of $x^\dagger$.
- a-posteriori parameter choice rules where $\alpha = \alpha(y_\delta, \delta)$.
- noise level free ([23]: error free) parameter choice rules where $\alpha = \alpha(y_\delta)$, i.e., α is independent of the noise level and depends only on the data.

So strictly speaking, a regularisation method is always defined as a family of regularisation operators together with a parameter choice rule, for a precise definition see [23].

The last one of the parameter choice rules might look as a very appealing choice, as it does not require knowledge on the noise level (which might not be available). The crux with this, however, is that such *a noise free parameter choice for ill-posed problems will never give rise to a convergent regularisation method in the worst case*, i.e., if the problem is ill-posed, there is always a noise such that for any regularisation method together with a noise level free parameter choice rule the regularised solution will not converge to the true one even though the noisy data converge to the exact one:

$$R_{\alpha(y_\delta)} y_\delta \not\to x^\dagger \text{ if } y_\delta \to y.$$

This result is often referred to as the Bakushinskii veto [2]. It is the reason why in the deterministic theory one is bounded to use a-priori or a-posteriori parameter choice rules. Note that a similar result holds in the stochastic case for the Ky Fan and the Prokhorov metrics [35]. On the contrary, for the Gaussian noise case described above it is indeed possible to use noise level free parameter choice rules. Furthermore, in [50] it was shown that excluding a so-called smooth noise, one can again obtain convergence even with a noise level free method.

Next we should indicate what we mean by convergence: The regularised solution should converge to the true one if the noisy data converge to the exact data (or the noise level tends to 0). More precisely, a regularisation method with a parameter choice rule $\alpha(y_\delta, \delta)$ converges if for all $x^\dagger$

$$\lim_{\delta \to 0} \|x_\alpha^\delta - x^\dagger\|_X \to 0 \quad \forall\, y_\delta : \|y_\delta - y\|_Y \le \delta.$$

This is often called the worst-case convergence because we ask for the convergence for all noisy data below the noise level.

For the stochastic case, an average case error can be used: For instance, if n is a random noise, we define average case convergence as

$$\lim_{\delta \to 0} \mathbb{E}\|x_\alpha^\delta - x^\dagger\|_X^2 \to 0 \quad \forall\, y_\delta : \mathbb{E}\|y_\delta - y\|_Y^2 = \delta^2$$

where in this definition, n is assumed to be a Gaussian noise with finite variance. (This definition can be modified if n is a generalised random process such as the white noise). Note that the convergence analysis for the worst case and the average case have quite different aspects (as we have seen for the Bakushinskii veto) although the general theme of ill-posedness is central in both of them.

A convergent regularisation method has all the properties we want: We can compute a regularised solution in a stable way and the regularised solution is close to the true one if the noise level is sufficiently small. It is of interest to further quantify what "close" means here: Can we find estimates ("convergence rates") for the error $\|x_\alpha^\delta - x^\dagger\|_X$ in terms of the noise level of the following type

$$\|x_\alpha^\delta - x^\dagger\|_X \leq f(\delta) \quad \forall\, x^\dagger \tag{3.2}$$

for some function f (e.g., of the Hölder type $f(\delta) = \delta^\tau$). Again there is an important negative result: For ill-posed problems and any regularisation method there cannot be a continuous function with $f(0) = 0$ such that the uniform estimate holds. In other words: *for ill-posed problems convergence can be arbitrarily slow* [59]. Such a result is rather disappointing as we can never be sure if a computed regularised solution has anything to do with the true one. Note that (3.2) is impossible because it is an estimate uniform in $x^\dagger$. Because of convergence it is immediate that for any $x^\dagger$ we can find such a function $f(\delta)$. We simply cannot give a uniform bound for all such solutions.

The remedy in this situation is to impose additional conditions on the exact solution $x^\dagger$. It is a rule of thumb for many algorithms that a smoother solution gives rise to faster convergence. This is also the case here. If we a-priori assume that the solution is smoother than just $x^\dagger \in X$ it is possible to show for many regularisation methods convergence rates, i.e.,

$$\|x_\alpha^\delta - x^\dagger\|_X \leq f(\delta) \quad \forall\, x^\dagger \in X^\mu. \tag{3.3}$$

Here X^μ denotes a set of a certain smoothness. This smoothness has to be related to the operator F. For linear problems the question how such a smoothness class looks like is answered completely: For instance the set

$$X^\mu = \{x \,|\, x = (F^*F)^\mu w,\ w \in X\}$$

for some $\mu > 0$ gives rise to convergence rates $f(\delta) = \delta^{\frac{\mu}{2\mu+1}}$ for many regularisation methods [23]. The condition that $x^\dagger$ is in the range of the operator $(F^*F)^\mu$ is called a (Hölder) *source condition*. If F is a smoothing operator (e.g., it takes L^2-functions to functions in the Sobolev space H^s), such a condition means that the exact solution is in an appropriate Sobolev space. Hence a source condition can be seen as an abstract smoothness condition. It is possible to extend this convergence rate analysis to the nonlinear case (see below).

The source condition is not a condition that can be tested if the solution is not known, it has to be postulated. Nevertheless the convergence rate analysis is quite useful not only because it gives uniform bounds but also because it allows to classify problems according to their ill-posedness. If an operator F is highly smoothing, the condition in the set X^μ will be very restrictive and we have to expect slow convergence in general. On the other hand if F is only lightly smoothing, we can expect faster convergence for more solutions than in the first case. Related to this is the *degree of ill-posedness* of an inverse problem [23].

3.1 Tikhonov regularisation

Let us now turn to the most prominent example of a regularisation method for nonlinear problems: the Tikhonov regularisation. We have already pointed out the main principle of regularisation: Approximate an ill-posed problem by a well-posed one and solve this problem instead of the original one.

The idea of Tikhonov regularisation [60] starts with the least squares formulation in (2.2). We have already indicated that solving equation (2.1) in a least squares sense does not lead to a well-posed problem or a stable algorithm. In Tikhonov regularisation one therefore stabilises the least squares problem by adding a suitable penalty term: Instead of solving (2.1) or (2.2) we consider now minimising the Tikhonov functional

$$J(x) := \|F(x) - y_\delta\|_Y^2 + \alpha\|x\|_X^2 \tag{3.4}$$

where X and Y are Hilbert spaces, $\alpha > 0$ is a fixed regularisation parameter, and y_δ indicates the possibly noisy data.

As an approximate solution to our ill-posed problem we look for a minimiser of the Tikhonov functional

$$x_\alpha^\delta := \operatorname{argmin}_{x \subset \mathcal{D}(F)} J(x). \tag{3.5}$$

We argued that adding a penalty term in (3.4) helps to stabilise the problem. The question arises if this is true, i.e., if the problem of finding a minimiser (3.5) is a well-posed problem at all. As the reader might guess this is true but needs some mild conditions that have to be satisfied:

1. For noise free data y there exists an exact solution
$$F(x) = y.$$
2. There exists a minimum norm least squares solution of (2.1).
3. $F : \mathcal{D}(F) \subset X \to Y$ is continuous.
4. F is weakly sequentially closed, i.e., for any sequence $x_n \subset \mathcal{D}(F)$ the weak convergence of x_n to x (in X) and the weak convergence of $F(x_n)$ to y (in Y) imply that $x \in \mathcal{D}(F)$ and $F(x) = y$.

The first two conditions have to be postulated. If our model is correct, i.e., for exact data there exists a true solution and if this solution is unique, these conditions are satisfied. The second condition is used to cover the case of nonunique solutions.

The last two conditions are assumptions on the forward operator and they have to be shown for a specific problem. F satisfies these conditions if $\mathcal{D}(F)$ is closed and convex (and hence weakly closed) and F is the composition of a linear compact and a continuous mapping.

Under these assumptions it is not difficult to show [23, 25]:

Theorem 3.1. *For any $y_\delta \in Y$, (3.5) admits a solution x_α^δ.*

So the Tikhonov regularisation is well-defined, and the first criterion of the Hadamard well-posedness is met for (3.5). Remember that the third criterion of well-posedness was concerned with stability: If y_δ converges to some element y_δ', then the regularised solutions x_α^δ should converge as well to the solution of (3.5) with y_δ replaced by y_δ'. Under the weak assumptions on F, only a subsequence-type of continuity can be established [23, 25]:

Theorem 3.2. *Let the general assumptions hold. Let $\alpha > 0$ and let y_k and x_k be sequences such that $y_k \to y_\delta$ and x_k is a minimiser of* (3.4) *with y_δ replaced by y_k. Then x_k has a convergent subsequence and the limit of every convergent subsequence is a minimiser of* (3.4).

This theorem established the fact that the regularised solution depends continuously (in an appropriate sense) on the data y_δ. The reason why subsequences are needed in the proof is the nonuniqueness of minimisers of (3.4). If we additionally impose that a minimiser to (3.4) is unique, we can conclude that x_k converges to it (and not only a subsequence) in the previous theorem. Showing this uniqueness is, however, a rather tedious work, and usually one is satisfied with subsequence convergence.

The two preceding theorems basically show that computing the regularised solution is a well-posed problem: The most important conclusion is that x_α^δ depends stably on the data. This of course is only true if $\alpha > 0$. Thus we have replaced the original ill-posed least squares problem by a well-posed one.

The central point of interest is now convergence: Does the regularised solution converge to the true one in the sense of the previous section (i.e., when the noise level tends to 0)? This is the essence of the third main theorem on Tikhonov regularisation [23, 25]:

Theorem 3.3. *Let the general assumptions hold. Let $y_\delta \in Y$ and $\|y - y_\delta\|_Y \leq \delta$. Let $\alpha(\delta)$ be such that $\alpha(\delta) \to 0$ and $\frac{\delta^2}{\alpha(\delta)} \to 0$ as $\delta \to 0$. Then every sequence $x_{\alpha(\delta_k)}^{\delta_k}$ of minimisers* (3.5) *has a convergent subsequence as $\delta_k \to 0$. Moreover, the limit is a minimum norm least squares solution. If, in addition, this solution is unique,*

$$\lim_{\delta_k \to 0} x_{\alpha(\delta_k)}^{\delta_k} = x^\dagger .$$

This is the main convergence theorem: If the noise level tends to 0, the regularised solution converges to the 'true' one. As we have already pointed out, the regularisation parameter has to be related to the noise level. The conditions in the theorem mean that the regularisation parameter must not go too fast to 0 as the noise level vanishes.

Theorem 3.3 is the 'working horse' for applying Tikhonov regularisation to any specific problem. The main requirements for an application of Tikhonov regularisation is that the general assumptions hold. This has to be shown for any specific problem.

The convergence theorem does not specify how fast the regularised solution will converge to the true solution. As we have said above, without additional conditions convergence can be arbitrarily slow. So in order to find the speed of convergence some additional assumptions have to be made. As in the linear case, source conditions are the appropriate assumptions. Apart of these, one additionally has to impose differentiability of the forward operator in the nonlinear case.

The following theorem establishes convergence rates for the Tikhonov regularisation for nonlinear problems [23, 25]:

Theorem 3.4. *Let the general assumptions hold. Let $\mathcal{D}(F)$ be convex, $y_\delta \in Y$ and $\|y - y_\delta\|_Y \le \delta$. Let the following conditions hold:*

- *F is Fréchet-differentiable.*
- *There exists a constant $\gamma \ge 0$ such that $\|F'(x) - F'(x^\dagger)\|_Y \le \gamma\|x - x^\dagger\|_X$ for all $x \in \mathcal{D}(F)$ in a sufficiently large ball around $x^\dagger$.*
- *There exists a $\omega \in Y$ such that $x^\dagger = F'(x^\dagger)^*\omega$.*
- *$\gamma\|\omega\|_Y < 1$.*

Then for the choice $\alpha \sim \delta$ we obtain

$$\|x_\alpha^\delta - x^\dagger\|_X = O(\sqrt{\delta}) \quad \text{and} \quad \|F(x_\alpha^\delta) - F(x^\dagger)\|_Y = O(\delta).$$

The third condition is the source condition which holds if $x^\dagger$ is in the range of $F'(x^\dagger)^*$. It can be interpreted as an abstract smoothness condition.

There are many generalisations of this convergence rate theorem, for instance, one can postulate weaker source conditions and get weaker results (see [23, Theorem 10.7]). Moreover the problem can be viewed in Hilbert scales which allows to find further convergence rates [23]. There is no need for the a-priori parameter choice $\alpha \sim \delta$ but a-posteriori type of choices can be made, for instance, the popular Morozov discrepancy principle [23] or the balancing principle [54, 57, 53]. Let us mention that all the convergence and convergence rates can be generalised to the stochastic setting using metrics in probability spaces [35].

When applying Tikhonov regularisation there are some choices to be made: Most important is the choice of the regularisation term $\|x\|_X^2$. Note that the space X and its norm was rather arbitrary. The only condition that is required is that F satisfies the assumptions of the previous theorem as an operator from X to Y. Quite often a Sobolev space norm is taken for X. Note that if F satisfies the conditions of the theorem, it remains true if X is taken with a stronger norm. For a stronger norm it is usually easier to show the conditions on F. On the other hand we always have to make sure that the exact solution is in X, so one usually has to bear in mind that $\|x^\dagger\|_X < \infty$.

The choice of the regularisation term can to some extent be derived if one adopts the Bayesian point of view. The regularisation term is directly related to the postulated prior distribution of the exact solution. So the regularisation term reflects which space we believe the exact solution to be in.

The main work in applying Tikhonov regularisation and the convergence (rates) theorem has to be done in showing the conditions on F. This restricts to some extent the choice of the space X: the forward mapping F has to be continuous and weakly closed on X. Moreover, if we want to establish convergence rates, we additionally have to show that F is differentiable and the Lipschitz-type continuity of F' in the second condition.

The results so far hold in the Hilbert space case: Both X and Y are assumed to be Hilbert spaces. Recently, there has been some extension of this theory to the more complicated Banach space case. The main reason why Banach spaces are needed is to cover more general regularisation terms. For some problems, for instance, involving measures [12] or sparse solutions [30], it is convenient not to use Hilbert space norms but general convex functionals as regularisation terms. We briefly outline some theoretical results in this field. For instance, in [38] the following functional was considered

$$J(x) = \|F(x) - y_\delta\|_Y^p + \alpha R(x) \tag{3.6}$$

where $1 \leq p < \infty$, F is a mapping between Banach spaces X and Y, and R is a convex proper functional on X.

The general theory is based on the following assumptions:

- X and Y are Banach spaces and there are topologies τ_X and τ_Y, respectively, which are weaker than the norm topology.
- $\|\cdot\|_Y$ is sequentially lower semicontinuous with respect to the topology τ_Y.
- F is continuous with respect to the topologies τ_X and τ_Y.
- $R : X \to \mathbb{R}$ is proper convex and τ_X-lower semicontinuous.
- $\mathcal{D}(F)$ is closed with respect to the τ_X-topology and $\mathcal{D}(F) \cap \mathcal{D}(R) \neq \emptyset$.
- For any $\alpha > 0$ and $M > 0$ the level sets $\{x : J(x) \leq M\}$ are sequentially compact with respect to τ_X.

These assumptions are satisfied in the Hilbertian case if F satisfies the standard assumptions and one choose τ_X and τ_Y as the weak topologies. Typically also in the Banach space case τ_X and τ_Y are taken as weak or weak* topologies. Under these assumptions, there exists a minimiser, the minimiser is stable (in the τ_X-topology) with respect to the data noise and as $\alpha \to 0$ appropriately the minimiser converges to the R-minimal solution (again in the τ_X-topology). So all the results of Theorems 3.1, 3.2 and 3.3 carry over to the Banach space case. Also convergence rates (in the Bregman distance) can be shown. The mentioned results are shown in [38]. Further extensions to this theory can be found in [30, 52, 33, 55]. As before, the theorems can be used to find convergence and convergence rates for a specific problem when one can verify the assumptions for the specific operator F, the specific choice of Banach spaces, and regularisation functionals.

4 Calibration of option price models via regularisation

In this section we give a survey on the use of regularisation techniques in calibration problems in mathematical finance. We concentrate on calibration of option price models. The identification of an unknown local volatility in the Dupire model and model parameters in the jump diffusion models have been studied in the literature. In the following sections, we summarise the main ideas of the Dupire, the Lévy and the local Lévy option price models and review how regularisation has been utilised for solving corresponding calibration problems. Using the Dupire model as an example we demonstrate how the general theory of Tikhonov regularisation can be employed in calibration problems by emphasising the steps needed for the theoretical results. Even though more sophisticated asset price models than the geometric Brownian motion are in use in practice, the Dupire model serves as a benchmark for the theoretical and numerical analysis of calibration problems in financial mathematics.

4.1 Dupire model

An option is a contract that gives the owner the right to buy or to sell a specified amount of a particular underlying asset at a fixed price (the strike price) within a fixed period of time (before or at the maturity date). In the Dupire model the dynamics of the price of the underlying asset is described by a geometric Brownian motion, i.e., the price S_t is a stochastic process defined by the stochastic differential equation

$$\mathrm{d}S_t = S_t(\mu\,\mathrm{d}t + \sigma(t, S_t)\,\mathrm{d}W_t), \quad 0 < t \le T$$

where $\mu \in \mathbb{R}$ is a constant drift and W_t is a standard Wiener process. In the Black–Scholes model [6] the volatility σ was assumed to be a constant but in the Dupire model the local volatility can depend on both the time and the price.

Suppose that the market is liquid and free of arbitrage and transaction costs. The price $C(t, S; T, K)$ of a European call option, i.e., the right to buy the asset with the strike price K at the maturity T, then satisfies the Black–Scholes parabolic PDE

$$\begin{aligned} &\frac{\partial C}{\partial t} + \frac{1}{2}\sigma^2(t, S)S^2\frac{\partial^2 C}{\partial S^2} + rS\frac{\partial C}{\partial S} - rC = 0, \quad 0 < t \le T,\ S > 0,\\ &C(t, 0; T, K) = 0,\\ &C(T, S; T, K) = (S - K)^+, \quad S > 0, \end{aligned} \tag{4.1}$$

where r is the constant interest rate on a riskless investment. If the volatility σ is a constant, equation (4.1) admits an analytic solution (the famous Black–Scholes formula [6]). As a function of K and T, the price of a European call option satisfies the Dupire equation [19]

$$\begin{aligned} &\frac{\partial C}{\partial T} = \frac{1}{2}\sigma^2(T, K)K^2\frac{\partial^2 C}{\partial K^2} - rK\frac{\partial C}{\partial K}, \quad T > t,\ K > 0,\\ &C(t, S; T, 0) = S, \quad T > t,\\ &C(t, S; t, K) = (S - K)^+, \quad K > 0, \end{aligned} \tag{4.2}$$

where S is the spot price of the underlying asset at the time t. The option pricing problem is to define the price of a European call option when the local volatility σ is a known function, by solving equations (4.1) and/or (4.2).

The calibration problem of the Dupire option price model is to identify the local volatility function $\sigma(t, S)$ such that the theoretical prices $C(t, S; T, K)$ given by (4.1) and (4.2) coincide with the observed prices $C^*(T, K)$ of the European call options for all (given) strike prices K and maturities T. As a parameter identification problem the calibration problem is an inverse problem. The option pricing problem is the corresponding direct problem.

The possible ill-posedness of the calibration problem is an essential issue. Since the local volatility is a function and hence usually an infinite-dimensional object, the data required for the unique solvability of the calibration problem have to be continuous on the strike price K and/or the maturity T (see, e.g., [8, 9, 10] for uniqueness results for the time-independent volatilities, i.e., $\sigma(t, S) = \sigma(S)$). In practice, data are always discrete both in strike prices and maturities. For each underlying asset the prices of European call options are given only for few strike prices and maturities. In addition, the prices of European call options are not known accurately but the bid-ask spread can be seen as a noise in the data. To assure the existence and the uniqueness the solution of the calibration problem needs to be defined in the least squares sense.

The main source of ill-posedness for an inverse problem is that the forward mapping F from the parameter to the data does not have a continuous inverse. The ill-posedness of the calibration problem has been studied in the literature. The existing ill-posedness results can be split into the cases where the unknown local volatility is assumed either to be space-independent [34, 39], i.e., $\sigma(t, S) = \sigma(t)$, time-independent [32], i.e., $\sigma(t, S) = \sigma(S)$, or dependent on both variables [16, 20, 21]. In these theoretical results, prices of the European call options are supposed to be known either for a fixed strike price K but all maturities T, i.e., $C^*(T, K) = C^*(T)$, for a fixed maturity T but all strike prices K, i.e., $C^*(T, K) = C^*(K)$, or for all strike prices K and maturities T, respectively. In the space-independent case, the Black–Scholes equation (4.1) has a solution in a closed form. Hence the exact definition of the forward mapping F can be given. In the other two cases, the parameter-to-solution mapping is defined via the Dupire equation (4.2). In the references above, the parameter and the data spaces were selected to be suitable function spaces, either Hilbert or Banach spaces. Mostly, the ill-posedness was proven by showing that the forward mapping is a compact operator. As was mentioned in Section 2 a compact operator cannot have a continuous inverse.

Due to the ill-posedness a stable way for solving the calibration problem is to utilise some regularisation method. Here, we concentrate on Tikhonov regularisation. Since the calibration problem is a nonlinear inverse problem, the theory of Tikhonov regularisation summarised in Section 3.1 can be applied to the problem. In the Hilbert space setting if the forward mapping F fulfils the general assumptions given in Section 3.1 for the appropriate parameter and data spaces, the minimisation of the Tikhonov functional (3.4) is a well-posed problem by Theorems 3.1 and 3.2. Furthermore, the minimiser of the Tikhonov functional converges to the minimum norm least squares solution of the calibration problem as the noise level tends to zero according to Theorem 3.3. A convergence rate result for the Tikhonov regularised solution is obtained by showing that the forward mapping F satisfies the assumptions of Theorem 3.4. Hence the main theoretical task in applying Tikhonov regularisation to the calibration problem is to study the properties of the parameter-to-solution mapping F. The parameter and the data spaces need to be chosen such a way that the forward mapping fulfils the assumptions of Section 3.1. In more general setting, e.g. in Banach spaces, also the penalty functional R has to be taken into account (see (3.6)).

Theoretical results concerning Tikhonov regularisation and the calibration problem, including convergence and convergence rate results, have been published in the literature. Both space-independent [34, 39], time-independent [20, 32, 44, 45], and general local volatilities [16, 20, 21] with corresponding continuous data have been considered. In [16] also the convergence of the Tikhonov regularised solution for discrete data was examined with rates for general local volatilities. In the references above, the penalty term in Tikhonov regularisation was mainly given by a norm in a Hilbert or a Banach space (mainly Sobolev spaces were used) but in [39] the maximum entropy regularisation (see, e.g., [26, 27]) was considered. In the space-independent case, the least square minimisation problem was regularised by the functional

$$E(a,\bar{a}) = \int_0^T \left(a(t)\ln\frac{a(t)}{\bar{a}(t)} + \bar{a}(t) - a(t) \right) \,\mathrm{d}t,$$

called the cross entropy of a relative to the prior $\bar{a}$, where $a(t) = \sigma^2(t)$ for all $0 \le t \le T$ and $\bar{a} \in L^1(0,T)$ such that $\bar{a}(t) \ge c > 0$ for almost all $0 \le t \le T$.

In numerical implementations of Tikhonov regularisation for the calibration problem, the unknown local volatility has to be represented by finite degrees of freedom. One possibility is to assume that the local volatility is described by a finite number of parameters. In [43] the local volatility was supposed to be a space-time spline with a finite number of nodal points. A less restrictive way is to discretise the local volatility in a suitable grid and to take the unknown to be the values of the local volatility in the grid points [7, 17, 20, 21, 34, 39, 51]. In practice the data are discrete. Therefore both the least squares and the penalty terms in the Tikhonov functional have to be replaced by discrete versions. In the minimisation of the Tikhonov functional, the calculation of the forward mapping is needed. For the space-independent local volatilities, the parameter-to-solution mapping is known in a closed form and hence the theoretical option prices can be computed in a straight-forward manner by discretisation [34, 39]. For the other two cases, the Black–Scholes or the Dupire equations need to be solved numerically, e.g., by finite difference and/or finite element methods [20, 21, 51] or by trinomial tree discretisation [17]. For minimising the Tikhonov functional, gradient based minimisation methods have been employed, e.g., steepest decent or quasi-Newton techniques (see the references above). Numerical tests have mainly been done with simulated data but real data were used in [7, 17, 51]. For obtaining a regularised solution close to the true solution the regularisation parameter has to be selected properly. In the references above, known parameter choice rules, either a-posteriori or noise level free rules, were utilised to choose the regularisation parameter.

We want to point out that in the Dupire framework, calibration problems where the given data are prices of more complicated options than European call options, have also been treated by using the regularisation theory. In [17, 40] the calibration of American call options, i.e., the right to buy an asset with a strike price at any time up to a maturity date, was studied as a Tikhonov regularised problem mainly from the computational point of view.

Note that there are few earlier review papers of the calibration problem of the Dupire option price model. In [9] the main emphasis was on the theoretical study of the corresponding inverse problem whereas in [61] the focus was on the use of the regularisation theory.

4.2 Lévy model

One of the disadvantages of the Dupire option price model is that the model does not allow jumps in the price of the underlying asset. A generalisation of the Dupire model is the jump diffusion model. The dynamics of the underlying asset is modeled, under a risk-neutral measure Q, as an exponential of a Lévy process:

$$S_t = e^{rt}e^{X_t}$$

where $r > 0$ is the interest rate. The process X_t is a Lévy process with characteristic triplet (σ, γ, ν) where $\sigma > 0$ is called the volatility, $\gamma \in \mathbb{R}$ the drift, and the Lévy measure ν is a positive measure on $\mathbb{R}$ verifying

$$\nu(\{0\}) = 0 \quad \text{and} \quad \int_{-\infty}^{\infty} \min(1, x^2)\, \nu(\mathrm{d}x) < \infty.$$

The Lévy measure ν gives the expected number of jumps of the process X_t per time unit. Since Q is a risk-neutral probability measure, e^{X_t} is a martingale and hence the drift is uniquely defined by the volatility and the Lévy measure:

$$\gamma = -\frac{\sigma^2}{2} - \int_{-\infty}^{\infty} (e^y - 1 - y 1_{\|y\| \le 1})\, \nu(\mathrm{d}y).$$

The price of a European call option with the strike price K and the maturity T fulfils the partial integro-differential equation [12]

$$\begin{aligned} &\frac{\partial C}{\partial t}(t, S) + rS\frac{\partial C}{\partial S}(t, S) + \frac{\sigma^2}{2} S^2 \frac{\partial^2 C}{\partial S^2}(t, S) - rC(t, S) \\ &\quad + \int_{-\infty}^{\infty} \left[C(t, Se^y) - C(t, S) - S(e^y - 1)\frac{\partial C}{\partial S}(t, S) \right] \nu(\mathrm{d}y) = 0 \end{aligned} \tag{4.3}$$

for $0 \le t < T$ and $S > 0$ with the terminal condition

$$C(T, S) = (S - K)^+, \quad S > 0.$$

Equation (4.3) consists of the Black–Scholes PDE (cf. (4.1)) and an integral term related to the Lévy measure. Note that the volatility σ is a constant like in the Black–Scholes model.

The calibration problem for the Lévy model is to identify the parameters (σ, ν) such that the theoretical option prices given, e.g., by (4.3) coincide with the observed option prices. Different sources of the ill-posedness of the calibration problem were pointed out and shown by examples in [14]. Due to the ill-posedness, for solving the calibration problem regularisation is required.

In [12, 13, 14, 15] the weighted least squares problem for option prices was regularised by the relative entropy of the risk-neutral measure Q with respect to a prior measure Q_0. For risk-neutral exponential Lévy models, the relative entropy is given by

$$\begin{aligned} H(\nu) = &\frac{T}{2\sigma^2} \left(\int_{-\infty}^{\infty} (e^x - 1)\, (\nu - \nu_0)(\mathrm{d}x) \right)^2 \\ &+ T \int_{-\infty}^{\infty} \left(\frac{\mathrm{d}\nu}{\mathrm{d}\nu_0} \ln \frac{\mathrm{d}\nu}{\mathrm{d}\nu_0} + 1 - \frac{\mathrm{d}\nu}{\mathrm{d}\nu_0} \right) \nu_0(\mathrm{d}x) \end{aligned}$$

where σ is the common volatility for both risk-neutral measures and ν_0 is the Lévy measure related to the prior Q_0. Since the prior Q_0 defines the volatility, only the Lévy measure ν needs to be calibrated according to given option prices. Possible choices for an appropriate prior measure were discussed in [12, 13]. The convergence of the relative entropy regularised solution as the noise in the data tends to zero was examined in [13] in the case where the Lévy measure is a finite sum of point measures and in [14] for general Lévy measures. Convergence rate results do not exist in the literature.

The numerical implementation of the relative entropy regularisation method was presented in [12, 13, 15]. To discretise the problem, the unknown Lévy measure was modeled by a finite sum of point measures. The forward mapping can be defined by using the characteristic function of the Lévy process and the Fourier transform, not only by equation (4.3). Hence option prices needed in the Tikhonov functional can be calculated by the fast Fourier transform (FFT). The choice of appropriate weights in the least squares functional was discussed in the papers. The suitable regularisation parameter was determined by the Morozov discrepancy principle. The corresponding Tikhonov functional was then minimised by a gradient descent method. Both simulated and real data were used in the numerical tests.

Calibration of the Lévy model was also studied in [3] where regularisation of the calibration problem was done in the spectral domain by cutting off high frequencies. Observation noise was assumed to be stochastic instead of the deterministic bid-ask spread. Exact minimax rates of convergence were obtained and it was shown that the proposed spectral estimators are rate optimal. The method was numerically tested with simulated data.

4.3 Local Lévy model

Even though in the Lévy model the price of the underlying asset can have jumps, both the volatility and the Lévy measure cannot vary over the time or with the price of the asset. The Lévy model can be generalised by the local Lévy model. Let the asset price S_t has the risk-free dynamics

$$\begin{aligned} S_t = S_0 + \int_0^t r S_{s-} \, \mathrm{d}s + \int_0^t \sigma(s, S_{s-}) S_{s-} \, \mathrm{d}W_s \\ + \int_0^t \int_{\mathbb{R}} S_{s-}(e^x - 1) \left(m_{(S_{s-},s)}(\mathrm{d}x, \mathrm{d}s) - \mu_{(S_{s-},s)}(\mathrm{d}x, \mathrm{d}s) \right) \end{aligned}$$

where r is the riskless interest rate, σ is the local volatility function, W_t is a standard Wiener process, m is an integer-valued random measure associated to the jumps of S_t independent of W_t, and μ is the compensator of m. The process S_t can equivalently be presented as an exponential of an inhomogeneous Markov process X_t of Lévy type, i.e.,

$$S_t = S_0 e^{rt} e^{X_t}$$

where e^{X_t} is a martingale. We assume that the compensator μ has the form

$$\mu_{(S,t)}(\mathrm{d}x, \mathrm{d}t) = a(t, S)\nu(\mathrm{d}x)\mathrm{d}t$$

where $a(t, S)$ is the local speed function and ν is a Radon measure satisfying

$$\nu(\{0\}) = 0 \quad \text{and} \quad \int_{-\infty}^{\infty} \frac{x^2}{1 + x^2} \, \nu(\mathrm{d}x) < \infty.$$

This assumption means that the distribution of jumps remains unchanged over time while the arrival rate varies in the time and the asset price. Note that ν is a Lévy measure.

The price $C(T, K)$ of a European call option as a function of the strike price K and the maturity T fulfils the partial integro-differential equation [11]

$$\begin{aligned}\frac{\partial C}{\partial T}(T,K) &= \frac{1}{2}\sigma^2(T,K)K^2\frac{\partial^2 C}{\partial K^2}(T,K) - rK\frac{\partial C}{\partial K}(T,K) \\ &\quad + \int_0^\infty Y\frac{\partial^2 C}{\partial K^2}(T,Y)a(T,Y)\psi\left(\log\frac{K}{Y}\right)\,\mathrm{d}Y\end{aligned} \tag{4.4}$$

for all $K > 0$ and $T > 0$ where

$$\psi(z) = \begin{cases} \int_{-\infty}^{z}(e^z - e^x)\,\nu(\mathrm{d}x) & \text{for } z < 0, \\ \int_z^\infty (e^x - e^z)\,\nu(\mathrm{d}x) & \text{for } z > 0 \end{cases}$$

with the initial value

$$C(0,K) = (S_0 - K)^+ \quad \text{for all } K > 0$$

and the boundary condition

$$C(T,0) = S_0 \quad \text{for all } T > 0.$$

Note that the PDE part of equation (4.4) is equal to the Dupire equation (4.2) and the integral term depends only on the parameters a and ν. The calibration problem for the local Lévy model is to identify the parameters (σ, a, ν) such that the theoretical option prices given by (4.4) coincide with the observed option prices.

In [49] it was assumed that the only unknown parameter is the local speed function a. The parameter-to-solution mapping was defined by a PIDE in logarithmic variables related to equation (4.4). The ill-posedness of the calibration problem was shown and the source of the ill-posedness was discussed. It was pointed out that the calibration problem for the local Lévy model is more ill-posed than that for the Dupire model. The local speed function was calibrated by Tikhonov regularisation using a Sobolev penalty term. The convergence and convergence rate results for the Tikhonov regularised solution were obtained. In the numerical implementation the forward mapping was discretised by finite differences and the integral term by a midpoint rule. In the selection of the regularisation parameter the discrepancy principle was used. The Tikhonov functional was minimised by a Gauss–Newton method. Numerical tests were done for both simulated and real data.

In [48] the complete calibration problem was studied. Like in [49], the forward mapping was defined by a PIDE in logarithmic variables. In Tikhonov regularisation, Sobolev space penalty functionals were used for the local volatility and the local speed function whereas for the Lévy measure several choices of a regularising functional were proposed. The convergence of the Tikhonov regularised solution was shown and a possible source condition for the speed of convergence was mentioned.

4.4 Further remarks

Another generalisation of the Dupire model are stochastic volatility models where the local volatility in the Dupire framework is modeled by a stochastic process, not as a function of the time and the asset price like in Section 4.1. In [1, 58] the stochastic volatility was calibrated by minimising the relative entropy functional constrained by the observed prices of a European call option. In these papers the calibration problem was viewed as a stochastic control problem which is closely related to the regularisation point of view. By Hull and White [41], the price of a European call option with a stochastic volatility can be given by the expectation of Black–Scholes option prices over the distribution of the quadratic variation of the stochastic volatility. In [29] the quadratic variation was calibrated by using Tikhonov regularisation.

In addition to the regularisation theory, the Bayesian approach to inverse problems can be used to solve ill-posed problems in a stable way. For a comprehensive introduction into the topic see [47]. The theory of Bayesian inversion is not fully developed, especially in infinite-dimensional spaces, but some convergence results similar to the ones in the regularisation theory have been recently published in [36, 37, 56]. In the financial mathematics context, the Bayesian inversion theory was applied to calibrate the quadratic variation of the stochastic volatility in [46].

Robust calibration using Tikhonov regularisation is not restricted only to option price models. In fact, Tikhonov regularisation has been employed for other calibration problems in mathematical finance as well. For applications in the calibration of interest rates models, see, e.g., [5, 18, 22]. The main message of this article is that the regularisation theory is a general technique which is applicable to various range of parameter identification/calibration problems.

Bibliography

[1] M. Avellaneda, C. Friedman, R. Holmes, and D. Samperi, *Calibrating volatility surfaces via relative-entropy minimization*, Appl. Math. Finance 4 (1997), pp. 37–64.

[2] A. Bakushinskii, *Remarks on choosing a regularization parameter using the quasioptimality and ratio criterion*, Comput. Math. Math. Phys. 24 (1984), pp. 181–182.

[3] D. Belomestny and M. Reiss, *Spectral calibration of exponential Lévy models*, Finance Stoch. 10 (2006), pp. 449–474.

[4] A. Binder, H. W. Engl, C. W. Groetsch, A. Neubauer, and O. Scherzer, *Weakly closed nonlinear operators and parameter identification in parabolic equations by Tikhonov regularization.*, Appl. Anal. 55 (1994), pp. 215–234.

[5] A. Binder, H. W. Engl, and A. Schatz, *Advanced Numerical Techniques for Financial Engineering*, Derivatives Week XII (2003), pp. 6–7.

[6] F. Black and M. Scholes, *The pricing of options and corporate liabilities*, J. Polit. Econ. 81 (1973), pp. 637–654.

[7] J. N. Bodurtha and M. Jermakyan, *Non-parametric estimation of an implied volatility surface*, J. Comput. Finance 2 (1999), pp. 29–60.

[8] I. Bouchouev and V. Isakov, *The inverse problem of option pricing*, Inverse Problems 13 (1997), pp. L11–L17.

[9] ———, *Uniqueness, stability and numerical methods for inverse problems that arises in financial markets*, Inverse Problems 15 (1999), pp. R95–R116.

[10] I. Bouchouev, V. Isakov, and N. Valdivia, *Recovery of volatility coefficient by linearization*, Quant. Finance 2 (2002), pp. 257–263.

[11] P. Carr, H. Geman, D. B. Madan, and M. Yor, *From local volatility to local Lévy models*, Quant. Finance 4 (2004), pp. 581–588.

[12] R. Cont and P. Tankov, *Financial Modelling With Jump Processes*, Chapman & Hall/CRC, Boca Raton, U. S. A., 2004.

[13] ———, *Non-parametric calibration of jump-diffusion option pricing models*, J. Comput. Finance 7 (2004), pp. 1–49.

[14] ———, *Retrieving Lévy processees from option prices: regularization of an ill-posed inverse problem*, SIAM J. Control Optim. 45 (2006), pp. 1–25.

[15] R. Cont, P. Tankov, and E. Voltchkova, *Option pricing models with jumps: integro-differential equations and inverse problems*, European Congress on Computational Methods in Applied Sciences and Engineering (P. Neittaanmäki, T. Rossi, S. Korotov, E. Onate, J. Périaux, and D. Knörzer, eds.), ECCOMAS, 2004.

[16] S. Crépey, *Calibration of the local volatility in a generalized Black-Sholes model using Tikhonov regularization*, SIAM J. Math. Anal. 34 (2003), pp. 1183–1206.

[17] ———, *Calibration of the local volatility in a trinomial tree using Tikhonov regularization*, Inverse Problems 19 (2003), pp. 91–127.

[18] A. d'Aspremont, *Interest Rate Model Calibration Using Semidefinite Programming*, Appl. Math. Finance 3 (2003), pp. 183–213.

[19] B. Dupire, *Pricing with a smile*, RISK 7 (1994), pp. 18–20.

[20] H. Egger and H. W. Engl, *Tikhonov regularization applied to the inverse problem of option pricing: convergence analysis and rates*, Inverse Problems 21 (2005), pp. 1027–1045.

[21] H. Egger, T. Hein, and B. Hofmann, *On decoupling of volatility smile and term structure in inverse oprion pricing*, Inverse Problems 22 (2006), pp. 1247–1259.

[22] H. W. Engl, *Calibration problems – A inverse problems view*, WILMOTT magazine (July 2007), pp. 16–20.

[23] H. W. Engl, M. Hanke, and A. Neubauer, *Regularization of Inverse Problems*, Kluwer Academic Publisher, Dordrecht, the Netherlands, 1996.

[24] H. W. Engl, A. Hofinger, and S. Kindermann, *Convergence rates in the Prokhorov metric for assessing uncertainty in ill-posed problems*, Inverse Problems 21 (2005), pp. 399–412.

[25] H. W. Engl, K. Kunisch, and A. Neubauer, *Convergence rates for Tikhonov regularisation of nonlinear ill-posed problems*, Inverse Problems 5 (1989), pp. 523–540.

[26] H. W. Engl and G. Landl, *Convergence rates for maximum entropy regularization*, SIAM J. Numer. Anal. 30 (1993), pp. 1509–1536.

[27] ———, *Maximum entropy regularization of nonlinear ill-posed problems*, Proceedings of the First World Congress of Nonlinear Analysis (Berlin, Germany) (V. Lakshmikantham, ed.), vol. I, de Gruyter, 1996, pp. 513–525.

[28] T. Felici and H. W. Engl, *On shape optimization of optical waveguides using inverse problem techniques*, Inverse Problems 17 (2001), pp. 1141–1162.

[29] P. Friz and J. Gatheral, *Valuation of volatility derivatives as an inverse problem*, Quant. Finance 5 (2005), pp. 531–542.

[30] M. Grasmair, M. Haltmeier, and O. Scherzer, *Sparse regularization with l^q penalty term*, Inverse Problems 24 (2008), p. 055020, (13 pp).

[31] J. Hadamard, *Lectures on Cauchy's problem in linear partial differential equations*, Yale University Press, New Haven, U. S. A., 1923.

[32] T. Hein, *Some analysis of Tikhonov regularization for the inverse problems of option pricing in the price-dependent case*, Z. Anal. Anwendungen 24 (2005), pp. 593–609.

[33] T. Hein, *Tikhonov regularization in Banach spaces – improved convergence rates results*, Inverse Problems 25 (2009), p. 035002, (18 pp).

[34] T. Hein and B. Hofmann, *On the nature of ill-posedness of an inverse problem arising in option pricing*, Inverse Problems 19 (2003), pp. 1319–1338.

[35] A. Hofinger, *Ill-posed Problems: Extending the Deterministic Theory to a Stochastic Setup*, Trauner Verlag, Linz, Austria, 2006, (Doctoral Thesis).

[36] A. Hofinger and H. K. Pikkarainen, *Convergence rates for the Bayesian approach to linear inverse problems*, Inverse Problems 23 (2007), pp. 2469–2484.

[37] ———, *Convergence rates for linear inverse problems in the presence of an additive normal noise*, Stoch. Anal. Appl. 27 (2009), pp. 240–257.

[38] B. Hofmann, B. Kaltenbacher, C. Poeschl, and O. Scherzer, *A convergence rates result for Tikhonov regularization in Banach spaces with non-smooth operators*, Inverse Problems 23 (2007), pp. 987–1010.

[39] B. Hofmann and R. Krämer, *On maximum entropy regularization for a specific inverse problem of option pricing*, J. Inverse Ill-Posed Probl. 13 (2005), pp. 41–63.

[40] J. Huang and J.-S. Pang, *A mathematical programming with equilibrium constraints approach to the implied volatility surface of American options*, J. Comput. Finance 4 (2000), pp. 21–56.

[41] J. Hull and A. White, *The pricing of options with stochastic volatilities*, J. Finance 42 (1987), pp. 281–300.

[42] V. Isakov, *Inverse Problems for Partial Differential Equations*, Springer-Verlag, New York, U. S. A., 2006.

[43] N. Jackson, E. Süli, and S. Howison, *Computation of deterministic volatility surfaces*, J. Comput. Finance 2 (1999), pp. 5–32.

[44] L. Jiang, Q. Chen, L. Wang, and J. E. Zhang, *A new well-posed algorithm to recover implied local volatility*, Quant. Finance 3 (2003), pp. 451–457.

[45] L. Jiang and Y. Tao, *Identifying the volatility of underlying assets from option prices*, Inverse Problems 17 (2001), pp. 137–155.

[46] R. Kaila, *The Integrated Volatility Implied by Option Prices, A Bayesian Approach*, TKK Mathematics, Espoo, Finland, 2008, (Doctoral Thesis).

[47] J. P. Kaipio and E. Somersalo, *Statistical and Computational Inverse Problems*, Springer-Verlag, Berlin, Germany, 2005.

[48] S. Kindermann and P. Mayer, *On the calibration of local jump-diffusion market models*, (2008), submitted.

[49] S. Kindermann, P. Mayer, H. Albrecher, and H. W. Engl, *Identification of the local speed function in a Lévy model for option pricing*, J. Integral Equations Appl. 20 (2008), pp. 161–200.

[50] S. Kindermann and A. Neubauer, *On the convergence of the quasi-optimality criterion for (iterated) Tikhonov regularization*, Inverse Probl. Imaging 2 (2008), pp. 291–299.

[51] R. Lagnado and S. Osher, *A technique for calibrating derivative security pricing models: numerical solution of inverse problems*, J. Comput. Finance 1 (1997), pp. 13–25.

[52] D. A. Lorenz, *Convergence rates and source conditions for Tikhonov regularization with sparsity constraints*, J. Inverse Ill-Posed Probl. 16 (2008), pp. 463–478.

[53] S. Lu, S. V. Pereverzev, and R. Ramlau, *An analysis of Tikhonov regularization for nonlinear ill-posed problems under a general smoothness assumption*, Inverse Problems 23 (2007), pp. 217–230.

[54] P. Mathé and S. V. Pereverzev, *Geometry of linear ill-posed problems in variable Hilbert spaces*, Inverse Problems 19 (2003), pp. 789–803.

[55] A. Neubauer, *On enhanced convergence rates for Tikhonov regularization of nonlinear problems in Banach spaces*, Inverse Problems (2009), p. 065009, (10 pp).

[56] A. Neubauer and H. K. Pikkarainen, *Convergence results for the Bayesian inversion theory*, J. Inverse Ill-Posed Probl. 16 (2008), pp. 601–613.

[57] S. Pereverzev and E. Schock, *On the adaptive selection of the parameter in regularization of ill-posed problems*, SIAM J. Numer. Anal. 43 (2005), pp. 2060–2076.

[58] D. Samperi, *Calibrationg a diffusion pricing model with uncertain volatility: regularization and stability*, Math. Finance 12 (2002), pp. 71–87.

[59] E. Schock, *Approximate solution of ill-posed equations: arbitrarily slow convergence vs. superconvergence*, Constructive Methods for the Practical Treatment of Integral Equations (G. Hämmerlin and K. H. Hoffmann, eds.), Birkhäuser, Basel, Switzerland, 1985, pp. 234–243.

[60] A. N. Tikhonov and V. B. Glasko, *An approximate solution of Fredholm integral equations of the first kind*, Ž. Vyčisl. Mat. i Mat. Fiz. 4 (1964), pp. 564–571.

[61] J. P. Zubelli, *Inverse problems in finance: A short survey of calibration techniques*, Proceedings of the 2nd Brazilian Conference on Statistical Modelling in Insurance and Finance (Maresias, Brazil) (N. Kolev and P. Morettin, eds.), Institute of Mathematics and Statistics, University of São Paulo, 2005, pp. 64–75.

Author information

Stefan Kindermann, Industrial Mathematics Institute, Johannes Kepler University Linz, Altenbergerstrasse 69, A-4040 Linz, Austria.
Email: stefan.kindermann@jku.at

Hanna K. Pikkarainen, Johann Radon Institute for Computational and Applied Mathematics (RICAM), Austrian Academy of Sciences, Altenbergerstrasse 69, A-4040 Linz, Austria.
Email: hanna.pikkarainen@ricam.oeaw.ac.at

Radon Series Comp. Appl. Math **8**, 245–273

Optimal consumption and investment with bounded downside risk measures for logarithmic utility functions

Claudia Klüppelberg and Serguei Pergamenshchikov

Abstract. We investigate optimal consumption problems for a Black–Scholes market under uniform restrictions on Value-at-Risk and Expected Shortfall for logarithmic utility functions. We find the solutions in terms of a dynamic strategy in explicit form, which can be compared and interpreted. This paper continues our previous work, where we solved similar problems for power utility functions.

Key words. Black–Scholes model, capital-at-risk, expected shortfall, logarithmic utility, optimal consumption, portfolio optimisation, utility maximisation, value-at-risk.

AMS classification. primary: 91B70, 93E20, 49K30; secondary: 49L20, 49K45

1 Introduction

One of the principal questions in mathematical finance is the optimal investment/consumption problem for continuous time market models. By applying results from stochastic control theory, explicit solutions have been obtained for some special cases (see e.g. Karatzas and Shreve [9], Korn [11] and references therein).

With the rapid development of the derivatives markets, together with margin trading on certain financial products, the exposure to losses of investments into risky assets can be considerable. Without a careful analysis of the potential danger, the investment can cause catastrophic consequences such as, for example, the recent crisis in the "Société Générale".

To avoid such situations the Basel Committee on Banking Supervision suggested some measures for the assessment of market risks. It is widely accepted that the *Value-at-Risk* (VaR) is a useful summary risk measure (see, Jorion [7] or Dowd [4]). We recall that the VaR is the maximum expected loss over a given horizon period at a given confidence level. Alternatively, the *Expected Shortfall* (ES) or *Tail Condition Expectation* (TCE) measures also the expected loss given the confidence level is violated.

In order to satisfy the Basel committee requirements, portfolios have to control the level of VaR or (the more restrictive) ES throughout the investment horizon. This leads to stochastic control problems under restrictions on such risk measures.

Our goal in this paper is the optimal choice of a dynamic portfolio subject to a risk limit specified in terms of VaR or ES uniformly over the investment interval $[0, T]$.

In Klüppelberg and Pergamenshchikov [10] we considered the optimal invest-

Second author: This work was supported by the European Science Foundation through the AMaMeF programme.

ment/consumption problem with uniform risk limits throughout the investment horizon for power utility functions. In that paper also some interpretation of VaR and ES besides an account of the relevant literature can be found. Our results in [10] have interesting interpretations. We have, for instance, shown that for power utility functions with exponents less than one, the optimal constrained strategies are riskless for sufficiently small risk bounds: they recommend consumption only. On the contrary, for the (utility bound of a) linear utility function the optimal constrained strategies recommend to invest everything into risky assets and consume nothing.

In this paper we investigate the optimal investment/consumption problem for logarithmic utility functions again under constraints on uniform versions of VaR and ES over the whole investment interval $[0,T]$. Using optimisation methods in Hilbert functional spaces, we find all optimal solutions in explicit form. It turns out that the optimal constrained strategies are the unconstrained ones multiplied by some coefficient which is less than one and depends on the specific constraints.

Consequently, we can make the main recommendation: *To control the market risk throughout the investment interval* $[0,T]$ *restrict the optimal unconstrained portfolio allocation by specific multipliers (given in explicit form in* (3.6) *for the VaR constraint and in* (3.26) *for the ES constraint).*

Our paper is organised as follows. In Section 2 we formulate the problem. We define the Black–Scholes model for the price processes and present the wealth process in terms of an SDE. We define the cost function for the logarithmic utility function and present the admissible control processes. We also present the unconstrained consumption and investment problem of utility maximisation for logarithmic utility. In Sections 3 we consider the constrained problems. Section 3.1 is devoted to a risk bound in terms of Value-at-Risk, whereas Section 3.2 discusses the consequences of a risk bound in terms of Expected Shortfall. Auxiliary results and proofs are postponed to Section 4. We start there with material needed for the proofs of both regimes, the Value-at-Risk and the ES risk bounds. In Section 4.1 all proofs of Section 3.1 can be found, and in Section 4.2 all proofs of Section 3.2. Some technical lemmas are postponed to the Appendix, again divided in two parts for the Value-at-Risk regime and the ES regime.

2 Formulating the problem

2.1 The model and first results

We work in the same framework of self-financing portfolios as in Klüppelberg and Pergamenshchikov in [10], where the financial market is of Black–Scholes type consisting of one *riskless bond* and several *risky stocks* on the interval $[0,T]$. Their respective prices $S_0 = (S_0(t))_{0\le t\le T}$ and $S_i = (S_i(t))_{0\le t\le T}$ for $i = 1,\dots,d$ evolve according to the equations:

$$\begin{cases} \mathrm{d}S_0(t) = r_t\, S_0(t)\,\mathrm{d}t\,, & S_0(0) = 1\,,\\ \mathrm{d}S_i(t) = S_i(t)\,\mu_i(t)\,\mathrm{d}t + S_i(t)\sum_{j=1}^d \sigma_{ij}(t)\,\mathrm{d}W_j(t)\,, & S_i(0) > 0\,. \end{cases} \tag{2.1}$$

Here $W_t = (W_1(t), \ldots, W_d(t))'$ is a standard d-dimensional Wiener process in $\mathbb{R}^d$; $r_t \in \mathbb{R}$ is the *riskless interest rate*; $\mu_t = (\mu_1(t), \ldots, \mu_d(t))'$ is the vector of *stock-appreciation rates* and $\sigma_t = (\sigma_{ij}(t))_{1\le i,j\le d}$ is the matrix of *stock-volatilities*. We assume that the coefficients $(r_t)_{0\le t\le T}$, $(\mu_t)_{0\le t\le T}$ and $(\sigma_t)_{0\le t\le T}$ are deterministic cadlag functions. We also assume that the matrix σ_t is not degenerated for all $0 \le t \le T$.

We denote by $\mathcal{F}_t = \sigma\{W_s, s \le t\}$, $t \ge 0$, the filtration generated by the Brownian motion (augmented by the null sets). Furthermore, $|\cdot|$ denotes the Euclidean norm for vectors and the corresponding matrix norm for matrices and prime denotes the transposed. For $(y_t)_{0\le t\le T}$ square integrable over the fixed interval $[0, T]$ we define $\|y\|_T = (\int_0^T |y_t|^2 \,\mathrm{d}t)^{1/2}$.

The portfolio process $(\pi_t = (\pi_1(t), \ldots, \pi_d(t))')_{0\le t\le T}$ represents the fractions of the wealth process invested into the stocks. The consumption rate is denoted by $(v_t)_{0\le t\le T}$. Then (see [10] for details) the wealth process $(X_t)_{0\le t\le T}$ is the solution to the SDE

$$\mathrm{d}X_t = X_t\,(r_t + y_t'\,\theta_t - v_t)\,\mathrm{d}t + X_t\,y_t'\,\mathrm{d}W_t\,, \quad X_0 = x > 0\,, \tag{2.2}$$

where

$$\theta_t = \sigma_t^{-1}(\mu_t - r_t\,\mathbf{1}) \quad \text{with} \quad \mathbf{1} = (1, \ldots, 1)' \in \mathbb{R}^d\,,$$

and we assume that

$$\int_0^T |\theta_t|^2 \,\mathrm{d}t < \infty\,.$$

The control variables are $y_t = \sigma_t'\pi_t \in \mathbb{R}^d$ and $v_t \ge 0$. More precisely, we define the $(\mathcal{F}_t)_{0\le t\le T}$-progressively measurable control process as $\nu = (y_t, v_t)_{t\ge 0}$, which satisfies

$$\int_0^T |\,y_t\,|^2 \,\mathrm{d}t < \infty \quad \text{and} \quad \int_0^T v_t \,\mathrm{d}t < \infty \quad \text{a.s..} \tag{2.3}$$

In this paper we consider logarithmitic utility functions. Consequently, we assume throughout that

$$\int_0^T (\ln v_t)_- \mathrm{d}t < \infty \quad \text{a.s.,} \tag{2.4}$$

where $a_- := -\min(a, 0)$.

To emphasise that the wealth process (2.2) corresponds to some control process ν we write X^ν. Now we describe the set of control processes.

Definition 2.1. A stochastic control process $\nu = (\nu_t)_{0\le t\le T} = ((y_t, v_t))_{0\le t\le T}$ is called *admissible*, if it is $(\mathcal{F}_t)_{0\le t\le T}$-progressively measurable with values in $\mathbb{R}^d \times \mathbb{R}_+$, satisfying integrability conditions (2.3)–(2.4) such that the SDE (2.2) has a unique strong a.s. positive continuous solution $(X_t^\nu)_{0\le t\le T}$ for which

$$\mathbb{E}\left(\int_0^T (\ln(v_t X_t^\nu))_- \,\mathrm{d}t + (\ln X_T^\nu)_-\right) < \infty\,.$$

We denote by $\mathcal{V}$ the class of all *admissible control processes*.

For $\nu \in \mathcal{V}$ we define the cost function

$$J(x,\nu) := \mathbb{E}_x \left(\int_0^T \ln\left(v_t\, X_t^\nu\right)\, dt \,+\, \ln X_T^\nu \right). \tag{2.5}$$

Here $\mathbb{E}_x$ is the expectation operator conditional on $X_0^\nu = x$.

We recall a well-known result, henceforth called the *unconstrained problem*:

$$\max_{\nu \in \mathcal{V}} J(x,\nu)\,. \tag{2.6}$$

To formulate the solution we set

$$\omega(t) = T - t + 1 \quad \text{and} \quad \widehat{r}_t = r_t + \frac{|\theta_t|^2}{2}\,, \qquad 0 \le t \le T\,.$$

Theorem 2.2 (Karatzas and Shreve [9], Example 6.6, p. 104). *The optimal value of $J(x,\nu)$ is given by*

$$\max_{\nu \in \mathcal{V}} J(x,\nu) \;=\; J(x,\nu^*) \;=\; (T+1)\,\ln\frac{x}{T+1} + \int_0^T \omega(t)\,\widehat{r}_t\, dt\,.$$

The optimal control process $\nu^* = (y_t^*, v_t^*)_{0\le t\le T} \in \mathcal{V}$ is of the form

$$y_t^* \,=\, \theta_t \quad \textit{and} \quad v_t^* \,=\, \frac{1}{\omega(t)}\,, \tag{2.7}$$

where the optimal wealth process $(X_t^*)_{0\le t\le T}$ is given as the solution to

$$dX_t^* \,=\, X_t^*\,\left(r_t + |\theta_t|^2 - v_t^*\right)\, dt \,+\, X_t^*\,\theta_t'\, dW_t\,, \quad X_0^* = x\,, \tag{2.8}$$

which is

$$X_t^* \;=\; x\,\frac{T+1-t}{T+1}\,\exp\Big(\int_0^t \widehat{r}_u\, du + \int_0^t \theta_u'\, dW_u \Big)\,.$$

Note that the optimal solution (2.7) of problem (2.6) is deterministic, and we denote in the following by $\mathcal{U}$ the set of deterministic functions $\nu = (y_t, v_t)_{0\le t\le T}$ satisfying conditions (2.3) and (2.4).

For the above result we can state that

$$\max_{\nu \in \mathcal{V}} J(x,\nu) = \max_{\nu \in \mathcal{U}} J(x,\nu)\,.$$

Intuitively, it is clear that to construct financial portfolios in the market model (2.1) the investor can invoke only information given by the coefficients $(r_t)_{0\le t\le T}$, $(\mu_t)_{0\le t\le T}$ and $(\sigma_t)_{0\le t\le T}$ which are deterministic functions.

Then for $\nu \in \mathcal{U}$, by Itô's formula, equation (2.2) has solution

$$X_t^\nu \;=\; x\,\mathcal{E}_t(y)\, e^{R_t - V_t + (y,\theta)_t}\,,$$

with $R_t = \int_0^t r_u \mathrm{d}u$, $V_t = \int_0^t v_u \mathrm{d}u$, $(y,\theta)_t = \int_0^t y'_u \, \theta_u \mathrm{d}u$ and the stochastic exponential

$$\mathcal{E}_t(y) = \exp\Big(\int_0^t y'_u \mathrm{d}W_u - \frac{1}{2}\int_0^t |y_u|^2 \mathrm{d}u \Big) .$$

Therefore, for $\nu \in \mathcal{U}$ the process $(X_t^\nu)_{0\le t\le T}$ is positive, continuous and satisfies

$$\sup_{0\le t\le T} \mathbb{E}\,|\ln X_t^\nu| < \infty .$$

This implies that $\mathcal{U} \subset \mathcal{V}$. Moreover, for $\nu \in \mathcal{U}$ we can calculate the cost function (2.5) explicitly as

$$\begin{aligned} J(x,\nu) = (T+1)\ln x &+ \int_0^T \omega(t) \left(r_t + y'_t\theta_t - \frac{1}{2}|y_t|^2 \right) \mathrm{d}t \\ &+ \int_0^T (\ln v_t - V_t)\mathrm{d}t \, - V_T . \end{aligned} \tag{2.9}$$

3 Optimisation with constraints: main results

3.1 Value-at-Risk constraints

As in Klüppelberg and Pergamenshchikov [10] we use as risk measures the modifications of Value-at-Risk and Expected Shortfall introduced in Emmer, Klüppelberg and Korn [5], which reflect the capital reserve. For simplicity, in order to avoid non-relevant cases, we consider only $0 < \alpha < 1/2$.

Definition 3.1 (Value-at-Risk (VaR)). For a control process ν and $0 < \alpha \le 1/2$ define the *Value-at-Risk (VaR)* by

$$\mathrm{VaR}_t(\nu,\alpha) := x\,e^{R_t} - Q_t , \quad t \ge 0 ,$$

where for $t \ge 0$ the quantity $Q_t = \inf\{z \ge 0 : \mathbb{P}(X_t^\nu \le z) \ge \alpha\}$ is the α-quantile of X_t^ν.

Note that for every $\nu \in \mathcal{U}$ we find

$$Q_t = x \exp\Big(R_t - V_t + (y,\theta)_t - \frac{1}{2}\|y\|_t^2 - |q_\alpha|\|y\|_t \Big) , \tag{3.1}$$

where q_α is the α-quantile of the standard normal distribution.

We define the *level risk function* for some coefficient $0 < \zeta < 1$ as

$$\zeta_t = \zeta\, x\, e^{R_t} , \quad t \in [0,T] . \tag{3.2}$$

The coefficient $\zeta \in (0,1)$ introduces some risk aversion behaviour into the model. In that sense it acts similarly as a utility function does. However, ζ has a clear interpretation, and every investor can choose and understand the influence of the risk bound ζ as a proportion of the riskless bond investment.

We consider the maximisation problem for the cost function (2.9) over strategies $\nu \in \mathcal{U}$ for which the Value-at-Risk is bounded by the level function (3.2) over the interval $[0,T]$; i.e.

$$\max_{\nu\in\mathcal{U}} J(x,\nu) \quad \text{subject to} \quad \sup_{0\le t\le T} \frac{\mathrm{VaR}_t(\nu,\alpha)}{\zeta_t} \le 1\,. \tag{3.3}$$

To formulate the solution of this problem we define

$$G(u,\lambda) := \int_0^T \frac{(\omega(t)+\lambda)^2}{(\lambda|q_\alpha| + u(\omega(t)+\lambda))^2}|\theta_t|^2\,\mathrm{d}t\,, \quad u\ge 0\,, \lambda\ge 0\,. \tag{3.4}$$

Moreover, for fixed $\lambda > 0$ we denote by

$$\rho(\lambda) = \inf\{u\ge 0\,:\, G(u,\lambda)\le 1\}\,, \tag{3.5}$$

if it exists, and set $\rho(\lambda) = +\infty$ otherwise. For a proof of the following lemma see A.1.

Lemma 3.2. *Assume that* $|q_\alpha| > \|\theta\|_T > 0$ *and*

$$0\le\lambda\le\lambda_{\max} = \frac{k_1 + \sqrt{k_2\left(q_\alpha^2 - \|\theta\|_T^2\right) + k_1^2}}{q_\alpha^2 - \|\theta\|_T^2}\,,$$

where $k_1 = \|\sqrt{\omega}\theta\|_T^2$ *and* $k_2 = \|\omega\theta\|_T^2$. *Then the equation* $G(\cdot,\lambda) = 1$ *has the unique positive solution* $\rho(\lambda)$. *Moreover,* $0 < \rho(\lambda) < \infty$ *for all* $0 \le \lambda < \lambda_{\max}$, *and* $\rho(\lambda_{\max}) = 0$.

Now for $\lambda \ge 0$ fixed and $0 \le t \le T$ we define the weight function

$$\tau_\lambda(t) = \frac{\rho(\lambda)(\omega(t)+\lambda)}{\lambda|q_\alpha| + \rho(\lambda)(\omega(t)+\lambda)}\,. \tag{3.6}$$

Here we set $\tau_\lambda(\cdot)\equiv 1$ for $\rho(\lambda) = +\infty$. It is clear, that for every fixed $\lambda \ge 0$,

$$0\le\tau_\lambda(T)\le\tau_\lambda(t)\le 1\,,\quad 0\le t\le T\,. \tag{3.7}$$

To take the VaR constraint into account we define

$$\Phi(\lambda) = |q_\alpha|\|\tau_\lambda\theta\|_T + \frac{1}{2}\|\tau_\lambda\theta\|_T^2 - \|\sqrt{\tau_\lambda}\theta\|_T^2\,. \tag{3.8}$$

Denote by Φ^{-1} the inverse of Φ, provided it exists. A proof of the following lemma is given in A.1.

Lemma 3.3. *Assume that* $\|\theta\|_T > 0$ *and*

$$0 < \zeta < 1 - e^{-|q_\alpha| \|\theta\|_T + \|\theta\|_T^2/2} . \tag{3.9}$$

Then for all $0 \le a \le -\ln(1-\zeta)$ *the inverse* $\Phi^{-1}(a)$ *exists. Moreover,*

$$0 \le \Phi^{-1}(a) < \lambda_{\max} \quad \textit{for} \quad 0 < a \le -\ln(1-\zeta)$$

and $\Phi^{-1}(0) = \lambda_{\max}$.

Now set

$$\phi(\kappa) := \Phi^{-1}\left(\ln(1-\kappa)/(1-\zeta)\right), \quad 0 \le \kappa \le \zeta, \tag{3.10}$$

and define the investment strategy

$$\widetilde{y}_t^\kappa := \theta_t \tau_{\phi(\kappa)}(t), \quad 0 \le t \le T. \tag{3.11}$$

To introduce the optimal consumption rate we define

$$v_t^\kappa = \frac{\kappa}{T - t\kappa} \tag{3.12}$$

and recall that for

$$\kappa = \kappa_0 = \frac{T}{T+1}$$

the function v_t^κ coincides with the optimal unconstrained consumption rate $1/\omega(t)$ as defined in (2.7).

It remains to fix the parameter κ. To this end we introduce the cost function

$$\Gamma(\kappa) = \ln(1-\kappa) + T \ln \kappa + \int_0^T \omega(t) \, |\theta_t|^2 \left(\tau_{\phi(\kappa)}(t) - \frac{1}{2}\tau^2_{\phi(\kappa)}(t)\right) \mathrm{d}t . \tag{3.13}$$

To choose the parameter κ we maximise Γ:

$$\gamma = \gamma(\zeta) = \operatorname*{argmax}_{0 \le \kappa \le \zeta} \Gamma(\kappa) . \tag{3.14}$$

With this notation we can formulate the main result of this section.

Theorem 3.4. *Assume that* $\|\theta\|_T > 0$. *Then for all* $\zeta > 0$ *satisfying* (3.9) *and for all* $0 < \alpha < 1/2$ *for which*

$$|q_\alpha| \ge 2\,(T+1)\,\|\theta\|_T , \tag{3.15}$$

the optimal value of $J(x, \nu)$ *for problem* (3.3) *is given by*

$$J(x, \nu^*) = A(x) + \Gamma\left(\gamma(\zeta)\right), \tag{3.16}$$

where

$$A(x) = (T+1) \ln x + \int_0^T \omega(t) r_t \, dt - T \ln T \tag{3.17}$$

and the optimal control $\nu^* = (y_t^*, v_t^*)_{0\le t\le T}$ *is of the form*

$$y_t^* = \widetilde{y}_t^\gamma \quad and \quad v_t^* = v_t^\gamma\,. \tag{3.18}$$

The optimal wealth process is the solution of the SDE

$$dX_t^* = X_t^* \,(r_t - v_t^* + (y_t^*)'\,\theta_t)\,dt + X_t^*\,(y_t^*)'\,dW_t\,, \quad X_0^* = x\,,$$

given by

$$X_t^* = x\,\mathcal{E}_t(y^*)\,\frac{T-\gamma(\zeta)t}{T}\,e^{R_t - V_t + (y^*,\theta)_t}\,, \quad 0\le t\le T\,.$$

The following corollary is a consequence of (2.9).

Corollary 3.5. *If* $\|\theta\|_T = 0$*, then for all* $0<\zeta<1$ *and for all* $0<\alpha<1/2$

$$y_t^* = 0 \quad and \quad v_t^* = v_t^\gamma$$

with $\gamma = \text{argmax}_{0\le\kappa\le\zeta}(\ln(1-\kappa) + T\ln\kappa) = \min(\kappa_0,\zeta)$*. Moreover, the optimal wealth process is the deterministic function*

$$X_t^* = x\,\frac{T-\min(\kappa_0,\zeta)\,t}{T}\,e^{R_t}\,, \quad 0\le t\le T\,.$$

In the next corollary we give some sufficient condition, for which the investment process equals zero (the optimal strategy is riskless). This is the first marginal case.

Corollary 3.6. *Assume that* $\|\theta\|_T > 0$ *and that* (3.9) *holds. If* $0<\zeta<\kappa_0$ *and*

$$|q_\alpha| \ge (1+T)\|\theta\|_T\left(2 + \frac{\zeta(T+1)}{(1-\zeta)T-\zeta}\right)\,, \tag{3.19}$$

then $\gamma=\zeta$ *and the optimal solution* $\nu^* = (y_t^*, v_t^*)_{0\le t\le T}$ *is of the form*

$$y_t^* = 0 \quad and \quad v_t^* = v_t^\zeta\,.$$

Moreover, the optimal wealth process is the deterministic function

$$X_t^* = x\,\frac{T-\zeta t}{T}\,e^{R_t}\,, \quad 0\le t\le T\,.$$

Below we give some sufficient conditions, for which the solution of optimisation problem (3.3) coincides with the unconstrained solution (2.7). This is the second marginal case.

Theorem 3.7. *Assume that*

$$\zeta > 1 - \frac{1}{T+1}\,e^{-|q_\alpha|\|\theta\|_T + \|\theta\|_T^2/2}\,. \tag{3.20}$$

Then for all $0<\alpha<1/2$ *for which* $|q_\alpha| \ge \|\theta\|_T$*, the solution of the optimisation problem* (3.3) *is given by* (2.7)–(2.8).

3.2 Expected shortfall constraints

Our next risk measure is an analogous modification of the *Expected Shortfall* (ES).

Definition 3.8 (Expected Shortfall (ES)). For a control process ν and $0 < \alpha \le 1/2$ define

$$m_t(\nu,\alpha) = \mathbb{E}_x \left(X_t^\nu \,|\, X_t^\nu \le Q_t \right) , \quad t \ge 0 ,$$

where Q_t is the α-quantile of X_t^ν given by (3.1). The *Expected Shortfall (ES)* is then defined as

$$ES_t(\nu,\alpha) = x e^{R_t} - m_t(\nu,\alpha) , \quad t \ge 0 .$$

Again for $\nu \in \mathcal{U}$ we find

$$m_t(\nu,\alpha) = x\, F_\alpha \left(\|y\|_t \right) e^{R_t - V_t + (y,\theta)_t} ,$$

where

$$F_\alpha(z) = \frac{1}{\int_{|q_\alpha|}^\infty e^{-t^2/2}\,\mathrm{d}t} \int_{|q_\alpha|+z}^\infty e^{-t^2/2}\,\mathrm{d}t .$$

We consider the maximisation problem for the cost function (2.5) over strategies $\nu \in \mathcal{U}$ for which the Expected Shortfall is bounded by the level function (3.2) over the interval $[0,T]$, i.e.

$$\max_{\nu\in\mathcal{U}} J(x,\nu) \quad \text{subject to} \quad \sup_{0\le t\le T} \frac{ES_t(\nu,\alpha)}{\zeta_t} \le 1 . \tag{3.21}$$

We proceed similarly as for the VaR-constraint problem (3.3). Define

$$G_1(u,\lambda) := \int_0^T \frac{(\omega(t)+\lambda)^2}{\left(\lambda\,\psi_\alpha(u) + u(\omega(t)+\lambda)\right)^2} |\theta_t|^2\,\mathrm{d}t , \quad u \ge 0, \lambda \ge 0 . \tag{3.22}$$

where

$$\psi_\alpha(u) = \frac{1}{\varphi(u+|q_\alpha|)} - u \quad \text{with} \quad \varphi(y) = e^{\frac{y^2}{2}} \int_y^\infty e^{-\frac{t^2}{2}}\,\mathrm{d}t . \tag{3.23}$$

It is well known and easy to prove that

$$\frac{1}{y} - \frac{1}{y^3} \le \varphi(y) \le \frac{1}{y} , \quad y > 0 . \tag{3.24}$$

This means that $\psi_\alpha(u) \ge |q_\alpha|$ for all $u \ge 0$, which implies for every fixed $\lambda \ge 0$ that $G_1(u,\lambda) \le G(u,\lambda)$ for all $u \ge 0$. Moreover, similarly to (3.5) we define

$$\rho_1(\lambda) = \inf\{u \ge 0 \,:\, G_1(u,\lambda) \le 1\} . \tag{3.25}$$

Since G_1 has similar behaviour as G, the following lemma is a modification of Lemma 3.2. Its proof is analogous to the proof of Lemma 3.2.

Lemma 3.9. *Assume that* $|q_\alpha| > \|\theta\|_T > 0$ *and*

$$0 \le \lambda \le \lambda^1_{\max} = \frac{k_1 + \sqrt{k_2\left(\psi^2_\alpha(0) - \|\theta\|^2_T\right) + k_1^2}}{\psi^2_\alpha(0) - \|\theta\|^2_T}\,,$$

where k_1 *and* k_2 *are given in Lemma 3.2. Then the equation* $G_1(\cdot,\lambda) = 1$ *has the unique positive solution* $\rho_1(\lambda)$. *Moreover,* $0 < \rho_1(\lambda) < \infty$ *for* $0 \le \lambda < \lambda^1_{\max}$ *and* $\rho_1(\lambda^1_{\max}) = 0$.

Now for $\lambda \ge 0$ fixed and $0 \le t \le T$ we define the weight function

$$\varsigma_\lambda(t) = \frac{\rho_1(\lambda)\,(\omega(t)+\lambda)}{\lambda\,\psi_\alpha(\rho_1(\lambda)) + \rho_1(\lambda)\,(\omega(t)+\lambda)}\,, \tag{3.26}$$

and we set $\varsigma_\lambda(\cdot) \equiv 1$ for $\rho_1(\lambda) = +\infty$. Note that for every fixed $\lambda \ge 0$,

$$0 \le \varsigma_\lambda(T) \le \varsigma_\lambda(t) \le 1\,, \quad 0 \le t \le T\,. \tag{3.27}$$

To take the ES constraint into account we define

$$\Phi_1(\lambda) \;=\; -\|\sqrt{\varsigma_\lambda}\theta\|^2_T - \ln F_\alpha\left(\|\varsigma_\lambda\theta\|_T\right)\,. \tag{3.28}$$

Denote by Φ_1^{-1} the inverse of Φ_1 provided it exists. The proof of the next lemma is given in Section A.2.

Lemma 3.10. *Assume that* $\|\theta\|_T > 0$ *and*

$$0 < \zeta < 1 - F_\alpha\left(\|\theta\|_T\right)\, e^{\|\theta\|^2_T}\,. \tag{3.29}$$

Then for all $0 \le a \le -\ln(1-\zeta)$ *the inverse* Φ_1^{-1} *exists and* $0 \le \Phi_1^{-1}(a) < \lambda_{\max}$ *for* $0 < a \le -\ln(1-\zeta)$ *and* $\Phi_1^{-1}(0) = \lambda^1_{\max}$.

Now, similarly to (3.5) we set

$$\phi_1(\kappa) = \Phi_1^{-1}\left(\ln(1-\kappa)/(1-\zeta)\right)\,, \quad 0 \le \kappa \le \zeta\,, \tag{3.30}$$

and define the investment strategy

$$\widetilde{y}^{1,\kappa}_t = \theta_t \varsigma_{\phi_1(\kappa)}(t)\,, \quad 0 \le t \le T\,. \tag{3.31}$$

We introduce the cost function

$$\Gamma_1(\kappa) = \ln(1-\kappa) + T\ln\kappa + \int_0^T \omega(t)\,|\theta_t|^2 \left(\varsigma_{\phi_1(\kappa)}(t) - \frac{1}{2}\varsigma^2_{\phi_1(\kappa)}(t)\right)\mathrm{d}t\,. \tag{3.32}$$

To fix the parameter κ we maximise Γ_1:

$$\gamma_1 = \gamma_1(\zeta) = \operatorname*{argmax}_{0\le\kappa\le\zeta}\, \Gamma_1(\kappa)\,. \tag{3.33}$$

With this notation we can formulate the main result of this section.

Theorem 3.11. *Assume that* $\|\theta\|_T > 0$. *Then for all* $\zeta > 0$ *satisfying* (3.29) *and for all* $0 < \alpha < 1/2$ *satisfying*

$$|q_\alpha| \geq \max(1, 2(T+1)\|\theta\|_T) \tag{3.34}$$

the optimal value of $J(x,\nu)$ *for the optimisation problem* (3.21) *is given by*

$$J(x,\nu^*) = A(x) + \Gamma_1(\gamma_1(\zeta)),$$

where the function A *is defined in* (3.17) *and the optimal control* $\nu^* = (y_t^*, v_t^*)_{0\leq t\leq T}$ *is of the form (recall the definition of* v_t^κ *in* (3.12))

$$y_t^* = \widetilde{y}_t^{1,\gamma_1} \quad \text{and} \quad v_t^* = v_t^{\gamma_1}. \tag{3.35}$$

The optimal wealth process is the solution to the SDE

$$dX_t^* = X_t^*(r_t - v_t^* + (y_t^*)'\theta_t)\,dt + X_t^*(y_t^*)'\,dW_t, \quad X_0^* = x,$$

given by

$$X_t^* = x\,\mathcal{E}_t(y^*)\,\frac{T-\gamma_1(\zeta)t}{T}\,e^{R_t - V_t + (y^*,\theta)_t}, \quad 0 \leq t \leq T.$$

Corollary 3.12. *If* $\|\theta\|_T = 0$, *then the optimal solution of problem* (3.21) *is given in Corollary 3.5.*

Similarly to the optimisation problem with VaR constraint we observe two marginal cases. Note that the following corollary is again a consequence of (2.9).

Corollary 3.13. *Assume that* $\|\theta\|_T > 0$ *and that* (3.29) *and* (3.34) *hold. Then* $\gamma_1 = \zeta$ *and the assertions of Corollary 3.6 hold with* γ *replaced by* γ_1.

Theorem 3.14. *Assume that*

$$\zeta > 1 - \frac{1}{T+1} F_\alpha(\|\theta\|_T)\,e^{\|\theta\|_T^2}. \tag{3.36}$$

Then for all $0 < \alpha < 1/2$ *for which* $|q_\alpha| > \max(1, \|\theta\|_T)$ *the solution of problem* (3.21) *is given by* (2.7)–(2.8).

3.3 Conclusion

If we compare the optimal solutions (3.18) and (3.35) with the unconstrained optimal strategy (2.7), then the risk bounds force investors to restrict their investment into the risk assets by multiplying the unconstrained optimal strategy by the coefficients given in (3.11) and (3.14) for VaR constraints and (3.31) and (3.33) for ES constraints. The impact of the risk measure constraints enter into the portfolio process through the risk level ζ and the confidence level α.

4 Auxiliary results and proofs

In this section we consider maximisation problems with constraints for the two terms of (2.9):

$$I(V) := \int_0^T (\ln v_t - V_t)\mathrm{d}t \quad \text{and} \quad H(y) := \int_0^T \omega(t)\Big(y_t'\theta_t - \frac{1}{2}|y_t|^2\Big)\mathrm{d}t\,. \tag{4.1}$$

We start with a result concerning the optimisation of $I(\cdot)$, which will be needed to prove results from both Sections 3.1 and 3.2.

Let $\mathcal{W}[0,T]$ be the set of differentiable functions $f : [0,T] \to \mathbb{R}$ having positive cadlag derivative $\dot{f}$ satisfying condition (2.4). For $b > 0$ we define

$$\mathcal{W}_{0,b}[0,T] = \{f \in \mathcal{W}[0,T] : f(0) = 0 \quad \text{and} \quad f(T) = b\}\,. \tag{4.2}$$

Lemma 4.1. *Consider the optimisation problem*

$$\max_{f\in\mathcal{W}_{0,b}[0,T]} I(f)\,.$$

The optimal value of I is given by

$$I^*(b) = \max_{f\in\mathcal{W}_{0,b}[0,T]} I(f) = I(f^*) = -T\ln T - T\ln\frac{e^b}{e^b-1}\,, \tag{4.3}$$

with optimal solution

$$f^*(t) = \ln\frac{Te^b}{Te^b - t(e^b-1)}\,, \quad 0 \le t \le T\,. \tag{4.4}$$

Proof. Firstly, we consider this optimisation problem in the space $\mathcal{C}^2[0,T]$ of two times continuously differentiable functions on $[0,T]$:

$$\max_{f\in\mathcal{W}_{0,b}[0,T]\cap\mathcal{C}^2[0,T]} I(f)\,,$$

By variational calculus methods we find that it has solution (4.3); i.e.

$$\max_{f\in\mathcal{W}_{0,b}[0,T]\cap\mathcal{C}^2[0,T]} I(f) = I(f^*)\,.$$

where the optimal solution f^* is given in (4.4).

Take now $f \in \mathcal{W}_{0,b}[0,T]$ and suppose first that its derivative

$$\dot{f}_{\min} = \inf_{0\le t\le T} \dot{f}(t) > 0\,.$$

Let Υ be a positive two times differentiable function on $[-1,1]$ such that $\int_{-1}^1 \Upsilon(z)\,\mathrm{d}z = 1$, and set $\Upsilon(z) := 0$ for $|z| \ge 1$. We can take, for example,

$$\Upsilon(z) = \begin{cases} \frac{1}{\int_{-1}^1 \exp\left(-\frac{1}{1-v^2}\right)\mathrm{d}v} \exp\left(-\frac{1}{1-z^2}\right) & \text{if } |z| \le 1\,, \\ 0 & \text{if } |z| > 1\,. \end{cases}$$

By setting $\dot{f}(t) = \dot{f}(0)$ for all $t \le 0$ and $\dot{f}(t) = \dot{f}(T)$ for all $t \ge T$, we define an approximating sequence of functions by

$$\upsilon_n(t) = n \int_{\mathbb{R}} \Upsilon(n(u-t))\, \dot{f}(u)\, \mathrm{d}u = \int_{-1}^{1} \Upsilon(z)\, \dot{f}\left(t + \frac{z}{n}\right)\, \mathrm{d}z\,.$$

It is clear that $(\upsilon_n)_{n\ge 1} \in \mathcal{C}^2[0,T]$. Moreover, we recall that $\dot{f}$ is cadlag, which implies that it is bounded on $[0,T]$; i.e.

$$\sup_{0\le t\le T} \dot{f}(t) := \dot{f}_{\max} < \infty\,,$$

and its discontinuity set has Lebesgue measure zero. Therefore, the sequence $(\upsilon_n)_{n\ge 1}$ is bounded; more precisely,

$$0 < \dot{f}_{\min} \le \upsilon_n(t) \le \dot{f}_{\max} < \infty\,, \quad 0 \le t \le T\,, \tag{4.5}$$

and $\upsilon_n \to \dot{f}$ as $n \to \infty$ for Lebesgue almost all $t \in [0,T]$. Therefore, by the Lebesgue convergence theorem we obtain

$$\lim_{n\to\infty} \int_0^T |\upsilon_n(t) - \dot{f}(t)|\, \mathrm{d}t = 0\,.$$

Moreover, inequalities (4.5) imply

$$|\ln \upsilon_n| \le \ln\left(\max(\dot{f}_{\max}\,, 1)\right) + \left|\ln\left(\min(\dot{f}_{\min}\,, 1)\right)\right|\,.$$

Therefore, $f_n(t) = \int_0^t \upsilon_n(u)\, \mathrm{d}u$ belongs to $\mathcal{W}_{0,b_n}[0,T] \cap \mathcal{C}^2[0,T]$ for $b_n := \int_0^T \upsilon_n(u)\, \mathrm{d}u$. It is clear that

$$\lim_{n\to\infty} I(f_n) = I(f) \quad \text{and} \quad \lim_{n\to\infty} b_n = b\,.$$

This implies that

$$I(f) \le I^*(b)\,,$$

where $I^*(b)$ is defined in (4.3).

Consider now the case, where $\inf_{0\le t\le T} \dot{f}(t) = 0$. For $0 < \delta < 1$ we consider the approximation sequence of functions

$$\dot{f}_\delta(t) = \max(\delta\,, \dot{f}(t)) \quad \text{and} \quad f_\delta(t) = \int_0^t \dot{f}_\delta(u)\, \mathrm{d}u\,, \quad 0 \le t \le T\,.$$

It is clear that $f_\delta \in \mathcal{W}_{0,b_\delta}[0,T]$ for $b_\delta = \int_0^T \dot{f}_\delta(t)\, \mathrm{d}t$. Therefore, $I(f_\delta) \le I^*(b_\delta)$. Moreover, in view of the convergence

$$\lim_{\delta\to 0} \int_0^T \left(\dot{f}_\delta(t) - \dot{f}(t)\right) \mathrm{d}t = 0$$

we get $\limsup_{\delta\to 0} I(f_\delta) \le I^*(b)$. Moreover, note that

$$|I(f_\delta) - I(f)| \le \int_{A_\delta} (\ln\delta - \ln \dot f(t))\,\mathrm{d}t + T\int_{A_\delta}\left(\delta - \dot f(t)\right)\,\mathrm{d}t$$
$$\le \int_{A_\delta} (\ln \dot f(t))_-\,\mathrm{d}t + \delta T\,\Lambda(A_\delta)\,,$$

where $A_\delta = \{t \in [0,T] : 0 \le \dot f(t) \le \delta\}$ and $\Lambda(A_\delta)$ is the Lebesgue measure of A_δ. Moreover, by the definition of $\mathcal{W}[0,T]$ in (4.2) the Lebesgue measure of the set $\{t \in [0,T] : \dot f(t) = 0\}$ equals zero and $\int_0^T (\ln \dot f_t)_-\,\mathrm{d}t < \infty$. This implies that $\lim_{\delta\to 0}\Lambda(A_\delta) = 0$ and hence

$$\lim_{\delta\to 0} I(f_\delta) = I(f)\,,$$

i.e. $I(f) \le I^*(b)$. □

In order to deal with H as defined in (4.1) we need some preliminary result. As usual, we denote by $\mathcal{L}_2[0,T]$ the Hilbert space of functions y satisfying the square integrability condition in (2.3).

Define for $y \in \mathcal{L}_2[0,T]$ with $\|y\|_T > 0$

$$\overline{y}_t = y_t/\|y\|_T \quad\text{and}\quad l_y(h) = \|y+h\|_T - \|y\|_T - (\overline{y}, h)_T\,. \tag{4.6}$$

We shall need the following lemma.

Lemma 4.2. *Assume that $y \in \mathcal{L}_2[0,T]$ and $\|y\|_T > 0$. Then for every $h \in \mathcal{L}_2[0,T]$ the function $l_y(h) \ge 0$.*

Proof. Obviously, if $h \equiv ay$ for some $a \in \mathbb{R}$, then $l_y(h) = (|1+a| - 1 - a)\|y\|_T \ge 0$. Let now $h \not\equiv ay$ for all $a \in \mathbb{R}$. Then

$$l_y(h) = \frac{2(y\,,\,h)_T + \|h\|_T^2}{\|y+h\|_T + \|y\|_T} - (\overline{y}, h)_T = \frac{\|h\|_T^2 - (\overline{y}, h)_T((\overline{y}, h)_T + l_y(h))}{\|y+h\|_T + \|y\|_T}\,.$$

It is easy to show directly that for all h

$$\|y+h\|_T + \|y\|_T + (\overline{y}, h)_T \ge 0$$

with equality if and only if $h \equiv ay$ for some $a \le -1$.

Therefore, if $h \not\equiv ay$, we obtain

$$l_y(h) = \frac{\|h\|_T^2 - (\overline{y}\,,\,h)_T^2}{\|y+h\|_T + \|y\|_T + (\overline{y}, h)_T} \ge 0\,.$$

□

4.1 Results and proofs of Section 3.1

We introduce the constraint $K : \mathcal{L}_2[0,T] \to \mathbb{R}$ as

$$K(y) := \frac{1}{2}\|y\|_T^2 + |q_\alpha|\,\|y\|_T - (y,\theta)_T \tag{4.7}$$

For $0 < a \le -\ln(1-\zeta)$ we consider the following optimisation problems

$$\max_{y\in\mathcal{L}_2[0,T]} H(y) \quad \text{subject to} \quad K(y) = a \tag{4.8}$$

Proposition 4.3. *Assume that the conditions of Lemma 3.3 hold. Then the optimisation problem* (4.8) *has the unique solution* $y^* = \widetilde{y}^a = \theta_t \tau_{\lambda_a}(t)$ *with* $\lambda_a = \Phi^{-1}(a)$.

Proof. According to Lagrange's method we consider the following unconstrained problem

$$\max_{y\in\mathcal{L}_2[0,T]} \Psi(y,\lambda)\,, \tag{4.9}$$

where $\Psi(y,\lambda) = H(y) - \lambda K(y)$ and $\lambda \in \mathbb{R}$ is the Lagrange multiplier. Now it suffices to find some $\lambda \in \mathbb{R}$ for which the problem (4.9) has a solution, which satisfies the constraint in (4.8). To this end we represent Ψ as

$$\Psi(y,\lambda) = \int_0^T (\omega(t)+\lambda)\left(y_t'\theta_t - \frac{1}{2}|y_t|^2\right)\mathrm{d}t \;-\; \lambda\,|q_\alpha|\,\|y\|_T\,.$$

It is easy to see that for $\lambda < 0$ the maximum in (4.9) equals $+\infty$; i.e. the problem (4.8) has no solution. Therefore, we assume that $\lambda \ge 0$. First we calculate the Fréchet derivative; i.e. the linear operator $D_y(\cdot,\lambda) : \mathcal{L}_2[0,T] \to \mathbb{R}$ defined for $h \in \mathcal{L}_2[0,T]$ as

$$D_y(h,\lambda) = \lim_{\delta\to 0} \frac{\Psi(y+\delta h,\lambda) - \Psi(y,\lambda)}{\delta}\,.$$

For $\|y\|_T > 0$ we obtain

$$D_y(h,\lambda) = \int_0^T (d_y(t,\lambda))' h_t\,\mathrm{d}t$$

with

$$d_y(t,\lambda) = (\omega(t)+\lambda)(\theta_t - y_t) - \lambda|q_\alpha|\,\overline{y}_t\,.$$

If $\|y\|_T = 0$, then

$$D_y(h,\lambda) = \int_0^T (\omega(t)+\lambda)\,\theta_t'\,h_t\,\mathrm{d}t - \lambda|q_\alpha|\,\|h\|_T\,.$$

Define now

$$\Delta_y(h,\lambda) = \Psi(y+h,\lambda) - \Psi(y,\lambda) - D_y(h,\lambda)\,. \tag{4.10}$$

We have to show that $\Delta_y(h, \lambda \leq 0$ for all $y, h \in \mathcal{L}_2[0,T]$. Indeed, if $\|y\|_T = 0$ then

$$\Delta_y(h,\lambda) = -\frac{1}{2}\int_0^T (\omega(t)+\lambda)\,|h_t|^2\,\mathrm{d}t \leq 0\,.$$

If $\|y\|_T > 0$, then

$$\Delta_y(h,\lambda) = -\frac{1}{2}\int_0^T (\omega(t)+\lambda)\,|h_t|^2\,\mathrm{d}t - \lambda\,|q_\alpha|\,l_y(h) \leq 0\,,$$

by Lemma 4.2 for all $\lambda \geq 0$ and for all $y, h \in \mathcal{L}_2[0,T]$.

To find the solution of the optimisation problem (4.9) we have to find $y \in \mathcal{L}_2[0,T]$ such that

$$D_y(h,\lambda) = 0 \qquad \text{for all} \qquad h \in \mathcal{L}_2[0,T]\,. \tag{4.11}$$

First notice that for $\|\theta\|_T > 0$, the solution of (4.11) can not be zero, since for $y = 0$ we obtain $D_y(h,\lambda) < 0$ for $h = -\theta$. Consequently, we have to find an optimal solution to (4.11) for y satisfying $\|y\|_T > 0$. This means we have to find a non-zero $y \in \mathcal{L}_2[0,T]$ such that

$$d_y(t,\lambda) = 0\,.$$

One can show directly that for $0 \leq \lambda < \lambda_{\max}$ the unique solution of this equation is given by

$$y_t^\lambda := \theta_t \tau_\lambda(t)\,, \tag{4.12}$$

where $\tau_\lambda(t)$ is defined in (3.6). Note that in view of the second part of Lemma 3.2 and definition (3.6) the function $y^\lambda \not\equiv 0$ for every $0 \leq \lambda < \lambda_{\max}$.

It remains to choose the Lagrage multiplier λ so that it satisfies the constraint in (4.8). To this end note that

$$K(y^\lambda) = \Phi(\lambda)\,.$$

Under the conditions of Lemma 3.3 the inverse of Φ exists and for $0 < a \leq -\ln(1-\zeta)$ the inverse function $0 \leq \Phi^{-1}(a) < \lambda_{\max}$. Thus $y^{\lambda_a} \not\equiv 0$ with $\lambda_a = \Phi^{-1}(a)$ is the solution of the problem (4.8). □

We are now ready to proof the main results in Section 3.1. The auxiliary lemmas are proved in A.1.

Proof of Theorem 3.4. In view of the representation (2.9) and definitions (4.1), we can rewrite the cost function as

$$J(x,\nu) = (T+1)\ln x + \int_0^T \omega(t) r_t \mathrm{d}t + \ln(1-\kappa) + I(V) + H(y)\,, \tag{4.13}$$

where $\kappa = 1 - e^{-V_T}$.

We start to maximise $J(x,\nu)$ by maximising I over all functions V. To this end we fix the last value of the consumption process, by setting $V_T = -\ln(1-\kappa)$ for some parameter $0 \leq \kappa < 1$ which will be chosen later. By Lemma 4.1 we find that

$$I(V) \leq I(V^\kappa) = -T\ln T + T\ln\kappa\,,$$

where

$$V_t^{\kappa} = \int_0^t v^{\kappa}(t)\mathrm{d}t = \ln \frac{T}{T - \kappa t}, \quad 0 \le t \le T. \tag{4.14}$$

Define now

$$L_t(\nu) = (y, \theta)_t - \frac{1}{2}\|y\|_t^2 - V_t - |q_\alpha| \, \|y\|_t, \quad 0 \le t \le T,$$

and note that condition (3.3) is equivalent to

$$\inf_{0 \le t \le T} L_t(\nu) \ge \ln (1 - \zeta). \tag{4.15}$$

Firstly, we consider the bound in (4.15) only at time $t = T$:

$$L_T(\nu) \ge \ln (1 - \zeta).$$

Recall definition (4.7) of K and choose the function V as V^{κ} as in (4.14). Then we can rewrite the bound for $L_T(\nu)$ as a bound for K and obtain

$$K(y) \le \ln(1 - \kappa)/(1 - \zeta), \quad 0 \le \kappa \le \zeta.$$

To find the optimal investment strategy we need to solve the optimisation problem (4.8) for $0 \le a \le \ln(1 - \kappa)/(1 - \zeta)$. By Proposition 4.3 for $0 < a \le -\ln(1 - \zeta)$

$$\max_{y \in \mathcal{L}_2[0,T],\, K(y) = a} H(y) = H(\widetilde{y}^a) := C(a), \tag{4.16}$$

where the solution $\widetilde{y}^a$ is defined in Proposition 4.3. Note that the definitions of the functions H and $\widetilde{y}^a$ imply

$$C(a) = \int_0^T \omega(t)\Big(\tau_{\lambda_a}(t) - \frac{1}{2}\tau_{\lambda_a}^2(t)\Big) |\theta_t|^2 \mathrm{d}t \quad \text{with} \quad \lambda_a = \Phi^{-1}(a).$$

To consider the optimisation problem (4.8) for $a = 0$ we observe that

$$K(y) \ge \|y\|_T \left(|q_\alpha| - \|\theta\|_T\right) + \frac{1}{2}\|y\|_T^2 \ge 0,$$

provided that $|q_\alpha| > \|\theta\|_T$ (which follows from (3.15)). Thus, there exists only one function for which $K(y) = 0$, namely $y \equiv 0$. Furthermore, by Lemma 3.2 $\rho(\lambda_{\max}) = 0$ and, therefore, definition (3.6) implies

$$\tau_{\lambda_{\max}}(\cdot) \equiv 0 \quad \text{and} \quad y^{\lambda_{\max}} \equiv 0. \tag{4.17}$$

In view of Lemma 3.3 we get $\Phi^{-1}(0) = \lambda_{\max}$, therefore, $y^{\Phi^{-1}(0)} = y^{\lambda_{\max}} \equiv 0$; i.e. y^{λ_a} with $\lambda_a = \Phi^{-1}(a)$ is the solution of the optimisation problem (4.8) for all $0 \le a \le -\ln(1 - \zeta)$. Now we calculate the derivative of $C(\cdot)$ as

$$\dot{C}(a) = \dot{\lambda}_a \int_0^T \omega(t) \left(1 - \tau_{\lambda_a}(t)\right) |\theta_t|^2 \, \tau_1(t, \lambda_a) \, \mathrm{d}t,$$

where

$$\tau_1(t,\lambda) = \frac{\partial \tau_\lambda(t)}{\partial \lambda}\,. \tag{4.18}$$

Since $\dot\lambda_a = 1/\dot\Phi(\lambda_a)$, by Lemma A.1, the derivative $\dot C(a)$ is positive. Therefore,

$$\max_{0\le a\le \ln(1-\kappa)/(1-\zeta)} C(a) = C\left(\frac{\ln(1-\kappa)}{1-\zeta}\right),$$

and we choose $a = \ln(1-\kappa)/(1-\zeta)$ in (4.16).

Now recall the definitions (3.11) and (3.12), the representation (4.13) and set $\nu^\kappa = (\widetilde{y}^\kappa_t\,,\, v^\kappa_t)_{0\le t\le T}$. Thus for $\nu \in \mathcal{U}$ with $V_T = -\ln(1-\kappa)$ we have

$$J(x,\nu) \le J(x,\nu^\kappa) = A(x) + \Gamma(\kappa)\,.$$

It is clear that (3.14) gives the optimal value for the parameter κ.

To finish the proof we have to verify condition (4.15) for the strategy ν^* defined in (3.18). Indeed, we have

$$\begin{aligned} L_t(\nu^*) &= (y^*,\theta)_t - \frac{1}{2}\|y^*\|_t^2 - |q_\alpha|\,\|y^*\|_t - \int_0^t v_s^*\mathrm{d}s \\ &=: -\int_0^t g(u)\,\mathrm{d}u - \int_0^t v_s^*\mathrm{d}s\,, \end{aligned}$$

where

$$g(t) = \tau_t^*|\theta_t|^2\left(|q_\alpha|\,\chi(t) - 1 + \frac{\tau_t^*}{2}\right) \quad\text{and}\quad \chi(t) = \frac{\tau_t^*}{2\sqrt{\int_0^t(\tau_s^*)^2|\theta_s|^2\,\mathrm{d}s}}\,.$$

We recall $\phi(\kappa)$ from (3.10) and γ from (3.14), then $\tau_t^* = \tau_{\phi(\gamma)}(t)$. Definition (3.6) implies

$$\chi(t) \ge \frac{\tau_{\phi(\gamma)}(T)}{2\tau_{\phi(\gamma)}(0)\|\theta\|_T} \ge \frac{1+\phi(\gamma)}{2\|\theta\|_T\,(1+T+\phi(\gamma))} \ge \frac{1}{2\|\theta\|_T\,(1+T)}\,.$$

Therefore, condition (3.15) guarantees that $g(t)\ge 0$ for $t\ge 0$, which implies

$$L_t(\nu^*) \ge L_T(\nu^*) = \ln(1-\zeta)\,.$$

This concludes the proof of Theorem 3.4. □

Proof of Corollary 3.6. Consider now the optimisation problem (3.14). To solve it we have to find the derivative of the integral in (3.13)

$$B(\kappa) := \int_0^T \omega(t)\,|\theta_t|^2\left(\tau_{\phi(\kappa)}(t) - \frac{1}{2}\tau^2_{\phi(\kappa)}(t)\right)\mathrm{d}t\,.$$

Indeed, we have with $\phi(\kappa)$ as in (3.10),

$$\dot B(\kappa) = \int_0^T \omega(t)|\theta_t|^2\,\left(1-\tau_{\phi(\kappa)}(t)\right)\,\frac{\partial}{\partial\kappa}\,\tau_{\phi(\kappa)}(t)\,\mathrm{d}t\,.$$

Obviously,

$$\frac{\partial}{\partial \kappa}\,\tau_{\phi(\kappa)}(t) = \tau_1(t,\phi(\kappa))\,\dot{\phi}(\kappa)\,, \tag{4.19}$$

where $\tau_1(t,\lambda)$ is defined in (4.18). By the definition of ϕ in (3.10) $\Phi(\phi(\kappa)) = \ln(1-\kappa)/(1-\zeta)$, we have

$$\dot{\phi}(\kappa) = -\frac{1}{\dot{\Phi}(\phi(\kappa))(1-\kappa)}\,.$$

Therefore,

$$\dot{B}(\kappa) = -\frac{1}{1-\kappa}\,\widetilde{B}(\phi(\kappa)) \quad \text{with} \quad \widetilde{B}(\lambda) = \frac{\int_0^T \omega(t)|\,(1-\tau_\lambda(t))\,\tau_1(t,\lambda)\,|\theta_t|^2\,\mathrm{d}t}{\dot{\Phi}(\lambda)}\,.$$

We calculate now the derivative of Φ as

$$\dot{\Phi}(\lambda) = \int_0^T \widehat{\tau}(t,\lambda)\,\tau_1(t,\lambda)\,|\theta_t|^2\mathrm{d}t\,, \tag{4.20}$$

where

$$\widehat{\tau}(t,\lambda) = \frac{|q_\alpha|\tau_\lambda(t)}{\|\tau_\lambda\theta\|_T} - 1 + \tau_\lambda(t)\,.$$

By inequality (A.1), $\widehat{\tau}(t,\lambda) > 0$ and, moreover, in view of Lemma A.1, we have $\tau_1(t,\lambda) \le 0$. Therefore, taking representation (4.20) into account, we obtain

$$\widetilde{B}(\lambda) = \frac{\int_0^T \omega(t)\,(1-\tau_\lambda(t))\,|\tau_1(t,\lambda)|\,|\theta_t|^2\,\mathrm{d}t}{\int_0^T \widehat{\tau}(t,\lambda)\,|\tau_1(t,\lambda)|\,|\theta_t|^2\mathrm{d}t}\,.$$

Moreover, using the lower bound (A.1) we estimate

$$\widetilde{B}(\lambda) < \frac{(1+T)^2\,\|\theta\|_T}{|q_\alpha| - (T+1)\,\|\theta\|_T} =: B_{\max}\,. \tag{4.21}$$

Condition (3.19) for $0 < \zeta < \kappa_0$ implies that

$$B_{\max} \le \left(\frac{1}{\zeta} - 1\right) T - 1\,.$$

Thus for $0 \le \kappa \le \zeta < \kappa_0$ we obtain

$$\dot{\Gamma}(\kappa) > \frac{T}{\kappa} - \frac{1}{1-\kappa}\,(1+B_{\max}) \ge \frac{T}{\zeta} - \frac{1}{1-\zeta}\,(1+B_{\max}) \ge 0\,.$$

This implies $\gamma = \zeta$, i.e. $\phi(\gamma) = \Phi^{-1}(0)$. Therefore, by Lemma 3.3, $\phi(\gamma) = \lambda_{\max}$. Therefore, we conclude from (4.17) that $y_t^* = \tau_{\lambda_{\max}}(t)\theta_t = 0$ for all $0 \le t \le T$. □

Proof of Theorem 3.7. It suffices to verify condition (4.15) for the strategy $\nu^* = (y_t^*, v_t^*)_{0 \le t \le T}$ with $y_t^* = \theta_t$ and $v_t^* = 1/\omega(t)$ for $t \in [0, T]$. It is easy to show that condition (3.20) implies that $L_T(\nu^*) \geq \ln(1-\zeta)$. Moreover, for $0 \le t \le T$ we can represent $L_t(\nu^*)$ as

$$L_t(\nu^*) = -\int_0^t g_s^* \mathrm{d}s - \int_0^t v_s^* \, \mathrm{d}s \,,$$

where

$$g_t^* = \left(\frac{|q_\alpha|}{\|\theta\|_t} - 1 \right) \frac{|\theta_t|^2}{2} \geq \left(\frac{|q_\alpha|}{\|\theta\|_T} - 1 \right) \frac{|\theta_t|^2}{2} \geq 0$$

since we have assumed that $|q_\alpha| \geq \|\theta\|_T$. Therefore, $L_t(\nu^*)$ is decreasing in t; i.e. $L_t(\nu^*) \geq L_T(\nu^*)$ for all $0 \le t \le T$. This implies the assertion of Theorem 3.7. □

4.2 Results and proofs of Section 3.2

Next we introduce the constraint

$$K_1(y) := -(y, \theta)_T - \ln F_\alpha\left(\|y\|_T\right). \tag{4.22}$$

For $0 < a \le -\ln(1-\zeta)$ we consider the following optimisation problems

$$\max_{y \in \mathcal{L}_2[0,T]} H(y) \quad \text{subject to} \quad K_1(y) = a\,. \tag{4.23}$$

The following result is the analog of Proposition 4.3.

Proposition 4.4. *Assume that the conditions of Lemma 3.10 hold. Then the optimisation problem* (4.23) *has the unique solution* $y_t^* = \widetilde{y}_t^{1,a} = \theta_t \varsigma_{\lambda_{1,a}}(t)$ *with* $\lambda_{1,a} = \Phi_1^{-1}(a)$.

Proof. As in the proof of Proposition 4.3 we use Lagrange's method. We consider the unconstrained problem

$$\max_{y \in \mathcal{L}_2[0,T]} \Psi_1(y, \lambda)\,, \tag{4.24}$$

where $\Psi_1(y,\lambda) = H(y) - \lambda K_1(y)$ and $\lambda \geq 0$ is the Lagrange multiplier. Taking into account the definition of F_α in (4.22), and setting $f_\alpha = \ln F_\alpha$, we obtain the representation

$$\Psi_1(y,\lambda) = \int_0^T \left((\omega(t) + \lambda)\, \theta_t' \, y_t - \frac{\omega(t)}{2} |y_t|^2 \right) \mathrm{d}t + \lambda \, f_\alpha\left(\|y\|_T\right).$$

Its Fréchet derivative is given by

$$D_{1,y}(h,\lambda) = \lim_{\delta \to 0} \frac{\Psi_1(y + \delta h, \lambda) - \Psi_1(y,\lambda)}{\delta}\,.$$

It is easy to show directly that for $\|y\|_T > 0$

$$D_{1,y}(h,\lambda) = \int_0^T (d_{1,y}(t,\lambda))' h_t \, \mathrm{d}t\,,$$

where

$$d_{1,y}(t,\lambda) = (\omega(t)+\lambda)\theta_t - \omega(t)\,y_t + \lambda \dot{f}_\alpha(\|y\|_T)\,\overline{y}_t\,,$$

and $\dot{f}_\alpha(\cdot)$ denotes the derivative of $f_\alpha(\cdot)$.

If $\|y\|_T = 0$, then

$$D_{1,y}(h,\lambda) = \int_0^T (\omega(t)+\lambda)\,\theta'_t\,h_t\,\mathrm{d}t + \lambda\,\dot{f}_\alpha(0)\|h\|_T\,.$$

We set now

$$\Delta_{1,y}(h,\lambda) = \Psi_1(y+h,\lambda) - \Psi_1(y,\lambda) - D_{1,y}(h,\lambda)\,, \tag{4.25}$$

and show that $\Delta_{1,y}(h,\lambda) \le 0$ for all $y, h \in \mathcal{L}_2[0,T]$. Indeed, if $\|y\|_T = 0$, then

$$\Delta_{1,y}(h,\lambda) = -\frac{1}{2}\int_0^T \omega(t)\,|h_t|^2\,\mathrm{d}t + \lambda\left(f_\alpha(\|h\|_T) - \dot{f}_\alpha(0)\|h\|_T\right)\,.$$

Recalling the definition of φ in (3.23) and setting $x_1 = |q_\alpha| + x$, the derivatives of f_α are given by

$$\dot{f}_\alpha(x) = -\frac{1}{\varphi(x_1)} \quad \text{and} \quad \ddot{f}_\alpha(x) = -\frac{1 - x_1\,\varphi(x_1)}{\varphi^2(x_1)} \le 0\,. \tag{4.26}$$

The last inequality follows directly from the right inequality in (3.24). Therefore, taking into account that $f_\alpha(0) = 0$ we get $f_\alpha(x) \le \dot{f}_\alpha(0)x$ for all $x \ge 0$. Thus for $\lambda \ge 0$ we have $\Delta_{1,y}(h,\lambda) \le 0$ in the case when $\|y\|_T = 0$.

Let now $\|y\|_T > 0$ and $\overline{y} = y/\|y\|_T$. Then

$$\Delta_{1,y}(h,\lambda) = -\frac{1}{2}\int_0^T \omega(t)\,|h_t|^2\,\mathrm{d}t + \lambda\,\delta_{1,y}(h)\,,$$

where

$$\delta_{1,y}(h) = f_\alpha(\|y+h\|_T) - f_\alpha(\|y\|_T) - \dot{f}_\alpha(\|y\|_T)\,(\overline{y},h)_T\,.$$

Moreover, by Taylor's formula and denoting by $\ddot{f}_\alpha$ the second derivative of f_α, we get

$$\delta_{1,y}(h) = \dot{f}_\alpha(\|y\|_T)\,l_y(h) + \frac{1}{2}\,\ddot{f}_\alpha\,(\vartheta)\,(\|y+h\|_T - \|y\|_T)^2\,,$$

where $l_y(\cdot)$ is defined in (4.6) and

$$\min(\|y\|_T\,,\,\|y+h\|_T) \le \vartheta \le \max(\|y\|_T\,,\,\|y+h\|_T)\,.$$

Now the last inequality in (4.26) and Lemma 4.2 imply that $\Delta_{1,y}(h,\lambda) \le 0$ for all $\lambda \ge 0$ and $y, h \in \mathcal{L}_2[0,T]$. The solution of the optimisation problem (4.24) is given by $y \in \mathcal{L}_2[0,T]$ such that

$$D_{1,y}(h,\lambda) = 0 \quad \text{for all} \quad h \in \mathcal{L}_2[0,T]\,. \tag{4.27}$$

Notice that for $\|\theta\|_T > 0$ the solution (4.27) can not be zero, since for $y = 0$ we obtain $D_{1,y}(h, \lambda) < 0$ for $h = -\theta$. Therefore, we have to solve equation (4.27) for y with $\|y\|_T > 0$, equivalently, we have to find a non-zero function in $\mathcal{L}_2[0, T]$ satisfying

$$d_{1,y}(t, \lambda) = 0\,.$$

One can show directly that for $0 \le \lambda \le \lambda^1_{\max}$ the solution of this equation is given by

$$y_t^{1,\lambda} = \varsigma_\lambda(t)\theta_t\,, \tag{4.28}$$

where $\varsigma_\lambda(t)$ is defined in (3.26). Now we have to choose the parameter λ to satisfy the constraint in (4.23). Note that

$$K_1(y^{1,\lambda}) = \Phi_1(\lambda)\,.$$

Under the conditions of Lemma 3.10 the inverse of Φ_1 exists. Therefore, the function $y^{1,\lambda_a} \not\equiv 0$ with $\lambda_a = \Phi_1^{-1}(a)$ is the solution of the optimisation problem (4.23). □

Proof of Theorem 3.11. Define

$$L_{1,t}(\nu) = (y, \theta)_t \,-\, V_t \,+\, f_\alpha\left(\|y\|_t\right)\,, \quad 0 \le t \le T\,, \tag{4.29}$$

with $f_\alpha = \ln F_\alpha$.

First note that the risk bound in the optimisation problem (3.21) is equivalent to

$$\inf_{0\le t\le T} L_{1,t}(\nu) \ge \ln\,(1-\zeta)\,, \tag{4.30}$$

As in the proof of Theorem 3.4 we start with the constraint at time $t = T$:

$$L_{1,T}(\nu) \ge \ln\,(1-\zeta)\,.$$

Taking the definition of K_1 in (4.22) into account and choosing $V = V^\kappa$ as in (4.14) we rewrite this inequality as

$$K_1(y) \le \ln(1-\kappa)/(1-\zeta)\,, \quad 0 \le \kappa \le \zeta\,.$$

To find the optimal strategy we use the optimisation problem (4.23), extending the range of a to $0 \le a \le \ln(1-\kappa)/(1-\zeta)$. In Proposition 4.4 we established that for each $0 < a \le -\ln(1-\zeta)$

$$\max_{y\in\mathcal{L}_2[0,T]\,,\,K_1(y)=a} H(y) = H(\widetilde{y}^{1,a}) =: C_1(a)\,, \tag{4.31}$$

where the optimal solution $\widetilde{y}^{1,a}$ is defined in Proposition 4.4. We observe that

$$C_1(a) = \int_0^T \omega(t)|\theta_t|^2\left(\varsigma_{\lambda_{1,a}}(t) - \frac{1}{2}\varsigma^2_{\lambda_{1,a}}(t)\right)\,\mathrm{d}t \quad \text{with} \quad \lambda_{1,a} = \Phi_1^{-1}(a)\,.$$

To study the optimisation problem (4.23) for $a = 0$ note that

$$K_1(y) \ge k_{\min}(\|y\|_T) \quad \text{with} \quad k_{\min}(x) = -x\|\theta\|_T - f_\alpha(x)\,, \quad x \ge 0\,.$$

Moreover,

$$\dot{k}_{\min}(x) = \frac{1}{\varphi(|q_\alpha| + x)} - \|\theta\|_T\,, \quad x \geq 0\,,$$

and by the right inequality in (3.24) we obtain for $|q_\alpha| > \|\theta\|_T$ (which follows from condition (3.15))

$$\dot{k}_{\min}(x) \geq |q_\alpha| + x - \|\theta\|_T > 0\,, \quad x \geq 0,.$$

Therefore, $k_{\min}(x) > k_{\min}(0) = 0$ for all $x > 0$ and $k_{\min}(x) = 0$ if and only if $x = 0$. This means that only $y \equiv 0$ satisfies $K_1(y) = 0$. Moreover, in view of Lemma 3.9 and Lemma 3.10, as in the proof of Theorem 3.4, we obtain $\widetilde{y}^{1,0} \equiv 0$. Therefore, the function $\widetilde{y}^{1,a}$ is the solution of (4.23) for all $0 \leq a \leq -\ln(1-\zeta)$.

To choose the parameter $0 \leq a \leq \ln(1-\kappa)/(1-\zeta)$ we calculate the derivative of $C_1(a)$ as

$$\dot{C}_1(a) = \dot{\lambda}_{1,a} \int_0^T \omega(t)|\theta_t|^2 \left(1 - \varsigma_{\lambda_{1,a}}(t)\right) \varsigma_1(t, \lambda_{1,a})\,\mathrm{d}t\,,$$

where

$$\varsigma_1(t,\lambda) = \frac{\partial}{\partial\lambda}\varsigma_\lambda(t)\,. \tag{4.32}$$

We recall that $\dot{\lambda}_{1,a} = 1/\dot{\Phi}_1(\lambda_{1,a})$ with $\lambda_{1,a} = \Phi_1^{-1}(a)$. Therefore, by Lemma A.2, the derivative $\dot{C}_1(a) > 0$. This implies

$$\max_{0\leq a\leq \ln(1-\kappa)/(1-\zeta)} C_1(a) = C_1\left(\frac{\ln(1-\kappa)}{1-\zeta}\right).$$

So in (4.31) we take $a = \ln(1-\kappa)/(1-\zeta)$.

Recalling the notation $\widetilde{y}_t^{1,\kappa} = \theta_t \varsigma_{\phi_1(\kappa)}(t)$ from (3.31) we set $\nu^{1,\kappa} = (\widetilde{y}_t^{1,\kappa}\,,\, v_t^\kappa)_{0\leq t\leq T}$. Then, for $\nu \in \mathcal{U}$ with $V_T = -\ln(1-\kappa)$,

$$J(x,\nu) \leq J(x, \overline{\nu}^\kappa) = A(x) + \Gamma_1(\kappa)\,.$$

It is clear that (3.33) gives the optimal value for the parameter κ.

To finish the proof we have to verify condition (4.30) for the strategy ν^* as defined in (3.35). To this end, with $\phi_1(\kappa) = \Phi_1^{-1}\big(\ln(1-\kappa)/(1-\zeta)\big)$, we set

$$\varsigma_t^* = \varsigma_{\phi_1(\gamma_1)}(t) \quad \text{and} \quad \chi_1(t) = \frac{\varsigma_t^*}{2\|\varsigma^*\theta\|_t}\,.$$

With this notation we can represent the function $L_{1,t}(\nu^*)$ in the following integral form

$$L_t(\nu^*) = -\int_0^t g_1(u)\,\mathrm{d}u - \int_0^t v_s^* \mathrm{d}s\,,$$

where

$$g_1(t) = \varsigma_t^* |\theta_t|^2 \left(\frac{\widetilde{f}_\alpha(t)\chi_1(t)}{2} - 1\right) \quad \text{with} \quad \widetilde{f}_\alpha(t) = -\dot{f}_\alpha\left(\|\varsigma^*\theta\|_t\right).$$

Note that definition (3.26) and the inequalities (3.27) imply

$$\chi_1(t) \ge \frac{\varsigma_T^*}{2\varsigma_0^* \|\theta\|_t} \ge \frac{1+\phi_1(\gamma_1)}{2\|\theta\|_t\,(1+T+\phi_1(\gamma_1))} \ge \frac{1}{2\|\theta\|_T\,(1+T)}\,.$$

Moreover, from the right inequality in (3.24) we obtain

$$\widetilde{f}_\alpha(t) = \frac{1}{\varphi\left(|q_\alpha| + \|\varsigma^*\theta\|_t\right)} \ge |q_\alpha| + \|\varsigma^*\theta\|_t \ge |q_\alpha|\,.$$

Therefore, condition (3.15) implies that $g_1(t) \ge 0$, i.e.

$$L_{1,t}(\nu^*) \ge L_{1,T}(\nu^*) = \ln(1-\zeta)\,.$$

This concludes the proof of Theorem 3.11. □

Proof of Corollary 3.13. Consider now the optimisation problem (3.33). To solve this we have to calculate the derivative of the integral in (3.32)

$$B_1(\kappa) := \int_0^T \omega(t)\,|\theta_t|^2 \left(\varsigma_{\phi_1(\kappa)}(t) - \frac{1}{2}\varsigma^2_{\phi_1(\kappa)}(t)\right)\mathrm{d}t\,.$$

We obtain

$$\dot{B}_1(\kappa) = \dot{\phi}_1(\kappa) \int_0^T \omega(t)|\theta_t|^2\,(1-\varsigma(t))\,\varsigma_1(t,\phi_1(\kappa))\,\mathrm{d}t\,,$$

where $\varsigma_1(t,\lambda)$ is defined in (4.32). We recall the definition of ϕ_1 in (3.30). Therefore,

$$\dot{B}_1(\kappa) = -\frac{1}{1-\kappa}\,\widetilde{B}_1(\Phi_1^{-1}(\kappa))$$

with

$$\widetilde{B}_1(\lambda) = \frac{\int_0^T \omega(t)\,(1-\varsigma_\lambda(t))\,\varsigma_1(t,\lambda)\,|\theta_t|^2\mathrm{d}t}{\dot{\Phi}_1(\lambda)}\,.$$

By Lemma A.2, $\varsigma_1(t,\lambda) \le 0$, therefore, taking representation (A.5) into account, we obtain

$$\widetilde{B}_1(\lambda) = \frac{\int_0^T \omega(t)\,(1-\varsigma_\lambda(t))\,|\varsigma_1(t,\lambda)|\,|\theta_t|^2\,\mathrm{d}t}{\int_0^T \eta(t,\lambda)\,|\varsigma_1(t,\lambda)|\,|\theta_t|^2\mathrm{d}t}\,.$$

Moreover, with the lower bound (A.6) we can estimate $\widetilde{B}_1(\lambda)$ as in in (4.21), i.e.

$$\widetilde{B}_1(\lambda) \le B_{\max}\,.$$

The remainding proof is the same as the proof of Corollary 3.13. □

Proof of Theorem 3.14. We have to verify condition (4.30) for the strategy $\nu^* = (y_t^*, v_t^*)_{0\le t\le T}$ with $y_t^* = \theta_t$ and $v_t^* = 1/\omega(t)$ for $t \in [0,T]$.

First note that condition (3.36) implies

$$L_{1,T}(\nu^*) \ge \ln(1-\zeta)\,.$$

Moreover, for $0 \le t \le T$ we can represent the function $L_{1,t}(\nu^*)$ as

$$L_{1,t}(\nu^*) = \|\theta\|_t^2 + f_\alpha(\|\theta\|_t) - V_t^* = -\int_0^t l_s^* \mathrm{d}s - \int_0^t v_s^* \,\mathrm{d}s\,,$$

where

$$l_t^* = \left(\frac{1}{\varphi(|q_\alpha| + \|\theta\|_t)} - 1\right) |\theta_t|^2\,.$$

Therefore, by the right inequality in (3.24) we obtain

$$l_t^* \ge (|q_\alpha| + \|\theta\|_t - 1)\; |\theta_t|^2 \ge (|q_\alpha| - 1)\; |\theta_t|^2$$

and by condition (3.34) we get $l_t^* > 0$ for $0 \le t \le T$, therefore, $L_{1,t}(\nu^*)$ is decreasing in t, i.e. for $0 < t \le T$

$$L_{1,t}(\nu^*) \ge L_{1,T}(\nu^*) \ge \ln(1 - \zeta)\,.$$

This concludes the proof of Theorem 3.14. □

5 Appendix

A.1 Results for Section 3.1

Proof of Lemma 3.2. Since $G(u, \lambda)$ is for fixed λ decreasing to 0 in u, equation $G(u, \lambda) = 1$ has a positive solution if and only if $G(0, \lambda) \ge 1$. But this is equivalent to $k_2 + 2\lambda k_1 - \lambda^2(|q_\alpha|^2 - \|\theta\|_T^2) \ge 0$, which gives the upper bound for λ. Moreover, taking into account that $G(0, \lambda_{\max}) = 1$ we obtain from definition (3.5) that $\rho(\lambda_{\max}) = 0$. □

Next we prove some properties of Φ and τ_λ.

Lemma A.1. *The function $\tau_\lambda(t)$ is continuously differentiable in λ for $0 \le \lambda \le \lambda_{\max}$, and the partial derivative* (4.18) *is negative for all $0 \le t \le T$. Moreover, under the condition* (3.15) *the derivative $\dot{\Phi}(\lambda) < 0$ for $0 \le \lambda \le \lambda_{\max}$.*

Proof. First note that

$$\tau_1(t, \lambda) = -|q_\alpha| \frac{(\rho(\lambda)\omega(t) - \lambda\dot{\rho}(\lambda)(\omega(t) + \lambda))}{(\lambda|q_\alpha| + \rho(\lambda)(\omega(t) + \lambda))^2}\,.$$

By the definition of $\rho(\lambda)$ in (3.5) we get $G(\rho(\lambda), \lambda) = 1$ for $0 \le \lambda \le \lambda_{\max}$. Therefore,

$$\dot{\rho}(\lambda) = -\frac{G_2(\rho(\lambda), \lambda)}{G_1(\rho(\lambda), \lambda)}$$

with

$$G_1(u, \lambda) = \frac{\partial G(u, \lambda)}{\partial u} \quad \text{and} \quad G_2(u, \lambda) = \frac{\partial G(u, \lambda)}{\partial \lambda}\,.$$

The definition of G in (3.4) implies that

$$G_1(u,\lambda) = -2\int_0^T \frac{(\omega(t)+\lambda)^3}{(\lambda|q_\alpha| + u(\omega(t)+\lambda))^3}\,|\theta_t|^2\,\mathrm{d}t$$

and

$$G_2(u,\lambda) = -2|q_\alpha|\int_0^T \frac{\omega(t)(\omega(t)+\lambda)}{(\lambda|q_\alpha| + u(\omega(t)+\lambda))^3}\,|\theta_t|^2\,\mathrm{d}t\,.$$

Therefore, for all $0 \le \lambda \le \lambda_{\max}$ and $0 \le t \le T$

$$\dot\rho(\lambda) < 0 \quad \text{and} \quad \tau_1(t,\lambda) < 0\,.$$

Consider now the derivative of Φ given in (4.20). To find a lower bound for $\dot\Phi$ note that by the inequalities (3.7)

$$\frac{\tau_\lambda(t)}{\|\tau_\lambda\theta\|_T} \ge \frac{\tau(T,\lambda)}{\tau(0,\lambda)\|\theta\|_T} \ge \frac{1}{(T+1)\|\theta\|_T}\,.$$

Therefore,

$$\widehat{\tau}(t,\lambda) \ge \frac{|q_\alpha|}{(T+1)\|\theta\|_T} - 1 \tag{A.1}$$

and by condition (3.15) $\widehat{\tau}(t,\lambda) > 0$ for $0 \le t \le T$ and $0 \le \lambda \le \lambda_{\max}$, i.e. $\dot\Phi(\lambda) < 0$. □

Proof of Lemma 3.3. Taking into account that $\tau_0(\cdot) \equiv 1$ we get

$$\Phi(0) = |q_\alpha|\|\theta\|_T - \frac{1}{2}\|\theta\|_T^2\,.$$

Moreover, condition (3.9) implies $\Phi(0) > -\ln(1-\zeta)$. The second part of Lemma 3.2, the definitions (3.6) and (3.8) imply immediately that $\Phi(\lambda_{\max}) = 0$. Therefore, in view of Lemma A.1 the inverse $\Phi^{-1}(a)$ exists for $0 \le a \le -\ln(1-\zeta)$. Moreover, $0 \le \Phi^{-1}(a) < \lambda_{\max}$, for $0 < a \le -\ln(1-\zeta)$ and $\Phi^{-1}(0) = \lambda_{\max}$. □

A.2 Results for Section 3.2

We present some properties of $\Phi_1(\lambda)$ and ς_λ.

Lemma A.2. *The function $\varsigma_\lambda(t)$ is continuously differentiable in λ for all $0 \le \lambda \le \lambda^1_{\max}$ and the partial derivative* (4.32) *is negative for all $0 \le t \le T$. Moreover, under condition* (3.15) *the derivative $\dot\Phi_1(\lambda) < 0$ for $0 \le \lambda \le \lambda^1_{\max}$.*

Proof. First note that

$$\varsigma_1(t,\lambda) = -\frac{(\omega(t)+\lambda)\psi_\alpha(\lambda)\rho_1}{((\omega(t)+\lambda)\rho_1 + \psi_\alpha(\lambda))^2}\left(\frac{\omega(t)}{\omega(t)+\lambda} - \lambda\dot\rho_1\,\Omega_\alpha(\rho_1)\right) \tag{A.2}$$

where $\rho_1 = \rho_1(\lambda)$ is defined in (3.25) and

$$\Omega_\alpha(\rho_1) = \frac{\psi_\alpha(\rho_1) - \rho_1 \dot{\psi}_\alpha(\rho_1)}{\rho_1 \psi_\alpha(\rho_1)} .$$

Note that we can represent the numerator as

$$\psi_\alpha(\rho_1) - \rho_1 \dot{\psi}_\alpha(\rho_1) = \frac{\varphi(y)\,(1 + y(y - |q_\alpha|)) - (y - |q_\alpha|)}{\varphi^2(y)}$$

with $y = |q_\alpha| + \rho_1$. Therefore, the left inequality in (3.24) implies

$$\begin{aligned}\varphi(y)\,(1 + y(y - |q_\alpha|)) - (y - |q_\alpha|) &\geq (1 + y(y - |q_\alpha|))\left(\frac{1}{y} - \frac{1}{y^3}\right) - (y - |q_\alpha|)\\ &= \frac{y|q_\alpha| - 1}{y^3} \geq \frac{q_\alpha^2 - 1}{y^3},\end{aligned}$$

and by condition (3.34) we obtain

$$\Omega_\alpha(\rho_1) \geq 0 \quad \text{for} \quad \rho_1 \geq 0 .$$

Let us now calculate $\dot{\rho}_1$. To this end note that definition (3.25) implies

$$G_1(\rho_1(\lambda), \lambda) = 1 \quad \text{for all} \quad 0 \leq \lambda \leq \lambda^1_{\max} .$$

Therefore,

$$\dot{\rho}_1(\lambda) = -\frac{G_{1,2}(\rho_1(\lambda), \lambda)}{G_{1,1}(\rho_1(\lambda), \lambda)}$$

with

$$G_{1,1}(u, \lambda) = \frac{\partial G_1(u, \lambda)}{\partial u} \quad \text{and} \quad G_{1,2}(u, \lambda) = \frac{\partial G_1(u, \lambda)}{\partial \lambda} .$$

The definition of G_1 in (3.22) implies that

$$G_{1,1}(u, \lambda) = -2 \int_0^T \frac{(\omega(t) + \lambda)^2(\lambda(\dot{\psi}_\alpha(u) + 1) + \omega(t))}{(\lambda \psi_\alpha(u) + u(\omega(t) + \lambda))^3}\, |\theta_t|^2 \, \mathrm{d}t \tag{A.3}$$

and

$$G_{1,2}(u, \lambda) = -2\psi_\alpha(u) \int_0^T \frac{\omega(t)\,(\omega(t) + \lambda)}{(\lambda \psi_\alpha(u) + u(\omega(t) + \lambda))^3}\, |\theta_t|^2 \, \mathrm{d}t . \tag{A.4}$$

Taking into account that

$$\dot{\psi}_\alpha(u) + 1 = \frac{1 - (|q_\alpha| + u)\varphi(|q_\alpha| + u)}{\varphi^2(|q_\alpha| + u)},$$

we obtain from the right inequality in (3.24)

$$\dot{\psi}_\alpha(x) + 1 \geq 0 \quad \text{for all} \quad x \geq 0 .$$

Therefore, for all $0 \le \lambda \le \lambda^1_{\max}$ and $0 \le t \le T$

$$\dot{\rho}_1(\lambda) < 0 \quad \text{and} \quad \varsigma_1(t,\lambda) < 0\,.$$

Let us calculate now the derivative of Φ_1. We obtain

$$\dot{\Phi}_1(\lambda) = \int_0^T \eta(t,\lambda)\,\varsigma_1(t,\lambda)\|\theta\|^2_T \mathrm{d}t\,, \tag{A.5}$$

where

$$\eta(t,\lambda) = -\frac{\dot{f}_\alpha(a_\lambda)\,\varsigma_\lambda(t)}{a_\lambda} - 1 = \frac{1}{\varphi\left(|q_\alpha| + a_\lambda\right)}\,\frac{\varsigma_\lambda(t)}{a_\lambda} - 1$$

with $a_\lambda = \|\varsigma_\lambda\theta\|_T$. In view of the inequalities (3.27) we obtain

$$\frac{\varsigma_\lambda(t)}{a_\lambda} = \frac{\varsigma_\lambda(t)}{\|\varsigma_\lambda\theta\|_T} \ge \frac{\varsigma_\lambda(T)}{\varsigma_\lambda(0)\|\theta\|_T} \ge \frac{1}{(T+1)\|\theta\|_T}\,.$$

Therefore, by the right inequality in (3.24) and the condition (3.15)

$$\eta(t,\lambda) \ge \frac{|q_\alpha| + a_\lambda}{(T+1)\|\theta\|_T} - 1 \ge \frac{|q_\alpha|}{(T+1)\|\theta\|_T} - 1 > 0 \tag{A.6}$$

for $0 \le t \le T$ and $0 \le \lambda \le \lambda^1_{\max}$. □

Proof of Lemma 3.10. Similarly to the proof of Lemma 3.3 we observe that condition (3.29) implies

$$\Phi_1(0) = -\|\theta\|_T - f_\alpha(\|\theta\|_T) > -\ln(1-\zeta)\,.$$

Moreover, $\Phi_1(\lambda^1_{\max}) = 0$ since $\rho_1(\lambda^1_{\max}) = 0$. This means that $\phi_1(0) = \lambda^1_{\max}$.

In view of Lemma A.2 $\Phi_1(\cdot)$ is strictly decreasing on $[0, \lambda^1_{\max}]$. Therefore, Φ_1^{-1} exists for all $0 \le a \le -\ln(1-\zeta)$ such that

$$0 \le \phi_1(a) < \lambda^1_{\max} \quad \text{for} \quad 0 < a \le -\ln(1-\zeta)$$

and $\phi_1(0) = \lambda^1_{\max}$. □

Bibliography

[1] Artzner, P., Delbaen, F., Eber, J.-M. and Heath, D.: Coherent measures of risk. *Math. Finance.* **9**, 203–228 (1999)

[2] Basak, S. and Shapiro, A. (1999) Value at Risk based risk management: optimal policies and asset prices. *Review of Financial Studies.* **14**(2), 371–405 (1999)

[3] Cuoco, D., He, H. and Isaenko, S. (2005) Optimal dynamic trading strategies with risk limits. *Working paper*

[4] Dowd, K. (1998) *Beyond Value at Risk: the New Science of Risk Management.* Wiley, London.

[5] Emmer, S., Klüppelberg, C. and Korn, T. (2001) Optimal portfolios with bounded Capital-at-Risk. *Math. Finance* **11**, 365–384.

[6] Gabih, A., Grecksch, W. and Wunderlich, R. (2005) Dynamic portfolio optimisation with bounded shortfall risks. *Stoch. Anal. Appl.* **23**, 579–594

[7] Jorion, P. (2001) *Value at Risk.* McGraw-Hill, New York.

[8] Karatzas, I. and Shreve, S.E. (1988) *Brownian Motion and Stochastic Calculus.* Springer, Berlin.

[9] Karatzas, I. and Shreve, S.E. (1998) *Methods of Mathematical Finance.* Springer, Berlin.

[10] Klüppelberg, C. and Pergamenshchikov, S. (2008) Optimal consumption and investment with bounded downside risk for power utility functions. Invited book contribution. Available at *www-m4.ma.tum.de/Papers/*

[11] Korn, R. (1997) *Optimal Portfolios.* World Scientific, Singapore.

[12] Yiu, K.F.C. (2004) Optimal portfolios under a value-at-risk constrain. *J. Econom. Dynam. Control* **28** (7), 1317–1334.

Author information

Claudia Klüppelberg, Center for Mathematical Sciences, Technische Universität München, 85747 Garching, Germany.
Email: ckl u@ma.tum.de

Serguei Pergamenshchikov, Laboratoire de Mathématiques Raphaël Salem, UMR 6085 CNRS, Université de Rouen, Avenue de l'Université, BP.12, 76801 Saint Etienne du Rouvray, France.
Email: serge.pergamenchtchikov@univ-rouen.fr

Radon Series Comp. Appl. Math **8**, 275–301

A review of some recent results on Malliavin Calculus and its applications

Kohatsu-Higa Arturo and Yasuda Kazuhiro

Abstract. We review some of the recent developments of Malliavin Calculus and its applications with some focus in Finance. In particular, we discuss the finite difference methods which lead in a generalised form to kernel density estimation methods. We compare this method in relation with the Malliavin Calculus method and in particular with the Malliavin–Thalmaier formula. We finish by giving a short review of other developments in the area.

Key words. Multidimensional density function, Malliavin calculus, the Malliavin–Thalmaier formula, greeks.

AMS classification. 60H07, 60H35, 60J60, 62G07, 65C05

1 Brief introduction to Malliavin Calculus

Let $(\Omega, \mathcal{F}, P; \mathcal{F}_t)$ be a filtered probability space. Here $\{\mathcal{F}_t\}$ satisfies the usual conditions. That is, it is right-continuous and $\mathcal{F}_0$ contains all the P-negligible events in $\mathcal{F}$. Suppose that H is a real separable Hilbert space whose norm and inner product are denoted by $\|\cdot\|_H$ and $\langle\cdot,\cdot\rangle_H$ respectively (in this article, we usually have $H = L^2([0,T],\mathbb{R}^d)$). Let $W(h)$ denote a Wiener process on H.

We denote by $C_p^\infty(\mathbb{R}^n)$ the set of all infinitely differentiable functions $f : \mathbb{R}^n \to \mathbb{R}$ such that f and all of its partial derivatives have at most polynomial growth.

Let $\mathcal{S}$ denote the class of *smooth* random variables of the form

$$F = f(W(h_1), \ldots, W(h_n)), \tag{1.1}$$

where $f \in C_p^\infty(\mathbb{R}^n)$, $h_1, \ldots, h_n \in H$, and $n \geq 1$.

If F has the form (1.1) we define its derivative DF as the H-valued random variable given by

$$DF = \sum_{i=1}^n \frac{\partial f}{\partial x_i}(W(h_1), \ldots, W(h_n))h_i.$$

We will denote the domain of D in $L^p(\Omega)$ by $\mathcal{D}^{1,p}$. This space is the closure of the class of smooth random variables $\mathcal{S}$ with respect to the norm

$$\|F\|_{1,p} = \left\{\mathbb{E}\left[|F|^p\right] + \mathbb{E}\left[\|DF\|_H^p\right]\right\}^{\frac{1}{p}}.$$

We can define the iteration of the operator D in such a way that for a smooth random variable F, the derivative D^kF is a random variable with values on $H^{\otimes k}$. Then for every $p \geq 1$ and $k \in \mathbb{N}$ we introduce a seminorm on $\mathcal{S}$ defined by

$$\|F\|_{k,p}^p = \mathbb{E}\big[|F|^p\big] + \sum_{j=1}^{k} \mathbb{E}\big[\|D^j F\|_{H^{\otimes j}}^p\big].$$

For any real $p \geq 1$ and any natural number $k \geq 0$, we will denote by $\mathcal{D}^{k,p}$ the completion of the family of smooth random variables $\mathcal{S}$ with respect to the norm $\|\cdot\|_{k,p}$. Note that $\mathcal{D}^{j,p} \subset \mathcal{D}^{k,q}$ if $j \geq k$ and $p \geq q$.

Consider the intersection

$$\mathcal{D}^\infty = \bigcap_{p\geq 1}\bigcap_{k\geq 1} \mathcal{D}^{k,p}.$$

Then $\mathcal{D}^\infty$ is a complete, countably normed, metric space.

We will denote by D^* *the adjoint of the operator* D as an unbounded operator from $L^2(\Omega)$ into $L^2(\Omega; H)$. That is, the domain of D^*, denoted by $\mathrm{Dom}(D^*)$, is the set of H-valued square integrable random variables u such that

$$|\mathbb{E}[\langle DF, u\rangle_H]| \leq c\|F\|_2,$$

for all $F \in \mathcal{D}^{1,2}$, where c is some constant depending on u (here $\|\cdot\|_2$ denotes the $L^2(\Omega)$-norm).

Suppose that $F = (F_1, \ldots, F_d)$ is a random vector whose components belong to the space $\mathcal{D}^{1,1}$. We associate with F the following random symmetric nonnegative definite matrix:

$$\gamma_F = \big(\langle DF_i, DF_j\rangle_H\big)_{1\leq i,j\leq d}.$$

This matrix is called the *Malliavin covariance matrix* of the random vector F.

Definition 1.1. We will say that the random vector $F = (F_1, \ldots, F_d) \in (\mathcal{D}^\infty)^d$ is nondegenerate if the matrix γ_F is invertible a.s. and

$$(\det \gamma_F)^{-1} \in \bigcap_{p\geq 1} L^p(\Omega). \tag{1.2}$$

In what follows, we always assume $G \in \mathcal{D}^\infty$, $F = (F_1, \ldots, F_d) \in (\mathcal{D}^\infty)^d$ is d-dimensional nondegenerate random variable. Therefore the integration by parts formulas will always hold (see Nualart [39], Proposition 2.1.4, p. 100 or Sanz [47], Proposition 5.4, p. 67 and formula (1.3) below). For other references, see [49].

1.1 Three methods to compute densities of random variables on Wiener space

1.1.1 The classical integration by parts formula

Let $F = (F_1, \ldots, F_d)$ be a nondegenerate random vector and G a smooth random variable. We denote by $p_{F,G} = \mathbb{E}\,[G/F = x]\, p_{F,1}(x)$, where $p_{F,1}(x) \equiv p_F(x)$ denotes the density of F. Then there exists a random variable $H_{(1,2,\ldots,d)}(F;1) \in L^p(\Omega)$ for any $p > 2$ such that

$$p_{F,G}(\hat{x}) = \mathbb{E}\left[\prod_{i=1}^{d} \mathbf{1}_{[0,\infty)}(F_i - \hat{x}_i) H_{(1,2,\ldots,d)}(F;G)\right], \tag{1.3}$$

where $\mathbf{1}_{[0,\infty)}(x)$ denotes the indicator function. In fact, for $i = 2, \ldots, d$,

$$H_{(1)}(F;1) := \sum_{j=1}^{d} \delta\Big(G(\gamma_F^{-1})^{1j} DF_j\Big),$$

$$H_{(1,\ldots,i)}(F;1) := \sum_{j=1}^{d} \delta\Big(H_{(1,\ldots,i-1)}(F;G)(\gamma_F^{-1})^{ij} DF_j\Big). \tag{1.4}$$

Here δ denotes the adjoint operator of the Malliavin derivative operator D and γ_F the Malliavin covariance matrix of F.

In particular, we remark that δ is an extension of the Itô integral that also integrates non-adapted processes and is usually called the Skorohod integral. The definition of $H_{(1,\ldots,i)}(F;1)$ in iterative form in (1.4) shows that in order to compute this expression one requires the calculation of i-iterated stochastic integrals.

1.1.2 The finite difference or kernel density method

The finite difference (FD) method consists in computing the approximate derivative of any distribution function in order to obtain the density function. This introduces the choice of a parameter in order to compute the approximate derivative. This is a particular case of the kernel density estimation method. In fact, this method requires the choice of a kernel function K and a sufficiently small $h > 0$ (usually called the bandwidth or the tuning parameter) which gives as an approximation

$$\frac{1}{N}\sum_{j=1}^{N} \frac{1}{h^d} K\left(\frac{F^j - \hat{x}}{h}\right) G^j \tag{1.5}$$

where (F^j, G^j), $j = 1, \ldots, N$ denotes N independent copies of (F, G) obtained by simulation.

First, we remark that the classical finite difference method is obtained with the choice $K(x) = 2^{-d}\prod_{i=1}^{d} \mathbf{1}_{[-1,1]^d}(x)$. The theory of kernel density estimation deals with the statistical problem of given some data (F^j, G^j), $j = 1, \ldots, N$ what is the "optimal" way of choosing the kernel K and the tunning parameter h. The theory of kernel density estimation is quite vast and we are not able to give a fair account of the theory but it seems that the multidimensional case $d > 1$ is less well understood than the one dimensional case.

In the multidimensional case, one may use multiplicative type of kernels. The order of the bias is of order h^2 if the kernel is symmetric and regular in some sense (say Gaussian type kernels). The variance diverges in the order of $\left(Nh^d\right)^{-1}$. For more information on this method, see e.g. [48] or [55].

1.1.3 Malliavin–Thalmaier representation of multidimensional density functions

We represent the delta function by

$$\delta_0(x) = \Delta Q_d(x) \quad (x \in \mathbb{R}^d,\ d \geq 2),$$

in the following sense. If f is a smooth function then the solution of the Poisson equation $\Delta u = f$ is given by the convolution $Q_d * f$ (see e.g. [20]).

Definition 1.2. Given the $\mathbb{R}^d$-valued random vector F and the $\mathbb{R}$-valued random variable G, a multi-index α and a power $p \geq 1$ we say that there is an integration by parts formula (IBP formula) in Malliavin sense if there exists a random variable $H_\alpha(F;G) \in L^p(\Omega)$ such that

$$IP_{\alpha,p}(F,G):\ \mathbb{E}\left[\frac{\partial^\alpha}{\partial x^\alpha} f(F)G\right] = \mathbb{E}\Big[f(F)H_\alpha(F;G)\Big] \quad \text{for all } f \in C_0^{|\alpha|}(\mathbb{R}^d). \tag{1.6}$$

Related to the Malliavin–Thalmaier formula [38], Bally and Caramellino [8], have obtained the following result.

Proposition 1.3 (Bally, Caramellino [8]). *Suppose that for some $p > 1$*

$$\sup_{|a|\leq R} \mathbb{E}\left[\left|\frac{\partial}{\partial x_i} Q_d(F-a)\right|^{\frac{p}{p-1}} + \Big|Q_d(F-a)\Big|^{\frac{p}{p-1}}\right] < \infty \quad \text{for all } R>0,\ a \in \mathbb{R}^d. \tag{1.7}$$

(i) *If $IP_{i,p}(F;G)$ $(i = 1,\dots,d)$ holds then the law of F is absolutely continuous with respect to the Lebesgue measure on $\mathbb{R}^d$ and the density p_F is represented as*

$$p_F(x) = \mathbb{E}\left[\sum_{i=1}^d \frac{\partial}{\partial x_i} Q_d(F-x)H_{(i)}(F;G)\right]. \tag{1.8}$$

(ii) *If $IP_{\alpha,p}(F;G)$ holds for every multi-index α with $|\alpha| \leq m+1$ then $p_F \in C^m(\mathbb{R}^d)$ and for every multi-index ρ with $|\rho| \leq m$ one has*

$$\frac{\partial^\rho}{\partial x^\rho} p_F(x) = \mathbb{E}\left[\sum_{i=1}^d \frac{\partial}{\partial x_i} Q_d(F-x)H_{(i,\rho)}(F;G)\right].$$

The heuristic idea of the above proof is to use the integration by parts formula in Malliavin sense as follows

$$\begin{aligned} p_F(x) = \mathbb{E}\Big[\Delta Q_d(F-x)G\Big] &= \sum_{i=1}^d \mathbb{E}\left[\frac{\partial^2}{\partial x_i^2} Q_d(F-x)G\right] \\ &= \mathbb{E}\left[\sum_{i=1}^d \frac{\partial}{\partial x_i} Q_d(F-x)H_{(i)}(F;G)\right]. \end{aligned}$$

Next we impose conditions to assure that the assumptions of Proposition 1.3 are satisfied.

Corollary 1.4. *If $G \in \mathcal{D}^\infty$, $F = (F_1, \dots, F_d) \in (\mathcal{D}^\infty)^d$ is a nondegenerate random vector, then the probability density function of the random vector F is*

$$p_F(x) = \mathbb{E}\left[\sum_{i=1}^{d} \frac{\partial}{\partial x_i} Q_d(F - x) H_{(i)}(F; G)\right].$$

1.1.4 Theoretical comparison of the methods

The method of kernel density estimation is the oldest method of the three methods introduced above and the one that has been used by practitioners for a long time. The method is easy to implement and various standard recommendations are available on the choice of kernels K and the tuning parameter h.

The classical integration by parts formula (1.3) attracted the attention of practitioners as it allows in principle the calculation of density functions using Monte Carlo simulations without any bias (see e.g. [22] and [23]).

In comparison with kernel density estimation methods this method does not require any tunning as there is no bias. In exchange, the estimator obtained by integration by parts involve in general d iterated stochastic integrals and its calculation is not available for all models. Furthermore, the estimator obtained by integration by parts has a constant variance which tends to be big and one needs to use variance reduction methods.

In the one dimensional case a thorough comparison between the classical integration by parts formula and the kernel density estimation method can be found in [29]. When the dimension is bigger than one, one can try to compute the d-iterated Skorohod integrals but this becomes cumbersome as d increases. Furthermore as stochastic integrals have to be approximated by their Riemann sum counterparts the error increases. Nevertheless one can still write the system of linear equations satisfied by the higher order derivatives and try to use this structure in order to improve the system simulation (see e.g. [17]).

Another alternative that is in between the classical integration by parts and the kernel density estimation method is the Malliavin–Thalmaier formula (1.8).

The significance of the Malliavin–Thalmaier formula is clear. Instead of the d-iterated stochastic integrals which appear in (1.3) we have instead only one stochastic integral. The problem with the above formula is that the expectation is well defined in the sense of duality. That is, $\frac{\partial}{\partial x_i} Q_d(F - x) \in L^p(\Omega)$ for any $p < 2$ and $H_{(i)}(F; G) \in L^q(\Omega)$ for $p^{-1} + q^{-1} = 1$. Therefore the variance of the Malliavin–Thalmaier estimator is infinite.

In fact, we have for some constant A_d that

$$\frac{\partial}{\partial x_i} Q_d(x) := A_d \frac{x_i}{|x|^d}.$$

Therefore, we have to resort again to kernel density estimation methods. We will see later that the order of degeneration of these estimators is milder in comparison with estimators of the type (1.5).

2 Error estimation for the Malliavin–Thalmaier formula

In order to avoid the explosion of the variance of the Malliavin–Thalmaier estimator, we have proposed the use of a kernel density type alternative to this estimator, using instead of Q, we define

$$\frac{\partial}{\partial x_i} Q_d^h(x) := A_d \frac{x_i}{|x|_h^d},$$

where $|\cdot|_h$ is defined as

$$|x|_h := \sqrt{\sum_{i=1}^d x_i^2 + h} \quad (h > 0,\ x \in \mathbb{R}^d).$$

Then we define the approximation to the density function of F as;

$$p_{F,G}^h(x) := \mathbb{E}\left[\sum_{i=1}^d \frac{\partial}{\partial x_i} Q_d^h(F - x) H_{(i)}(F; G)\right]. \tag{2.1}$$

Note that clearly, $Q_d = Q_d^0$. We now give the Central Limit Theorem associated with the proposed approximation.

Theorem 2.1. *Let Z be a random variable with standard normal distribution and let $(F^{(j)}, G^{(j)}) \in (\mathcal{D}^\infty)^d \times \mathcal{D}^\infty$, $j \in \mathbb{N}$ be a sequence of independent identically distributed random vectors.*

(i) When $d = 2$, set $n = \left\lfloor \frac{C}{h \ln \frac{1}{h}} \right\rfloor$ and $N = \left\lfloor \frac{C^2}{h^2 \ln \frac{1}{h}} \right\rfloor$ for some positive constant C fixed throughout. Then as $h \to 0$

$$n\left(\frac{1}{N}\sum_{j=1}^N \sum_{i=1}^2 \frac{\partial}{\partial x_i} Q_2^h(F^{(j)} - \hat{x}) H_{(i)}(F; G)^{(j)} - p_{F,G}(\hat{x})\right) \Longrightarrow \sqrt{C_3^{\hat{x}}} Z - C_1^{\hat{x}} C, \tag{2.2}$$

where $H_{(i)}(F; G)^{(j)}$, $i = 1, \ldots, d$, $j = 1, \ldots, N$, denotes the weight obtained in the j-th independent simulation (the same that generates $F^{(j)}$ and $G^{(j)}$).

(ii) When $d \geq 3$, set $n = \left\lfloor \frac{C}{h \ln \frac{1}{h}} \right\rfloor$ and $N = \left\lfloor \frac{C^2}{h^{\frac{d}{2}+1} (\ln \frac{1}{h})^2} \right\rfloor$ for some positive constant C fixed throughout. Then as $h \to 0$

$$n\left(\frac{1}{N}\sum_{j=1}^N \sum_{i=1}^d \frac{\partial}{\partial x_i} Q_d^h(F^{(j)} - \hat{x}) H_{(i)}(F; G)^{(j)} - p_{F,G}(\hat{x})\right) \Longrightarrow \sqrt{C_4^{\hat{x}}} Z - C_1^{\hat{x}} C. \tag{2.3}$$

This result clearly also gives the asymptotic bias and variance of the estimators. In fact the bias is of the order

$$p_{F,G}(\hat{x}) - p^h_{F,G}(\hat{x}) = C_1^{\hat{x}} h \ln \frac{1}{h} + C_2^{\hat{x}} h + o(h), \tag{2.4}$$

Note that this bias is almost of the same order as in the kernel density estimation method. The asymptotic $L^2(\Omega)$-error is of the order for $d = 2$,

$$\mathbb{E}\left[\left(\sum_{i=1}^{2} \frac{\partial}{\partial x_i} Q_2^h(F - \hat{x}) H_{(i)}(F;G) - p_{F,G}(\hat{x})\right)^2\right] = C_3^{\hat{x}} \ln \frac{1}{h} + O(1) \quad (\hat{x} \in \mathbb{R}^d),$$

and for $d \geq 3$,

$$\mathbb{E}\left[\left(\sum_{i=1}^{d} \frac{\partial}{\partial x_i} Q_d^h(F - \hat{x}) H_{(i)}(F;G) - p_{F,G}(\hat{x})\right)^2\right]$$
$$= C_4^{\hat{x}} \frac{1}{h^{\frac{d}{2}-1}} + o\left(\frac{1}{h^{\frac{d}{2}-1}}\right) \quad (\hat{x} \in \mathbb{R}^d).$$

All the above constants $C_i^{\hat{x}}$ have explicit expressions that depend on the density itself. Note that the order of explosion of the $L^2(\Omega)$-error is reduced in comparison with the classical kernel density estimation methods.

3 Financial application

When computing the greek of any option, the instability of the calculation comes from the irregularities of the payoff function. In Fournié et al. it was shown how to deal with the problem. One essentially divides the payoff function in two parts

$$F = F_1^h + F_2^h$$

The first function F_1^h is a smooth function which depends on a smoothing parameter h and the second localises the irregularity of the payoff. In the second we apply the previous integration by parts formula and in the first one uses a direct simulation method. The question on the choice of the parameter h remains although in Fournié et al. the authors seem to suggest that this is not an important issue. Nevertheless note that this is also a tunning problem. In financial applications, one could use the classical integration by parts as follows.

$$\begin{aligned}
\frac{\partial}{\partial \mu}\mathbb{E}[f(F^\mu)] &= \int_{\mathbb{R}^d} f(x) \frac{\partial}{\partial \mu} p_F(\mu, x) dx \\
&= E\left[f(F^\mu) \sum_{j=1}^{d} H_{(j)}\left(F^\mu, \frac{\partial}{\partial \mu} F^{\mu,j}\right)\right].
\end{aligned}$$

The classical application to Greek calculations is for the case when f involves a step function.

Another possibility hinted by the Malliavin–Thalmaier formula is

$$f(x) = \int f(y) \sum_{i=1}^{d} \frac{\partial^2 Q_d}{\partial x_i^2}(y-x)dy$$

Therefore one can use any of the following alternative expressions (under certain regularity conditions, for details see [32])

$$\begin{aligned} \frac{\partial}{\partial \mu}\mathbb{E}[f(F^\mu)] &= \sum_{j=1}^{d} \mathbb{E}\left[f(F^\mu) H_{(j)}\left(F^\mu, \frac{\partial F^{\mu,j}}{\partial \mu}\right)\right] \\ &= \sum_{i,j=1}^{d} \mathbb{E}\left[\int f(y) \frac{\partial^2 Q_d}{\partial x_i \partial x_j}(y - F^\mu) dy H_{(i)}\left(F^\mu, \frac{\partial F^{\mu,j}}{\partial \mu}\right)\right]. \end{aligned} \tag{3.1}$$

In some cases, the above representation for the Greeks gives a variance reduction effect. In fact, if we consider Delta of a digital put option with two assets;

$$\frac{\partial}{\partial S_0^1} e^{-rT} E^Q \left[\mathbf{1}\left(0 \le S_T^1 \le K_1\right) \mathbf{1}\left(0 \le S_T^2 \le K_2\right)\right],$$

then a method by Fournié et al. [22] without localisation gives the following expression;

$$e^{-rT} E^Q \left[\mathbf{1}\left(0 \le S_T^1 \le K_1\right) \mathbf{1}\left(0 \le S_T^2 \le K_2\right) H_{(1)}\left(S_T^1, S_T^2; \frac{\partial S_T^1}{\partial S_0^1}\right)\right]. \tag{3.2}$$

On the other hand, an expression of this new method gives as follows;

$$\begin{aligned} e^{-rT} \Bigg\{ & E^Q \left[g_1\left(S_T^1, S_T^2\right) H_{(1)}\left(S_T^1, S_T^2; \frac{\partial S_T^1}{\partial S_0^1}\right)\right] \\ & + E^Q \left[g_2\left(S_T^1, S_T^2\right) H_{(2)}\left(S_T^1, S_T^2; \frac{\partial S_T^1}{\partial S_0^1}\right)\right] \Bigg\}, \end{aligned} \tag{3.3}$$

where we can explicitly calculate the integral parts of (3.1) to obtain that

$$\begin{aligned} g_1(x,y) &:= \frac{1}{2\pi}\left\{ \arctan\frac{y}{x} - \arctan\frac{y-K_2}{x} - \arctan\frac{y}{x-K_1} + \arctan\frac{y-K_2}{x-K_1}\right\}, \\ g_2(x,y) &:= \frac{1}{4\pi} \ln \frac{(x^2+y^2)((x-K_1)^2+(y-K_2)^2)}{((x-K_1)^2+y^2)(x^2+(y-K_2)^2)}. \end{aligned}$$

If we assume that the assets follow the Black–Scholes model, then (3.3) gives a variance reduction effect, compared with (3.2). In Kohatsu–Higa, Yasuda [31], we can find the simulation results where we conclude that the variance of (3.3) is about a third of variance of (3.2). This issue needs to be further studied.

4 Estimation of the optimal value of h

4.1 About optimal h

In this section, we give an ad-hoc method to compute a quasi-optimal value of h using similar ideas as in kernel density estimation and the central limit theorem obtained in Theorem 2.1. We consider the $L^2(\Omega)$ error of approximation;

$$\mathbb{E}\left[\left\{\frac{1}{N}\sum_{j=1}^{N}\left(\sum_{i=1}^{d}\frac{\partial}{\partial x_i}Q_d^h\left(F^{(j)}-\hat{x}\right)H_{(i)}(F;1)^{(j)}\right)-p_F\left(\hat{x}\right)\right\}^2\right]. \tag{4.1}$$

From Theorem 2.1 and the comments following it, we have for $d=2$,

$$\begin{aligned}(4.1) =& \frac{1}{N}\left\{C_3^{\hat{x}}\ln\frac{1}{h}+O(1)\right\}+\left(1-\frac{1}{N}\right)\left\{C_1^{\hat{x}}h\ln\frac{1}{h}+C_2^{\hat{x}}h+o(h)\right\}^2\\ &+\frac{2}{N}\left\{C_1^{\hat{x}}h\ln\frac{1}{h}+C_2^{\hat{x}}h+o(h)\right\}p_F\left(\hat{x}\right)-\frac{1}{N}p_F\left(\hat{x}\right)^2,\end{aligned}$$

and if $d\geq 3$,

$$\begin{aligned}(4.1) =& \frac{1}{N}\left\{C_4^{\hat{x}}\frac{1}{h^{\frac{d}{2}-1}}+o\left(\frac{1}{h^{\frac{d}{2}-1}}\right)\right\}+\left(1-\frac{1}{N}\right)\left\{C_1^{\hat{x}}h\ln\frac{1}{h}+C_2^{\hat{x}}h+o(h)\right\}^2\\ &+\frac{2}{N}\left\{C_1^{\hat{x}}h\ln\frac{1}{h}+C_2^{\hat{x}}h+o(h)\right\}p_F\left(\hat{x}\right)-\frac{1}{N}p_F\left(\hat{x}\right)^2.\end{aligned}$$

Then we select the leading terms from the above equations to find a trade-off relation between the small bias and the exploding L^2-error;

$$g(h):=\begin{cases}\dfrac{1}{N}C_3^{\hat{x}}\ln\dfrac{1}{h}+\left(C_1^{\hat{x}}\right)^2h^2, & d=2,\\ \dfrac{1}{Nh^{\frac{d}{2}-1}}C_4^{\hat{x}}+\left(C_1^{\hat{x}}\right)^2h^2, & d\geq 3.\end{cases}$$

Note that the intervention of the sample size becomes crucial in the above equation: the right choice of N will make the variance of the estimator converge to 0.

By considering the minimum value of $g(h)$, finally we obtain the following asymptotic optimal value for h;

$$h=\begin{cases}\sqrt{\dfrac{C_3^{\hat{x}}}{2N(C_1^{\hat{x}})^2}}, & d=2,\\ \left\{\dfrac{d-2}{4N}\dfrac{C_4^{\hat{x}}}{(C_1^{\hat{x}})^2}\right\}^{\frac{2}{2+d}}, & d\geq 3.\end{cases} \tag{4.2}$$

The problem with the above theoretical formula is that it requires the knowledge of the constants $C_i^{\hat{x}}$.

4.2 Calculation of constants C_1^x, C_3^x and C_4^x

Here we give a heuristic idea on how to obtain the constants $C_i^{\hat{x}}$ for $i = 1, 3, 4$ using pilot simulation. From our CLT result, we have

$$n\left(\frac{1}{N}\sum_{j=1}^{N}\sum_{i=1}^{d}\frac{\partial}{\partial x_i}Q_d^h\left(F^{(j)}-\hat{x}\right)H_{(i)}\left(F;1\right)^{(j)}-p_F\left(\hat{x}\right)\right) \Longrightarrow \sqrt{C_a^{\hat{x}}}Z - C_1^{\hat{x}}C, \tag{4.3}$$

where $C_a^{\hat{x}} = C_3^{\hat{x}}$ if $d = 2$ and $C_a^{\hat{x}} = C_4^{\hat{x}}$ if $d \geq 3$. Let $Y_{\hat{x}}^{h,N}$ be the left hand side of (4.3). Therefore we consider the following approximation

$$Y_{\hat{x}}^{h,N} \approx \sqrt{C_a^{\hat{x}}}Z - C_1^{\hat{x}}C.$$

Then using that Z follows a standard Normal distribution, we have the following approximations;

$$\mathbb{E}\left[Y_{\hat{x}}^{h,N}\right] \approx \mathbb{E}\left[\sqrt{C_a^{\hat{x}}}Z - C_1^{\hat{x}}C\right] = -C_1^{\hat{x}}C, \tag{4.4}$$

$$\mathbb{E}\left[\left(Y_{\hat{x}}^{h,N}\right)^2\right] \approx \mathbb{E}\left[\left(\sqrt{C_a^{\hat{x}}}Z - C_1^{\hat{x}}C\right)^2\right] = C_a^{\hat{x}} + \left(C_1^{\hat{x}}C\right)^2. \tag{4.5}$$

The computation of constants is done by first fixing the values of h and N in test simulations, this gives the value of the constant C and n according to the relation in the CLT (Theorem 2.1).

We use these test Monte–Carlo simulations in order to approach the mean and the variance in (4.4) and (4.5). In practice, one obtains a stable result for $C_a^{\hat{x}}$, but the result of $C_1^{\hat{x}}$ is unstable if one uses all the choices of h and N in the pilot simulations. This is due to the fact that when the value of C becomes too small then the above procedure is not good to obtain the value of $C_1^{\hat{x}}$ as the error terms become bigger than the quantity to be estimated. To obtain a stable estimation method, besides deleting (or avoiding) the simulations with small values of C, we additionally consider the following approximating procedure for $C_1^{\hat{x}}$.

$$\frac{1}{M}\sum_{k=1}^{M}Y_{\hat{x},(k)}^{h,N} \approx \sqrt{C_a^{\hat{x}}}\frac{1}{\sqrt{M}}\tilde{Z} - C_1^{\hat{x}}C^{h,N},$$

where let $\tilde{Z}$ be a random variable with the standard normal distribution. Now if we try this test simulation L times using different values of h, then we have

$$\frac{1}{L}\sum_{l=1}^{L}\left(\frac{1}{M}\sum_{k=1}^{M}Y_{\hat{x},(k)}^{h(l),N}\right) \approx -\frac{1}{L}\sum_{l=1}^{L}C_1^{\hat{x}}C^{h(l),N} = -C_1^{\hat{x}}\frac{1}{L}\sum_{l=1}^{L}C^{h(l),N}.$$

Therefore we obtain $C_1^{\hat{x}}$ as follow;

$$C_1^{\hat{x}} \approx -\frac{\sum_{l=1}^{L}\left(\frac{1}{M}\sum_{k=1}^{M}Y_{\hat{x},(k)}^{h(l),N}\right)}{\sum_{l=1}^{L}C^{h(l),N}}.$$

Remark 4.1. Once we obtain the constant $C_1^{\hat{x}}$, we can modify the approximation as follows;

$$\dot{p}_F^h(\hat{x}) = p_F^h(\hat{x}) + C_1^{\hat{x}} h \ln \frac{1}{h}.$$

Then from the bias error (2.4), we can improve the bias of the error;

$$p_F(\hat{x}) - \dot{p}_F^h(\hat{x}) = p_F(\hat{x}) - \left(p_F^h(\hat{x}) + C_1^{\hat{x}} h \ln \frac{1}{h} \right) = C_2^{\hat{x}} h + o(h).$$

5 Numerical results

In this section, we give a short report on some simulation results on the following models: the multidimensional Black–Scholes model and two factor models in finance: the Heston model [24] and the double volatility Heston model [21], [25].

5.1 The multidimensional Black–Scholes model

We consider the following d-dimensional Black–Scholes model; for $i = 1, \dots, d$,

$$\frac{dS_t^i}{S_t^i} = \mu_i dt + \sum_{j=1}^d \sigma_j^i dW_t^j, \quad S_0^i = s^i,$$

where μ_i and σ_j^i, $i, j = 1, \dots, d$ are constants, s^i, $i = 1, \dots, d$ is a positive constant, and $W = \{W_t = (W_t^1, \dots, W_t^d)\}_{t \geq 0}$ is a d-dimensional standard Brownian motion, whose components are independent of each other. As it is well known, the joint density of the random vector $S_T = (S_T^1, \dots, S_T^d)$ is the lognormal density which can be written explicitly. and $\Sigma^{-1} = (\sigma_{ij}^{-1})_{i,j=1,\dots,d}$ is the inverse matrix of Σ.

We can also represent the density $p_{S_T}(x)$ through the Malliavin–Thalmaier formula.

Lemma 5.1. *Let $F = S_T$ be a nondegenerate random vector. Then the density p_{S_T} can be expressed as*

$$p_{S_T}(x) = A_d \sum_{i=1}^d \mathbb{E}\left[\frac{S_T^i - x_i}{|S_T - x|^d} \sum_{j=1}^d (-1)^{i+j} \frac{\det(\Sigma_i^j)}{\det(\Sigma)} \left\{ \frac{W_T^j}{S_T^i} + \frac{\sigma_j^i T}{S_T^i} \right\} \right], \tag{5.1}$$

for $x \in \mathbb{R}^d$, where Σ_i^j, $i, j = 1, \dots, d$, is a $(d-1) \times (d-1)$-matrix obtained from Σ by deleting row j and column i.

For more details on the above lemma, see Kohatsu–Higa and Yasuda [32].

Hence we have the following approximation of (5.1); for $x \in \mathbb{R}^d$,

$$p_{S_T}^h(x) := A_d \sum_{i=1}^d \mathbb{E}\left[\frac{S_T^i - x_i}{|S_T - x|_h^d} \sum_{j=1}^d (-1)^{i+j} \frac{\det(\Sigma_i^j)}{\det(\Sigma)} \left\{ \frac{W_T^j}{S_T^i} + \frac{\sigma_j^i T}{S_T^i} \right\} \right]. \tag{5.2}$$

Now we provide a short summary of results in case $d = 2$. The simulation result through the classical representation is unstable and does not work well (unless variance reduction methods are applied (e.g. see [30], [28] and [12]), because of the appearance of a double-Skorohod integral for the Malliavin weight. Compared with the classical method, the Malliavin–Thalmaier formula (5.1) works better since it does not involve double Skorohod integral. But the density approximation exhibited unexpected peaks, which are due to the unstable behaviour of $\frac{\partial}{\partial x_i} Q_d$. This instability also appears when the density estimation is magnified locally. To improve these instability, we use the approximation formula (5.2). In fact, this approximation although slightly biased in comparison with the Malliavin–Thalmaier formula (5.1), behaves smoothly. For more details and graphs, see Kohatsu–Higa and Yasuda [33].

5.2 Heston model

In this section, we provide some simulation results for the Heston model [24];

$$\begin{aligned} dS_t &= \mu S_t dt + \sqrt{v_t} S_t dW_t^{(1)}, \\ dv_t &= -\gamma(v_t - \theta)dt + \kappa\sqrt{v_t} dW_t^{(2)}, \end{aligned}$$

where $\mu,\ \gamma,\ \theta,\ \kappa$ are constants with $\gamma\theta \geq \frac{\kappa^2}{2}$ (see Lamberton, Lapeyre [35]) and $W_t^{(1)},\ W_t^{(2)}$ are standard Brownian motions with $\mathbb{E}[W_t^{(1)} W_t^{(2)}] = \rho t$.

We introduce a new standard Brownian motion Z, which is independent of $W_t^{(2)}$ and $W_t^{(1)} = \rho W_t^{(2)} + \sqrt{1-\rho^2} Z_t$. We also change variables. Set $X_t := \ln(S_t/S_0) - \mu t$ and $u_t := a v_t$. Then from Itô's formula, we have the following dynamics;

$$\begin{aligned} dX_t &= -\frac{u_t}{2a} dt + \sqrt{\frac{u_t}{a}} \left\{ \rho dW_t^{(2)} + \sqrt{1-\rho^2} dZ_t \right\}, \\ du_t &= -\gamma(u_t - a\theta)dt + \sqrt{a}\kappa\sqrt{u_t} dW_t^{(2)}. \end{aligned} \tag{5.3}$$

As the exact value of the joint density value of (X_t, u_t) is unknown, we estimate the value by using the Malliavin–Thalmaier formula, the approximated version and the finite difference algorithm applied to the Kolmogorov equation.

Set $F := (F_1, F_2) := (X_t, u_t)$ for fixed $t > 0$. First we give the Malliavin–Thalmaier formula for this model. For $x = (x_1, x_2) \in \mathbb{R}^2$, we have

$$p_F(x) = \frac{1}{2\pi} \mathbb{E}\left[\sum_{i=1}^{2} \frac{F_i - x_i}{|F - x|^2} H_{(i)}(F; 1) \right], \tag{5.4}$$

where

$$H_{(1)}(F;1) := \frac{\sqrt{a}}{\sqrt{1-\rho^2}t}\int_0^t \frac{1}{\sqrt{u_s}}dZ_s,$$

$$H_{(2)}(F;1) := \frac{1}{t}\left\{A - B\right\},$$

$$A := \frac{1}{\sqrt{a}\kappa e(t)}\int_0^t \frac{e(s)}{\sqrt{u_s}}dW_s^{(2)} + \frac{1}{2e(t)}\int_0^t \frac{e(s)}{u_s}ds$$
$$+ \frac{a\kappa^2}{8e(t)}\int_0^t s\frac{e(s)}{u_s^2}ds - \frac{\sqrt{a}\kappa}{4e(t)}\int_0^t s\frac{e(s)}{u_s^{\frac{3}{2}}}dW_s^{(2)},$$

$$B := \frac{\rho}{\kappa\sqrt{a(1-\rho^2)}e(t)}\int_0^t \frac{e(s)}{\sqrt{u_s}}dZ_s - \frac{1}{2\sqrt{a(1-\rho^2)}e(t)}\int_0^t e(r)\int_0^r \frac{1}{\sqrt{u_s}}dZ_s dr$$
$$+ \frac{\rho}{2\sqrt{1-\rho^2}e(t)}\int_0^t \frac{e(r)}{\sqrt{u_r}}\int_0^r \frac{1}{\sqrt{u_s}}dZ_s dW_r^{(2)} + \frac{1}{2e(t)}\int_0^t \frac{e(r)}{\sqrt{u_r}}\int_0^r \frac{1}{\sqrt{u_s}}dZ_s dZ_r,$$

$$e(t) := \exp\left(-\gamma t - \frac{a\kappa^2}{8}\int_0^t \frac{1}{u_r}dr + \frac{\sqrt{a}\kappa}{2}\int_0^t \frac{1}{\sqrt{u_r}}dW_r^{(2)}\right).$$

And our approximation is given as follows;

$$p_F^h(x) = \frac{1}{2\pi}\mathbb{E}\left[\sum_{i=1}^2 \frac{F_i - x_i}{|F-x|_h^2}H_{(i)}(F;1)\right]. \tag{5.5}$$

All the stochastic integrals appearing in the above formulas are approximated using the corresponding Riemann sums. This obviously introduces a further error of approximation in the above formulas. We will compare the above approximation values with the following deterministic method.

Finite difference method applied to the associated Kolmogorov equation

Next we give the corresponding forward Kolmogorov equation of the model (5.3);

$$\frac{\partial p_t}{\partial t} = \gamma p_t + \gamma(u - a\theta)\frac{\partial p_t}{\partial u} + \frac{u}{2a}\frac{\partial p_t}{\partial x} + \rho\kappa u\frac{\partial^2 p_t}{\partial x \partial u}$$
$$+ \rho\kappa\frac{\partial p_t}{\partial x} + \frac{u}{2a}\frac{\partial^2 p_t}{\partial x^2} + \frac{a\kappa^2 u}{2}\frac{\partial^2 p_t}{\partial u^2} + a\kappa^2\frac{\partial p_t}{\partial u}. \tag{5.6}$$

The initial condition is the Dirac delta function;

$$p_0(x,u) = \delta_0(x)\delta_0(u - u_0).$$

When we compute the approximative solution to equation (5.6), we use the following explicit scheme;

$$\begin{aligned}\frac{P^{k+1}_{i,j}-P^k_{i,j}}{\Delta t} &= \gamma P^k_{i,j} + \left(\frac{u_j}{2a}+\rho\kappa\right)\frac{P^k_{i+1,j}-P^k_{i-1,j}}{2\Delta x}\\ &\quad + \left(\gamma(u_j-a\theta)+a\kappa^2\right)\frac{P^k_{i,j+1}-P^k_{i,j-1}}{2\Delta u}\\ &\quad + \frac{u_j}{2a}\frac{P^k_{i+1,j}-2P^k_{i,j}+P^k_{i-1,j}}{(\Delta x)^2}\\ &\quad + \rho\kappa u_j\frac{P^k_{i,j+1}+P^k_{i-1,j}-P^k_{i-1,j+1}-P^k_{i,j}}{\Delta x\Delta u}\\ &\quad + \frac{a\kappa^2 u_j}{2}\frac{P^k_{i,j+1}-2P^k_{i,j}+P^k_{i,j-1}}{(\Delta u)^2}, \end{aligned} \tag{5.7}$$

where $P^k_{i,j} := p_{t_k}(x_i, u_j|u_0)$ and $\Delta t, \Delta x, \Delta u > 0$. In order to achieve a stable simulation (positivity of the density) in the negative correlation case, we use the forward difference method w.r.t. x and the backward difference method w.r.t. u for the term $\frac{\partial^2 p_t}{\partial x \partial u}$ [1]. The stability property also requires some particular relation between the parameters, that is, assume that (i). $\Delta x = \Delta u$ is small enough, (ii). $\frac{(\Delta x)^2}{u_j(1+a\kappa^2+\rho\kappa)} \geq \Delta t$ under a restriction $c_1 \leq u_j \leq c_2$, (iii). $\frac{1}{2a} \geq -\rho\kappa$, (iv). $a\kappa \geq -2\rho$. [2]

Kernel density estimation method

We compare the density value to the kernel density method. Here we use the Gaussian kernel and all bandwidth sizes are the same. That is, for $F := (F_1, \ldots, F_d)$ and $x = (x_1, \ldots, x_d) \in \mathbb{R}^d$,

$$p_F(x) \approx \frac{1}{N}\sum_{j=1}^N \frac{1}{h^d}\prod_{i=1}^d \frac{1}{\sqrt{2\pi}}\exp\left(-\frac{(F_i^{(j)}-x_i)^2}{2h^2}\right), \tag{5.8}$$

where $F_i^{(j)}$, $i = 1, \ldots, d$, $j = 1, \ldots, N$ is a sequence of r.v.'s, copies of F_i.

To use (5.8), we have to decide how to choose the bandwidth size. To introduce this optimal choice and the calculations of constants as in Section 4, we consider the general case of KDE. Let $K : \mathbb{R} \to \mathbb{R}_+$ be a function with $\int_{\mathbb{R}} x^a K(x)dx = 0$ for $a = 1, 3$. And for $x = (x_1, \ldots, x_d) \in \mathbb{R}^d$, set

$$p^h_{KDE}(x) := \mathbb{E}\left[\frac{1}{h^d}\prod_{i=1}^d K\left(\frac{F_i - x_i}{h}\right)\right].$$

Then we have the following central limit theorem for kernel density estimations.

[1] When the correlation is positive, we have to use another approximation to achieve stability.

[2] If we use other approximation or consider a case $\rho \geq 0$, these relations vary.

Proposition 5.2. *Set* $h = (\frac{C^2}{N})^{\frac{1}{d+4}}$ *and* $n = \lfloor \frac{C}{h^2} \rfloor$, *where* C *is a positive constant. Let* Z *be a random variable with standard normal distribution. Then we have*

$$n\left(\frac{1}{N}\sum_{j=1}^{N}\frac{1}{h^d}\prod_{i=1}^{d}K\left(\frac{F_i^{(j)}-x_i}{h}\right)-p_F(x)\right)\Longrightarrow\sqrt{\dot{C}_2^x}Z-\dot{C}_1^x C,$$

where $F_i^{(j)}$, $i = 1, \ldots, d$, $j = 1, 2, \ldots$ *are an i.i.d. random variable of* F_i.

In fact, from Scott [48], we have that the bias error is

$$p_F(x) - p_{KDE}^h(x) = -h^2\frac{1}{2}\int_{\mathbb{R}} z^2K(z)dz\sum_{i=1}^{d}\frac{\partial^2 p_F(x)}{\partial x_i^2} + O\left(h^4\right) =: \dot{C}_1^x h^2 + O\left(h^4\right).$$

And also we obtain the L^2-error;

$$\begin{aligned}
&\mathbb{E}\left[\left\{\frac{1}{h^d}\prod_{i=1}^{d}K\left(\frac{F_i-x_i}{h}\right)-p_{KDE}^h(x)\right\}^2\right]\\
&= \frac{1}{h^d}p_F(x)\prod_{i=1}^{d}\int_{\mathbb{R}}K(z_i)^2dz_i + O\left(\frac{1}{h^{d-1}}\right)\\
&=: \dot{C}_2^x\frac{1}{h^d} + O\left(\frac{1}{h^{d-1}}\right).
\end{aligned}$$

Finally, we obtain an optimal bandwidth size from a calculation like in Section 4. Then we obtain the following asymptotic optimal size of the bandwidth

$$h = \left(\frac{d\dot{C}_2^x}{4N(\dot{C}_1^x)^2}\right)^{\frac{1}{d+4}}. \tag{5.9}$$

And we can calculate the constants $\dot{C}_1^x, \dot{C}_2^x$ through a pilot simulation as explained in Section 4.2 and following Proposition 5.2.

Using a KDE method on the Laplacian of the Poisson kernel

We also can estimate the density function through the Laplacian of the Poisson kernel. That is, for $\hat{x} \in \mathbb{R}^d$,

$$p_F(\hat{x}) = \mathbb{E}\left[\delta_0(F-\hat{x})\right] = \mathbb{E}\left[\sum_{i=1}^{d}\frac{\partial^2}{\partial x_i^2}Q(F-\hat{x})\right]. \tag{5.10}$$

If we simulate (5.10) directly, it is clear that the simulation will return either zero or an error.

Therefore we introduce the following approximation of (5.10); for $h > 0$,

$$p_{Poi}^h(\hat{x}) := \mathbb{E}\left[\sum_{i=1}^{d}\frac{\partial^2}{\partial x_i^2}Q_d^h(F-\hat{x})\right]. \tag{5.11}$$

We give a central limit theorem for (5.11).

Proposition 5.3. *Set* $N = \left\lfloor \frac{C^2}{h^{\frac{d+4}{2}} (\ln \frac{1}{h})^2} \right\rfloor$ *and* $n = \left\lfloor \frac{C}{h \ln \frac{1}{h}} \right\rfloor$, *where* C *is a positive constant. Let* $F^{(j)}$, $j \in \mathbb{N}$ *be an i.i.d. random variable of* F *and* Z *be a random variable with the standard normal distribution. Then as* $h \to 0$, *we have*

$$n\left(\frac{1}{N}\sum_{j=1}^{N}\sum_{i=1}^{d}\frac{\partial^2}{\partial x_i^2}Q_d^h\left(F^{(j)}-\hat{x}\right)-p_F\left(\hat{x}\right)\right) \Longrightarrow \sqrt{\hat{C}_3^{\hat{x}}}Z-\hat{C}_1^{\hat{x}}C.$$

The proof uses the following error estimations. First, the bias error is

$$p_F\left(\hat{x}\right)-p_{Poi}^h\left(\hat{x}\right)=\hat{C}_1^{\hat{x}}h\ln\frac{1}{h}+\hat{C}_2^{\hat{x}}h+o(h),$$

where $\hat{C}_1^{\hat{x}}$ and $\hat{C}_2^{\hat{x}}$ are some constants defined as $C_1^{\hat{x}}$ and $C_2^{\hat{x}}$ in the Malliavin–Thalmaier formula respectively. Next, the L^2-error;

$$\mathbb{E}\left[\left\{\sum_{i=1}^{d}\frac{\partial^2}{\partial x_i^2}Q_d^h\left(F-\hat{x}\right)-p_F\left(\hat{x}\right)\right\}^2\right]=\frac{1}{h^{\frac{d}{2}}}\hat{C}_3^{\hat{x}}+o\left(\frac{1}{h^{\frac{d}{2}}}\right),$$

where $\hat{C}_3^{\hat{x}}$ is some positive constant.

As before, we obtain the following optimal bandwidth

$$h=\left(\frac{d\hat{C}_3^{\hat{x}}}{4N\left(\hat{C}_1^{\hat{x}}\right)^2}\right)^{\frac{2}{d+4}}. \tag{5.12}$$

And we can calculate the constants $\hat{C}_1^{\hat{x}}, \hat{C}_3^{\hat{x}}$ through a pilot simulation as Section 4.2 and Proposition 5.3.

Numerical results

Now we give a survey of the simulation results on the model (5.3). We use the following parameters;

parameter		value
initial log stock price	S_0	100
(initial volatility)^2	v_0	0.1
scale parameter	a	3
expected return	μ	0.1
speed of mean reversion	γ	2
long term mean	θ	0.1
volatility of volatility process	κ	0.2
maturity	t	1

We estimate the density value at $(x, u) = (0, 0.3)$ (the initial point). We simulate two cases, the correlation $\rho = -0.1,\ -0.8$, through five methods, the Malliavin–Thalmaier formula, the approximated Malliavin–Thalmaier formula, the finite difference method applied to the Kolmogorov equation (only $\rho = -0.1$), the Gaussian kernel density estimation and the Laplacian of the Poisson kernel method. Then their density estimation and variances appear in Figures 5.1 and 5.2.

In Figure 5.1, we have computed two different approximations of the Kolmogorov equation. That is, $(\Delta x, \Delta u) = (0.02, 0.02)$ for "PDE 1" and $(\Delta x, \Delta u) = (0.01, 0.01)$ for "PDE 2". These results depend heavily on the approximation of the initial condition (the Dirac function), In order to achieve a stable simulation (positivity of the density), we need to restrict the region of u. Then the calculation looses small mass on the boundary of the region. Hence our results depend on these conditions. As in the case $\rho = -0.8$, (5.7) does not satisfy the stability conditions, we have not included them in Figure 5.3.

In Figure 5.1 and 5.3, we give simulation results using the Malliavin–Thalmaier formula and its approximation with the optimal value for h (using (4.2)). The number of time steps until maturity is 50, that is, $\Delta t = t/50 = 0.02$ and the number of the Monte–Carlo simulation changes from 10^4 to 10^6. From these graphs, we can say that the approximative Malliavin–Thalmaier formula 2.1 performs well in comparison to the other approximative density values (Figure 5.1 and 5.3). In Figures 5.2 and 5.4, we can see that the variance of the approximated Malliavin–Thalmaier formula is stable and about a half of the variance of the Malliavin–Thalmaier formula without h.

Note that even if we use the optimal size of the bandwidth h, the variance of KDE is comparatively larger than the other methods. Compared with KDE, the Poisson kernel method works better. To reduce L^2-error, the optimal size of the parameter h becomes slightly big, then we find that the numerical results have somewhat large bias errors in Figure 5.1 and 5.3.

In Tables 5.1 and 5.2, we give the constant values from the central limit theorems obtained using pilot simulations. Here we first simulate through each method by using from $h = 0.1$ to $h = 10^{-10}$ and $N = 10^5$. The cases in which the value of C is too small are removed from further consideration. This gives a narrow range of h where the pilot simulation are carried out. For these we use $N = 10^5$ and $M = 100$. The results of the calculations appear in Tables 5.1 and 5.2. In Table 5.3, we give the optimal size of the parameter h for the case $N = 10^6$ by using the constants from Tables 5.1 and 5.2. And we also give the simulation times for each method. In this respect there is no big difference among the methods.

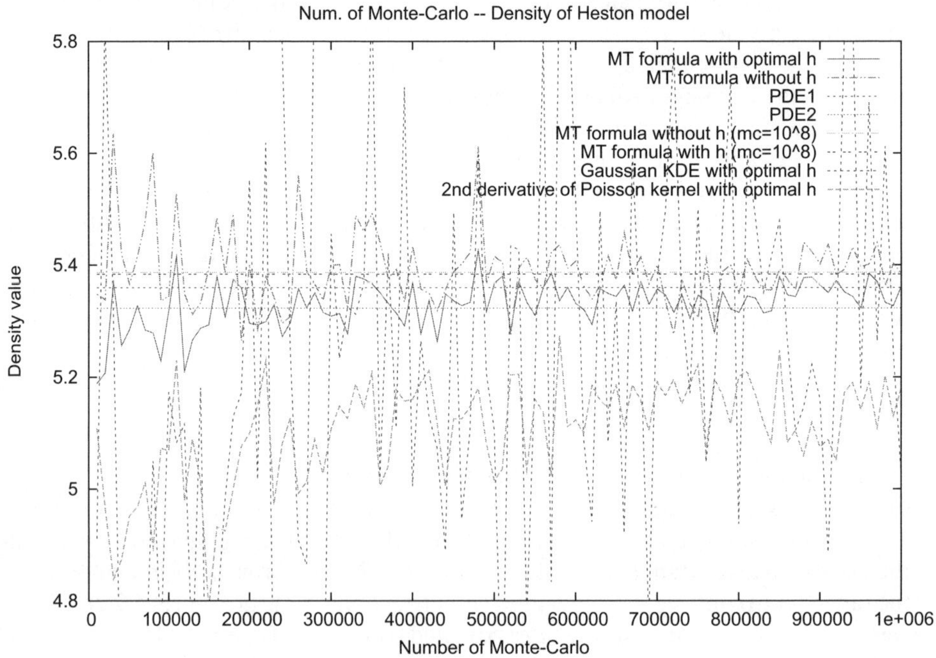

Figure 5.1. Number of MC simulations and density estimates of the Heston model ($\rho = -0.1$)

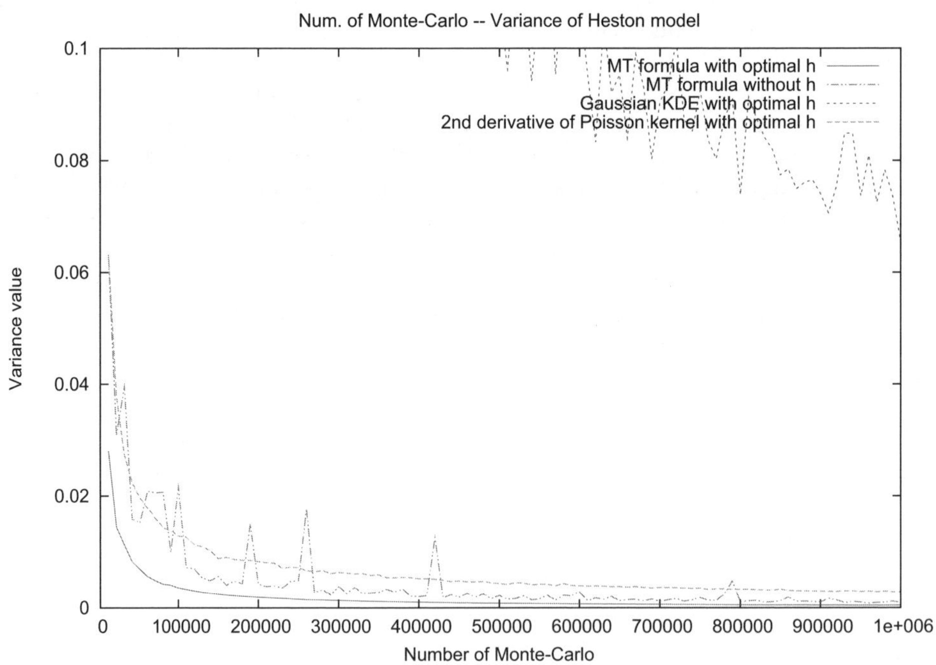

Figure 5.2. Number of MC simulations and variance of the density estimates for Heston model ($\rho = -0.1$)

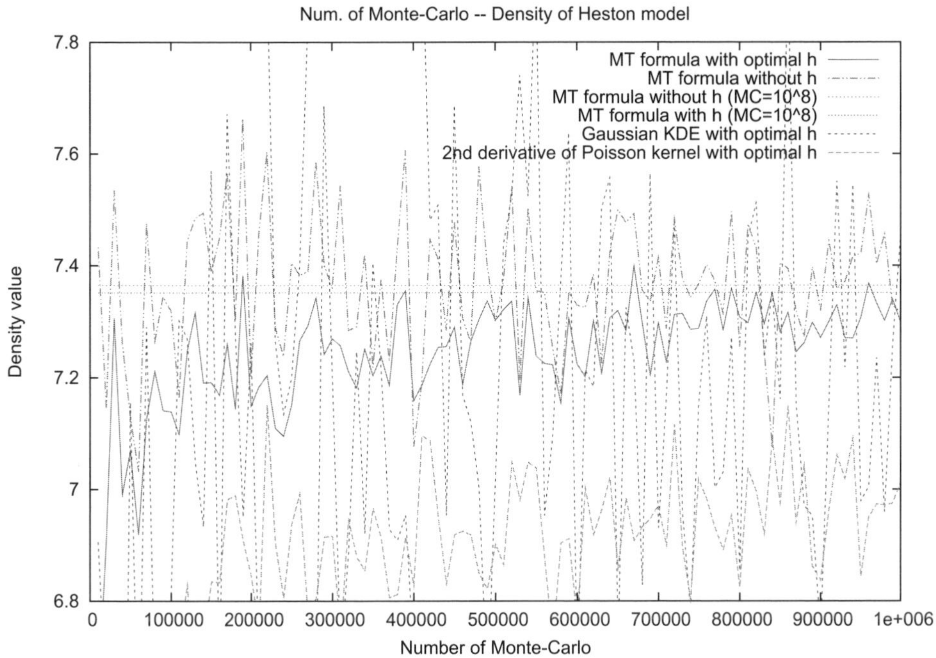

Figure 5.3. Number of MC simulations and density estimates for the Heston model ($\rho = -0.8$)

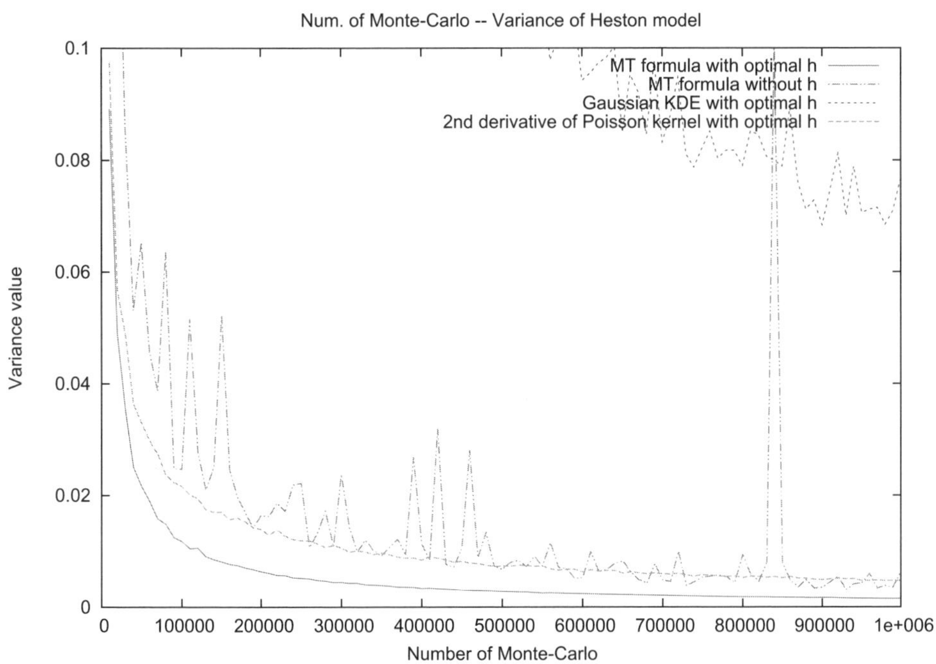

Figure 5.4. Number of MC simulations and variance of the density estimates for the Heston model ($\rho = -0.8$)

Method	Bias error	$\rho = -0.1$	$\rho = -0.8$
MT formula	$C_1^{\hat{x}} h \ln \frac{1}{h} + C_2^{\hat{x}} h + o(h)$	$C_1^{\hat{x}} = 97.2983$	$C_1^{\hat{x}} = 273.0708762$
KDE	$\dot{C}_1^{x} h^2 + O(h^4)$	$\dot{C}_1^{x} = 30258018$	$\dot{C}_1^{x} = 9209822.1$
Poisson	$\hat{C}_1^{\hat{x}} h \ln \frac{1}{h} + \hat{C}_2^{\hat{x}} h + o(h)$	$\hat{C}_1^{\hat{x}} = 195.2020997$	$\hat{C}_1^{\hat{x}} = 274.6290929$

Table 5.1. Bias error and constants computed using pilot simulations for the Heston model

Method	L^2-error	$\rho = -0.1$	$\rho = -0.8$
MT formula	$C_3^{\hat{x}} \ln \frac{1}{h} + O(1) \quad (d = 2)$ $C_4^{\hat{x}} \frac{1}{h^{\frac{d}{2}-1}} \quad (d \geq 3)$	$C_3^{\hat{x}} = 42.741$	$C_3^{\hat{x}} = 159.642$
KDE	$\dot{C}_2^{x} \frac{1}{h^d} + O(\frac{1}{h^{d-1}})$	$\dot{C}_2^{x} = 372966$	$\dot{C}_2^{x} = 92540.7$
Poisson	$\hat{C}_3^{\hat{x}} \frac{1}{h^{\frac{d}{2}}} + o(\frac{1}{h^{\frac{d}{2}}})$	$\hat{C}_3^{\hat{x}} = 0.555882$	$\hat{C}_3^{\hat{x}} = 0.598472$

Table 5.2. L^2-error and constants computed using pilot simulations for the Heston model

Method	Optimal size of h	$\rho = -0.1$	$\rho = -0.8$
MT formula	$\left(\frac{C_3^{\hat{x}}}{2N (C_1^{\hat{x}})^2} \right)^{\frac{1}{2}} \quad (d = 2)$ $\left\{ \frac{d-2}{4N} \frac{C_4^{\hat{x}}}{(C_1^{\hat{x}})^2} \right\}^{\frac{2}{2+d}} \quad (d \geq 3)$	4.75119×10^{-5}	3.27177×10^{-5}
KDE	$\left(\frac{d \dot{C}_2^{x}}{4N (\dot{C}_1^{x})^2} \right)^{\frac{1}{d+4}}$	0.0024256	0.0028585
Poisson	$\left(\frac{d \hat{C}_3^{\hat{x}}}{4N (\hat{C}_1^{\hat{x}})^2} \right)^{\frac{2}{d+4}}$	0.000193937	0.000158309

Table 5.3. Optimal bandwidth h for the Heston model ($d = 2$ and $N = 10^6$)

Method	$N = 10^4$	$N = 10^5$	$N = 10^6$
MT formula	0.406	3.978	39.312
KDE	0.296	2.933	27.456
Poisson	0.312	3.37	28.143

Table 5.4. Computation time for the Heston model (in seconds)

5.3 Double volatility Heston model

In this section, we consider a 3-dimensional case, the double volatility Heston model [25] which is a special case of [21], given by

$$\begin{aligned} dS_t &= \mu S_t dt + \sqrt{v_t} S_t dW_t^{(1)} + \sqrt{u_t} S_t dW_t^{(2)}, \\ dv_t &= \gamma(\theta - v_t)dt + \kappa\sqrt{v_t} dB_t^{(1)}, \\ du_t &= \alpha(\beta - u_t)dt + \tau\sqrt{u_t} dB_t^{(2)}, \end{aligned}$$

where $\mu, \gamma, \theta, \kappa, \alpha, \beta, \tau$ are constants with $\gamma\theta \geq \frac{\kappa^2}{2}$ and $\alpha\beta \geq \frac{\tau^2}{2}$, and $W_t^{(1)}, W_t^{(2)}, B_t^{(1)}$, $B_t^{(2)}$ are standard Brownian motions with $\mathbb{E}[W_t^{(1)} B_t^{(1)}] = \rho_1 t$ and $\mathbb{E}[W_t^{(2)} B_t^{(2)}] = \rho_2 t$ $(-1 \leq \rho_1, \rho_2 \leq 1)$ and others are independent of each other.

Then we introduce Brownian motions $Z_t^{(1)}$ and $Z_t^{(2)}$,

$$W_t^{(1)} = \rho_1 B_t^{(1)} + \sqrt{1-\rho_1^2} Z_t^{(1)}, \quad \text{and} \quad W_t^{(2)} = \rho_2 B_t^{(2)} + \sqrt{1-\rho_2^2} Z_t^{(2)}.$$

where $B^{(1)} \bigsqcup Z^{(1)}$, $B^{(2)} \bigsqcup Z^{(2)}$ and $Z^{(1)} \bigsqcup Z^{(2)}$ where $\bigsqcup$ stands for independence of processes. And set $X_t := \ln(S_t/S_0) - \mu t$, $V_t := a_1 v_t$ and $U_t := a_2 u_t$. Then we have

$$\begin{aligned} dX_t &= -\frac{1}{2}\left(\frac{V_t}{a_1} + \frac{U_t}{a_2}\right) dt + \frac{\rho_1}{\sqrt{a_1}} \sqrt{V_t} dB_t^{(1)} + \frac{\sqrt{1-\rho_1^2}}{a_1} \sqrt{V_t} dZ_t^{(1)} \\ &\quad + \frac{\rho_2}{\sqrt{a_2}} \sqrt{U_t} dB_t^{(2)} + \frac{\sqrt{1-\rho_2^2}}{\sqrt{a_2}} \sqrt{U_t} dZ_t^{(2)}, \\ dV_t &= \gamma\left(a_1\theta - V_t\right) dt + \sqrt{a_1}\kappa\sqrt{V_t} dB_t^{(1)}, \\ dU_t &= \alpha\left(a_2\beta - U_t\right) dt + \sqrt{a_2}\tau\sqrt{U_t} dB_t^{(2)}. \end{aligned} \tag{5.13}$$

Through usual calculations for weights $H_{(i)}(X_t, V_t, U_t; 1)$, $i = 1, 2, 3$, we obtain the Malliavin–Thalmaier formula. Although the weights are too long to write here (See Kohatsu–Higa and Yasuda [34]) the computational complexity is the same as in the previous example. Then we compare the density value and variance through some methods as the Heston model.

Numerical results

We use the following parameters;

Parameter	Notation	Value
Correlation	(ρ_1, ρ_2)	$(0.2, -0.15)$
Scale parameters	(a_1, a_2)	$(1, 1)$
Speed of mean-reversion	(γ, α)	$(2, 1.5)$
Long term mean	(θ, β)	$(0.2, 0.15)$
Volatility of volatility process	(κ, τ)	$(0.2, 0.15)$
Initial value of volatility process	(V_0, U_0)	$(0.2, 0.15)$
Initial value of log-price	X_0	0
Maturity	t	1
Time step size	Δt	$1/200 = 0.005$

The density estimates are carried at the point $(x, v, u) = (0, 0.2, 0.15)$ (the initial point).

From Figure 5.5, we arrive at conclusions similar to those related to the Heston model case. The KDE method has a large bias and variance even if we use the optimal bandwidth size. The bias error of the Poisson kernel method is larger than the corresponding biases of the Malliavin–Thalmaier formula without and with h. Variance of the approximated Malliavin–Thalmaier formula with the optimal h is much smaller than the variances of the other methods. We can easily find that the Malliavin–Thalmaier formula (without h) have some singular values in Figure 5.6. But the approximated version is stable and has smaller variance.

Expressions of the Malliavin weights are similar to ones of the Heston model. But computation time is longer than the Heston case, since a problem appears when one performs the simulation of the two volatility processes (the CIR model), for which we need a precise approximation. The time step size $\Delta t = 0.005$ is smaller than the Heston case. Therefore this issue has to be taken into account in the final result.

6 Conclusions and further comments

In this article we have only concentrated on the integration by parts formula in the setting of Wiener spaces and we have compared the kernel density methods with integration by parts methods. In [18], an interesting mixed approach is introduced, although some of the results do not seem encouraging it may lead to new ideas for new simulation methods.

There is also another tendency to obtain the infinite dimensional integration by parts formula as a limit of finite dimensional integration by parts. This is the point of view of [15] which also shows that there are various other integration by parts formulae that can be obtained beside the classical ones. This approach can also be used theoretically as shown in [5] and [53]. This has also lead to interesting results in the jump driven stochastic differential equations.

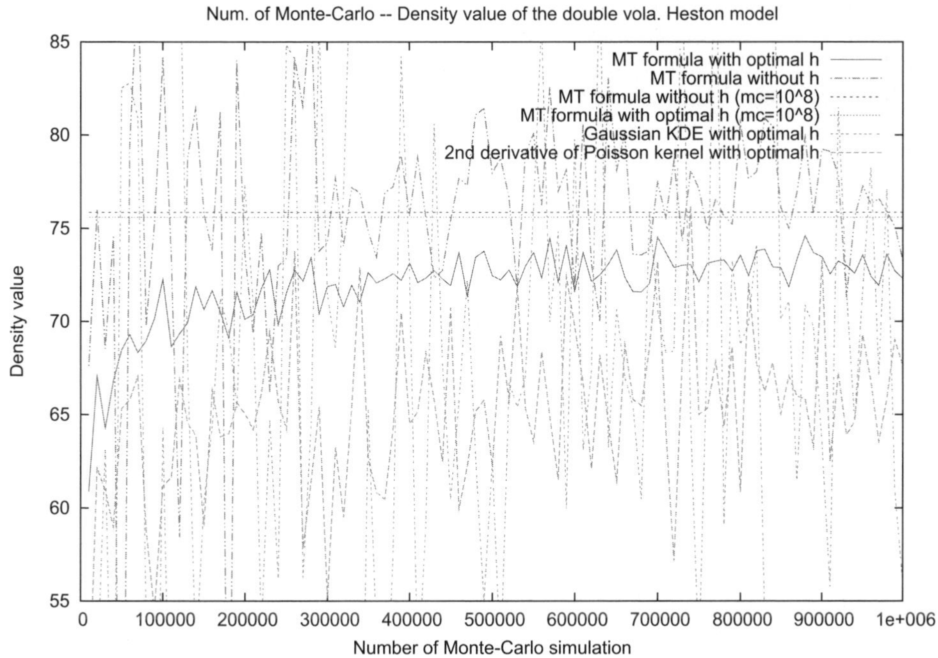

Figure 5.5. Number of MC simulations and estimation of the density for the double volatility Heston model

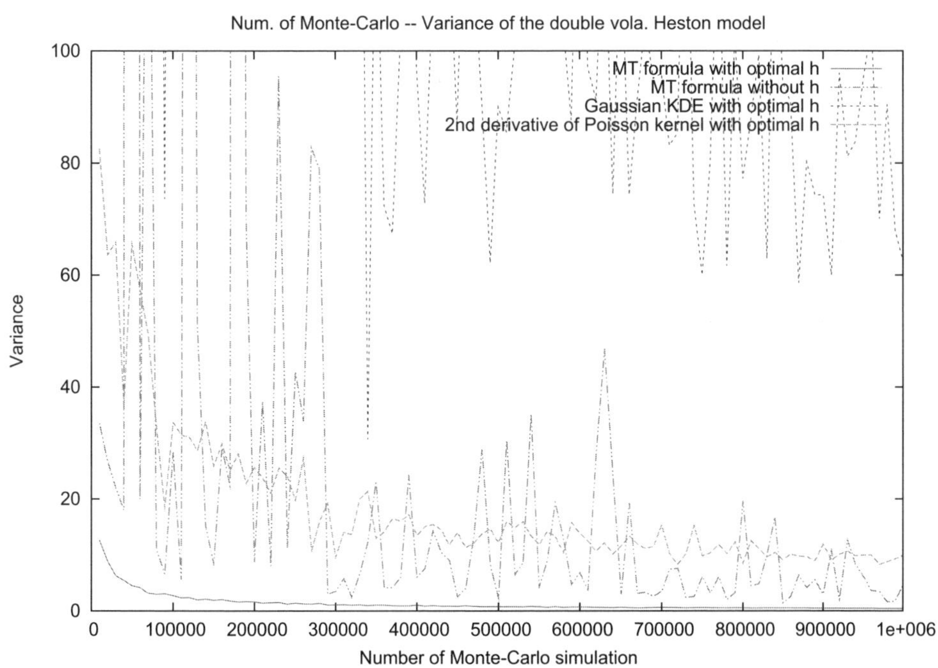

Figure 5.6. Number of MC simulations and variance of the density estimates for the double volatility Heston model

There is an increasing literature dealing with the integration by parts formula in the setting of Lévy driven stochastic differential equations. In the early 90's this became a hot topic of research leading to articles and books (see the references, [11], [13], [40], [45], [46], [14], [51] and [52]). There are various approaches that lead to different integration by parts formula depending which variable one uses to base the integration by parts. Some use the jump distribution, other the jump times and other are based in other variables. There is not a unified approach as in the Wiener case. In most cases, as in the case of the Wiener space, the interest is in proving the existence and smoothness of densities for solutions of stochastic differential equations with jumps.

There is another approach centred in the chaos decompositions. See for example, [37], [36]. This approach leads to a definition of derivative but its consequences for densities of random variables have been largely ignored. Also, in this setting, it becomes hard to verify that the solution of stochastic differential equations with jumps are differentiable.

In the past few years various authors have studied the application of this methodology in finance and insurance. Leading to similar studies of greeks in Finance. See, e.g. [6], [7], [16], [43], [19] and [27].

Another issue that has raised recent interest is the application of the asymptotic expansion theory developed by S. Watanabe on Wiener space (see [56], [41], [42]) and recently extended to the Poisson space case. These formulas found an application in statistics in the form of Berry-Essen type expansions (see [54] or [50]). In Finance this has lead to approximative formulas for option pricing. In particular, there has been a recent development of expansion formulas using greeks (see [9] and [10]). This formulas seem to have an application in the calibration problem. Although we still seem far from solving this difficult problem from the practical point of view.

Partial approaches that do not seem to lead to a clear expansion but give approximative formulas can be found in [1], [2] and[3]. We also remark that there are various other competitive approaches using partial differential equations or a combination of probabilistic arguments and analytic ones. For this, see eg. [4], [26] and [44].

Bibliography

[1] E. Alós, C.-O. Ewald, *Malliavin differentiability of the Heston Volatility and applications to option pricing*, Adv. in Appl. Probab. 40 (1) (2008), pp. 144–162.

[2] ———, J. Leon, and J. Vives, *On the short-time behavior of the implied volatility for jump-diffusion models with stochastic volatility*, Finance Stoch. 11 (4) (2007), pp. 571–589.

[3] ———, *A generalization of the Hull and White formula with applications to option pricing approximation*, Finance Stoch. 10 (3) (2006), pp. 353–365.

[4] F. Antonelli and S. Scarlatti. Pricing options under stochastic volatility. A power series approach. Preprint.

[5] V. Bally, *An elementary introduction to Malliavin calculus*, INRIA RR-4718 (2003), http://www.inria.fr/rrrt/rr-4718.html.

[6] ———, M.-P. Bavouzet, and M. Messaoud, *Integration by parts formula for locally smooth*

laws and applications to sensitivity computations, Ann. Appl. Probab. 17 (1) (2007), pp. 33–66.

[7] M.-P. Bavouzet, and M. Messaoud, *Computation of Greeks using Malliavin's calculus in jump type market models*, Electronic Journal of Probability 11 (10) (2006), pp. 276–300.

[8] ———, and L. Caramellino, *Lower bounds for the density of Ito processes under weak regularity assumptions*, working paper.

[9] E. Benhamou, E. Gobet, and M. Miri, *Smart expansion and fast calibration for jump diffusion*, preprint, http://papers.ssrn.com/sol3/papers.cfm?abstract_id=1079627.

[10] ———, ———, and ———, *Closed forms for European options in a local volatility model*, preprint, http://papers.ssrn.com/sol3/papers.cfm?abstract_id=1275872.

[11] K. Bichteler, J. Gravereaux, and J. Jacod, *Malliavin calculus for processes with jumps*, Stochastics Monographs, 2. Gordon and Breach Science Publishers, New York,1987. Math. Review 100847

[12] B. Bouchard, I. Ekeland, and N. Touzi, *On the Malliavin approach to Monte Carlo approximation of conditional expectations*, Finance Stoch. 8 (1) (2004), pp. 45–71.

[13] E. Carlen, and E. Pardoux, *Differential calculus and integration by parts on Poisson space*, In Stochastics, Algebra and Analysis in Classical and Quantum Dynamics, Kluwer, 1990, pp. 63–73.

[14] T. Cass, *Smooth densities for solutions to stochastic differential equations with jumps*, doi:10.1016/j.spa.2008.07.005.

[15] N. Chen, and P. Glasserman, *Malliavin greeks without Malliavin Calculus*, Stochastic Processes and their Applications, 117 (2007), pp. 1689–1723.

[16] M. Davis, and M. Johansson, *Malliavin Monte Carlo Greeks for jump diffusions*, Stochastic Process. Appl., 1 (2006), pp. 101–129.

[17] J. Detemple, R., Garcia and M., Rindisbacher. *Representation formulas for Malliavin derivatives of diffusion processes*, Finance Stoch. 9 (2005), pp. 349–367.

[18] R. Elie, J.-D. Fermanian, and N. Touzi, *Kernel estimation of Greek weights by parameter randomization*, Annals of Applied Probability, 17 (2007), pp. 1399–1423.

[19] Y. El-Khatib, and N. Privault, *Computations of Greeks in a market with jumps via the Malliavin calculus*, Finance Stoch. 8 no. 2 (2004), pp. 161–179.

[20] L. C. Evans, *Partial differential equations*, Graduate studies in Mathematics, Vol. 19, American Mathematical Society, 1998.

[21] J. Fonseca, and M. Grasselli, *Wishart Multi-Deimensional Stochastic Volatility*, preprint. (http://www.riskturk.com/ec2/submitted/IMEWISHART.pdf)

[22] E. Fournié, J. M. Lasry, J. Lebuchoux, P. L. Lions, and N. Touzi, *Applications of Malliavin calculus to Monte Carlo methods in finance*, Finance Stoch. 3 (4) (1999), pp. 391–412.

[23] E. Fournié, J. M. Lasry, J. Lebuchoux and P. L. Lions, *Applications of Malliavin calculus to Monte Carlo methods in finance II*, Finance Stoch. 5 (2) (1999), pp. 201–236.

[24] S. L. Heston, *A closed-form solution for options with stochastic volatility with applications to bond and currency options*, The Review of Financial Studies Vol. 6 No. 2 (1993), pp. 327–343.

[25] D. Kainth, and N. Saravanamuttu, *Modelling the FX Skew*, presentation slide. (http://www.quarchome.org/FXSkew2.ppt)

[26] J. Kampen, A. Kolodko, J.G.M. Schoenmakers. Monte Carlo Greeks for financial products via approximative transition densities. SIAM J. Sci. Comput. 31(1) 1–22 (2008).

[27] R. Kawai, and A. Takeuchi, *Greeks formulae for an asset price dynamics model with gamma processes*, submitted.

[28] A. Kebaier, and A. Kohatsu-Higa, *An optimal control variance reduction method for density estimation*, Stochastic Processes and their Applications, Vol. 118, 12, (2008), pp. 2143–2180.

[29] A. Kohatsu-Higa, and M. Montero, *Malliavin Calculus in Finance*, Handbook of Computational Finance, Birkhauser, 2004.

[30] A. Kohatsu-Higa and R. Pettersson. Variance Reduction Methods for simulations of densities on Wiener space. SIAM J. Numerical Analysis. 40, 431–450, 2002.

[31] ———, and K. Yasuda, *Estimating multidimensional density functions for random variables in Wiener space*, C. R. Math. Acad. Sci. Paris 346 5-6 (2008), pp. 335–338.

[32] ———, ———, *Estimating multidimensional density functions using the Malliavin-Thalmaier formula*, to appear in SIAM Journal of Numerical Analysis.

[33] ———, ———, *Simulation of multidimensional density functions through the Malliavin-Thalmaier formula and its application to finance*, submitted.

[34] ———, ———, *Heston-type density estimation through the Monte-Carlo method and its application to Greeks calculation*, in preparation.

[35] D. Lamberton, and B. Lapeyre, *Introduction to stochastic calculus applied to finance*, Chapman & Hall, 1996.

[36] J. Leon, J.L. Sole, F. Utzet and J. Vives. On Levy processes, Malliavin Calculus and market models with jumps. Finance and Stoch. 6, 197-225 (2006).

[37] A. Løkka, *Martingale representation of functionals of Lévy processes*, Stochastic Anal. Appl. 22 no. 4 (2004), pp. 867–892.

[38] P. Malliavin, and A. Thalmaier, *Stochastic calculus of variations in mathematical finance*, Springer Finance, Springer-Verlag, Berlin, 2006.

[39] D. Nualart, *The Malliavin calculus and related topics (Second edition)*, Probability and its Applications (New York), Springer-Verlag, Berlin, 2006.

[40] ———, and J. Vives, *A duality formula on the Poisson space and some applications*, Seminar on Stochastc Analysis, Random Fields and Applications, Progr. Probab. 36 (1995), pp. 205–213.

[41] Y. Osajima, *The Asymptotic Expansion Formula of Implied Volatility for Dynamics SABR Model and FX Hybrid Model*, preprint, hhtp://papers.ssrn.com/sol3/papers.cfm?abstract_id=965265.

[42] ———, *General Asymptotics of Wiener Functions and Applilcation to Mathematical Finance*, preprint, hhtp://papers.ssrn.com/sol3/papers.cfm?abstract_id=1019587.

[43] E. Petrou, *Malliavin Calculus in Lévy spaces and Applications to Finance*, Electronic Journal of Probability, 13 (2008), pp. 852–879.

[44] A. Pascucci, F. Corielli, *Parametrix approximation of diffusion transition densities*, AMS Acta, Universitá di Bologna (2008), preprint.

[45] J. Picard, *Formules de dualité sur l'espace de Poisson*, Ann. Inst. H. Poincaré Probab. Statist. 32 4 (1996), pp. 509–548.

[46] J. Picard, *On the existence of smooth densities for jump processes*, Probab. Theory Related Fields 105 4 (1996), pp. 481–511.

[47] M. Sanz-Solé, *Malliavin Calculus with applications to stochastic partial differential equations*, EPFL Press, 2005.

[48] D. W. Scott, *Multivariate Density Estimation: Theory, Practice, and Visualization*, Wiley, New York, 1992.

[49] I. Shigekawa, *Stochastic Analysis*, Translations of Mathematical Monographs, AMS, 2004.

[50] A. Takahashi, and M. Yoshida, *Monte Carlo simulation with asymptotic method*, preprint (2002), J. Japan Statist. Soc. 35 (2005), pp. 171–203.

[51] A. Takeuchi, *The Malliavin calculus for SDE with jumps and the partially hypoelliptic problem*, Osaka J. Math. 39 (2002), pp. 523–559.

[52] ———, *The Bismut-Elworthy-Li type formulae for stochastic differential equations with jumps*, submitted.

[53] J. Teichmann, *Stochastic evolution equations in infinite dimension with applications to term structure problems*, Lecture note (2005), http://www.fam.tuwien.ac.at/ jteichma/leipzigparislinz080605.pdf.

[54] M. Uchida, and N. Yoshida, *Asymptotic expansion for small diffusions applied to option pricing*, Statist. Infer. Stochast. Process, 7 (2004), pp. 189–223.

[55] M. P. Wand, and M. C. Jones, *Kernel Smoothing*, Chapman & Hall, 1995.

[56] S. Watanabe, *Analysis of Wiener functionals (Malliavin calculus) and its applications to heat kernels*, Ann. Probab. 15 1 (1987), pp. 1–39.

Author information

Kohatsu-Higa Arturo, Osaka University. Graduate School of Engineering Sciences. Division of Mathematical Science for Social Systems. 1-3 Machikaneyama, Toyonaka, Osaka 560-8531, Japan.
Email: kohatsu@sigmath.es.osaka-u.ac.jp

Yasuda Kazuhiro, Hosei University. Faculty of Science and Engineering. Department of Industrial and Systems Engineering. 3-7-2, Kajino-cho, Koganei-shi, Tokyo, 184-8584, Japan.
Email: k_yasuda@hosei.ac.jp

The numeraire portfolio in discrete time: existence, related concepts and applications

Ralf Korn and Manfred Schäl

Abstract. We survey the literature on the numeraire portfolio, explain its relation to various other concepts in financial mathematics and present two applications in insurance mathematics and portfolio optimization.

Key words. Numeraire portfolio, value preserving portfolio, growth optimal portfolio, benchmark approach, minimal martingale measure, NUIP condition.

AMS classification. 90A09, 91B28, 91B62, 93E20, 62P05

1 Introduction and summary

An important subject of financial mathematics is adequate pricing of financial derivatives, in particular options. In the modern theory (see e.g. Duffie 1992), the historical concept based on expectations of discounted quantities (the present value principle) is replaced by the concept of deflators, numeraires (inverse deflators) or the application of the present value principle after a change of measure. In this paper we focus on the concept of the numeraire portfolio, present its definition, its relation to various valuation concepts and its role in important applications. When the value process of a numeraire portfolio is used as a discount process, the relative value processes of all other portfolios with respect to it will be martingales or at least supermartingales (see Vasicek 1977, Long 1990, Artzner 1997, Bajeux-Besnainou and Portait 1997, Korn & Schäl 1999, Schäl 2000a, Becherer 2001, Platen 2001, 2006, Christensen and Larsen 2007, Karatzas & Kardaras 2007).

We will study a financial market with small investors which is free of arbitrage opportunities but incomplete (although we will see that much is valid under a weaker assumption than the no arbitrage assumption). Then in discrete time, one has several choices for an equivalent martingale measure (EMM) needed to value derivatives. In continuous time an EMM exists under more restrictive conditions. It is known (see Harrison & Kreps 1979) that each EMM corresponds to a consistent price system. Thus in incomplete markets, no preference-independent pricing of financial derivatives is possible. In the present paper, the unique martingale measure Q^* is studied which is defined by the concept of the numeraire portfolio (see Korn & Korn 2001, Section 3.7). The choice of Q^* can be justified by a change of numeraire in place of a change of measure. Uniqueness is obtained by the fact that the EMM after the change of numeraire should be the original real-world probability measure.

It is known that in many cases one can get a numeraire portfolio from the growth optimal portfolio (GOP) which maximises the expected utility when using the log-utility. Utility optimisation is now a classical subject. Recent papers with the log-utility are Goll & Kallsen 2000, Kallsen 2000, Goll and Kallsen 2003.

When looking for a numeraire portfolio (in the strict martingale sense), we are interested in optimal portfolios which can be chosen from the interior of the set of admissible portfolios. Also for more general utilities, optimal 'interior' portfolios can be used to define equivalent martingale measures (see Karatzas & Kou 1996, Schäl 2000a,b).

In order to get full equivalence of a numeraire portfolio and a GOP, one has to generalise the concept by defining a weak numeraire portfolio introduced by Becherer 2001 under the name 'numeraire portfolio'. Such a portfolio defines a supermartingale measure in the above sense.

The paper is laid out as follows. We consider a discrete-time market. It turns out that all the ideas can be explained in a simple one-period model starting in 0 and finishing at the time-horizon $T = 1$. In fact, for a log-utility investor, the optimal strategy is myopic even for market models where optimal power-utility strategies are not guaranteed to be myopic (see Hakansson & Ziemba, 1995). Given the solution to a one-period model, the form of the optimal strategy for a multi-period model is obvious. Therefore we will restrict to such a (0,1)-period. Then strategies ξ and portfolios π can be described by d-dimensional vectors. In fact when considering general semi-martingale models, it is sufficient (in most passages) to replace the inner products $\xi^\top \Delta S$ or $\pi^\top R$ by stochastic integrals $\xi \cdot S$ or $\pi \cdot R$, where S describes the prices and R the cumulative returns.

Except for the restriction to a $(0, 1)$-period, we try to choose the framework as general as possible where the recent paper by Karatzas & Kardaras 2007 will serve as a model. In particular, we accept the framework with general convex constraints.

We then consider various valuation and optimisation concepts that are directly related to the numeraire portfolio. Among them are the GOP, the benchmark portfolio, the value preserving portfolio and of course the valuation with the help of EMMs. This is followed by existence considerations for (weak) numeraire portfolios. Finally, we give two important applications of the numeraire portfolios in insurance mathematics and in portfolio optimisation.

2 The one-period market setting

On the market an investor can observe the prices of $1 + d$ assets at the dates $t = 0, 1$ which are described by S_t^0 and $S_t = (S_t^1, \ldots, S_t^d)^\top t = 0, 1$. [For any vector x we write $x^\top$ for the transposed vector and $x^\top y$ for the inner product of $x, y \in \mathbb{R}^d$ thought of as column vectors.]

Hence our time horizon will be $T = 1$. Then S_0^0 and S_0 are deterministic, S_1^0 is a random variable, S_1 is a random vector on a probability space (Ω, F, P) and S_t^0 is positive.

One of these assets will play a special role for which we will choose S^0. But any other component S^k can be chosen in place of S^0. An important situation will be the case where the asset with price S^0 describes the bank account (or money market) and the other d assets are stocks. This is a very useful interpretation and we will use it. The interpretation of S^0 as money market leads to further convenient interpretations. But remember that, mathematically, all price components will satisfy the same assumptions.

Given an initial capital $V_0 > 0$, one can invest in the assets described by S by choosing some $\xi \in \mathbb{R}^d$ which describes the *strategy* in the present simple case with $T = 1$. The number ξ^k represents the number of shares for stock k bought and held by the investor at time 0. The total amount invested in stocks is $\xi^\top S_0 = \sum_{k=1}^d \xi^k S_0^k$. For satisfying the self-financing condition, the remaining wealth of the initial value V_0, namely $\xi^0 := V_0 - \xi^\top S_0$ is invested in the bank account. Then $V_0 = V_0^\xi = \sum_{k=0}^d \xi^k S_0^k$. Upon defining $\Delta X := X_1 - X_0$ for X being defined for $t = 0, 1$, the value V_1^ξ of ξ at time 1 is described by

$$\Delta V^\xi = \xi^0 \Delta S^0 + \xi^\top \Delta S \left(= \sum_{k=0}^d \xi^k (S_1^k - S_0^k) \right). \tag{2.1}$$

Upon defining discounted quantities $\breve{S}_t = (\breve{S}_t^1, \dots, \breve{S}_t^d)^\top$ and $\breve{V}_t$ by

$$\breve{S}_t^k := S_t^k / S_t^0, \; \breve{V}_t^\xi := V_t^\xi / S_t^0, \tag{2.2}$$

we easily obtain

$$\Delta \breve{V}^\xi = \xi^\top \Delta \breve{S}. \tag{2.3}$$

This simple relation is the mathematical reason for using "discounted" quantities. Since we might as well work in discounted terms, from now on we assume that $S_t^0 \equiv 1$ as is common in Mathematical Finance (see Harrison & Kreps 1979). Then $\Delta S^0 \equiv 0$ and one can dispense with ξ^0. Starting with capital $V_0 = x > 0$ and investing according to strategy ξ, the investor's value at time 1 is $V_1^\xi(x) := x + \xi^\top \Delta S$. For any $V_0 = x > 0$ and any strategy ξ, $V_1^\xi(x) = x + \xi^\top \Delta S$ is called *admissible* if $V_1^\xi(x) \geq 0$. The *return* R^k for stock k is defined by

$$S_1^k = S_0^k \cdot (1 + R^k). \tag{2.4}$$

Then we can write $V_1^\xi(x) = x \cdot (1 + \sum_{k=1}^d (\xi^k S_0^k / V_0) R^k)$. Defining $\pi \in \mathbb{R}^d$ as the vector with components $\pi^k = \xi^k S_0^k / V_0$, π^k signifies the proportion of V_0 invested in stock k and we have

$$V_1^\xi(x) = x \cdot (1 + \pi^\top R) =: x \cdot V_1^\pi. \tag{2.5}$$

The equivalent of "$V_1^\xi(x) > 0$ (≥ 0)", for $x > 0$, is "$V_1^\pi = 1 + \pi^\top R > 0$ (≥ 0)". This simple representation is the reason for our *restriction* to the case $x = 1$ in the sequel where we write V_1^π in place of $V_1^\xi(1)$. By use of π, admissibility is independent of the initial wealth x and thus easier to handle.

We will now introduce constraints, where Karatzas & Kardaras 2007, Kardaras 2006 will serve as a model. For the sake of motivation, we will start with the following example.

Example A. The case where the investor is prevented from selling stock short or borrowing from the bank can be describe by $\xi^k \geq 0$, $1 \leq k \leq d$, and $\xi^0 := V_0 - \xi^\top S_0 \geq 0$.

This condition is equivalent to $\pi^k \geq 0$, $1 \leq k \leq d$, and $\pi^0 := 1 - \sum_{k=1}^d \pi^k \geq 0$. By setting $C := \{\pi \in \mathbb{R}^d : \pi^k \geq 0 \text{ and } \sum_{k=1}^d \pi^k \leq 1\}$, the prohibition of short sales and borrowing is translated into the requirement $\pi \in C$.

Definition 2.1. Consider an arbitrary convex closed set $C \subset \mathbb{R}^d$ with $0 \in C$. The admissible value V_1^π is called C-constrained, if $\pi \in C$. Here the following set

$$\check{C} := \cap_{a>0}\, aC \tag{2.6}$$

is called the *set of cone points* (or recession cone) of C.

Note in particular that the "safe" portfolio $\pi = 0$ is always admissible.

Example A (continuation). Here we have $aC = \{a\pi \in \mathbb{R}^d : \pi^k \geq 0 \text{ and } \sum_{k=1}^d \pi^k \leq 1\} = \{\vartheta \in \mathbb{R}^d : \vartheta^k \geq 0 \text{ and } \sum_{k=1}^d \vartheta^k \leq a\}$. This leads to the relation $\check{C} = \{0\} \subset \mathbb{R}^d$.

The following example describes the positivity constraints for admissibility.

Example B (Natural Constraints).

$$C := \Theta := \{\vartheta \in \mathbb{R}^d;\ 1 + \vartheta^\top R \geq 0 \text{ a.s.}\} = \{\vartheta \in \mathbb{R}^d;\ 1 + \vartheta^\top z \geq 0 \,\forall z \in Z\},$$

where Z is the support of R, i.e. the smallest closed subset B of $\mathbb{R}^d$ such that $P[R \in B] = 1$.

The representation of Θ by means of Z is easily proved (see Korn and Schäl, 1999 Lemma 4.3a).

We use "$\geq$"in place of "$>$" in the definition of Θ to keep the set Θ closed. Then $aC = \{a\pi \in \mathbb{R}^d;\ 1 + \pi^\top R \geq 0 \text{ a.s.}\} = \{\vartheta \in \mathbb{R}^d;\ a + \vartheta^\top R \geq 0 \text{ a.s.}\}$ and $\check{C} = \cap_{a>0}\, aC = \{\vartheta \in \mathbb{R}^d;\ \vartheta^\top R \geq 0 \text{ a.s.}\}$.

The requirement of admissibility of V_1^π is exactly what corresponds to π being Θ-constrained.

Consider the special case $d = 1$ and the no-arbitrage condition: $-\alpha, \beta \in Z$ for some $\alpha, \beta > 0$. Then again $\check{C} = \{0\} \subset \mathbb{R}^1$. We shall always assume that C is enriched with the natural constraints, i.e. $C \subset \Theta$. Otherwise, we can replace C by $C \cap \Theta$.

Example C. The case where the investor is prevented from selling stock short but not from borrowing from the bank can be described by $\xi^k \geq 0$, $1 \leq k \leq d$. This condition is equivalent to $\pi^k \geq 0, 1 \leq k \leq d$. By setting $C := \{\pi \in \mathbb{R}^d : \pi^k \geq 0\}$, the prohibition of short sales is translated into the requirement $\pi \in C$. Here C is a cone and thus we get $\check{C} = C = aC$ for $a > 0$.

In the sequel we will write

$$\Pi := \{\pi \in C;\ 1 + \pi^\top R > 0 \text{ a.s.}\}\,. \tag{2.7}$$

The elements of Π will be called *portfolios*; we make this distinction with the corresponding notion of strategy, denoted by ξ.

Lemma 2.2. *For $\rho \in C$ and $\vartheta \in \check{C}$ we have $\rho + \vartheta \in C$.*

Proof (See Karatzas & Kardaras 2007). We know that $a\vartheta \in C$ for all $a > 0$. Then $(1 - \frac{1}{a})\rho + \frac{1}{a} a\,\vartheta = (1 - \frac{1}{a})\rho + \vartheta \in C$ by the convexity of C. But C is also closed, and so $\rho + \vartheta \in C$. □

3 Weak numeraire portfolio

In general, by "numeraire" one understands any strictly positive random variable Y such that it acts as an "inverse deflator $D = Y^{-1}$", e.g. a stochastic discount factor, for the values V_1^π. Then we see our investment according to portfolio π relative to Y, giving us a wealth of V_1^π / Y. There Y may not even be generated by a portfolio.

Definition 3.1. A portfolio $\rho \in \Pi$ will be called *weak numeraire portfolio*, if for the relative value defined as V_1^π / V_1^ρ one has: $\mathbb{E}\,[V_1^\pi / V_1^\rho] \leq 1\ (= V_0^\pi / V_0^\rho)$ for every portfolio π.

The qualifier "weak" is used because we have "$\leq$" in place of "$=$" in the definition above. Since $0 \in \Pi$, one has $\mathbb{E}\,[1/V_1^\rho] \leq 1\ (= 1/V_0^\rho)$. Thus V_t^π / V_t^ρ and $1/V_t^\rho$ are positive supermartingales. The definition in this form first appears in Becherer 2001.

Proposition 3.2. *If ρ_1 and ρ_2 are weak numeraire portfolios, then $V_1^{\rho_1} = V_1^{\rho_2}$ a.s.*

Proof. We have both $\mathbb{E}[V_t^{\rho_1} / V_t^{\rho_2}] \leq 1$ and $\mathbb{E}[V_t^{\rho_2} / V_t^{\rho_1}] \leq 1$ which implies that $V_1^{\rho_1} = V_1^{\rho_2}$ a.s. □

Therefore the value generated by weak numeraire portfolios is unique. Moreover $\rho_1^\top R = \rho_2^\top R$ a.s. In this sense, the weak numeraire portfolio is unique, too.

Of course, if ρ satisfies the requirements of the definition above, V_1^ρ can act as a numeraire in the sense of this discussion. For a weak numeraire portfolio ρ, V_1^ρ is in a sense the best tradable benchmark: whatever anyone else is doing, it looks as a supermartingale (decreasing in the mean) through the lens of relative value to V_1^ρ.

An obvious example for a numeraire would be $Y_1 = S_1^0$ before assuming $S_t^0 \equiv 1$. Obviously the relative values do not depend on the discount factor since $\breve{V}_1^\pi / \breve{V}_1^\rho = V_1^\pi / V_1^\rho = Y^{-1} V_1^\pi / Y^{-1} V_1^\rho$. Now we again see that there was no loss of generality in considering discounted values. It will turn out that ρ satisfies certain optimality properties. Thus, when using V_t^ρ as inverse deflator in place of the classical S_t^0, we take into account that an investment in the bank account may be far from optimal.

The following relation will be used for sake of motivation (see Kardaras 2006):

$$V_1^\pi / V_1^\rho = (1 + \rho^\top R)^{-1} (1 + \pi^\top R) = 1 + (\pi - \rho)^\top R^\rho$$

where $R^\rho = (1 + \rho^\top R)^{-1} R$ is the return in an auxiliary market. Therefore the relative value can be seen as the usual value generated by investing in the auxiliary market. If

$\mathbb{E}[\pi^\top R] \in [-1,\infty]$ is called the *rate of return* or *drift rate*, then

$$\begin{aligned} r(\pi|\rho) &:= \mathbb{E}[(\pi-\rho)^\top R^\rho] \\ &= \mathbb{E}[(1+\rho^\top R)^{-1}\cdot(\pi-\rho)^\top R] = (\pi-\rho)^\top \mathbb{E}[(1+\rho^\top R)^{-1} R] \end{aligned} \tag{3.1}$$

is the rate of return of the relative value process V_t^π / V_t^ρ. Since $\mathbb{E}[V_1^\pi/V_1^\rho] = 1 + r(\pi|\rho)$, we now obtain the following lemma.

Lemma 3.3. *ρ is a weak numeraire portfolio if and only if*

$$r(\pi|\rho) \leq 0 \textit{ for every } \pi \in \Pi\,. \tag{3.2}$$

It is obvious that if (3.2) is to hold for C, then it must also hold for the closed convex hull of C, so it was natural to assume that C is closed and convex if we want to find the portfolio ρ.

The market may show some degeneracies. This has to do with linear dependence that some stocks might exhibit and which are not excluded. As a consequence, there may be seemingly different portfolios producing exactly the same value. Thus they should then be treated as equivalent. To formulate this notion, consider two different portfolios π_1 and π_2 producing exactly the same value, i.e. $\pi_1^\top R = \pi_2^\top R$ a.s. Now $(\pi_2-\pi_1)^\top R = 0$ a.s. is equivalent to $(\pi_2-\pi_1)^\top z$ for all $z \in Z$ where Z is again the support of R. Let $\mathcal{L}$ be the smallest linear space in $\mathbb{R}^d$ containing Z and $\mathcal{L}^\perp = \{\vartheta \in \mathbb{R}^d; \vartheta \perp \mathcal{L}\}$ its orthogonal complement.

Lemma 3.4. *(a) $\pi_1^\top R = \pi_2^\top R$ a.s. is equivalent to $\pi_2 - \pi_1 \in \mathcal{L}^\perp$.*

(b) $\vartheta \in \mathbb{R}^d \setminus \mathcal{L}^\perp$ if and only if $P[\vartheta^\top R \neq 0] > 0$.

Two portfolios π_1 and π_2 satisfy $\pi_2 - \pi_1 \in \mathcal{L}^\perp$ if and only if $V_1^{\pi_1} = V_1^{\pi_2}$a.s. It is convenient to assume that $\mathcal{L}^\perp \subset C$. So the investor should have at least the freedom to do nothing; that is, if an investment leads to absolutely no profit or loss, one should be free to make it. In the non-degenerate case $\mathcal{L} = \mathbb{R}^d$ this just becomes $0 \in C$. The natural constraints Θ can easily be seen to satisfy this requirement as well as the requirements of closedness and convexity.

Definition 3.5. Let us define the set I of *arbitrage opportunities* to be the set of portfolios ϑ such that

$$P[\vartheta^\top R > 0] > 0 \text{ and } P[\vartheta^\top R \geq 0] = 1\,,$$

i.e., the set of portfolios $\vartheta \in \mathbb{R}^d \setminus \mathcal{L}^\perp$ such that $\vartheta^\top R \geq 0$ a.s.

4 The NUIP condition $I \cap \check{C} = \emptyset$

The condition $I \cap \check{C} = \emptyset$ will play an important role and will be called *No Unbounded Increasing Profit* (NUIP) condition as by Karatzas and Kardaras 2007. The qualifier "increasing" stems from the fact that $\vartheta^\top R \geq 0$ a.s. for $\vartheta \in I$. The qualifier "unbounded" reflects the following fact:

Suppose that $\vartheta \in I \cap \check{C}$ and $V_1^\vartheta = 1 + \vartheta^\top R$ where $\vartheta^\top R \geq 0$ a.s. and $P[\vartheta^\top R > 0] > 0$. Since $\vartheta \in \check{C}$, we know that $a\vartheta \in C$ for all $a > 0$. Moreover $a\vartheta^\top R \geq 0$ a.s. and $\{a\vartheta^\top R, a > 0\}$ is unbounded on the set $\vartheta^\top R > 0$ with positive measure.

Now suppose that $\vartheta \in I \cap \check{C}$ and ρ is a weak numeraire portfolio, then

$$\mathbb{E}[V_1^{a\vartheta}/V_1^\rho] = \mathbb{E}[1/V_1^\rho] + a\mathbb{E}[\vartheta^\top R/V_1^\rho] \text{ where } \mathbb{E}[\vartheta^\top R/V_1^\rho] > 0\,.$$

Thus $\mathbb{E}[V_1^{a\vartheta}/V_1^\rho]$ is unbounded in a, in particular $\mathbb{E}[V_1^{a\vartheta}/V_1^\rho] > 1$ for large a which is a contradiction. Therefore we can obtain the following result:

Proposition 4.1. *The NUIP condition $I \cap \check{C} = \emptyset$ is necessary for the existence of a weak numeraire portfolio.*

Note in particular that the NUIP condition is far weaker than the no arbitrage condition.

Example A (continuation). In the case $C := \{\pi \in \mathbb{R}^d : \pi^k \geq 0 \text{ and } \sum_k \pi^k \leq 1\}$, we know that $\check{C} = \{0\} \subset \mathbb{R}^d$.

Since $I \subset \mathbb{R}^d \setminus \mathcal{L}^\perp$, in particular $0 \notin I$, the NUIP condition $I \cap \check{C} = \emptyset$ is always satisfied.

Example B (Natural Constraints continuation). In the case $C := \Theta := \{\vartheta \in \mathbb{R}^d; 1 + \vartheta^\top R \geq 0 \text{ a.s.}\}$ we have $\check{C} = \{\vartheta \in \mathbb{R}^d; \vartheta^\top R \geq 0 \text{ a.s.}\} \supset I$. Here the NUIP condition $I \cap \check{C} = \emptyset$ amounts to the no arbitrage condition $I = \emptyset$.

Example C (continuation). In the case $C := \{\pi \in \mathbb{R}^d : \pi^k \geq 0\}$ we have $\check{C} = C$. Here the NUIP condition $I \cap \check{C} = \emptyset$ amounts to the no arbitrage condition $I \cap C = \emptyset$.

We now present an example where $\mathbb{E}[\log\, V_1^\pi] = \infty$ for nearly all π, but where V_1^π/V_1^ϑ is bounded in π for nearly all ϑ and where a unique numeraire portfolio exists.

Example D (see Kardaras 2006). Consider the case where $d = 1$ and

$$P[R \in dx] \propto \left(1_{(-1,1]} + x^{-1}(\log\{1+x\})^{-2} \cdot 1_{(1,\infty)}\right) dx\,.$$

Since the support Z of R is $[-1, \infty)$, we have $\Theta = [0, 1] =: C$. Now the expected log-utility is

$$\mathbb{E}[\log\, V_1^\pi] = \mathbb{E}[\log(1 + \pi R)] = \infty \text{ for } \pi \in (0, 1]$$

since $\int_1^\infty \log(1 + \pi x)x^{-1}(\log(1 + x))^{-2}dx = \infty$ which easily follows by use of the substitution $y = \log(1 + x)$. Obviously $\mathbb{E}[\log\, V_1^\pi] = 0$ for $\pi = 0$.

However if we consider relative values $V_1^\pi/V_1^\vartheta = \frac{1+\pi R}{1+\vartheta R}$, then V_1^π/V_1^ϑ is bounded since

$$\min\left(\frac{\pi}{\vartheta}, \frac{1-\pi}{1-\vartheta}\right) \leq V_1^\pi/V_1^\vartheta \leq \max\left(\frac{\pi}{\vartheta}, \frac{1-\pi}{1-\vartheta}\right)\,.$$

Moreover, if we fix $\vartheta \in (0,1)$ and define

$$g(\pi) = \mathbb{E}[\log(V_1^\pi / V_1^\vartheta)] = \mathbb{E}\left[\log\left(\frac{1+\pi R}{1+\vartheta R}\right)\right],$$

then we obtain for $\pi \in (0,1)$

$$g'(\pi) = \mathbb{E}\left[\frac{R}{1+\pi R}\right]$$

where $g'(0+) = \infty$ and $g'(1-) = -\infty$. Therefore there exists a unique $\rho \in (0,1)$ such that $g'(\rho) = 0$. As a consequence we obtain the relation

$$\mathbb{E}[V_1^\pi / V_1^\rho] = \mathbb{E}\left[\frac{1+\pi R}{1+\rho R}\right] = 1 + (\pi - \rho)\mathbb{E}\left[\frac{R}{1+\rho R}\right] = 1$$

for any $\pi \in \Theta$. Then ρ will be called a numeraire portfolio (in the strict sense). The portfolio ρ is computed by Kardaras as $\rho \cong .916$.

Although the expected log-utility is infinite, the numeraire portfolio does not put all the weight on the stock. Finally we know that ρ is the unique portfolio such that

$$\mathbb{E}[\log(V_1^\rho / V_1^\rho)] = \sup_{\pi \in \Pi} \mathbb{E}[\log(V_1^\pi / V_1^\rho)] = 0\,.$$

5 The weak numeraire portfolio and the growth-optimal portfolio

Definition 5.1. (a) A portfolio $\rho \in \Pi$ is *log-optimal* if $\mathbb{E}[\log\ V_1^\pi] \le \mathbb{E}[\log\ V_1^\rho]$ for every $\pi \in \Pi$.

(b) A portfolio $\rho \in \Pi$ will be called *growth optimal portfolio* (GOP) [or *relatively log-optimal*] if $\mathbb{E}[\log(V_1^\pi / V_1^\rho)] \le 0$ for every $\pi \in \Pi$.

The present concept of GOP is used e.g. by Christensen & Larsen 2007, the name (relatively) log-optimal is used e.g. by Karatzas and Kardaras 2007. Of course, if the portfolio ρ is log-optimal with $\mathbb{E}[\log\ V_1^\rho] < \infty$, then ρ is also a GOP and we will prefer the notation GOP in that case. The two notions coincide if $\sup_{\pi\in\Pi} \mathbb{E}[\log\ V_1^\pi] < \infty$. In the Example D above, this condition fails and almost every portfolio is log-optimal. But we have existence of a unique numeraire portfolio which is the unique GOP.

Theorem 5.2. *A portfolio is a weak numeraire portfolio if and only if it is a GOP.*

Note that this result shows in particular that the existence of a weak numeraire portfolio implies the existence of a GOP and vice versa.

Proof of Theorem 5.2. (See Becherer 1999, Christensen and Larsen 2007, Bühlmann and Platen 2003.)

(i) Suppose ρ is numeraire portfolio. Then we have by Jensen's inequality

$$\mathbb{E}[\log(V_1^\pi/V_1^\rho)] \leq \log\left(\mathbb{E}[V_1^\pi/V_1^\rho]\right) \leq \log\, 1 = 0\,.$$

(ii) Suppose that ρ is GOP and π is an arbitrary portfolio. Then $V_1^\varepsilon := (1-\varepsilon)V_1^\rho$ $+\varepsilon V_1^\pi$ is the value of some portfolio where $V_1^\varepsilon - V_1^\rho = \varepsilon(V_1^\pi - V_1^\rho)$. From $1 - t^{-1} \leq \log\, t$ for $t > 0$ we obtain

$$0 \geq \varepsilon^{-1}\cdot \mathbb{E}[\log(V_1^\varepsilon/V_1^\rho)] \geq \varepsilon^{-1}\cdot \mathbb{E}[(V_1^\varepsilon - V_1^\rho)/V_1^\varepsilon] = \mathbb{E}[(V_1^\pi - V_1^\rho)/V_1^\varepsilon].$$

From

$$-2 \leq 2\frac{x-y}{x+y} \leq \frac{x-y}{(1-\varepsilon)y+\varepsilon x} \uparrow \frac{x}{y} - 1 \text{ for } \frac{1}{2} \geq \varepsilon \downarrow 0 \text{ (where } x, y > 0)$$

we finally get $\mathbb{E}[V_1^\pi/V_1^\rho] \leq 1$ from the monotone convergence theorem.

□

Proposition 5.3. *The NUIP condition $I \cap \check{C} = \emptyset$ is necessary for the existence of a GOP.*

Proof. Suppose that ρ is a GOP and suppose that $\vartheta \in I \cap \check{C}$. Since $\vartheta \in I$, we know that $\vartheta^\top R \geq 0$ a.s. and $P[\vartheta^\top R > 0] > 0$. Now we conclude from Lemma 2.2 that $\rho+\vartheta \in C$ where $\mathbb{E}[\vartheta^\top R/V_1^{\rho+\vartheta}] > 0$ and thus $\mathbb{E}[\log(V_1^\rho/V_1^{\rho+\vartheta})] \leq \log\, \mathbb{E}[V_1^\rho/V_1^{\rho+\vartheta}] = \log\{1 - \mathbb{E}[\vartheta^\top R/V_1^{\rho+\vartheta}]\} < 0$. Now we have a contradiction to the optimality of ρ. □

However, one can also directly derive Proposition 5.3 from Theorem 5.2 and Proposition 4.1 without a proof.

6 Existence of weak numeraire portfolios

In this section we will show that the NUIP condition is also sufficient for the existence of a weak numeraire portfolio. This in particular shows that valuation via discounting by the wealth process of a weak numeraire can even be performed in situations where the no arbitrage condition is not satisfied. This was already emphasised by Platen 2006. For getting the existence result we need some technical notations and results.

Definition 6.1. For $f : \mathbb{R}^d \mapsto (0,1]$ we write $f \in \mathcal{F}$ if $\mathbb{E}[f(R)\cdot \log(1+\|R\|)] < \infty$.

Example E. (see Kardaras 2006) We have $f_k \in \mathcal{F}$ and $f_k \uparrow 1$ for

$$f_k(x) := 1_{\{\|x\|\leq 1\}} + 1_{\{\|x\|>1\}}\cdot \|x\|^{-1/k}\,.$$

Under the no-arbitrage condition $I \cap \Theta = \emptyset$, one knows that $\Theta \cap \mathcal{L}$ is compact (see Korn and Schäl 1999) where Θ is defined in Example B. Under the weaker NUIP condition we need the following more technical lemma.

Lemma 6.2. *Assume $I \cap \check{C} = \emptyset$. Let $\mathcal{F}^*$ be some subset of $\mathcal{F}$ which is bounded from below in the following sense: there is some $f^* \in \mathcal{F}$ such that $f \geq f^*$ for all $f \in \mathcal{F}^*$.*

Let $R \subset C$ be a set of portfolios which are "not too bad" in the following sense: for every $\rho \in R$, $\rho \in \mathcal{L} \setminus \{0\}$, there exists some $f \in F^*$ such that the function $[0,1] \ni u \mapsto g_f(u\rho)$ is increasing where

$$g_f(\pi) := \mathbb{E}[\log(1 + \pi^\top R) \cdot f(R)] \; [\leq (\log \|\pi\|)^+ + \mathbb{E}[\log(1 + \|R\|) f(R)] < \infty] \,.$$

Then R is bounded.

The lemma is hidden in the proof of Theorem 3.15 in Karatzas and Kardaras 2007.

Proof by contradiction. Suppose there exists some sequence $(\rho_m, f_m) \subset R \times \mathcal{F}^*$ such that $\rho_m \in \mathcal{L} \cap C$ and $[0,1] \ni u \mapsto g_{f_m}(u\rho_m)$ is increasing where $\|\rho_m\| \to \infty$. Define $\xi_m := \|\rho_m\|^{-1}\rho_m$. We can assume that $\xi_m \to \xi$ for some $\xi \in \mathcal{L}$. We want to show that $\xi \in \check{C}$.

Choose any $a > 0$ and m_a such that $0 < u = a/\|\rho_m\| < 1$ for $m \geq m_a$. Then $a\xi_m = u\rho_m = u\rho_m + (1-u)0 \in C$ since C is convex. Moreover, since C is closed, we also have $a\xi \in C$. This proves $\xi \in \check{C} \cap \mathcal{L}$ with $\|\xi\| = 1$. Now for $u \in (0,1]$ we have

$$\begin{aligned} 0 &\leq \varepsilon^{-1}[g_{f_m}(u\rho_m) - g_{f_m}((1-\varepsilon)u\rho_m)] \\ &= \mathbb{E}\left[\varepsilon^{-1}\left(\log\left\{1 + u\,\rho_m^\top R\right\} - \log\left\{1 + (1-\varepsilon)u\,\rho_m^\top R\right\}\right) \cdot f_m(R)\right]. \end{aligned}$$

From the concavity of log we conclude that the integrand is decreasing for $\varepsilon \downarrow 0$. Since the expectation if finite for $\varepsilon = 1$, we apply the monotone convergence theorem and obtain

$$0 \leq \mathbb{E}\left[\frac{d}{du}\log\{1 + u\rho_m^\top R\} \cdot f_m(R)\right] = \mathbb{E}\left[(1 + u\rho_m^\top R)^{-1}\rho_m^\top R f_m(R)\right].$$

Again choose any $a > 0$ and m_a such that $0 < u = a/\|\rho_m\| < 1$ for $m \geq m_a$. Then

$$0 \leq \mathbb{E}[(1 + a\,\xi_m^\top R)^{-1}\xi_m^\top R\, f_m(R)] \text{ where } (1 + a\,\xi_m^\top R)^{-1}\xi_m^\top R\, f_m(R) \leq a^{-1}.$$

From Fatou's lemma we now obtain

$$\begin{aligned} a^{-1} &\geq \mathbb{E}\Big[(1 + a\xi^\top R)^{-1}\xi^\top R \,\overline{\lim_m}\, f_m(R)\Big] \\ &\geq \overline{\lim_m}\, \mathbb{E}\Big[(1 + a\xi_m^\top R)^{-1}\xi_m^\top R f_m(R)\Big] \geq 0 \end{aligned}$$

Since $\overline{\lim}\, f_m(R) \geq f^*(R) > 0$, we conclude from the first inequality that $1 + a\,\xi^\top R > 0$ a.s.

Now $a > 0$ was arbitrary, so we conclude that $\xi^\top R \geq 0$ a.s. where $\|\xi\| = 1$ and $\xi \in \mathcal{L}$. Therefore $P[\xi^\top R = 0] > 0$, otherwise $\xi \in \mathcal{L}^\perp$. Thus we finally have $\xi \in I$ and hence $\xi \in I \cap \check{C}$ which is a contradiction to our assumption. □

Theorem 6.3. *Under the NUIP assumption $I \cap \check{C} = \emptyset$, there exists a weak numeraire portfolio ρ.*

If $\mathbb{E}[\log(1 + \|R\|)] < \infty$, *then* ρ *is obtained as the unique solution of the following concave optimisation problem and thus the only GOP in* $C \cap \mathcal{L}$:

$$\rho = \arg \max_{\pi \in C \cap \mathcal{L}} g(\pi) \text{ where } g(\pi) := \mathbb{E}[\log(1 + \pi^\top R)].$$

Remark 6.4. In the general case, where the condition $\mathbb{E}[\log(1 + \|R\|)] < \infty$ does not hold, one can solve a sequence of optimisation problems and show that the corresponding solutions converge to the solution of the original problem, see below and Theorem 3.15 in Karatzas and Kardaras 2007.

Proof. We start with a sequence $(f_k) \subset \mathcal{F}$ where $f_k \uparrow 1$. The sequence can be chosen as in Example E above. Now define $g_k(\pi) = g_{f_k}(\pi) := \mathbb{E}[\log(1 + \pi^\top R) \cdot f_k(R)]$.

Then g_k is strictly concave on $C \cap \mathcal{L}$ and $-\infty \leq g_k(\pi) < +\infty$. Further set

$$0 \leq g_k^* := \sup_{\pi \in C} g_k(\pi) = \lim_{n \to \infty} g_k(\rho_{kn})$$

for some sequence $(\rho_{kn}) \subset C$. Since $g_k(\pi + \zeta) = g_k(\pi)$ for $\zeta \perp \mathcal{L}$ we can choose $\rho_{kn} \in \mathcal{L} \cap C$. Moreover, we may choose ρ_{kn} such that $g_k(\rho_{kn}) = \max_{0 \leq u \leq 1} g_k(u\rho_{kn}) \leq \sup_{\pi \in C} g_k(\pi)$.

Then by concavity, $u \mapsto g_k(u\rho_{kn})$ is increasing. From the preceding lemma we know that $R = (\rho_{kn})$ is bounded, in particular $g_k^* \in [0, \infty)$ and $g_k^* = g_k(\rho_k^*)$.

Now fix some k and assume *w.l.o.g.* that $\rho_{kn} \to \rho_k^*$ for some $\rho_k^* \in C$ where $g_k^* = g_k(\rho_k^*)$ since C is closed. Choose $\pi \in C$, then $[0, 1] \ni u \mapsto g_k(\rho_k^* + u(\pi - \rho_k^*))$ is real valued, concave. Since ρ_k^* is a maximum point, we conclude from the concavity that

$$0 \leq G(u) := \frac{1}{u}[g_k(\rho_k^*) - g_k(\rho_k^* + u(\pi - \rho_k^*))] \leq g_k(\rho_k^*) - g_k(\pi)$$

is increasing in $0 < u \leq 1$. From the monotone convergence theorem, we obtain for $u \downarrow 0$, again by concavity of log,

$$G(u) \downarrow E\left[-\frac{d}{du} \log\left\{1 + [\rho_k^* + u(\pi - \rho_k^*)]^\top R\}\right]\Big|_{u=0} \cdot f_k(R) \right].$$

Thus we get

$$\mathbb{E}[(\pi - \rho_k^*)^\top R \,/(1 + \rho_k^{*\top} R)] \leq 0.$$

Since we know that (ρ_k^*) is also bounded, we may assume that $\rho_k^* \to \rho$ for some $\rho \in C \cap \mathcal{L}$.

Now $(\pi - \rho_k^*)^\top R \,/(1 + \rho_k^{*\top} R) = (1 + \pi^\top R)/(1 + \rho_k^{*\top} R) - 1 \geq 1$.

Then we obtain from Fatou's lemma $r(\pi|\rho) \leq 0$, since

$$r(\pi|\rho) = \mathbb{E}[(\pi - \rho)^\top R \,/(1 + \rho^\top R)] = \mathbb{E}[\lim_k (\pi - \rho_k^*)^\top R \,/(1 + \rho_k^{*\top} R)] \leq \overline{\lim}\, \mathbb{E}[\cdots] \leq 0.$$

From Lemma 3.3 we finally conclude that ρ is a weak numeraire portfolio.

In the case where $\mathbb{E}[\log(1 + \|R\|) < \infty] < \infty$, we can choose $f_k \equiv 1$ for all k and thus $\rho_k = \rho$.

$$\text{Then } g^* := \sup_{\pi \in C} g(\pi) = g(\rho_k^*) = \max_{\pi \in C \cap \mathcal{L}} \text{ where } g(\pi) := \mathbb{E}[\log(1 + \pi^\top R)].$$

Since g is strictly concave on $C \cap \mathcal{L}$, the maximum point is unique. □

7 Deflators and value preserving portfolios

The concept of a deflator is important for the valuation of uncertain payment streams and is more general than that of a numeraire portfolio.

Definition 7.1. The class $\mathcal{D}$ of *supermartingale deflators* is defined as

$$\mathcal{D} := \{D \geq 0;\ D \text{ is a random variable with } \mathbb{E}[DV_1^\pi] \leq 1 (= V_0^\pi)\} \text{ for all portfolios } \pi\}.$$

Since $0 \in \Pi$, we know that $\mathbb{E}[D] \leq 1$ for $D \in \mathcal{D}$.

Corollary 7.2. *(a) A portfolio $\rho \in \Pi$ is a weak numeraire portfolio if and only if $(V_1^\rho)^{-1}$ is a supermartingale deflator.*

(b) $\mathbb{E}[\log\ V_1^\rho] = \inf_{D \in \mathcal{D}} \mathbb{E}[\log(D^{-1})]$.

The second property in (a) is introduced by Korn 1997 and called "ρ is *interest-oriented*". The property (b) of ρ can be seen as an optimal property dual to log-optimality.

Proof. (a) is clear by definition. (b) See Becherer 2001. $\mathbb{E}[\log(D^{-1})]$ makes sense since $\mathbb{E}[\log^-(D^{-1})] \leq \mathbb{E}[D] \leq 1$. Assume *w.l.o.g.* that the right hand in (b) is finite and $\mathbb{E}[\log(D^{-1})] \in \mathbb{R}$. Then $\mathbb{E}[\log\ V_1^\rho - \log(D^{-1})] = \mathbb{E}[\log(DV_1^\rho)] \leq \log\ \mathbb{E}[DV_1^\rho] = 0$. □

Definition 7.3 (Hellwig 1996). For $\pi \in \Pi$ and $D \in \mathcal{D}$, $V_D^\pi := D \cdot (1 + \pi^\top R)$ is called *present economic value* of π (at time 0) associated with $D \in \mathcal{D}$.

Since D is a supermartingale deflator, we always have $\mathbb{E}[V_D^\pi] \leq 1$ where 1 is here the initial value. Therefore the following definition is interesting:

Definition 7.4 (Hellwig 1996). A portfolio $\pi \in \Pi$ is called *value preserving* if $V_D^\pi \equiv 1$ a.s. for some $D \in \mathcal{D}$.

Theorem 7.5. *The following properties are equivalent:*

(1) π is value preserving w.r.t. the supermartingale deflator D;

(2) π is a weak numeraire portfolio and $D = (1 + \pi^\top R)^{-1}$.

Thus, by Theorem 7.5 existence of a value preserving portfolio is also related to the existence of a GOP (see Korn and Schäl 1998).

Proof. "(1) $\Rightarrow$ (2)" From $D \cdot (1 + \pi^\top R) = 1$ we get $D = (1 + \pi^\top R)^{-1}$ where $D \in \mathcal{D}$. Now Corollary 7.2 (a) applies. "(2) $\Rightarrow$ (1)" Again from Corollary 7.2 we know that $D = (1 + \pi^\top R)^{-1}$ is a deflator and $D \cdot (1 + \pi^\top R) = 1$. □

8 Fair portfolios and applications in actuarial valuation

Benchmarked portfolios and fair valuation is a concept that is suggested for use in actuarial valuation by Bühlman and Platen 2003. As ibidem we call V_1^π/V_1^ρ the *benchmarked value of portfolio* π if ρ is a weak numeraire portfolio and hence V_1^ρ is uniquely determined according to Proposition 3.2. Then we know that: $\mathbb{E}[V_1^\pi/V_1^\rho] \le 1$ for every portfolio π.

In financial valuations in competitive markets, a price is typically chosen such that seller and buyer have no systematic advantage or disadvantage. Let the random variable H be a contingent claim which is a possibly negative random payoff. Candidates for prices of H are $\mathbb{E}[DH]$ for some deflator $D \in \mathcal{D}$. For $H = V_1^\pi$ we thus have $\mathbb{E}[DH] \le 1$. For the case $\mathbb{E}[DH] < 1$, this could give an advantage to the seller of the portfolio π; its expected future benchmarked payoff is less than its present value. The only situation when buyers and sellers are equally treated is when the benchmarked price process V_t^π/V_t^ρ is a martingale, that means in our situation: $\mathbb{E}[V_1^\pi/V_1^\rho] = 1$.

Definition 8.1 (see Bühlmann and Platen 2003). A value process V_t, $t = 0, 1$, is called *fair* if its benchmarked value V_t/V_t^ρ is a martingale, i.e. if $\mathbb{E}[V_1/V_1^\rho] = V_0$ (since $V_0^\rho = 1$).

Let us consider a contingent claim H, which has to be paid at the maturity date 1. Let ρ be the weak numeraire portfolio. We choose the following pricing formula

$$pr(H) := \mathbb{E}[H/V_1^\rho] \tag{8.1}$$

which by definition is fair. In contrast to classical actuarial valuation principles no loading factor enters the valuation formula. For premium calculations in insurance business the use of a change of measure is explained in Delbaen & Haezendonck 1989. An important case arises when H is independent of the value V_1^ρ. Then we obtain

$$pr(H) = \mathbb{E}[H] \cdot \mathbb{E}[1/V_1^\rho]\,. \tag{8.2}$$

Here $P(0,1) = \mathbb{E}[1/V_1^\rho]$ is the fair price of the contingent claim $H' \equiv 1$ to be paid at the maturity date $T = 1$ and thus the *zero coupon bond* with maturity 1. Thus (8.2) is the classical actuarial pricing formula in the case of stochastic interest rates and $pr(H) := \mathbb{E}[H/V_1^\rho]$ is an extension to the more general case where dependence may occur.

For *equity-linked* or *unit-linked insurance contracts* we look again at a claim H payed at $T = 1$ where H has the following form: $H = U \cdot V_1^\pi$. Intuitively, H stands now for unit linked benefit and premium. Then H can be of either sign. The benefit at maturity is linked to some strictly positive reference portfolio V_1^π with given portfolio π. The insurance contract specifies the reference portfolio π and the random variable U depending on the occurrence of insured events during the period $(0, 1]$, for instance, death, disablement or accident.

These products offer the insurance company as well as the insurance customer advantages compared to traditional products. The insurance industry may benefit from offering more competitive products and the customer may benefit from higher yields in financial markets. Compared to classical insurance products, one distinguishing feature of unit-linked products is the random amount of benefit. But the traditional basis for pricing life insurance policies, the principle of equivalence, based on the idea that premiums and expenses should balance in the long run, does not deal with random benefits. Therefore, we have to use financial valuation theories together with elements of actuarial theory to price such products.

The *standard actuarial value* $pr_o(H)$ of the contingent claim $H = U \cdot V_1^\pi$ is determined by the properly defined liability of prospective reserve as

$$pr_o(H) = V_0^\pi \cdot \mathbb{E}[H/V_1^\pi] = \mathbb{E}[H \cdot V_0^\pi / V_1^\pi].$$

The standard actuarial methodology assumes that the insurer invests all payments in the reference portfolio π. Then one obtains for $pr_o(H)$, when expressed in units of the domestic currency, the expression

$$pr_o(H) = pr_o(U \cdot V_1^\pi) = V_0^\pi \cdot \mathbb{E}[U].$$

We observe the difference between $pr_o(H)$ and $pr(H)$. Hence the standard actuarial pricing and fair pricing will, in general, lead to different results. As one could see this is to be expected when the endowments depend on the numeraire portfolio. Indeed, let us assume that ρ is a numeraire portfolio (in the strict sense), then $\mathbb{E}[V_1^\pi/V_1^\rho] = V_0^\pi/V_0^\rho = V_0^\pi$ and we obtain

$$pr(U \cdot V_1^\pi) - pr_o(U \cdot V_1^\pi) = \mathrm{Cov}(U, V_1^\pi/V_1^\rho). \tag{8.3}$$

A similar formula is derived by Dijkstra 1998. Hence, the two prices coincide if and only if U and V_1^π/V_1^ρ are uncorrelated. Moreover, the sign of the difference is the sign of the covariance. This condition differs from the one given by Bühlmann and Platen 2003.

In many cases, the endowment H of the insurance contract will include a guaranteed (non-stochastic) amount $g(K)$ where K is the premium paid by the insured. Then the benefit at maturity is composed of the guaranteed amount plus a call option with exercise price $g(K)$ and with the reference portfolio as underlying assets. Then the fair premium is the solution to an equation in K and $g(K)$ (see Nielsen and Sandmann 1995).

9 Existence of numeraire portfolios

It seems to be a general agreement that S_t^k should be fair, since S_0^k is a fair price for $H = S_1^k$ for every $k \in \{0, 1, \ldots, d\}$. This leads to the requirement

$$\mathbb{E}[S_1^k/V_1^\rho] = S_0^k,\ 0 \le k \le d. \tag{9.1}$$

Definition 9.1. A portfolio $\rho \in \Pi$ will be called *numeraire portfolio* (in the strict sense), if the above condition (9.1) holds.

Proposition 9.2. *(a) If ρ is a numeraire portfolio, then we have for any strategy ξ and*

$$V_1^\xi(x) := x + \xi^\top \Delta S : \quad \mathbb{E}[V_1^\xi(x)/V_1^\rho] = x = V_0^\xi .$$

(b) A numeraire portfolio is a weak numeraire portfolio.

Proof. Set $V_t^\xi(x) = \sum_{k=0}^d \xi^k S_t^k$. Then we obtain $\mathbb{E}[V_1^\xi/V_1^\rho] = \sum_{k=0}^d \xi^k \mathbb{E}[S_1^k/V_1^\rho] = \sum_{k=0}^d \xi^k S_0^k = V_0$. In the present simple situation, where the horizon is $T = 1$ we do not have any integrability problems and we even get the martingale property. In the more general case we obtain the supermartingale property from the fact that each non-negative local martingale is a supermartingale. □

As in Lemma 3.3 we know that ρ is a numeraire portfolio if and only if $r(\pi|\rho) = 0$, where π is a unit vector or the zero vector in $\mathbb{R}^d$. There $r(\pi|\rho)$ is the directional derivative of $g(\pi) := \mathbb{E}[\log(1 + \pi^\top R)]$ at the point ρ in the direction of $\pi - \rho$ (if g is finite).

In general, we cannot expect to be able to compute the numeraire portfolio just by naively trying to solve the first-order condition $\nabla g(\rho) = r(0|\rho) = 0$, because sometimes this equation simply fails to have a solution. In this section, we make the following assumptions.

Assumption 9.3. $C = \Theta$ describes the natural constraints, $I = \emptyset$ which here is the NUIP condition, and integration of the log exists in the following sense: $\mathbb{E}[\log(1 + \|R\|)] < \infty$.

We now introduce another condition given in the following theorem proved in Schäl (1999, Theorem 4.15):

Theorem 9.4. *Let ρ be the only GOP in $\Theta \cap \mathcal{L}$ according to Theorem 6.3. Then, the condition*

$$\mathbb{E}[\, \vartheta^\top \cdot R/(1 + \vartheta^\top \cdot R)] < 0 \textit{ for all } \vartheta \in \partial\Theta \cap \mathcal{L}\,, \tag{9.2}$$

implies the first order condition: $\mathbb{E}[R^k/(1 + \rho^\top \cdot R)] = 0$ for $k = 1, \dots, d$.

Corollary 9.5. *Let ρ be defined as in the preceding theorem. Then, under condition (9.2), ρ is a numeraire portfolio (in the strict sense) and*

$$\mathbb{E}[\, 1/V_1^\rho] = 1. \tag{9.3}$$

Proof. We obtain from Theorem 9.4

$$1 = 1 - \sum_{k=1}^d \rho^k \mathbb{E}[R^k/(1 + \rho^\top \cdot R)] = 1 - \mathbb{E}[\rho^\top \cdot R/(1 + \rho^\top \cdot R)]\,,$$

which implies (9.3). Now we get for $0 \le k \le d$ with $R^0 \equiv 0$:

$$\mathbb{E}[S_1^k/V_1^\rho] = \mathbb{E}[S_0^k(1 + R^k)/(1 + \rho^\top \cdot R)] = S_0^k \mathbb{E}[1/(1 + \rho^\top \cdot R)] = S_0^k\,.$$

Thus ρ is a numeraire portfolio. □

Example F. *The one-dimensional case.*

Consider the case where $d = 1$ and R is bounded, then the support Z is a compact subset of $\mathbb{R}$. Set $-\alpha = \min\ Z$, $\beta = \max\ Z$. Then $\mathrm{conv}(Z) = [-\alpha, \beta]$. For the no-arbitrage condition we need $\alpha > 0$, $\beta > 0$.

Then condition (9.3) is satisfied if and only if

$$\mathbb{E}[\ R/(1 + \frac{1}{\alpha}R)] < 0 < \mathbb{E}[\ R/(1 - \frac{1}{\beta}R)]\,. \tag{9.4}$$

For a proof we have

$$\min_{z \in Z} 1 + \vartheta z = \min_{-\alpha \le z \le \beta} 1 + \vartheta z = 1 - \vartheta\alpha \text{ for } \vartheta > 0 \text{ and } = 1 + \vartheta\beta \text{ for } \vartheta < 0\,.$$

Hence, we know that $\Theta = [-\frac{1}{\beta}, \frac{1}{\alpha}]$ and $\partial\Theta = \{-\frac{1}{\beta}, \frac{1}{\alpha}\}$.

Then $E\left[\frac{\vartheta \cdot R}{1+\vartheta \cdot R}\right] = \vartheta \cdot E\left[\frac{R}{1+\vartheta \cdot R}\right] < 0$ for $\vartheta \in \partial\Theta$ if and only if (9.4) holds. In fact, the condition (9.4) is weak. It can be looked upon as a kind of no-arbitrage condition. The martingale case $\mathbb{E}[R] = 0$ is not interesting as we can choose $\vartheta = 0$ then. Let us suppose that $\mathbb{E}[R] > 0$. Then $\mathbb{E}[R/(1 - R/\beta)] \ge \mathbb{E}[R] > 0$ and the condition $\mathbb{E}[R/(1 + R/\alpha)] < 0$ requires that there should not be too little probability for negative values of R. The condition (9.4) can easily be proved to be also necessary for the first order condition.

We will give a sufficient condition for (9.2) which is far from being necessary, however.

Theorem 9.6. *If Ω or Z is finite, then the condition (9.2) is always satisfied and thus the statements of Corollary 9.5 hold true.*

Proof (See also Long 1990). If Ω is finite, then Z is finite. Choose $\vartheta \in \partial\Theta \cap \mathcal{L}$, then one obtains the following relation :

$$0 = \min_{z \in Z}(1 + \vartheta^\top \cdot z) = 1 + \vartheta^\top \cdot z_o \text{ for some } z_o \in Z\,.$$

Further, $\{R = z_o\} \subset \{1 + \vartheta^\top R = 0\} = \{\vartheta^\top \cdot R = -1\}$. Now

$$\begin{aligned}
\mathbb{E}[\vartheta^\top \cdot R/(1 + \vartheta^\top \cdot R)] &\le \mathbb{E}[1_{\{R=z_o\}} \cdot \vartheta^\top \cdot R/(1 + \vartheta^\top \cdot R)] \\
&\quad + \mathbb{E}[1_{\{\vartheta^\top \cdot R > 0\}} \cdot \vartheta^\top \cdot R/(1 + \vartheta^\top \cdot R)] \\
&\le -\mathbb{E}[1_{\{R=z_o\}}/(1 + \vartheta^\top \cdot R)] + 1 \\
&= -\infty, \text{ since } P[R = z_o] > 0\,.
\end{aligned}$$

□

For the theorem one can also use a result by Hakansson 1971 that the GOP can be chosen as an interior point. The theorem is generalised in (Korn and Schäl 1999, Theorem 4.22).

It is known that the existence of a growth-optimal portfolio will not imply the existence of a numeraire portfolio (see Becherer 2001). We will give an example.

Example G. We may restrict attention to the case $d = 1$ (see Example F). Let the distribution of R on $Z := [-1, 1]$ be given by

$$\mathbb{E}[g(R)] := \lambda \cdot \int_{-1}^{0} \frac{3}{2}(1 - z^2)g(z)dz + (1 - \lambda) \cdot \int_{0}^{1} \frac{3}{2}(1 - z^2)g(z)dz\,,$$

where we choose $\lambda > 0$ sufficient small, e.g. $\lambda = 1/12$. Then

$$\begin{aligned} \mathbb{E}[R] &:= \lambda \cdot \int_{-1}^{0} \frac{3}{2}(1 - z^2)z\,dz + (1 - \lambda) \cdot \int_{0}^{1} \frac{3}{2}(1 - z^2)z\,dz \\ &= (1 - 2\lambda) \int_{0}^{1} \frac{3}{2}(1 - z^2)z\,dz = \frac{3}{8}(1 - 2\lambda) > 0\,. \end{aligned}$$

[Obviously, by the choice of $\lambda = \lambda* = \frac{1}{2}$, one obtains an equivalent martingale measure (see below)]. Now set

$$f(\vartheta) := E\left[\frac{R}{1 + \vartheta \cdot R}\right],$$

then f is strictly decreasing on $\Theta := [-1, 1]$, where $f(-1) \geq f(\vartheta) \geq f(1)$ for $\vartheta \in \Theta$. Now

$$\begin{aligned} f(1) &= E\left[\frac{R}{1 + R}\right] \\ &= \lambda \int_{-1}^{0} \frac{3}{2}(1 - z)z\,dz + (1 - \lambda) \cdot \int_{0}^{1} \frac{3}{2}(1 - z)z\,dz \\ &= \frac{1}{4} - \frac{3}{2}\lambda > 0\,, \\ f(-1) &= E\left[\frac{R}{1 - R}\right] \\ &= \lambda \int_{-1}^{0} \frac{3}{2}(1 + z)z\,dz + (1 - \lambda) \cdot \int_{0}^{1} \frac{3}{2}(1 + z)z\,dz \\ &= \frac{5}{4} - \lambda > 0\,. \end{aligned}$$

Hence there is no $\vartheta \in \Theta$ such that $f(\vartheta) = 0$ and ϑ is hence a numeraire portfolio. On the other hand, we have $\infty > f(-1) \geq f(\vartheta) = \frac{d}{d\vartheta} E\left[\ln(1 + \vartheta \cdot R)\right] \geq f(1) > 0$ for $-1 < \vartheta < 1$.

Thus, we know that $\max_{\vartheta \in \Theta} \mathbb{E}[\ln(1 + \vartheta \cdot R)] = \mathbb{E}[\ln(1 + R)]$ and $\vartheta* = 1$ defines the GOP.

10 Equivalent martingale measures and the numeraire portfolio

A well-known candidate for a fair price of a financial derivative described by the contingent claim H is given by an EMM Q (defined below) with positive density dQ/dP according to

$$E_Q\left[H/S_1^0\right] = \mathbb{E}\left[\frac{dQ}{dP}H/S_1^0\right]$$

where $\frac{dQ}{dP}/S_1^0$ is a deflator (see Duffie 1992, p. 23).

Definition 10.1. A probability measure Q is an equivalent martingale measure (EMM), if Q has a (a.s.) positive density dQ/dP such that

$$E_Q[S_1^k/S_1^0] = \mathbb{E}\left[\frac{dQ}{dP}S_1^k/S_1^0\right] = S_0^k/S_0^0,\ 0 \le k \le d. \tag{10.1}$$

Here, we present the general property though we decided to consider only the case $S_t^0 \equiv 1$.

Proposition 10.2. *(a) A portfolio $\rho \in \Pi$ is a numeraire portfolio if and only if $1/V_1^\rho = dQ^*/dP$ for some EMM Q^*.*

(b) In the case of existence, an EMM Q^* implied by a numeraire portfolio in the sense of (a) is unique.

Proof. (a) We make use of (9.1). For the 'only if'-direction we get $\mathbb{E}[dQ^*/dP] = 1$ from $dQ^*/dP = 1/V_1^\rho > 0$ and $\mathbb{E}[S_1^k/V_1^\rho] = S_0^k$ for $k = 0$.

Part (b) follows from the uniqueness of V_1^ρ according to Proposition 3.2. □

From the "Fundamental Theorem of Asset Pricing" (see Back and Pliska 1990, Dalang et al. 1990, Schachermayer 1992, Rogers 1994, Jacod & Shiryaev 1998) we know that there exists an EMM if and only if the no arbitrage condition $I = \emptyset$ holds. If in addition the market is complete, then the EMM Q is known to be unique and we may consider $L^{-1} := (dQ/dP)^{-1}$ as a contingent claim. Upon making use of the definition of completeness, we obtain $L^{-1} = V^\xi(x)$ for some strategy ξ and some initial capital x. Then we obtain $x = \mathbb{E}[L\ V_1^\xi(x)] = \mathbb{E}[L\ L^{-1}] = 1$. Therefore we conclude that $V_1^\xi(x) = 1 + \rho^\top R$ where $\rho^k = \xi^k S_0^k$. From the preceding proposition we obtain the following result:

Corollary 10.3. *Let $C = \Theta$ describe the natural constraints. If the market is complete and free of arbitrage opportunities, then a numeraire portfolio (in the strict sense) exists.*

For the remainder of this section, we consider the case where $d = 1$ and (as in Example F):

$$\mathrm{conv}(Z) = [-\alpha, \beta] \text{ for some } \alpha,\ \beta > 0 \text{ with } -\alpha,\ \beta \in Z. \tag{10.2}$$

The minimal martingale measure was introduced by Föllmer and Schweizer 1991 in the context of option hedging and pricing in incomplete financial markets. By the discrete-time Girsanov transformation one obtains the minimal martingale Q^o according to $\frac{dQ}{dP}^o = b + a \cdot R$ (see Schweizer 1995). From the two conditions $\mathbb{E}[\frac{dQ}{dP}^o] = 1$ and $\mathbb{E}[\frac{dQ}{dP}^o R] = 0$, one can compute that

$$b = 1 + \{\mu/\sigma\}^2, \; a = -\mu/\sigma^2 \text{ where } \mu := \mathbb{E}[R] \text{ and } \sigma^2 := \mathrm{Var}[R]\,. \tag{10.3}$$

One difficulty with the Girsanav transformation in discrete time is that it may lead to a density with negative values. The resulting martingale measure is then called a *signed martingale measure*. However, in the case where $Z \subset \{d-1, 0, u-1\}$ for some $0 < d < 1 < u$, it is easy to see that $\frac{dQ}{dP}^o > 0$. On the other hand, we know from Theorem 9.6 and Corollary 9.5 that $\frac{dQ}{dP}^* = \{1 + \rho R\}^{-1} > 0$ always defines a (positive) martingale measure if Z is finite. Thus we know that the minimal martingale measure cannot coincide with the martingale measure Q^* induced by the numeraire portfolio if Q^o is not a positive measure but a signed measure. It can be shown that the two measures only coincide in a binomial model that means only for a complete market (according to Harrison & Pliska 1981 and Jacod & Shiryaev 1998).

A binomial model is characterised by the fact

$$R \in \{-\alpha, \beta\} \text{ a.s.} \tag{10.4}$$

Theorem 10.4. *Let Q^* be the measure defined by Proposition 10.2 and let Q^o be the minimal martingale measure. Then $Q^* = Q^o$ if and only if (10.4) holds.*

The proof is given in Korn and Schäl (1999, Theorem 5.18). The theorem is surprising because one always has $Q^* = Q^o$ in the important case of financial markets modeled by diffusion processes (see Becherer 2001, Korn 1998).

11 Portfolio optimisation and the numeraire portfolio

So far we mainly highlighted the role of the numeraire portfolio in valuation of uncertain payment streams. However, we already saw that the numeraire portfolio is closely related to the growth optimal portfolio. In this section, we generalise this and show that, for a wide class of portfolio optimisation problems, the numeraire portfolio is the main ingredient of their solution.

Definition 11.1. (a) A strictly concave function U on $(0, \infty)$ which is increasing, twice continuously differentiable and satisfies

$$U'(0+) = \infty, \quad U'(\infty) = 0 \tag{11.1}$$

is called a *utility function*.

(b) We call the optimisation problem

$$u(x) := \sup_{\pi\in\Pi} \mathbb{E}[U(V_1^\pi(x))], \text{ where } V_1^\pi(x) = x \cdot V_1^\pi, \tag{11.2}$$

the *portfolio problem* of an investor with initial value x.

Popular utility functions are $U(x) = \log\, x$ or $U(x) = \frac{1}{\gamma}x^\gamma$ for $\gamma < 1$. The portfolio problem can now be explicitly solved in a complete market setting:

Theorem 11.2. *Let ρ be the weak numeraire portfolio; define*

$$I(y) = (U')^{-1}(y),\ X(y) = \mathbb{E}[I(y/V_1^\rho)/V_1^\rho],\ Y(x) = X^{-1}(y)\,, \tag{11.3}$$

$$B = I(Y(x)/V_1^\rho) \text{ and assume} \tag{11.4}$$

$$X(y) < \infty\,. \tag{11.5}$$

(a) Then $\mathbb{E}[U(V_1^\pi(x))] \leq \mathbb{E}[U(B)]$ and $\mathbb{E}[B/V_1^\rho] = x$ for all admissible portfolios π.

(b) If the market is complete and ρ is chosen as the numeraire portfolio, then B is the optimal final value for the portfolio problem of an investor with initial wealth x.

Proof. Under the assumption (11.5) it can easily be shown (by dominated and/or monotone convergence) that $X(y)$ is strictly decreasing with $X(0) = \infty$, $X(\infty) = 0$. Thus, an inverse $Y(x)$ exists and one can define B as in (11.4). Further, by construction, B satisfies

$$\mathbb{E}[B/V_1^\rho] = x\,, \tag{11.6}$$

while for all other admissible portfolios π we have

$$\mathbb{E}[V_1^\pi(x)/V_1^\rho] \leq x\,. \tag{11.7}$$

The following property of a concave function

$$U(x) \leq U(y) + U'(y)(x-y),\ y, x > 0\,, \tag{11.8}$$

implies that

$$U(x) \leq U(I(y)) + y(x - I(y)),\ y, x > 0\,. \tag{11.9}$$

From (11.9), (11.4)–(11.7) we then obtain

$$\mathbb{E}[U(V_1^\pi(x))] \leq \mathbb{E}[U(B) + Y(x)(\mathbb{E}[V_1^\pi(x)/V_1^\rho] - \mathbb{E}[B/V_1^\rho]) \leq \mathbb{E}[U(B)]\,. \tag{11.10}$$

If the market is complete, then there exists a portfolio π^B and an initial value x^B generating the final payment of B, i.e. $V_1^{\pi^B}(x^B) = B$. Now, $x^B = \mathbb{E}[B/V_1^\rho] = x$ by (11.6), since ρ is a numeraire portfolio (see proof of Proposition 9.2). Thus π^B is a solution to the portfolio problem. □

Example H. (a) Note that for $U(x) = \log\, x$, we recover from Theorem 11.2 that we have

$$B = xV_1^\rho\,, \tag{11.11}$$

which restates the relation between the growth optimal and the numeraire portfolio.

(b) For $U(x) = \frac{1}{\gamma}x^\gamma$, Theorem 11.2 yields the optimal final wealth of

$$B = x(V_1^\rho)^{\gamma'}/\mathbb{E}[(V_1^\rho)^{\gamma\cdot\gamma'}] \text{ where } \gamma' = \frac{1}{1-\gamma}. \tag{11.12}$$

Remark 11.3. Portfolio optimisation problems in incomplete markets can be solved by duality methods in a similar way as in Kramkov and Schachermayer 1999. There, the problem is transformed to auxiliary markets which are complete. The portfolio problem in these markets is again solved with the help of the numeraire portfolio.

The optimisation problem (11.2) makes sense only if its value function u is finite. Due to the concavity of U, if $u(x) < +\infty$ for some $x > 0$, then $u(x) < +\infty$ for all $x > 0$ and u is continuous, concave and increasing. When we have $u(x) = \infty$ for some (equivalently, all) $x > 0$, there are two cases. Either the supremum in (11.2) is not attained, so there is no solution; or, in case there exists a portfolio with infinite expected utility, the concavity of U will imply that there will be infinitely many of them. We will show that one cannot do utility optimisation if the NUIP condition fails. We can use the same arguments as at the beginning of Section 4. Then $a\vartheta^\top R$ is unbounded for $a \to \infty$ on the set $\vartheta^\top R > 0$ where $P[\vartheta^\top R > 0] > 0$. Thus

$$u(x) \geq \lim_{a\to\infty} \mathbb{E}[U(xV_1^{a\vartheta})] = U(1)\cdot P[\vartheta^\top R = 0] + U(\infty)\cdot P[\vartheta^\top R > 0]$$

and we proved the following result (see Karatzas and Kardaras 2007 Prop. 4.19):

Proposition 11.4. *Assume that the NUIP condition fails. If $U(\infty) = \infty$ then $u(x) = \infty$ for all $x > 0$. If $U(\infty) < \infty$, then there is no solution.*

12 Additional remarks

1 Vasicek 1977 was perhaps the first who used the concept of a numeraire portfolio for an equilibrium characterisation of the term structure. In the language of Long 1990 and of the present paper, Vasicek constructed a numeraire portfolio investing in two assets: the short rate and a long rate.

2 By the use of the numeraire portfolio we can replace the change of measure $P \to Q$ where Q is an EMM by changing the numeraire $\{S_t^0\} \to \{V_t^\rho\}$ and sticking to the original probability measure P. There P models the 'true world'-probability which can be investigated by statistical methods. Long 1990, for example, studied the application of measuring abnormal stock returns by discounting NYSE-stock returns by empirical proxys of the numeraire portfolio.

3 Further properties and applications in the diffusion case, where the numeraire portfolio is mean-variance efficient and therefore related to the CAPM-theory, can also be found in Bajeux-Besnainou & Portait (97) and Johnson (96).

4 De Santis, Gerard and Ortu 2000 are interested in the case where no self-financing trading strategy has strictly positive value and introduce the concept of a generalised numeraire portfolio based on non-self-financing strategies.

5 A further advantage of the present discrete time market is the fact that there exists only one concept of no-arbitrage under the natural constraints (Example B). In particular, it cannot happen then as in continuous-time models that the weak numeraire portfolio exists but no equivalent martingale measure does. As mentioned above, a numeraire portfolio can be used for the purpose of pricing derivative securities. Platen 2002 argues that this can be done even in models where an equivalent martingale measure is absent and has developed a benchmark framework to do so (see Platen 2006).

6 Theorem 9.6 is generalised by Korn, Oertel and Schäl 2003 to a market modeled by a jump-diffusion process where the state space of the jumps is finite.

7 How to apply the results for the one-period model to the multi-period model is explained in Schäl 2000.

8 The concept of a numeraire portfolio (in the strict sense) is extended to financial markets with proportional transaction cost by Sass and Schäl 2009.

Bibliography

[1] P. Artzner (1997) *On the numeraire portfolio*. In: Mathematics of Derivative Securities, ed: M.A.H. Dempster and S.R. Pliska, Cambridge Univ. Press, pp. 216–226.

[2] F. Back and S.R. Pliska (1990) *On the fundamental theorem of asset pricing with an infinite state space*. J. of Mathematical Economics **20**, pp. 1–18.

[3] I. Bajeux-Besnainou, R. Portait (1997) *The numeraire portfolio: a new perspective on financial theory*. The European Journal of Finance **3**, pp. 291–309.

[4] D. Becherer (2001) *The numeraire portfolio for unbounded semimartingales*. Finance and Stochastics **5**, pp. 327–341.

[5] H. Bühlmann and E. Platen (2003) *A discrete time benchmark approach for insurance and finance*. ASTIN Bulletin **33**, pp. 153–172.

[6] M.M. Christensen and K. Larsen (2007) *No arbitrage and the growth optimal portfolio*. Stoch. Anal. Appl. **25**, pp. 255–280.

[7] M.M. Christensen and E. Platen (2005) *A general benchmark model for stochastic jumps*. Stochastic Analysis and Applications **23**, pp. 1017–1044.

[8] R.C. Dalang, A. Morton and W. Willinger (1990) *Equivalent martingale measures and no-arbitrage in stochastic securities market models*. Stochastics and Stochastic Reports **29**, pp. 185–201.

[9] F. Delbaen and J. Haezendonck (1989) *A martingale approach to premium calculation principles in an arbitrage free market*. Insur. Math. Econ. **8**, pp. 269–277.

[10] G. De Santis, B. Gerard, F. Ortu (2000) *Generalized Numeraire Portfolios*. Working paper, University of California, Anderson Graduate School of Management.

[11] T. Dijkstra (1998) *On numeraires and growth-optimum portfolios*. Working paper, University of Groningen.

[12] D. Duffie (1992) *Dynamic Asset Pricing Theory*. Princeton University Press.

[13] H. Föllmer and M. Schweizer (1991) *Hedging of contingent claims under incomplete information*, In: M.H.A. Davis and R.J. Elliot (eds.) "Applied Stochastic Analysis", Stochastic Monographs. **5**, Gordon and Breach, London, pp. 389–414.

[14] T. Goll, J. Kallsen (2000) *Optimal portfolios for logarithmic utility*. Stochastic Processes Appl. **89**, pp. 31–48.

[15] T. Goll and J. Kallsen (2003) *A complete explicit solution to the log-optimal portfolio problem*. Advances in Applied Probability **13**, pp. 774–779.

[16] N.H. Hakansson (1971) *Optimal entrepreneurial decisions in a completely stochastic environment*. Management Science **17**, pp. 427–449.

[17] N.H. Hakansson and W.T. Ziemba (1995) *Capital Growth Theory*. In: R. Jarrow et al. Handbook in Operations Research & Management Science, Volume 9, Finance. Amsterdam: North Holland.

[18] J.M. Harrison and D.M. Kreps (1979) *Martingales and arbitrage in multiperiod securities markets*. J. Economic Theory **20**, pp. 381–408.

[19] J.M. Harrison and S.R. Pliska (1981) *Martingales and stochastic integrals in the theory of continuous trading*. Stoch. Processes Appl. **11**, pp. 215–260.

[20] K. Hellwig (1996) *Portfolio selection under the condition of value preservation*. Review of Quantitative Finance and Accounting **7**, pp. 299–305.

[21] J. Jacod and A.N. Shiryaev (1998) *Local martingales and the fundamental asset pricing theorems in the discrete-time case*. Finance Stochast. **3**, pp. 259–273.

[22] B.E. Johnson (1996) *The pricing properties of the optimal growth portfolios: extensions and applications*. Working paper, Stanford University.

[23] J. Kallsen (2000) *Optimal portfolios for exponential Lévy processes*. Math. Methods Op. Res. **51**, pp. 357–374.

[24] I. Karatzas and C. Kardaras (2007) *The numéraire portfolio in semimartingale financial models*. Finance Stochast. **11**, pp. 447–493.

[25] C. Kardaras (2006) *The numéraire portfolio and arbitrage in semimartingale models of financial markets*. PhD dissertation, Columbia University.

[26] I. Karatzas and S.G. Kou (1996) *On the pricing of contingent claims under constraints*. Ann. Appl. Probab. **6**, pp. 321–369.

[27] R. Korn (1997) *Value preserving portfolio strategies in continuous-time models*. Math. Methods Op. Res. **45**, pp. 1–43.

[28] R. Korn (1997) *Optimal portfolios*. World Scientific, Singapore.

[29] R. Korn (1998) *Value preserving portfolio strategies and the minimal martingale measure*. Math. Methods Op. Res. **47**, pp. 169–179.

[30] R. Korn (2000) *Value preserving portfolio strategies and a general framework for local approaches to optimal portfolios*. Mathematical Finance **10**, pp. 227–241.

[31] R. Korn and E. Korn (2001) *Option pricing and portfolio optimization*, American Mathematical Society, Providence.

[32] R. Korn, F. Oertel and M. Schäl (2003) *The numeraire portfolio in financial markets modeled by a multi-dimensional jump diffusion process*. Decis. Econom. Finance **26**, pp. 153–166.

[33] R. Korn and M. Schäl (1999) *On value preserving and growth optimal portfolios*, Math. Methods Op. Res. **50**, pp. 189–218.

[34] D. Kramkov and W. Schachermayer (1999) *The asymptotic elasticity of utility functions and optimal investment in incomplete markets*. The Annals of Applied Probability **9**, pp. 904–950.

[35] J. Long (1990) *The numeraire portfolio*. J. Finance **44**, pp. 205–209.

[36] J.A. Nielsen and K. Sandmann (1995) *Equity-linked life insurance: A model with stochastic interest rates*. Insurance: Mathematics and Economics **16**, pp. 225–253.

[37] E. Platen (2001) *A minimal financial market model*. In: Trends in Mathematics. Birkhäuser Verlag, pp. 293–301.

[38] E. Platen (2002) *Arbitrage in continuous complete markets*. Adv. Appl. Probab. **34**, pp. 540–558.

[39] E. Platen (2006) *A benchmark approach to finance*. Mathematical Finance **16**, pp. 131–151.

[40] S.R. Pliska (1997) *Introduction to Mathematical Finance*. Blackwell Publisher, Malden, USA, Oxford, UK.

[41] L.C.G. Rogers (1994) *Equivalent martingale measures and no-arbitrage*. Stochastics and Stochastic Reports **51**, pp. 41–49.

[42] J. Sass and M. Schäl (2009) *The numeraire portfolio under proportional transaction costs*. Working paper.

[43] W. Schachermayer (1992) *A Hilbert space proof of the fundamental theorem of asset pricing in finite discrete time*. Insurance: Mathematics and Economics **11**, pp. 249–257.

[44] M. Schäl (1999) *Martingale measures and hedging for discrete-time financial markets*. Math. Oper. Res. **24**, pp. 509–528.

[45] M. Schäl (2000a) *Portfolio optimization and martingale measures*. Mathematical Finance **10**, pp. 289–304.

[46] M. Schäl (2000b) *Price systems constructed by optimal dynamic portfolios*. Math. Methods Op. Res. **51**, pp. 375–397.

[47] M. Schweizer (1995) *Variance-optimal hedging in discrete time*. Math. Oper. Res. **20**, pp. 1–32.

[48] O. Vasicek (1977) *An equilibrium characterization of the term structure*. J. Financial Economics **5**, pp. 177–188.

[49] T. Wiesemann (1996) *Managing a value-preserving portfolio over time*. European Journal of Operations Research **91**, pp. 274–283.

Author information

Ralf Korn, Fraunhofer-Institut für Techno- und Wirtschaftsmathematik, Fraunhofer-Platz 1, 67663 Kaiserslautern, Germany.
Email: ralf.korn@itwm.fraunhofer.de

Manfred Schäl, Inst. Angew. Math. Endenicher Allee 60, 53115 Bonn, Germany.
Email: schael@uni-bonn.de

A worst-case approach to continuous-time portfolio optimisation

Ralf Korn and Frank Thomas Seifried

Abstract. We survey the main ideas, results and methods behind the worst-case approach to portfolio optimisation in continuous time. This will cover the indifference approach, the HJB-system approach and the very recent martingale approach. We illustrate the difference to conventional portfolio optimisation with explicitly solved examples.

Key words. Optimal portfolios, worst-case approach, utility indifference, HJB-equation.

AMS classification. 93E20

1 Introduction

Stock price models that abandon the continuity of sample paths to include the possibility of asset price jumps have (re-)gained an enormous interest in recent years with the introduction of Lévy processes to financial mathematics (see the monograph [1] and their impressive list of references). Their main motivation is the inability of the standard geometric Brownian motion based models to explain large stock price moves, which are often observed at the markets. In particular, sudden price falls of the whole market, so-called crashes, are not incorporated into the standard continuous-path framework.

While many of those recently introduced Lévy process models exhibit a very good fit to observed market prices, they have the drawback that their analytical handling is not easy. Even more seriously, estimating the necessary input parameters from market data is not at all trivial, sometimes not even very stable from a statistical point of view. Motivated by this and also by the desire to be able to model market crashes, [3] introduced their so-called *crash model*. Its distinctive feature is that stock prices are assumed to follow geometric Brownian motions in *normal times*; at a *crash time* they suddenly fall by an unknown factor, which they assume to be bounded by an explicitly known constant. Besides the height of the crash, the time and the number of crashes up to a given time horizon are also unknown, but not explicitly modeled in a stochastic way. [3] obtain so-called *worst-case option prices* by figuring out the crash scenario that generates the worst case with respect to the option price.

In the context of portfolio optimisation, looking at the worst case is also an inter-

First author: Ralf Korn was supported by the Center for Mathematical and Computational Modeling $(CM)^2$ at the University of Kaiserslautern.
Second author: Frank Seifried was supported by the Center for Mathematical and Computational Modeling $(CM)^2$ at the University of Kaiserslautern.

esting alternative to the focus on expected utility or on the mean-variance criterion. Of course, such a consideration of the worst case needs two essential components: an exact definition of the worst case and a concept how this worst case enters the portfolio decisions. Examples for worst-case approaches that appeared in the continuous-time portfolio optimisation literature are [14], [12], [13] and [2]. These approaches typically focus on the parameters of asset prices, i.e. on the market coefficients. The worst case is then modeled as the parameter setting leading to an optimal portfolio process with the lowest utility. In [14] the market is explicitly regarded as an opponent to the investor that chooses the market coefficients. However, the price processes still remain diffusion processes. [13] formalises the idea by considering a whole set of probability measures that are candidates to govern the evolution of stock prices. In light of this setting, he determines the portfolio strategy that yields the highest lower bound for the expected utility from terminal wealth over all those possible probability measures.

In contrast to the approaches mentioned above, [8] have taken up the [3] framework. They focus on the uncertainty of the number, time and height of possible market crashes. By an indifference argument they show how to derive a characterisation of the worst-case optimal portfolio process. This approach is extended to a more general market setting by [6] and to problems including insurance risk processes in [5]. [7] relate the indifference approach of [8] to a system of inequalities that they call the HJB-system and thereby obtain optimality of the worst-case portfolio process in a wider class of strategies.

In this survey paper, we will focus on the worst-case approach in the sense of [8]. The indifference approach will be considered in Chapter 2, the HJB-systems approach will be the subject of Chapter 3, while a new approach (that we call the martingale approach) will be presented in Chapter 4. Some hints on open problems will close the paper.

2 The indifference approach to worst-case portfolio optimisation in the log-utility case

2.1 Motivation and model

In this section we consider the simplest case of the continuous-time worst-case market model introduced by [3] and taken up by [8]. We look at a market consisting of a riskless bond and one risky security whose price dynamics are given by

$$\begin{aligned} dP_0(t) &= P_0(t)\, r\, dt, \quad P_0(0) = 1 && (2.1)\\ dP_1(t) &= P_1(t)\, [b\, dt + \sigma\, dW(t)], \quad P_1(0) = p_1 && (2.2)\end{aligned}$$

with constant market coefficients $b > r$ and $\sigma \neq 0$ in *normal times*. At a so-called *crash time* τ, which is modeled as a stopping time, the stock price can suddenly fall by a relative amount k with $0 \leq k \leq k^* < 1$. Here, k^* is assumed to be the biggest possible crash height. Thus in a crash scenario (τ, k) we shall have

$$P_1(\tau) = (1-k)P_1(\tau-).$$

In this section we restrict ourselves to the case that at most one such crash can happen before the investment horizon T.

We assume the crash to be unknown a priori, but observable, so an investor will specify his actions completely by a pre-crash portfolio strategy $\underline{\pi}$ and a post-crash portfolio strategy $\overline{\pi}$, both of which we assume to be progressive processes. For ease of exposition, we assume throughout that pre-crash strategies are *bounded* and *continuous*. As an abbreviation we introduce $\pi := (\underline{\pi}, \overline{\pi})$. Then for a possible crash scenario (τ, k) the dynamics of the investor's wealth process $X = X^\pi = \{X^\pi(t) : t \in [0,T]\}$ are governed by the stochastic differential equation

$$\begin{aligned}
\frac{\mathrm{d}X^\pi(t)}{X^\pi(t)} &= (r + \underline{\pi}(t)(b-r))\,\mathrm{d}t + \underline{\pi}(t)\sigma\,\mathrm{d}W(t) \text{ on } [0,\tau), \quad X^\pi(0) = x \\
X^\pi(\tau) &= (1 - \underline{\pi}(\tau)k)X^\pi(\tau-) \\
\frac{\mathrm{d}X^\pi(t)}{X^\pi(t)} &= (r + \overline{\pi}(t)(b-r))\,\mathrm{d}t + \overline{\pi}(t)\sigma\,\mathrm{d}W(t) \text{ on } (\tau, T]
\end{aligned}$$

where $x > 0$ denotes the initial wealth. Thus, in accordance with the intended interpretation, the pre-crash strategy $\underline{\pi}$ is valid up to and including the crash time, whereas $\overline{\pi}$ is only applied starting immediately afterwards. All portfolio strategies π that guarantee a corresponding non-negative wealth process starting from an initial wealth of $x > 0$ form the class $\mathcal{A}(x)$ of admissible strategies with initial wealth x. If we consider only the time interval $[t, T]$, we use the obviously modified notation $\mathcal{A}(t, x)$ for the class of admissible strategies starting at time t with wealth $x > 0$.

Before we state the worst-case portfolio problem, we define $\tilde{X}^\pi = \{\tilde{X}^\pi(t) : t \in [0,T]\}$ as the wealth process in the standard crash-free market model given by equations (2.1), (2.2) that corresponds to the portfolio process π.

Definition 2.1 (Worst-Case Portfolio Problem). Let U be an $\mathbb{R}$-valued strictly concave, increasing and differentiable function. U will be called a utility function.

1. The problem

$$\sup_{\pi \in \mathcal{A}(x)} \inf_{0 \le \tau \le T,\ 0 \le k \le k^*} \mathbb{E}\left[U(X^\pi(T))\right] \tag{P}$$

with final wealth $X^\pi(T)$ in the case of a crash of size k at time τ given by

$$X^\pi(T) = (1 - \underline{\pi}(\tau)k)\,\tilde{X}^\pi(T)$$

is called the *worst-case portfolio problem* with *value function*

$$\nu^1(t,x) := \sup_{\pi \in \mathcal{A}(t,x)} \inf_{t \le \tau \le T,\ 0 \le k \le k^*} \mathbb{E}\left[U\left(X^\pi(T)\right)\right].$$

2. We denote by $\nu^0(t,x)$ the value function of the optimisation problem in the standard (crash-free) Black–Scholes setting; it is given by

$$\nu^0(t,x) = \sup_{\pi \in \mathcal{A}(t,x)} \mathbb{E}[U(\tilde{X}^\pi(T))].$$

To allow for explicit computations, we consider the special case of the logarithmic utility function

$$U(x) = \ln(x),\ x > 0$$

in this chapter. We then have the following representation of the value function in the Black–Scholes setting (see e.g. [4]):

$$\nu^0(t,x) = \ln(x) + r(T-t) + \frac{1}{2}\left(\frac{b-r}{\sigma}\right)^2 (T-t) \tag{2.3}$$

as well as the corresponding optimal portfolio process

$$\pi^* = \frac{b-r}{\sigma^2}\,. \tag{2.4}$$

We motivate the basic ideas of our worst-case concept by looking at two extreme strategies. Note first that the (worst-case) optimal post-crash strategy is π^*. This is simply due to the fact that this is the optimal portfolio process in the then relevant market. If we also chose the portfolio process π^* before the crash (provided that it satisfies $\pi^* < \frac{1}{k^*}$), the worst case would be a crash of maximal height k^* (recall that due to the assumption $b > r$ the log-optimal portfolio process is positive!). One can easily verify that the exact time of this crash would have no impact on the resulting final expected utility. It can therefore be obtained from the worst crash happening immediately and equals

$$\nu^0\left(t,(1-\pi^*k^*)x\right) = \ln(x) + r(T-t) + \frac{1}{2}\left(\frac{b-r}{\sigma}\right)^2 (T-t) + \ln(1-\pi^*k^*). \tag{2.5}$$

If, instead, we consider a very prudent investor that chooses $\underline{\pi}(t) = 0$ before the crash, the worst case for him is the no-crash scenario. To see this, note that a crash would not harm the investor; however, he could never switch to the strategy π^* after the crash (such a switch would result in a higher expected terminal utility!). Hence, he can never benefit from the knowledge that no further crash can happen. His corresponding final utility would simply be

$$\mathbb{E}\left[\ln\left(x\exp\left(r(T-t)\right)\right)\right] = \ln(x) + r(T-t). \tag{2.6}$$

Comparing the representations (2.5) and (2.6) one can draw the following conclusions:

- It depends on the investment time left $T-t$ which of the two extreme strategies yields a higher worst-case bound.
- While the first strategy takes too much risk (especially when the remaining investment time is small), the second one is too risk averse (especially when the remaining investment time is big). An optimal strategy should in a way balance this out.
- A portfolio process that consists of two constant parts $\underline{\pi}$ and $\overline{\pi}$ cannot be optimal with respect to the worst-case criterion.

2.2 Indifference strategies: characterisation and optimality

We take up the conclusions from the end of the preceding section and look for a portfolio process that attains a balance between good performance of the wealth process when no crash happens and a (just) acceptable loss in the crash scenario. For this we try to find a pre-crash portfolio process making us indifferent between the two scenarios:

- The worst crash happens immediately.
- No crash occurs at all.

Such a portfolio process $\hat{\pi} = (\hat{\underline{\pi}}, \pi^*)$ has to satisfy the following identity between the expected utilities corresponding to the two different scenarios:

$$\nu^0(t, (1-\hat{\underline{\pi}}(t)k^*)x) = \mathbb{E}^{t,x}\big[\ln\big(\tilde{X}^{\hat{\pi}}(T)\big)\big].$$

Applying Itô's formula to the right-hand side of this equality and using the explicit form of $\nu^0(t,x)$ on the left-hand side results in

$$\begin{aligned}
&\ln(x) + r(T-t) + \frac{1}{2}\Big(\frac{b-r}{\sigma}\Big)^2 (T-t) + \ln\big(1-\hat{\underline{\pi}}(t)k^*\big) \\
&= \ln(x) + r(T-t) + \mathbb{E}\left[\int_t^T \Big\{\hat{\underline{\pi}}(s)(b-r) - \frac{1}{2}\hat{\underline{\pi}}(s)^2\sigma^2\Big\}\,\mathrm{d}s\right] \\
&+\mathbb{E}\left[\int_t^T \hat{\underline{\pi}}(s)\sigma\,\mathrm{d}W(s)\right]. \qquad (2.7)
\end{aligned}$$

If we assume existence of a deterministic indifference portfolio process $\hat{\pi}$, the stochastic integral has mean zero and the expectation in front of the ds-integral can be dropped. Eliminating identical terms on both sides of equation (2.7) yields

$$\frac{1}{2}\Big(\frac{b-r}{\sigma}\Big)^2 (T-t) + \ln\big(1-\hat{\underline{\pi}}(t)k^*\big) = \int_t^T \Big\{\hat{\underline{\pi}}(s)(b-r) - \frac{1}{2}\hat{\underline{\pi}}(s)^2\sigma^2\Big\}\,\mathrm{d}s.$$

Assuming that $\hat{\underline{\pi}}$ is differentiable, differentiation of this identity with respect to t leads to the ordinary differential equation

$$\hat{\underline{\pi}}'(t) = -\frac{\sigma^2}{2k^*}\,[1-\hat{\underline{\pi}}(t)k^*]\,[\hat{\underline{\pi}}(t)-\pi^*]^2 \qquad (2.8)$$

while the obvious final condition

$$\hat{\underline{\pi}}(T) = 0 \qquad (2.9)$$

follows directly from (2.7). It is now straightforward to verify that there is a unique solution to equations (2.8) and (2.9). Even more, one can directly prove that the strategy determined by (2.8) and (2.9) solves the worst-case problem. The following result is taken from [8], but we will give a somewhat shorter proof.

Theorem 2.2 (Worst-Case Optimal Portfolio for Logarithmic Utility). *The portfolio process* $\hat{\pi} = (\hat{\underline{\pi}}, \pi^*)$ *determined by* (2.8), (2.9) *and* (2.4) *solves the worst-case investment problem* (P) *with logarithmic utility.*

Proof. Let $\hat{\underline{\pi}}$ be the unique pre-crash portfolio process determined by (2.8), (2.9).

1. We first show that the worst-case scenario for $\hat{\pi}$ is attained by a jump of maximum size k^* at any time $t \in [0,T]$. This obviously is the case if the corresponding expectation function

$$\hat{\nu}\left(t, X^{\hat{\pi}}(t)\right) = \nu^0\left(t, (1-\hat{\underline{\pi}}(t)k^*)X^{\hat{\pi}}(t)\right)$$

is a martingale. However, by the explicit form of $\nu^0(t,x)$ given in equation (2.3) and the fact that $\hat{\underline{\pi}}$ satisfies (2.8), (2.9), we obtain

$$\begin{aligned}
&\nu^0\left(t, (1-\hat{\underline{\pi}}(t)k^*)X^{\hat{\pi}}(t)\right)\\
&\quad= \ln\left(x\left(1-\hat{\underline{\pi}}(0)k\right)\right) + \left(r + \frac{1}{2}\left(\frac{b-r}{\sigma}\right)^2\right)T\\
&\qquad+ \int_0^t \frac{1}{1-\hat{\underline{\pi}}(s)k^*}\left[\hat{\underline{\pi}}(s)(b-r) - \frac{1}{2}\sigma^2\left(\hat{\underline{\pi}}(s)^2 + (\pi^*)^2\right)\right]\mathrm{d}s\\
&\qquad+ \int_0^t \sigma\hat{\underline{\pi}}(s)\,\mathrm{d}W(s)\\
&\quad= \ln\left(x\left(1-\hat{\underline{\pi}}(0)k\right)\right) + \left(r + \frac{1}{2}\left(\frac{b-r}{\sigma}\right)^2\right)T + \int_0^t \sigma\hat{\underline{\pi}}(s)\,\mathrm{d}W(s).
\end{aligned}$$

As the integrand of the stochastic integral is deterministic and bounded, the martingale property is established.

2. Let now $\pi = (\underline{\pi}, \pi^*)$ be an admissible portfolio process with a better worst-case performance than $\hat{\pi}$; without loss of generality suppose that the portfolio process π^* is used in the Black–Scholes setting after the crash. Due to continuity it must be constant in $t = 0$. Thus, to obtain a higher worst-case bound than $\hat{\pi}$, it must satisfy

$$\underline{\pi}(0) < \hat{\underline{\pi}}(0).$$

Further, as we have

$$\begin{aligned}
\mathbb{E}[\ln(\tilde{X}^{\pi}(T))] &= \ln(x) + \left(r + \frac{1}{2}\left(\frac{b-r}{\sigma}\right)^2\right)T + \int_0^T \mathbb{E}\left[\underline{\pi}(s)(b-r) - \frac{1}{2}\sigma^2\underline{\pi}(s)^2\right]\mathrm{d}s\\
&\le \ln(x) + \left(r + \frac{1}{2}\left(\frac{b-r}{\sigma}\right)^2\right)T\\
&\quad+ \int_0^T \left\{\mathbb{E}\left[\underline{\pi}(s)\right](b-r) - \frac{1}{2}\sigma^2\left(\mathbb{E}\left[\underline{\pi}(s)\right]\right)^2\right\}\mathrm{d}s
\end{aligned} \tag{2.10}$$

by Jensen's inequality, due to continuity of $\hat{\underline{\pi}}$ there has to be a smallest deterministic time $\bar{t} \in [0,T]$ with

$$\mathbb{E}\left[\underline{\pi}(\bar{t})\right] \ge \mathbb{E}\left[\hat{\underline{\pi}}(\bar{t})\right] = \hat{\underline{\pi}}(\bar{t})$$

if in the no-crash scenario the portfolio process π delivers a higher worst-case bound than $\hat{\pi}$. Note that due to the indifference construction $\hat{\pi}$ attains its worst-case bound also in the no-crash-scenario.

We now look at the worst-crash scenario at time $\bar{t}$. In this situation we obtain

$$\begin{aligned}
&\mathbb{E}\left[\ln\left(X^{\pi}(T)\right)\right] \\
&\quad = \mathbb{E}\left[\ln\left(X^{\pi}(\bar{t})\right) + \int_{\bar{t}}^{T} \left\{ r + \pi^*(b-r) + \frac{1}{2}\sigma^2(\pi^*)^2 \right\} \mathrm{d}s\right] \\
&\quad = \mathbb{E}\left[\ln(\tilde{X}^{\pi}(\bar{t}))\right] + \mathbb{E}\left[\ln\left(1-\underline{\pi}(\bar{t})k^*\right)\right] + \int_{\bar{t}}^{T} \left\{ r + \pi^*(b-r) + \frac{1}{2}\sigma^2(\pi^*)^2 \right\} \mathrm{d}s \\
&\quad \le \mathbb{E}\left[\ln(\tilde{X}^{\underline{\pi}}(\bar{t}))\right] + \ln\left(1-\mathbb{E}\left[\underline{\pi}(\bar{t})\right]k^*\right) + \int_{\bar{t}}^{T} \left\{ r + \pi^*(b-r) + \frac{1}{2}\sigma^2(\pi^*)^2 \right\} \mathrm{d}s \\
&\quad \le \mathbb{E}\left[\ln(\tilde{X}^{\underline{\hat{\pi}}}(\bar{t}))\right] + \ln\left(1-\left(\underline{\hat{\pi}}(\bar{t})\right)k^*\right) + \int_{\bar{t}}^{T} \left\{ r + \pi^*(b-r) + \frac{1}{2}\sigma^2(\pi^*)^2 \right\} \mathrm{d}s \\
&\quad = \mathbb{E}\left[\ln\left(X^{\hat{\pi}}(T)\right)\right].
\end{aligned}$$

Note that in the first inequality, we have used Jensen's inequality. The second inequality is a consequence of (2.10), the fact that for $\hat{\pi}$ (2.10) is satisfied with equality, and of course the defining property of $\bar{t}$. Hence, we arrive at a contradiction to the assumption that π attains a higher worst-case bound than $\hat{\pi}$. □

Remark 2.3 (Analysis of the Worst-Case Optimal Portfolio Process). From the explicit form of the differential equation (2.8) and (2.9) for the worst-case optimal pre-crash strategy $\hat{\pi}$, we can see that

$$0 \le \underline{\hat{\pi}}(t) \le \min\left\{\pi^*, \frac{1}{k^*}\right\} =: \pi_{\min} \text{ for } t \in [0,T].$$

More precisely, under the change of variable $t \mapsto T-t$ the differential equation (2.8), (2.9) takes the form

$$h'(t) = \frac{\sigma^2}{2k^*}\left[1-h(t)k^*\right]\left[h(t)-\pi^*\right]^2, \quad h(0)=0$$

with $\underline{\hat{\pi}}(t) = h(T-t)$. It is then clear that starting in 0, in particular below $\pi_{\min}$, h cannot cross either 0, π^* or $\frac{1}{k^*}$. Therefore, even in the case $\pi^* > \frac{1}{k^*}$, the worst-case optimal portfolio process avoids a negative wealth at any time.

As constant portfolio processes often play a very prominent role in portfolio optimisation, one might ask for the best constant portfolio process under the worst-case setting. As it is clear that the best constant portfolio process *after* the crash is $\overline{\pi}(t) = \pi^*$, we refer to a constant (worst-case) portfolio process as a pair of the form

$$\pi(t) = \left(\underline{\pi}(t), \overline{\pi}(t)\right) = (c, \pi^*) \text{ for all } t \in [0,T]$$

with c a constant. As shown in [8], the optimal constant c depends on the time horizon T. We therefore introduce the optimal constant portfolio process as a function of time

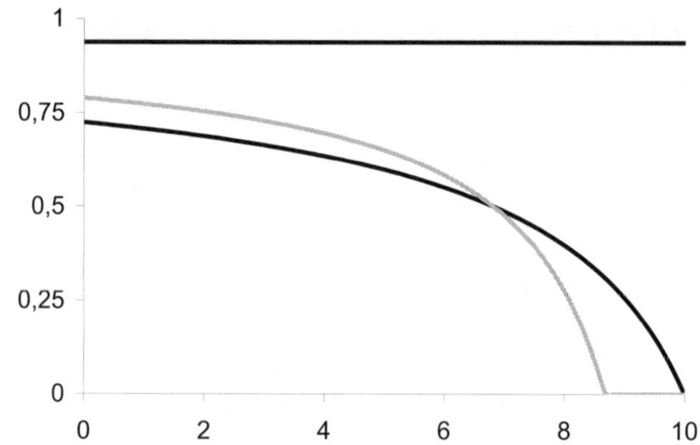

Figure 2.1. Worst-case optimal strategy (black lines) and worst-case constant best strategy as a function of the time horizon (grey line).

$c(t)$ (where the time variable actually denotes the time horizon) as

$$c(t) = \left(\frac{1}{2} \left(\frac{b-r}{\sigma^2} + \frac{1}{k^*} \right) - \sqrt{\frac{1}{4} \left(\frac{b-r}{\sigma^2} - \frac{1}{k^*} \right)^2 + \frac{1}{\sigma^2 t}} \right)^+ .$$

Obviously, this constant $c(t)$ converges towards $\pi_{\min}$ as $t \to \infty$. Note in particular that

$$c(t) = 0 \quad \Longleftrightarrow \quad b - r \leq \frac{k^*}{t},$$

i.e. if the "crash height per time unit" exceeds the excess return of the stock.

Example 2.4. To demonstrate the performance of the worst-case strategy together with the worst-case optimal constant portfolio process, we look at an example where we have chosen the following data:

$$r = 0.05,\ b = 0.20,\ \sigma = 0.4,\ k^* = 0.2,\ T = 10.$$

As long as no crash has happened, the worst-case optimal portfolio process $\hat{\pi}$ is given by the black curved line which shows $\underline{\hat{\pi}}$. After the jump the investor has to switch to the black line parallel to the x-axis with $\overline{\pi} = \pi^* = 0.9375$. For reasons of comparison, the grey line shows the optimal constant portfolio $c(T-t)$ which would be chosen if the portfolio problem started at time t. One clearly sees that the constant portfolio function c differs from the worst-case optimal portfolio $\underline{\hat{\pi}}$. It is below the worst-case optimal portfolio close to the investment horizon, and above it if the investment horizon is far away.

2.3 Indifference strategies: generalisations

The central result of the previous section can be generalised in various ways by simply using indifference arguments. Here we list some of them.

A finite number of possible crashes

In [6] we allow for more than one crash until the time horizon T. In such a situation of at most n crashes, a portfolio process is specified by an $(n+1)$-vector $\pi = (\pi_0, \dots, \pi_n)$ where π_j is the portfolio process that will be used by the investor if still at most j crashes can occur. Then an optimal portfolio process $\hat{\pi}$ exists and is given as the solution of the following sequence of ordinary differential equations for $j = 1, \dots, n$:

$$\begin{aligned} \hat{\pi}_0(t) &= \frac{b-r}{\sigma^2} \\ \hat{\pi}_j{}'(t) &= -\frac{\sigma^2}{2k^*}\left[1 - \hat{\pi}_j(t)k^*\right]\left[\hat{\pi}_j(t) - \hat{\pi}_{j-1}(t)\right]^2, \quad \hat{\pi}_j(T) = 0. \end{aligned}$$

Note that each such differential equation has a unique solution that satisfies

$$0 \le \hat{\pi}_j(t) \le \min\left\{\hat{\pi}_{j-1}(t), \frac{1}{k^*}\right\} \text{ for } t \in [0,T], \quad j = 1, \dots, n.$$

Indeed, the arguments used to ensure the corresponding properties in the one-crash setting are valid here, too.

More general utility functions

If instead of the logarithmic utility function we choose a general utility function U, then [6] and [5] contain verification results that are only valid under very restrictive assumptions. These assumptions are hard to verify, and they are by far not necessary conditions. However, by restricting to deterministic strategies it can be shown that similar differential equations as (2.8), (2.9) characterise the worst-case optimal *deterministic* portfolio process $\hat{\pi}$. In the case of the negative exponential utility function

$$U(x) = 1 - \exp(-\lambda x), \ x \in \mathbb{R}, \text{ for some } \lambda > 0$$

there is even a completely explicit result. By assuming $r = 0$, $b > r$ and allowing for a possibly negative wealth it is shown in [5] that we have:

Theorem 2.5 (Worst-Case Optimal Portfolio for Exponential Utility). *The optimal deterministic amount of money $A(t)$ invested in the stock before the crash is given by*

$$A(t) = A^* - \frac{2}{\lambda\sigma^2}\frac{\lambda k^*}{T - t + 2k^*/b} \quad \text{for } t \in [0,T]$$

while after the crash it is optimal to hold the amount of money

$$A^* = \frac{b}{\lambda\sigma^2}$$

in the stock.

Changing market conditions

As we have modeled the impact of a crash so far, its only consequence is a drop of the stock price. However, in real-world financial markets, the occurrence of a crash might have a more persistent effect. In [5] and [6] this is modeled by a change of the market coefficients after the crash. In such a situation, one can still insist on being indifferent between the worst possible crash and the no-crash scenario. Such a point of view is taken in [9]. However, under certain relations between the market situations before and after the crash, it is shown in [5] that one has to sacrifice indifference to obtain worst-case optimality.

In addition to the model considered so far, we assume that in the crash scenario (τ, k) *after* the crash the price dynamics are given by

$$\begin{aligned} \mathrm{d}P_0^1(t) &= P_0^1(t)\, r_1 \,\mathrm{d}t, \quad P_0^1(\tau) = P_0(\tau) \\ \mathrm{d}P_1^1(t) &= P_1^1(t)\,[b_1 \,\mathrm{d}t + \sigma_1 \,\mathrm{d}W(t)]\,, \quad P_1^1(\tau) = (1-k)P_1(\tau) \end{aligned}$$

with constant market coefficients r_1, b_1, and $\sigma_1 \neq 0$. To illustrate the possible new effect, we look again at the situation of Theorem 2.5, i.e. at the negative exponential utility case with $r = r_1 = 0$ and the notation

$$A^* = \frac{b}{\lambda\sigma^2}\,, \quad A_1^* = \frac{b_1}{\lambda\sigma_1^2}\,.$$

Note that after a crash we are in the new market. Thus, if we compare the crash-free scenario with a crash scenario we always have to use the value function in the crash-free scenario of the new market. Further, if it is more attractive to invest in the stock in the new market than in the original market, the possible loss caused by a crash might be overcompensated by the better market conditions in the new market. It can therefore be optimal *not* to insist on indifference. This is the content of the following theorem from [5]:

Theorem 2.6. *Under the assumptions of Theorem 2.5, $r = r_1 = 0$ and with the post-crash stock price dynamics given by*

$$\mathrm{d}P_1^1(t) = P_1^1(t)\,[b_1 \,\mathrm{d}t + \sigma_1 \,\mathrm{d}W(t)]$$

we have the following assertions:

a) For $A_1^* \leq A^*$ the results of Theorem 2.5 remain valid if we replace A^* by A_1^*.

b) For $A_1^* > A^*$ the optimal deterministic amount of money invested in the stock before the crash is given by

$$A(t) = \min\left\{A^*, A_1^* - \frac{2k^*}{\lambda\sigma_1^2(T-t) + 2k^*/A_1^*}\right\} \text{ for } t \in [0,T]. \tag{2.11}$$

The optimal amount of money invested into the stock after a crash equals A_1^*.

As part b) of the theorem shows, it can thus be better to invest optimally in the market before the crash than to insist on indifference. Following the (deterministic) indifference strategy before the crash would lead to a loss in terms of expected utility compared to A^* if no crash occurs. In the crash scenario, if there is still much time to the investment horizon T, equation (2.11) shows that the strategy A is below the indifference strategy and would thus also lead to a smaller loss.

3 HJB-systems for worst-case portfolio optimisation

The classical method to solve continuous-time portfolio problems is to apply the basic tool of continuous-time stochastic control theory, the Hamilton-Jacobi-Bellman equation (for short: HJB-equation). This approach has been introduced by Merton (see e.g. [10], [11]). Since then numerous papers have been written on this subject (see e.g. the monograph by [4]).

The purpose of this section is to introduce the approach of [7] who derive a system of inequalities that can be regarded as an analogue to the HJB-equation in the worst-case setting. The main achievement of the introduction of this *HJB-inequality system* is that one can prove that the optimal deterministic strategies derived in [5] and [6] are indeed optimal among all admissible portfolio processes.

The conceptually new aspect of [7] is the interpretation of the worst-case setting as a game between the market and the investor. While the market is "allowed" to choose a crash sequence, the investor chooses the portfolio process. The stock price dynamics are modeled by

$$dP_1(t) = P_1(t-)\left[b\,dt + \sigma\,dW(t) - k^*\,dN(t)\right], \quad P_1(0) = p_1.$$

Here, $N = \{N(t) : t \in [0,T]\}$ is a process that counts the number of jumps such that

$$N(t) = \#\{0 < s \le t : P_1(s) \ne P_1(s-)\} \text{ for } t \in [0,T]$$

and k^* is the (maximal) crash height. For simplicity, we always assume that a crash of maximum size happens (for more on this, see [7]). While in the indifference approach we simply ignored the modelling of jumps, we now assume that the market chooses a jump strategy N with a maximum number of jumps n and $N(t) - N(t-) \in \{0,1\}$. This strategy can also be characterised as a sequence of jump times $(\tau_1, \ldots, \tau_n)$. We denote by $\mathcal{B}(n)$ the class of crash scenarios with at most n jumps.

As before, we assume the portfolio process π to be adapted (now with respect to the filtration generated by the stock price and the counting process N, which models the investor's ability to know how many crashes can still occur!), and we suppose that portfolio processes take values in a subset A of $\mathbb{R}$. Further, we use the notation $\pi = (\pi_0, \ldots, \pi_n)$ where $\pi_j(t)$ denotes the part of the portfolio process that the investor chooses if still at most j crashes can occur. To apply standard arguments from stochastic control theory and to avoid a negative wealth due to a crash, we also assume

$$\mathbb{E}\left[\int_0^T |\pi_j(s)|^m\,ds\right] < \infty \text{ for } m = 1,2,\ldots, \quad \pi_j(t)k^* < -1, \quad j = 1,\ldots,n.$$

Then for a given "control" (π, N) the wealth process follows the dynamics

$$\begin{aligned} X^{(\pi,N)}(0) &= x, \\ \mathrm{d}X^{(\pi,N)}(t) &= X^{(\pi,N)}(t)\,[(r+\pi_j(t)(b-r))\,\mathrm{d}t + \sigma\,\mathrm{d}W(t)] \text{ on } (\tau_{j-1},\tau_j] \\ X^{(\pi,N)}(\tau_j) &= (1-\pi_j(\tau_j)k^*)\,X^{(\pi,N)}(\tau_j-), \quad j=1,\dots,n. \end{aligned}$$

We assume that the investor chooses a portfolio process to maximise worst-case expected utility of terminal wealth in the sense of the optimisation problem

$$\sup_{\pi\in\mathcal{A}(x)} \inf_{N\in\mathcal{B}(n)} \mathbb{E}\big[U(X^{(\pi,N)}(T))\big].$$

For $\nu \in C^{1,2}$ we define the differential operator $\mathcal{L}^\pi \nu$ by

$$\mathcal{L}^\pi \nu(t,x) := \nu_t(t,x) + \nu_x(t,x)(r+\pi(b-r))x + \frac{1}{2}\nu_{xx}(t,x)\pi^2\sigma^2 x^2$$

and for $n \in \mathbb{N}$ we define the *value function* $V^n(t,x)$ by

$$V^n(t,x) := \sup_{\pi\in\mathcal{A}(t,x)} \inf_{N\in\mathcal{B}(t,n)} \mathbb{E}^{t,x,n}\big[U(X^{(\pi,N)}(T))\big].$$

Here as above $\mathcal{A}(t,x)$ and $\mathcal{B}(t,n)$ denote, respectively, admissible strategies and possible crash sequences on $[t,T]$, given that the investor's wealth is x and n crashes are possible. With this notation we can now formulate the main result of this section.

Theorem 3.1 (Verification Theorem). *The worst-case optimisation problem can be solved via the following recursive system of HJB-equations.*

Step 0. *Assume that $\nu^0(t,x)$ is a polynomially bounded classical solution of*

$$0 = \sup_{\pi\in A}\left[\mathcal{L}^\pi \nu^0(t,x)\right], \quad \nu^0(T,x) = U(x)$$

and that

$$p(t,x) := \arg\sup_{\pi\in A}\left[\mathcal{L}^\pi \nu^0(t,x)\right]$$

is an admissible control function. Then we have

$$V^0(t,x) = \nu^0(t,x)$$

and the optimal control function exists and is given by

$$\pi_0^*(t) = p\big(t, X^{(\pi^*,N)}(t)\big).$$

Step n. *For $n \in \mathbb{N}$ and every function $\nu^n \in C^{1,2}$, define $A'_n(t,x)$ and $A''_n(t,x)$ by*

$$\begin{aligned} A'_n(t,x) &:= \left\{\pi \in A : \pi < \frac{1}{k^*},\ 0 \le \mathcal{L}^\pi \nu^n(t,x)\right\} \\ A''_n(t,x) &:= \left\{\pi \in A : \pi < \frac{1}{k^*},\ 0 \le \nu^{n-1}\left(t,(1-\pi k^*)x\right) - \nu^n(t,x)\right\}. \end{aligned}$$

Assume that there exists a polynomially bounded $C^{1,2}$-solution of

$$\begin{aligned}
0 &\leq \sup_{\pi \in A_n''(t,x)} \left[\mathcal{L}^\pi \nu^n(t,x)\right] \\
0 &\leq \sup_{\pi \in A_n'(t,x)} \left[\nu^{n-1}(t,(1-\pi k^*)x) - \nu^n(t,x)\right] \\
0 &= \sup_{\pi \in A_n''(t,x)} \left[\mathcal{L}^\pi \nu^n(t,x)\right] \sup_{\pi \in A_n'(t,x)} \left[\nu^{n-1}(t,(1-\pi k^*)x) - \nu^n(t,x)\right] \\
\nu^n(T,x) &= U(x)
\end{aligned}$$

and that

$$\begin{aligned}
p(t,x) &:= \arg \sup_{\pi \in A_n''(t,x)} \left[\mathcal{L}^\pi \nu^n(t,x)\right], \\
\theta(t,x) &:= \inf_{s \geq t} \left[\nu^{n-1}(s,(1-\pi k^*)x) - \nu^n(s,x) \leq 0\right]
\end{aligned}$$

is a pair of admissible control functions. Then

$$V^n(t,x) = \nu^n(t,x)$$

and the optimal control functions exist and are given by

$$\pi_n^*(t) = p\big(t, X^{(\pi^*,N)}(t)\big), \quad \tau_n^* = \theta\big(t, X^{(\pi^*,N)}(t)\big).$$

Remark 3.2 (Form of the HJB-System). The form of the HJB-system characterising the value functions $\nu^n(t,x)$ needs explanation, as it differs in certain aspects from the HJB-equation or HJB-inequalities of related problems. For this, note first that if we looked at the portfolio problem where the jump process is a Poisson process with constant intensity λ and jump size k^*, then the corresponding HJB-equation would read

$$\begin{aligned}
0 &= \sup_{\pi \in A} \tilde{\mathcal{L}}^\pi \nu^0(t,x) \\
&= \sup_{\pi \in A} \Big[\nu_t^0 + \frac{1}{2}\sigma^2\pi^2x^2\nu_{xx}^0 + \big(r + \pi(b-r)\big)x\nu_x^0 \\
&\qquad + \lambda\big(\nu^0(t,(1-\pi k^*)x) - \nu^0(t,x)\big)\Big] \\
&= \sup_{\pi \in A} \left[\mathcal{L}^\pi \nu^0(t,x) + \lambda\left(\nu^0(t,(1-\pi k^*)x) - \nu^0(t,x)\right)\right]. \qquad (3.1)
\end{aligned}$$

As for a utility function of class C^2 and $b > r$ the optimal portfolio process ought to be non-negative, we would expect from (3.1) that

$$0 \leq \sup_{\pi \in A} \left[\mathcal{L}^\pi \nu^0(t,x)\right] \qquad (3.2)$$

which also motivates this requirement for $\mathcal{L}^\pi \nu^n$ in the verification theorem. This inequality also characterises the set $A_n'(t,x)$. It further suggests that the investor should only search among those π that satisfy this inequality when he considers the optimal performance (with respect to the no-crash scenario). On the other hand, he should not give the market a chance to hit him more by a crash than necessary. Therefore, he ought to restrict π to those strategies that satisfy

$$\nu^n(t,x) \leq \nu^{n-1}\left(t,(1-\pi k^*)x\right) \tag{3.3}$$

which is the requirement that characterises the set $A''_n(t,x)$. The assumption that both inequalities (3.3) and (3.2) are strict would intuitively contradict the idea of ν^n being a value function, as it would not be in line with the form of the HJB-equation (3.1). This motivates the presence of the complementarity condition that (at least) one of the two inequalities always has to be satisfied with equality.

Remark 3.3 (Possible Generalisations). In [7] explicit examples are solved when the utility function is the negative exponential utility function or of the form

$$U(x) = \frac{1}{\gamma}x^\gamma, \ x > 0, \ \text{for some } \gamma < 1, \ \gamma \neq 0.$$

To give details of the rather technical way of solving the HJB-inequality systems in the verification theorem is beyond the scope (and length!) of this paper. However, we would like to point out that the HJB-system approach allows for an easy inclusion of a consumption rate process c, which is only subtracted from the wealth equation as a term $-c(t)dt$. It should therefore be possible to prove a suitably modified verification theorem and also to solve the corresponding HJB-inequality system for the special choices of the utility functions considered in [7].

4 A martingale approach to worst-case portfolio optimisation

In contrast to the dynamic programming approach, the martingale approach to the worst-case portfolio problem is based on martingale optimality arguments and the idea that the market acts as an opponent to the investor. In the following we briefly outline its main components: the Change-of-Measure Device, the Indifference-Optimality Principle, and the notion of an Indifference Frontier.

4.1 The Change-of-Measure device

We consider the worst-case portfolio problem (P) and assume that

$$U(x) = \frac{1}{\gamma}x^\gamma, \ x > 0, \ \text{with } \gamma < 1, \ \gamma \neq 0.$$

Moreover, suppose that if a crash occurs, it has maximum size k^*. We let Θ denote the class of $[0,T] \cup \{\infty\}$-valued stopping times and interpret the event $\{\tau = \infty\}$ as there being no crash at all. We recall that admissible strategies are assumed to be bounded and continuous before the crash. Then we may equivalently reformulate the worst-case portfolio problem (P) as the problem to optimally choose a *pre-crash* strategy so as to obtain

$$\sup_{\pi\in\mathcal{A}(x)} \inf_{\tau\in\Theta} \mathbb{E}\left[\nu^0\left(\tau,(1-\pi(\tau)k^*)X^\pi(\tau)\right)\right] \tag{P$_{pre}$}$$

where as above ν^0 denotes the value function of the post-crash optimisation problem, which is known explicitly:

$$\nu^0(t,x) = \frac{1}{\gamma}x^\gamma \exp\left(\left[\gamma r + \frac{1}{2}\left(\frac{b-r}{\sigma}\right)^2 \frac{\gamma}{1-\gamma}\right](T-t)\right). \tag{4.1}$$

This is intuitively completely obvious because no further crash can occur, and can be shown formally with the following trick:

Theorem 4.1 (Change-of-Measure Device). *Consider the classical optimal portfolio problem with random initial time τ and time-τ initial wealth ξ*

$$\sup_{\pi\in\mathcal{A}(\tau,\xi)} \mathbb{E}^{\tau,\xi}[U(\tilde{X}^\pi(T))] \tag{P$_{post}$}$$

where τ is a stopping time and $\mathcal{A}(\tau,\xi)$ denotes the corresponding class of admissible strategies on $[\tau,T]$. Then for any $\pi\in\mathcal{A}(\tau,\xi)$ we can write

$$U(\tilde{X}^\pi(T)) = U(\xi)\exp\left(\gamma\int_\tau^T \Phi(\pi(s))\,\mathrm{d}s\right) M_\pi(T) \tag{4.2}$$

with a martingale $M_\pi = \{M_\pi(t) : t\in[0,T]\}$ satisfying $M_\pi(\tau)=1$ and

$$\Phi(y) := r + (b-r)y - \frac{1}{2}(1-\gamma)\sigma^2 y^2.$$

Thus the optimal solution to problem (P$_{post}$) *is given by* $\pi^* = \frac{b-r}{(1-\gamma)\sigma^2}$.

Proof. The first part is a consequence of Itô's formula and Novikov's condition, making use of the boundedness assumption on π. To establish the second note that clearly π^* maximises Φ. Hence, if $\pi\in\mathcal{A}(\tau,\xi)$ is an arbitrary strategy, we have from (4.2) and the martingale property of M_π

$$\begin{aligned}
\mathbb{E}^{\tau,\xi}[U(\tilde{X}^\pi(T))] &= \mathbb{E}^{\tau,\xi}\left[U(\xi)\exp\left(\gamma\textstyle\int_\tau^T \Phi(\pi(s))ds\right) M_\pi(T)\right] \\
&\le \mathbb{E}^{\tau,\xi}\left[U(\xi)\exp\left(\gamma\textstyle\int_\tau^T \Phi(\pi^*)ds\right) M_\pi(T)\right] \\
&= \mathbb{E}^{\tau,\xi}\left[U(\xi)\exp\left(\gamma\textstyle\int_\tau^T \Phi(\pi^*)ds\right) M_{\pi^*}(T)\right] \\
&= \mathbb{E}^{\tau,\xi}[U(\tilde{X}^{\pi^*}(T))]
\end{aligned}$$

so π^* is optimal. □

The Change-of-Measure Device allows to transform the stochastic optimisation problem to a *pathwise* maximisation, quite similar to the log-case. Note that changing market coefficients are subsumed by the above framework, and that Theorem 4.1 also adapts immediately to situations with deterministic trading constraints.

4.2 Abstract indifference strategies

The form of (P_{pre}) suggests a reformulation of the worst-case portfolio problem as a zero-sum stochastic game; this is the motivation for the martingale approach. Let us consider an abstract controller-and-stopper game played between two players A (the controller) and B (the stopper). Player A controls a stochastic process

$$W = W^\lambda = \{W^\lambda(t) : t \in [0,T]\}$$

by choosing λ from a given class of admissible controls Λ, and player B decides on the duration of the game by choosing a stopping time $\tau \in \Theta$. The controller and stopper aim to maximise or minimise, respectively, the expectation

$$\mathbb{E}[W^\lambda(\tau)].$$

Assuming that player A has to choose his strategy first, he faces the problem to obtain

$$\sup_{\lambda\in\Lambda}\inf_{\tau\in\Theta}\mathbb{E}[W^\lambda(\tau)]. \tag{$P_{abstract}$}$$

Now if player A can choose his strategy $\hat{\lambda} \in \Lambda$ in such a way that $W^{\hat{\lambda}}$ is a martingale, then player B's actions become irrelevant to him because by optional stopping

$$\mathbb{E}[W^{\hat{\lambda}}(\sigma)] = \mathbb{E}[W^{\hat{\lambda}}(\tau)] \text{ for all stopping times } \sigma, \tau.$$

Thus it makes sense to call such a strategy $\hat{\lambda}$ an *indifference strategy*. The crucial benefit of indifference strategies is formulated in

Proposition 4.2 (Indifference-Optimality Principle). *If $\hat{\lambda}$ is an indifference strategy, and for all $\lambda \in \Lambda$ there exists a single $\tau \in \Theta$ such that $\mathbb{E}[W^{\hat{\lambda}}(\tau)] \geq \mathbb{E}[W^\lambda(\tau)]$, then $\hat{\lambda}$ is optimal for player A in* ($P_{abstract}$).

4.3 Optimality and the indifference frontier

In the framework of the previous section, observe that if we call player A the investor and player B the market, then setting $\Lambda := \mathcal{A}(x)$ and

$$W^\pi(t) := \nu^0(t,(1-\pi(t)k^*)X^\pi(t)) \text{ for } t \in [0,T] \text{ and } W^\pi(\infty) := \nu^0(T, X^\pi(T))$$

we obtain the worst-case portfolio problem (P_{pre}). Note also that the seemingly obvious terminal condition (2.9) is in fact a consequence of the martingale property of $W^{\hat{\pi}}$ between T and ∞. To construct an indifference strategy $\hat{\pi}$, one goes through the same calculation as in the first part of the proof of Theorem 2.2 to obtain the ordinary differential equation

$$\hat{\pi}'(t) = -\frac{\sigma^2}{2k^*}(1-\gamma)\big[1-\hat{\pi}(t)k^*\big]\big[\hat{\pi}(t)-\pi^*\big]^2, \quad \hat{\pi}(T)=0 \tag{4.3}$$

for $\hat{\pi}$, making use of the explicit form (4.1) of ν^0. Here and in the following, we assume for simplicity that market coefficients do not change after a crash; in particular one sees as in Remark 2.3 that $0 \le \hat{\pi}(t) \le \min\{\pi^*, \frac{1}{k^*}\}$ for all $t \in [0,T]$.

Lemma 4.3 (Indifference Frontier). *Let $\pi \in \mathcal{A}(x)$ be an admissible strategy, let $\hat{\pi}$ be determined by equation* (4.3)*, set $\sigma := \inf\{t : \pi(t) > \hat{\pi}(t)\}$ and define*

$$\tilde{\pi}(t) := \pi(t) \text{ if } t < \sigma \text{ and } \tilde{\pi}(t) := \hat{\pi}(t) \text{ if } t \ge \sigma.$$

Then $\tilde{\pi} \in \mathcal{A}(x)$ and the worst-case bound attained by $\tilde{\pi}$ is at least as big as that achieved by π.

Proof. Let τ be an arbitrary stopping time. By continuity we have $\tilde{\pi}(t) = \hat{\pi}(t)$ if $0 \le t \le \sigma$, and since $\hat{\pi}$ is an indifference strategy the process $W^{\tilde{\pi}}$ is a martingale on $[\sigma,T] \cup \{\infty\}$. Thus we obtain

$$\mathbb{E}\big[W^{\tilde{\pi}}(\tau)\big] = \mathbb{E}\big[W^{\tilde{\pi}}(\tau\wedge\sigma)\big] = \mathbb{E}\big[W^{\pi}(\tau\wedge\sigma)\big] \ge \inf_{\tau'\in\Theta} \mathbb{E}\big[W^{\pi}(\tau')\big].$$

Since τ is arbitrary, the conclusion follows. □

Remark 4.4. Lemma 4.3 implies that it suffices to search for optimal strategies which are dominated by the indifference strategy. Hence $\hat{\pi}$ represents a *frontier* which rules out too optimistic investment, i.e. a too great exposure to the risk of a crash.

Now it is not hard to see that the strategy $\hat{\pi}$ is worst-case optimal. Indeed, by the Change-of-Measure device (and the fact that Φ is a quadratic function) the indifference strategy yields an optimal performance for the no-crash scenario in the class of all strategies that remain below the Indifference Frontier. Hence, optimality follows from the Indifference-Optimality Principle.

Theorem 4.5 (Solution of the Worst-Case Portfolio Problem). *The optimal strategy in the pre-crash market for the worst-case portfolio problem* (P) *is given by the indifference strategy $\hat{\pi}$ determined from* (4.3)*. After the crash, the Merton strategy $\pi^* = \frac{b-r}{(1-\gamma)\sigma^2}$ is optimal.*

The indifference strategy has been verified to be optimal in [7] by means of the dynamic programming methods presented in the previous chapter. The martingale approach provides a simpler and more direct way to analyse the problem, as it focuses directly on the crucial notion of indifference.

4.4 Extensions

The approach outlined above applies to more general settings than that considered here. For instance we can consider general Lévy-driven asset price models, we can remove the continuity assumption imposed on admissible trading strategies, we can allow for changing market coefficients, and we can consider multiple crashes. Although this complicates the formal analysis, the concepts developed above remain valid and provide the key to solve the worst-case optimal portfolio problem. A detailed exposition of the martingale approach to worst-case portfolio problems in a more general framework is the subject of a forthcoming paper.

5 Conclusion and further aspects

The worst-case approach to continuous-time portfolio optimisation represents, on the one hand, a generalisation of the classical Merton setting, and on the other hand, an alternative to technically involved frameworks such as the Lévy process setting. Its main strength lies in the fact that for standard utility functions we can derive fully explicit optimal portfolio strategies. Their specific form is appealing, in particular the reduction of risky investments when the time horizon gets near while there is still crash risk. Of course, the strategies depend heavily on the assumed upper bound k^* for the jump height and on the maximum number of jumps n.

There are various ways to generalise the framework and the problems dealt with. We only mention a few of them:

- *More general price processes:* In fact, we mainly need the Black–Scholes type modelling framework in order to be able to explicitly solve the portfolio problem in the crash-free setting. Considering different models for which this is also possible is a direct generalisation. This is the subject of a forthcoming paper.
- *Including consumption:* The possibility of the investor to consume parts of his wealth during the investment time is a natural generalisation. The HJB-approach seems to be well-suited to deal with this aspect (as already mentioned above).
- *Dependence between jump heights:* So far, heights of previous crashes have no impact on the next crash. This memoryless behaviour of the market might be unrealistic. Again, this generalisation will be dealt with in forthcoming research.

We believe that there is a lot of potential in the worst-case approach from both the scientific and the application-oriented perspective.

Bibliography

[1] R. Cont, P. Tankov, *Financial Modelling with Jump Processes*, Chapman & Hall (2004).

[2] D. Hernández-Hernández, A. Schied, *Robust Utility Maximization in a Stochastic Factor Model*, Statistics & Decisions **24** (2006), pp. 109–125.

[3] P. Hua, P. Wilmott, *Crash Courses*, Risk **10** (1997), pp. 64–67.

[4] R. Korn, *Optimal Portfolios*, World Scientific (1997).

[5] R. Korn, *Worst-Case Scenario Investment for Insurers*, Insurance: Mathematics and Economics **36** (2005), pp. 1–11.

[6] R. Korn, O. Menkens, *Worst-Case Scenario Portfolio Optimization: a New Stochastic Control Approach*, Mathematical Methods of Operations Research **62** (2005), pp. 123–140.

[7] R. Korn, M. Steffensen, *On Worst-Case Portfolio Optimization*, SIAM Journal on Control and Optimization **46** (2007), pp. 2013–2030.

[8] R. Korn, P. Wilmott, *Optimal Portfolios under the Threat of a Crash*, International Journal of Theoretical and Applied Finance **5** (2002), pp. 171–187.

[9] O. Menkens, *Crash Hedging Strategies and Worst-Case Scenario Portfolio Optimization*, International Journal of Theoretical and Applied Finance **9** (2006), pp. 597–618.

[10] R. C. Merton, *Lifetime Portfolio Selection under Uncertainty: The Continuous-Time Case*, Review of Economics and Statistics **51** (1969), pp. 247–257.

[11] R. C. Merton, *Optimum Consumption and Portfolio Rules in a Continuous-Time Model*, Journal of Economic Theory **3** (1971), pp. 373–413.

[12] F. Riedel, *Optimal Stopping with Multiple Priors*, to appear in Econometrica (2009).

[13] A. Schied, *Optimal Investments for Robust Utility Functionals in Complete Market Models*, Mathematics of Operations Research **30** (2005), pp. 750–764.

[14] D. Talay, Z. Zheng, *Worst Case Model Risk Management*, Finance and Stochastics **6** (2002), pp. 517–537.

Author information

Ralf Korn, Department of Mathematics, University of Kaiserslautern, 67653 Kaiserslautern and Fraunhofer ITWM, Kaiserslautern, Germany.
Email: korn@mathematik.uni-kl.de

Frank Thomas Seifried, Department of Mathematics, University of Kaiserslautern, 67653 Kaiserslautern, Germany.
Email: seifried@mathematik.uni-kl.de

Radon Series Comp. Appl. Math **8**, 347–369

Time consistency and information monotonicity of multiperiod acceptability functionals

Raimund Kovacevic and Georg Ch. Pflug

Abstract. Time consistency is an often required property of functionals which measure the risk or the acceptability of financial processes. Based on elementary consistency properties for pairs of conditional acceptability mappings, we demonstrate how these properties translate into the multi-period setting. Moreover, we show that even in elementary cases time consistency may conflict with information monotonicity.

Key words. Time consistency, information monotonicity, multi-period risk measures, multi-period acceptability measures.

AMS classification. 91B28, 62P05

1 Introduction

Analysing risk inherent in stochastic financial payments or whole cash-flow processes and evaluating their favourability has become an important subject for mathematically oriented theoreticians, as well as for practitioners in different industries. Theoretical approaches reach from the Markowitz-model ([15]) and expected utility ([24, 1]) to the recent discussions regarding coherent risk measures ([4, 2] , [8]) and risk or acceptability functionals ([16]). On the other hand practice in banking and insurance was dominated by the introduction of value at risk based models.

Efforts have been made to extend theoretical approaches originally formulated in a static framework to multi-period settings. Therefore conditional and dynamic versions of the basic measures under consideration have been defined and analysed ([5, 7, 18, 22]). Recently many publications (e.g. [3, 6, 9, 12, 14, 17, 20]) explicitly deal with, or at least mention the subject of time consistency in multi-period valuation.

In the present paper we define and analyse the main concepts related to time consistency in a simple framework. We want to separate the core of the notion of time consistency from the more technical issues influenced by the respective viewpoints of the papers mentioned above. Finally those concepts are analysed in a genuine multi-period framework, including intermediate payments. Although we have in view the applicability of our concepts in multistage stochastic optimisation, compared with [23] we go back a step and analyse the properties of functionals and mappings suitable as objective functions or constraints rather than analysing time consistent decisions.

In the last section we will confront time consistency with another crucial property of multiperiod acceptability measures: information monotonicity ([16]). One might

First author: Supported by FWF

argue that this concept is even more fundamental for a multiperiod valuation than time consistency.

1.1 Basic framework

A *probability functional* is an extended real valued function on a set of random variables or random processes which are defined on some probability space $(\Omega, \mathcal{F}, \mathbb{P})$. If a functional is interpreted in the sense that higher values are preferable to lower values we call it a *monotonic* or *acceptability-type ([16]) functional.* We define the domain of the functional $\mathcal{A}$ as $\operatorname{dom} \mathcal{A} = \{X : \mathcal{A}(X) > -\infty\}$.

Definition 1.1.1. A probability functional $\mathcal{A}$ is called *centred at zero* if

$$\mathcal{A}(0) = 0$$

holds.

We concentrate on acceptability and not on risk in the following, but one should be aware that it is very easy to adapt all our statements to statements about risk, first of all changing the sign and moving from concavity to convexity. Besides, it should be clear that suitable acceptability measures implicitly account for the risk of a random variable or process in the sense that risk reduces acceptability.

Acceptability functionals are monotonic functionals with some additional properties:

Definition 1.1.2. A probability functional $\mathcal{A}(\cdot) : L_p(\Omega, \mathcal{F}, \mathbb{P}) \to \bar{\mathbb{R}} = R \cup \{-\infty\}$ is called an *acceptability functional*, if the following properties hold for all $X, Y \in L_p(\Omega, \mathcal{F}, \mathbb{P})$:

(A1) *Translation Equivariance.* $\mathcal{A}(X + c) = \mathcal{A}(X) + c$ holds for all constants c.

(A2) *Concavity.* $\mathcal{A}(\lambda \cdot X + (1 - \lambda) \cdot Y) \geq \lambda \cdot \mathcal{A}(X) + (1 - \lambda) \cdot \mathcal{A}(Y)$ holds for all $\lambda \in [0, 1]$.

(A3) *Monotonicity.* $X \leq Y \Rightarrow \mathcal{A}(X) \leq \mathcal{A}(Y)$.

For acceptability functionals it is not a heavy restriction to be centred at zero. If $\mathcal{A}(0) = c \neq 0$ we can easily switch to an acceptability mapping $\mathcal{A}'(X) = \mathcal{A}(X) - c$. Additionally most of relevant acceptability mappings fulfil the stronger condition $\mathcal{A}(c) = c$ for any $c \in \mathbb{R}$.

Although we do not deal with coherent risk functionals explicitly, it should be noted that ρ is a coherent risk functional if and only if $\mathcal{A} = -\rho$ is a positively homogeneous acceptability functional. Thus, acceptability functionals can be considered as a kind of generalisation of coherent measures. One reason why the use of non-homogeneous functionals is necessary in some situations is liquidity risk: For large financial positions it might be very risky to double their amount, because prices will be influenced, if liquidity is needed and the position has to be reduced to its original size.

We allow $p \in [1, \infty]$, but mention that $p = \infty$ is a very optimistic assumption in reality.

From the Fenchel–Moreau–Rockafellar Theorem ([19], Theorem 5) it follows that concave upper semicontinuous (u.s.c.) functionals can be represented in the following way:

$$\mathcal{A}(X) = \inf_{Z \in L_q(\Omega, \mathcal{F}, \mathbb{P})} \left\{ \mathbb{E}(X \cdot Z) - \mathcal{A}^+(Z) \right\}, \tag{1.1}$$

where $\frac{1}{p} + \frac{1}{q} = 1$ and $\mathcal{A}^+(Z) = \inf_{X \in L_p(\Omega, \mathcal{F}, \mathbb{P})} \{\mathbb{E}(X \cdot Z) - \mathcal{A}(X)\}$ is the concave conjugate or Fenchel–Moreau conjugate of the functional $\mathcal{A}$. Basically this means that the functional equals its biconjugate. This relation can be very useful for analysing the properties of a functional.

The second main class of mappings analysed in this paper are certainty equivalents, derived from expected utility. Such mappings are defined in the following way:

Definition 1.1.3. Let u be a (strictly monotonic, concave) function, interpreted as utility function. Then the related *expected utility* of a random variable X is given by the mapping

$$\mathcal{U}(X) = \mathbb{E}(u(X)).$$

The related *certainty equivalent* is given by

$$\mathcal{A}_u(X) = u^{-1}(\mathbb{E}(u(X))).$$

The special case $u(x) = -\exp(-x)$ leads to the *entropic mapping*

$$E_\gamma(X) = -\frac{1}{\gamma} \ln(\mathbb{E}(\exp(-\gamma X))).$$

Not all certainty equivalent functionals fulfil the concavity assumption (A2). It is well known that (A2) holds if and only if the function $h(t) = -\frac{u'(t)}{u''(t)}$ is concave ([11], 106 (i), p. 88 and [10], Lemma 8, p. 322). Note that $\frac{u'(t)}{u''(t)}$ is proportional to the reciprocal relative risk aversion coefficient of the utility function u.

Using the functional $I(\cdot) = (u')^{-1}(\cdot)$ we can characterise the concave conjugate of certainty equivalents $\mathcal{A}_u(\cdot)$ for differentiable utility functions u.

$$\mathcal{A}_u^+(Z) = \begin{cases} \mathbb{E}(Z \cdot I(c \cdot Z)) & \text{if } c = u'\left(u^{-1}(\mathbb{E}(u(I(c \cdot Z))))\right) \\ -\infty & \text{if no such } c \text{ can be found.} \end{cases}$$

If the acceptability of a random variable has to be valuated relative to some nontrivial information, represented by a σ-algebra $\mathcal{F}$, conditional monotonic mappings come into play.

Definition 1.1.4. For any space $L_p(\Omega, \mathcal{F}_1, \mathbb{P})$ the related extended space is given by $\overline{L}_p(\Omega, \mathcal{F}_1, \mathbb{P}) = L_p(\Omega, \mathcal{F}_1, \mathbb{P}) \cup \{-\infty\}$.

Here $\{-\infty\}$ represents the functions with (a.s.) constant value $-\infty$.

Acceptability functionals can be generalised to conditional acceptability mappings. These are mappings

$$\mathcal{A}(\cdot|\mathcal{F}_1) : \bar{L}_p(\Omega, \mathcal{F}, \mathbb{P}) \to \overline{L}_{p'}(\Omega, \mathcal{F}_1, \mathbb{P})$$

with $p \geq p'$ and the properties (A2), (A3) understood in the almost sure sense. The domain of such a mapping is given by $\operatorname{dom} \mathcal{A} = \{Y : \mathcal{A}(Y|\mathcal{F}_1) \neq -\infty\}$. This means that $\mathcal{A}(Y|\mathcal{F}_1)$ is either in $L_{p'}(\Omega, \mathcal{F}_1, \mathbb{P})$ or it is $-\infty$ a.s.

In this context, property (A1) has to be generalised in the following way:

Condition 1.1.5 (CA1). *$\mathcal{A}(Y + X|\mathcal{F}_1) = \mathcal{A}(Y|\mathcal{F}_1) + X$ holds for all $\mathcal{F}_1$-measurable random variables X.*

The idea of conditional mappings is that a random payment is valuated relative to some nontrivial information, which may arise at some future points of time. The spaces $L_p(\Omega, \mathcal{F}, \mathbb{P})$ together with the partial order $\leq$ a.s. are order complete Banach lattices.

With $Z \in \bar{L}_s(\Omega, \mathcal{F}, \mathbb{P}) \subseteq \bar{L}_q(\Omega, \mathcal{F}, \mathbb{P})$, $s \geq \frac{p \cdot p'}{p - p'}$ and the infimum related to the partial order $\leq$ a.s. it is possible ([13]) to define the concave conjugate of conditional acceptability mappings in a meaningful way: The conjugate is given by a mapping $\mathcal{A}^+ : \bar{L}_s(\Omega, \mathcal{F}, \mathbb{P}) \to \bar{L}_{p'}(\Omega, \mathcal{F}_1, \mathbb{P})$ with $\frac{1}{p} + \frac{1}{s} = \frac{1}{p'}$ and

$$\mathcal{A}^+(Z|\mathcal{F}_1) = \inf_{X \in \bar{L}_p(\Omega, \mathcal{F}, \mathbb{P})} \{\mathbb{E}(X \cdot Z|\mathcal{F}_1) - \mathcal{A}(X|\mathcal{F}_1)\}.$$

If the superdifferential

$$\partial \mathcal{A}(X_0|\mathcal{F}_1) = \\ \{Z \in \bar{L}_s(\Omega, \mathcal{F}, \mathbb{P}) : \mathcal{A}(X|\mathcal{F}_1) \leq \mathcal{A}(X_0|\mathcal{F}_1) + \mathbb{E}((X - X_0) \cdot Z|\mathcal{F}_1), \forall X \in \operatorname{dom} \mathcal{A}\}$$

is nonempty at X_0, a supergradient representation

$$\mathcal{A}(X_0|\mathcal{F}_1) = \mathcal{A}^{++}(X_0|\mathcal{F}_1)$$

holds for proper conditional acceptability mappings ([13], Theorem 4.2.7).

In particular it is possible to define conditional acceptability mappings, using acceptability functionals with known supergradient representation. This can be done by replacing any instance of an expectation in the supergradient representation with the suitable conditional expectation.

Similarly it is possible to define conditional certainty equivalents by replacing the expectation in Definition 1.1.3 with the suitable conditional expectation.

Definition 1.1.6. A conditional mapping $\mathcal{A}(\cdot|\mathcal{F}_1)$ is called *centred at zero* if

$$\mathcal{A}(0|\mathcal{F}_1) = 0$$

holds.

A basic example is the average value at risk (also known as tail value at risk, conditional value at risk or expected shortfall). For a random variable X with distribution function G it is defined as

$$AV@R_\alpha(X) = \frac{1}{\alpha} \int_0^\alpha G^{-1}(x)\, dx\,.$$

This functional fulfils the conditions (A1), (A2), (A3) and is positive homogeneous in addition.

It is well known that the conjugate representation of the average value at risk is given by

$$AV@R_\alpha(X) = \inf\left\{\mathbb{E}(X \cdot Z) : \mathbb{E}(Z) = 1,\ 0 \le Z \le \tfrac{1}{\alpha}\right\}.$$

This can be used to define the conditional average value at risk:

$$AV@R_\alpha(X|\mathcal{F}_1) = \inf\left\{\mathbb{E}(X \cdot Z|\mathcal{F}_1) : \mathbb{E}(Z|\mathcal{F}_1) = 1,\ 0 \le Z \le \tfrac{1}{\alpha} \text{ a.s.}\right\}.$$

2 Time consistency for pairs and related concepts

Let $(\Omega, \mathcal{F}, \mathbb{P})$ be a probability space and $\mathcal{F}_0 \subseteq \mathcal{F}_1 \subseteq \mathcal{F}$ a simple filtration related to three points of time $t_0 \le t_1 \le T$. We consider random variables X defined on this space, representing a financial payoff or the value of a property at time T, possibly discounted with a deterministic or stochastic deflator. We try to value the random variable at the beginning of the whole period. The filtration represents the information available at the points of time under consideration.

2.1 Time consistency

In the following $\mathcal{A}_0(\cdot|\mathcal{F}_0) : \bar{L}_p(\Omega, \mathcal{F}, \mathbb{P}) \to \bar{L}_{p_0}(\Omega, \mathcal{F}_0, \mathbb{P})$and $\mathcal{A}_1(\cdot|\mathcal{F}_1) : \bar{L}_p(\Omega, \mathcal{F}, \mathbb{P}) \to \bar{L}_{p_1}(\Omega, \mathcal{F}_1, \mathbb{P})$ denote conditional monotonic probability mappings. Both mappings might be conditional versions of the same functional $\mathcal{A}(\cdot)$, but this is not mandatory. It is possible to include unconditional functionals into the following considerations: If $\mathcal{F}_0$ is the trivial σ-algebra $\{\Omega, \emptyset\}$ then $\mathcal{A}_0(\cdot|\mathcal{F}_0) \in \overline{\mathbb{R}}$ and we call this functional "unconditional".

Definition 2.1.1. A pair $\mathcal{A}_0(\cdot|\mathcal{F}_0)$, $\mathcal{A}_1(\cdot|\mathcal{F}_1)$ is called *time consistent*, if for all $X, Y \in \bar{L}_p(\Omega, \mathcal{F}, \mu)$ the implication

$$\mathcal{A}_1(X|\mathcal{F}_1) \ge \mathcal{A}_1(Y|\mathcal{F}_1) \text{ a.s.} \implies \mathcal{A}_0(X|\mathcal{F}_0) \ge \mathcal{A}_0(Y|\mathcal{F}_0) \text{ a.s.}$$

holds.

This definition was used in Artzner et al. [3], where the focus was on a special class of conditional mappings, related to coherent functionals. One can also define weaker consistency properties.

Let χ_B and 1_B denote the following indicator functionals of the set B:

$$\chi_B(\omega) = \begin{cases} 0 & \text{if } \omega \in B \\ +\infty & \text{otherwise}, \end{cases}$$

$$1_B(\omega) = \begin{cases} 1 & \text{if } \omega \in B \\ 0 & \text{otherwise}. \end{cases}$$

Definition 2.1.2. A pair $\mathcal{A}_0(\cdot|\mathcal{F}_0)$, $\mathcal{A}_1(\cdot|\mathcal{F}_1)$ is called *acceptance consistent*, if for all $X \in \bar{L}_p(\Omega, \mathcal{F}, \mu)$ the inequality

$$\operatorname{ess\,inf}\, (\chi_B + \mathcal{A}_1(X|\mathcal{F}_1)) \leq \chi_B + \mathcal{A}_0(X|\mathcal{F}_0) \text{ a.s.}$$

holds for any set $B \in \mathcal{F}_0$. It is called *rejection consistent*, if

$$\operatorname{ess\,sup}\, (\chi_B + \mathcal{A}_1(X|\mathcal{F}_1)) \geq \chi_B + \mathcal{A}_0(X|\mathcal{F}_0) \text{ a.s.}$$

for any $B \in \mathcal{F}_0$.

The terms "acceptance consistent" and "rejection consistent" were first used by Weber [25], who defined $\mathcal{A}_0(\cdot), \mathcal{A}_1(X|\mathcal{F}_1)$ to be acceptance consistent if the implication $\mathcal{A}_1(X|\mathcal{F}_1) \geq 0$ a.s. $\Longrightarrow \mathcal{A}_0(X) \geq 0$ holds, and called them rejection consistent if $\mathcal{A}_1(X|\mathcal{F}_1) \leq 0$ a.s. $\Longrightarrow \mathcal{A}_0(X) \leq 0$ is valid. Weber based his analysis on positive homogeneous functionals and stated his definition in terms of acceptance and rejection sets. The consistency properties can be characterised equivalently in the following way:

Proposition 2.1.3. *Acceptance consistency is equivalent to*

$$\mathcal{A}_1(X|\mathcal{F}_1) \geq W \text{ a.s.} \Longrightarrow \mathcal{A}_0(X|\mathcal{F}_0) \geq W \text{ a.s.} \tag{2.1}$$

for any $\mathcal{F}_0$-measurable random variable W and rejection consistency is equivalent to

$$\mathcal{A}_1(X|\mathcal{F}_1) \leq W \text{ a.s.} \Longrightarrow \mathcal{A}_0(X|\mathcal{F}_0) \leq W \text{ a.s..} \tag{2.2}$$

for any $\mathcal{F}_0$-measurable random variable W.

For $W = 0$ this property was also called "weak time consistency" in [3].

Proof. We show the first part of the assertion.

Assume first acceptance consistency and let W be $\mathcal{F}_0$-measurable with $\mathcal{A}_1(X|\mathcal{F}_1) \geq W$ a.s.. Additionally let W_n be a sequence of simple functions

$$W_n = \sum_i \alpha_{n,i} \cdot 1_{B_{n,i}}$$

such that $W_n \uparrow W$. Since $\mathcal{A}_1(X|\mathcal{F}_1) \geq W_n$ it follows that for all i

$$\operatorname{ess\,inf}\, \left(\chi_{B_{n,i}} + \mathcal{A}_1(X|\mathcal{F}_1)\right) \geq \alpha_{n,i}\ ,$$

hence by acceptance consistency we have

$$\chi_{B_{n,i}} + \mathcal{A}_0(X|\mathcal{F}_0) \geq \alpha_{n,i}$$

i.e.

$$\mathcal{A}_0(X|\mathcal{F}_0) \geq W_n.$$

By letting n tend to infinity it follows that $\mathcal{A}_0(X|\mathcal{F}_0) \geq W$.

For the other direction let for all $B \in \mathcal{F}_0$

$$W_B = \operatorname{ess\,inf}\ (\chi_B + \mathcal{A}_1(X|\mathcal{F}_1)) \cdot 1_B + (-\infty) \cdot 1_{B^C}.$$

Then $\mathcal{A}_1(X|\mathcal{F}_1) \geq W_B$ and hence

$$\mathcal{A}_0(X|\mathcal{F}_0) \geq W_B = \operatorname{ess\,inf}\ (\chi_B + \mathcal{A}_1(X|\mathcal{F}_1)) \cdot 1_B + (-\infty) \cdot 1_{B^C}$$

which implies that

$$\chi_B + \mathcal{A}_0(X|\mathcal{F}_0) \geq \operatorname{ess\,inf}\ (\chi_B + \mathcal{A}_1(X|\mathcal{F}_1)).$$

□

The following lemma shows that time consistency implies weak time consistency or acceptability and rejection consistency.

Proposition 2.1.4. *If* $\mathcal{A}_0(\cdot|\mathcal{F}_0)$ and $\mathcal{A}_1(\cdot|\mathcal{F}_1)$ are centred at zero, *then time consistency implies both weak time consistency, i.e.*

$$\mathcal{A}_1(X|\mathcal{F}_1) \geq 0 \text{ a.s.} \Longrightarrow \mathcal{A}_0(X|\mathcal{F}_0) \geq 0 \text{ a.s.} \tag{2.3}$$

and

$$\mathcal{A}_1(X|\mathcal{F}_1) \leq 0 \text{ a.s.} \Longrightarrow \mathcal{A}_0(X|\mathcal{F}_0) \leq 0\,. \tag{2.4}$$

If $\mathcal{A}_0(\cdot|\mathcal{F}_0)$, $\mathcal{A}_1(\cdot|\mathcal{F}_1)$ *are translation equivariant in addition, time consistency implies both acceptability and rejection consistency.*

Proof. Let $\mathcal{A}_1(X|\mathcal{F}_1) \geq \mathcal{A}_1(0|\mathcal{F}_1) = 0$ a.s. From time consistency we have $\mathcal{A}_0(X|\mathcal{F}_0) \geq \mathcal{A}_0(0|\mathcal{F}_0) = 0$. The second implication (2.4) can be proved in a similar manner.

Consider now a $\mathcal{F}_0$-measurable random variable W with $\mathcal{A}_1(X|\mathcal{F}_1) \geq W$ a.s.. If translation equivariance holds, $\mathcal{A}_1(X - W|\mathcal{F}_1) \geq 0$ a.s. follows. Using (2.3) above we get $\mathcal{A}_0(X - W|\mathcal{F}_0) \geq 0$ and again by translation equivariance $\mathcal{A}_0(X) \geq W$. This condition is equivalent to acceptance consistency by Lemma 2.1.3. Rejection consistency can be proved analogously. □

Acceptance and rejection consistency are also implied by another strong condition, connected with martingale properties.

Definition 2.1.5. A pair of mappings $\mathcal{A}_0(\cdot|\mathcal{F}_0)$, $\mathcal{A}_1(\cdot|\mathcal{F}_1)$ is called

(i) a *submartingale pair*, if for all $X \in \operatorname{dom} \mathcal{A}$

$$\mathcal{A}_0(X|\mathcal{F}_0) \leq \mathbb{E}\left(\mathcal{A}_1(X|\mathcal{F}_1)|\mathcal{F}_0\right) \text{ a.s.} \tag{2.5}$$

(ii) a *supermartingale pair*, if for all $X \in \operatorname{dom} \mathcal{A}$

$$\mathcal{A}_0(X|\mathcal{F}_0) \geq \mathbb{E}\left(\mathcal{A}_1(X|\mathcal{F}_1)|\mathcal{F}_0\right) \text{ a.s..} \tag{2.6}$$

The pair is called *martingale pair* if it is both a submartingale pair and a supermartingale pair.

A special situation arises, when $\mathcal{A}_1(\cdot|\mathcal{F}_1)$ and $\mathcal{A}_0(\cdot|\mathcal{F}_0)$ are conditional versions of the same functional $\mathcal{A}(\cdot)$. In this case properties (i) and (ii) are called *compound convexity* and *compound concavity* ([16], Definition 2.11 and Proposition 2.12). It is well known that the $AV@R$ is compound convex ([16], Proposition 2.35).

One critical feature of martingale pair properties is that each of them implies the suitable weak consistency property.

Proposition 2.1.6. *(i) If a pair $\mathcal{A}_0(\cdot|\mathcal{F}_0))$, $\mathcal{A}_1(\cdot|\mathcal{F}_1)$ is a submartingale pair, it is rejection consistent.*

(ii) If a pair $\mathcal{A}_0(\cdot|\mathcal{F}_0))$, $\mathcal{A}_1(\cdot|\mathcal{F}_1)$ is a supermartingale pair, it is acceptance consistent.

(iii) If a pair $\mathcal{A}_0(\cdot|\mathcal{F}_0))$, $\mathcal{A}_1(\cdot|\mathcal{F}_1)$ is a martingale pair, it is both rejection and acceptance consistent.

Proof. We show the implication for a submartingale pair: Assume $\mathcal{A}_1(X|\mathcal{F}_1) \leq W$ and that W is $\mathcal{F}_0$.-measurable. Because the expectation is monotonic,

$$\mathbb{E}\left(\mathcal{A}_1(X|\mathcal{F}_1)|\mathcal{F}_0\right) \leq W$$

follows. Using the submartingale-pair property this means that

$$\mathcal{A}_0(Y|\mathcal{F}_0) \leq \mathbb{E}\left(\mathcal{A}_1(X|\mathcal{F}_1)|\mathcal{F}_0\right) \leq W.$$

Hence we have shown the implication $\mathcal{A}_1(X|\mathcal{F}_1) \leq W \Longrightarrow \mathcal{A}_0(Y|\mathcal{F}_0) \leq W$, which is equivalent to rejection consistency by Lemma 2.1.3.

Part (ii) can be shown by the same argument and (iii) follows, because a martingale pair is both a sub- and a supermartingale pair. □

2.2 Recursivity

Another concept, closely related to time consistency is recursivity.

Definition 2.2.1. The pair $\mathcal{A}_0(\cdot|\mathcal{F}_0)$, $\mathcal{A}_1(\cdot|\mathcal{F}_1)$ is called *recursive*, if for all $X \in \bar{L}_p(\Omega, \mathcal{F}, \mu)$ the equation

$$\mathcal{A}_0(X|\mathcal{F}_0) = \mathcal{A}_0\left(\mathcal{A}_1(X|\mathcal{F}_1)|\mathcal{F}_0\right) \text{ a.s.} \tag{2.7}$$

holds.

The pair $\mathcal{A}_0(\cdot|\mathcal{F}_1)$, $\mathcal{A}_1(\cdot|\mathcal{F}_1)$ is called *self-recursive* if (2.7) holds and $\mathcal{A}_1(\cdot|\mathcal{F}_1)$ and $\mathcal{A}_0(\cdot|\mathcal{F}_0)$ are conditional versions of the same functional $\mathcal{A}(\cdot)$.

It is a basic fact that recursivity is equivalent to time consistency under very general assumptions. The following property is critical in this context.

Definition 2.2.2. A mapping $\mathcal{A}_1(\cdot|\mathcal{F}_1)$ has the *weak projection property* if

$$\mathcal{A}_1(X|\mathcal{F}_1) = X \tag{2.8}$$

holds, whenever X is $\mathcal{F}_1$-measurable.

Let $\mathcal{W}$ denote the class of mappings with weak projection property. Simple examples for mappings in this class are:

Example 2.1.

a) Translation equivariant conditional monotonic mapping, centred at zero.

This is true because of $\mathcal{A}_1(X|\mathcal{F}_1) = \mathcal{A}_1(0 + X|\mathcal{F}_1) = \mathcal{A}_1(0|\mathcal{F}_1) + X = X$.

b) Conditional certainty equivalent mapping $u^{-1}\left(\mathbb{E}\left(u(\cdot)|\mathcal{F}_1\right)\right)$.

Such mappings are not translation equivariant in general. By the projection property of the conditional expectation

$$u^{-1}\left(\mathbb{E}\left(u(X)|\mathcal{F}_1\right)\right) = u^{-1}\left(u(X)\right) = X$$

holds, if X is $\mathcal{F}_1$-measurable.

Acceptability mappings are of type a) if they are centred at zero. It should be remembered that the Value at Risk – though not concave and therefore not a full acceptability mapping – has property a) and hence the weak projection property.

It should be clear that the class $\mathcal{W}$ contains more than just cases a) and b) above:

Example 2.2.

- Any convex combination $\sum_{i=1}^{n} \lambda_i \cdot \mathcal{A}_i(X|\mathcal{F}_1)$ of mappings $\mathcal{A}_i \in \mathcal{W}$ is a member of the class $\mathcal{W}$.
- If α is a parameter of an acceptability mapping $\mathcal{A}_\alpha(\cdot|\mathcal{F}_1) \in \mathcal{W}$, then the integral $\int_{-\infty}^{+\infty} \mathcal{A}_\alpha(\cdot|\mathcal{F}_1)\, dF(\alpha)$ is in $\mathcal{W}$ for any distribution function F.
- Any mapping $u^{-1}\left(\mathcal{A}\left(u(\cdot)|\mathcal{F}_1\right)\right)$ is in $\mathcal{W}$, if $\mathcal{A} \in \mathcal{W}$.

The weak projection property will be of basic interest for our analysis:

It is an important fact that for mappings with the weak projection property, recursivity and time consistency are equivalent.

Theorem 2.2.3. *A pair $\mathcal{A}_0(\cdot|\mathcal{F}_0)$, $\mathcal{A}_1(\cdot|\mathcal{F}_1)$, where $\mathcal{A}_1(\cdot|\mathcal{F}_1)$ possesses the weak projection property* 2.8 *and $\mathcal{A}_0(\cdot|\mathcal{F}_0)$ is monotonic, is time consistent if and only if the pair is recursive.*

Proof. Assume first that $\mathcal{A}_0(\cdot|\mathcal{F}_0)$, $\mathcal{A}_1(\cdot|\mathcal{F}_1)$ is a recursive pair. Using the monotonicity of $\mathcal{A}_0(\cdot|\mathcal{F}_0)$, from $\mathcal{A}_1(X|\mathcal{F}_1) \geq \mathcal{A}_1(Y|\mathcal{F}_1)$ a.s. we can infer

$$\mathcal{A}_0\left(\mathcal{A}_1(X|\mathcal{F}_1)|\mathcal{F}_0\right) \geq \mathcal{A}_0\left(\mathcal{A}_1(Y|\mathcal{F}_1)|\mathcal{F}_0\right).$$

Because of recursivity this implies $\mathcal{A}_0\left(X|\mathcal{F}_0\right) \geq \mathcal{A}_0\left(Y|\mathcal{F}_0\right)$.

For the other direction assume now that the pair is time consistent. Because of the weak projection property we have $\mathcal{A}_1\left(X|\mathcal{F}_1\right) = \mathcal{A}_1\left(\mathcal{A}_1(X|\mathcal{F}_1)|\mathcal{F}_1\right)$. Setting $Y = \mathcal{A}_1(X|\mathcal{F}_1)$ this means that both $\mathcal{A}_1\left(X|\mathcal{F}_1\right) \leq \mathcal{A}_1\left(Y|\mathcal{F}_1\right)$ and $\mathcal{A}_1\left(X|\mathcal{F}_1\right) \geq \mathcal{A}_1\left(Y|\mathcal{F}_1\right)$ hold.

Now if time consistency holds, we can conclude $\mathcal{A}_0(X|\mathcal{F}_0) \leq \mathcal{A}_0(Y|\mathcal{F}_0)$ and $\mathcal{A}_0(X|\mathcal{F}_0) \geq \mathcal{A}_0(Y|\mathcal{F}_0)$, which results in

$$\mathcal{A}_0(X|\mathcal{F}_0) = \mathcal{A}_0(Y|\mathcal{F}_0) = \mathcal{A}_0(\mathcal{A}_1(X|\mathcal{F}_1)|\mathcal{F}_0),$$

the recursivity equation. □

The connection between recursivity and time consistency was stated and proved in [3] (Theorem 5.1) in a more restricted context. The original definition of timc consistency is more and more replaced by the requirement of recursivity in newer papers ([14, 12]). We consider now some connections between time consistency and the martingale pair properties. We have seen that the combination of rejection and acceptance consistency is implied from both the martingale-pair property and (strong) time consistency, although in the latter case we need additional properties – e.g. translation equivariance. More generally we will see that the martingale-pair property is stronger than time consistency.

Corollary 2.2.4. *For a time consistent pair $\mathcal{A}_0(\cdot|\mathcal{F}_0)$, $\mathcal{A}_1(\cdot|\mathcal{F}_1)$ with translation equivariant $\mathcal{A}_1(\cdot|\mathcal{F}_1)$ and monotonic $\mathcal{A}_0(\cdot|\mathcal{F}_1)$ the assumption $\mathcal{A}_0(X|\mathcal{F}_0) \leq \mathbb{E}(X|\mathcal{F}_0)$ (strictness) implies the submartingale-pair property.*

Proof. By Theorem 2.2.3 recursivity and hence the inequality

$$\mathcal{A}_0(X|\mathcal{F}_0) \leq \mathcal{A}_0(\mathcal{A}_1(X|\mathcal{F}_1)|\mathcal{F}_0)$$

holds. Using the assumption of strictness we have $\mathcal{A}_0(X|\mathcal{F}_0) \leq \mathbb{E}(\mathcal{A}_1(X|\mathcal{F}_1)|\mathcal{F}_0)$, which is the submartingale-pair property. □

Proposition 2.2.5. *Any martingale pair is time consistent.*

Proof. Assume $\mathcal{A}_1(X|\mathcal{F}_1) \geq \mathcal{A}_1(Y|\mathcal{F}_1)$ a.s. Because expectation is monotonic we have $\mathbb{E}(\mathcal{A}_1(X|\mathcal{F}_1)|\mathcal{F}_0) \geq \mathbb{E}(\mathcal{A}_1(Y|\mathcal{F}_1)|\mathcal{F}_0)$. The martingale-pair property then gives $\mathcal{A}_0(X|\mathcal{F}_0) = \mathbb{E}(\mathcal{A}_1(X|\mathcal{F}_1)|\mathcal{F}_0) \geq \mathbb{E}(\mathcal{A}_1(Y|\mathcal{F}_1)|\mathcal{F}_0) = \mathcal{A}_0(Y|\mathcal{F}_0)$. □

Example 2.3. Expected conditional functionals are defined as $\mathbb{E}(\mathcal{A}_1(\cdot|\mathcal{F}_1))$. It is clear that – by definition – such a functional together with the mapping $\mathcal{A}_1(\cdot|\mathcal{F}_1)$ defines a martingale pair, which therefore is time consistent.

Example 2.4. Any certainty equivalent functional $u^{-1}(\mathbb{E}(u(\cdot)))$ is self-recursive and hence time consistent. A special case is the entropic functional $E_\gamma(X|\mathcal{F}_1)$.

3 Dynamic and multi-period functionals

So far we have considered only two periods and a single payoff at the end of the second period. Now we will generalise this to arbitrary filtrations. In addition we will also consider the valuation of processes, which includes intermediate cash-flows.

3.1 Dynamic acceptability mappings and multi-period acceptability functionals

For multi-period valuation we consider the following general setup: Time $t = 0, \dots, T$ is discrete. We restrict ourselves to a finite time horizon $T < \infty$. $\mathfrak{F} = (\mathcal{F}_0, \dots, \mathcal{F}_T)$ is a filtration with $\mathcal{F}_T = \mathcal{F}$.

As a first step into the multi-period framework we define dynamic acceptability mappings. Such mappings valuate one final payment at each intermediate time period.

Definition 3.1.1. A sequence $(\mathcal{A}_t(\cdot|\mathcal{F}_t))_{t\in\{0,\dots,T-1\}}$ of conditional monotonic mappings with $\mathcal{A}_t(\cdot|\mathcal{F}_t) : \bar{L}_p(\Omega, \mathcal{F}, \mathbb{P}) \to \bar{L}_{p_t}(\Omega, \mathcal{F}_t, \mathbb{P})$ for each t is called *dynamic monotonic mapping.*

If the constituent mappings are acceptability mappings we will call the sequence a *dynamic acceptability mapping.*

If X_T is a $\mathcal{F}_T$-measurable random variable, e.g. representing a final cash-flow, we can consider the sequence of random variables $\mathcal{A}_t(X_T|\mathcal{F}_t)$. These are adapted to the filtration $\mathfrak{F}$ and can be interpreted as the development of the valuation of the final cash-flow over time relative to the information available. If the underlying filtration can be represented as a tree and the random variable $X \in \bar{L}_p(\Omega, \mathcal{F}, \mathbb{P})$ as values sitting on the leaves of the tree, such a dynamic mapping will assign acceptability values to each node of the tree.

If $\mathcal{F}_0 = \{\Omega, \emptyset\}$ is the trivial σ-algebra the mapping $\mathcal{A}_0(\cdot|\mathcal{F}_0)$ represents an (unconditional) monotonic functional. Now we can generalise the notion of time consistency for dynamic mappings:

Definition 3.1.2. A dynamic monotonic mapping $(\mathcal{A}_t(\cdot|\mathcal{F}_t))_{t\in\{0,\dots,T-1\}}$ is called *time consistent* if each pair $\mathcal{A}_{t_1}(\cdot|\mathcal{F}_{t_1})$,$\mathcal{A}_{t_2}(\cdot|\mathcal{F}_{t_2})$ with $0 \le t_1 \le t_2 \le T$ is a time consistent pair.

In addition we define:

Definition 3.1.3. A dynamic monotonic mapping $(\mathcal{A}_t(\cdot|\mathcal{F}_t))_{t\in\{0,\dots,T-1\}}$ is called *recursive* if each pair $\mathcal{A}_{t_1}(\cdot|\mathcal{F}_{t_1})$,$\mathcal{A}_{t_2}(\cdot|\mathcal{F}_{t_2})$ with $0 \le t_1 \le t_2 \le T$ is a recursive pair. If in addition all conditional mappings are conditional versions of the same unconditional functional, we call the dynamic acceptability mapping *self-recursive.*

Clearly we have again equivalence between time consistency and recursivity:

Corollary 3.1.4. *A dynamic monotonic mapping* $(\mathcal{A}_t(\cdot|\mathcal{F}_t))_{t\in\{0,\dots,T-1\}}$*, where all the constituent mappings are monotonic and possess the weak projection property, is time consistent if and only if it is recursive.*

Proof. This is an application of Theorem 2.2.3. □

It is also possible to generalise Propositions 2.1.6 and 2.2.5 for dynamic mappings.

Corollary 3.1.5. *Let* $(\mathcal{A}_t(X|\mathcal{F}_t))_{t\in\{0,\ldots,T-1\}}$ *be a dynamic monotonic mapping, applied to* $\mathcal{F}$*-measurable random variables* X*.*

(1) *If* $(\mathcal{A}_t(X|\mathcal{F}_t))_{t\in\{0,\ldots,T-1\}}$ *is a martingale for any* X*, then it is time consistent.*

(2) *If* $(\mathcal{A}_t(X|\mathcal{F}_t))_{t\in\{0,\ldots,T-1\}}$ *is time consistent and all the mappings* $\mathcal{A}_t(\cdot|\mathcal{F}_t)$ *are strict, then* $(\mathcal{A}_t(X|\mathcal{F}_t))_{t\in\{0,\ldots,T-1\}}$ *will be a submartingale.*

Proof. These are applications of Proposition 2.2.5 and Corollary 2.2.4. □

3.2 Compositions

If we restrict ourselves to a finite time horizon there is an easy way to achieve time consistency by nesting monotonic mappings:

Definition 3.2.1. Let $(\mathcal{A}_t(X|\mathcal{F}_t))_{t\in\{0,\ldots,T-1\}}$ be a sequence of monotonic mappings $\mathcal{A}_t(\cdot|\mathcal{F}_t) : \bar{L}_{p_{t+1}}(\Omega, \mathcal{F}_{t+1}, \mathbb{P}) \to \bar{L}_{p_t}(\Omega, \mathcal{F}_t, \mathbb{P})$ and assume that each mapping $\mathcal{A}_t$ has the weak projection property (2.8).

The *dynamic composition* of the mappings is then defined recursively as

$$\mathcal{B}_{T-1}(\cdot|\mathcal{F}_{T-1}) = \mathcal{A}_{T-1}(\cdot|\mathcal{F}_{T-1})$$

for $T-1$ and

$$\mathcal{B}_{t-1}(\cdot|\mathcal{F}_{t-1}) = \mathcal{A}_{t-1}(\mathcal{B}_t(\cdot|\mathcal{F}_t)|\mathcal{F}_{t-1})$$

for $t < T-1$.

Remark 3.2.2. *Any dynamic composition is a recursive and hence time consistent dynamic monotonic mapping.*

Proof. Because each individual mapping $\mathcal{A}_t$ has the weak projection property, each nested mapping $\mathcal{B}_t$ will also possess this property. Now $\mathcal{B}_{t_1} = \mathcal{A}_{t_1} \circ \mathcal{A}_{t_{1+1}} \circ \ldots \circ \mathcal{A}_{t_2-1} \circ \mathcal{B}_{t_2}$ and $\mathcal{B}_{t_2}(X_T|\mathcal{F}_{t_2})$ is a $\mathcal{F}_{t_2}$-measurable random variable. Applying the weak projection property we have

$$\mathcal{B}_{t_1}\left(\mathcal{B}_{t_2}(X_T|\mathcal{F}_{t_2})\right) = \mathcal{A}_{t_1} \circ \mathcal{A}_{t_{1+1}} \circ \ldots \circ \mathcal{A}_{T-1}\left(\mathcal{B}_{t_2}(X_T)|\mathcal{F}_{t_2}\right) = \mathcal{A}_{t_1} \circ \mathcal{A}_{t_{1+1}} \circ \ldots \circ \mathcal{A}_{t_2-1}\left(\mathcal{B}_{t_2}(X_T)|\mathcal{F}_{t_2}\right) = \mathcal{B}_{t_1}\left(X_T|\mathcal{F}_{t_1}\right),$$

which is recursivity. This means that each pair $\mathcal{B}_{t_1}, \mathcal{B}_{t_2}$ is recursive. Hence the whole dynamic mapping is recursive and time consistent. □

In general, compositions are not self-recursive. Literature mostly concentrates on conditional mappings that are versions of the same unconditional functional. In particular self-recursive mappings are analysed ([12, 3]). Of course such functionals have very appealing mathematical properties but unfortunately they are very rare.

Kupper and Schachermayer ([14]) proved that in an a framework with infinite time and under additional assumptions regarding the filtrations involved, the only time consistent dynamic monotonic mappings are compositions of certainty equivalent mappings and the only translation equivariant and time consistent dynamic monotonic mapping is the entropic one.

Example 3.1. The nested $AV@R$, which is the dynamic composition of conditional $AV@R$s is compound convex ([16], Proposition 2.35), hence a submartingale pair and rejection consistent.

On the other hand it is well known that the average value at risk together with its conditional version is not acceptance consistent, hence neither time consistent (self-recursive) nor a martingale pair. This can easily be shown by a counterexample based on the following tree:

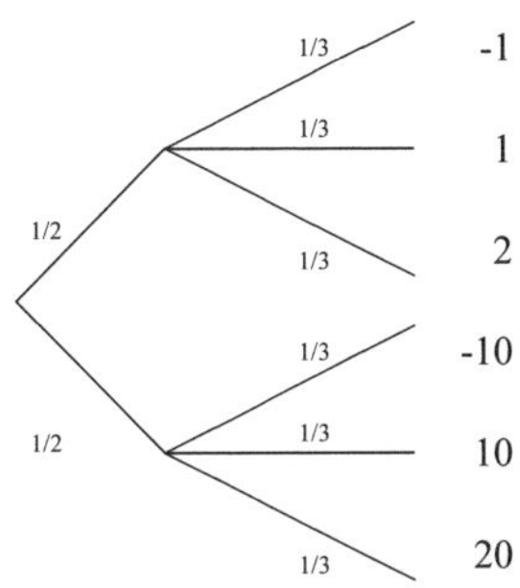

In this case the conditional $AV@R$ at level $\alpha = 2/3$ equals zero in each scenario but the unconditional $AV@R$ of the tree is negative and equals $-8/6$.

On the other hand the nested $AV@R$ together with the conditional $AV@R$ is time consistent by Corollary 3.1.4.

Example 3.2. The Value at risk, defined as the α-quantile of a distribution is not an acceptability mapping, because concavity is missing. Nevertheless it fulfils translation equivariance and is centred at zero. Therefore it possesses the weak projection property and any composition based on the value at risk will be time consistent.

3.3 Intermediate payoffs and multi-period functionals

For practical application we want to consider not only a single cash-flow at the end of our planning horizon. We also want take care of random cash-flows at intermediate periods in our valuation and decision.

Basically we want to jointly valuate a random vector $X = (X_1, \ldots, X_T)$ together with an information structure which represents the gain in information over time. The development of information is again modeled by a filtration $\mathfrak{F} = (\mathcal{F}_0, \ldots, \mathcal{F}_T)$. In addition we assume that at the beginning of the planning horizon no nontrivial information is available: $\mathcal{F}_0 = \{\Omega, \emptyset\}$.

In this setting it is possible to define dynamic multi-period monotonic mappings. Such mappings are constructed in order to evaluate a random income stream relative to nontrivial information available. This can be interpreted as a sequence of valuations of an income stream at future time points.

We write $X^{(t)} = (X_t, \ldots, X_T)$. If a mapping $\mathcal{A}_t$ is applied to a process that is zero at many time points we will for short write

$$\mathcal{A}_t\left(X_\tau; \mathfrak{F}\right) = \mathcal{A}_t\left((0, \ldots, 0, X_\tau, 0, \ldots, 0); \mathfrak{F}\right),$$

$$\mathcal{A}_t\left((X_\tau, X_\upsilon);\mathfrak{F}\right) = \mathcal{A}_t\left((0,\ldots,0,X_\tau,0,\ldots,0,X_\upsilon,0,\ldots,0);\mathfrak{F}\right)$$

etc.

Definition 3.3.1. A sequence of mappings

$$\mathcal{A}_t\left(\cdot;\mathfrak{F}\right) : \times_{j=t+1}^{T} \bar{L}_{p_j}\left(\Omega,\mathcal{F}_j,\mathbb{P}\right) \to \bar{L}_{p_t}\left(\Omega,\mathcal{F}_t,\mathbb{P}\right)$$

with the following properties is called a *dynamic multi-period monotonic mapping*:

(1) Any sequence $\mathcal{A}_t\left(X_\tau;\mathfrak{F}\right)$ with $t \le \tau \le T$ is a dynamic monotonic mapping.

(2) *Monotonicity*. If $X_\tau \ge Y_\tau$ a.s. for all $\tau \in \{t+1,\ldots,T\}$ then

$$\mathcal{A}_t\big(X^{(t+1)};\mathfrak{F}\big) \ge \mathcal{A}_t\big(Y^{(t+1)};\mathfrak{F}\big) \text{ a.s.}$$

The sequence is called a *dynamic multi-period acceptability mapping* if the following conditions hold in addition:

(3) *Translation equivariance.*

$$\mathcal{A}_t\big((X_{t+1},\ldots,X_\tau+c,\ldots,X_T);\mathfrak{F}\big) = \mathcal{A}_t\big((X_{t+1},\ldots,X_\tau,\ldots,X_T);\mathfrak{F}\big) + c \text{ a.s.}$$

(4) *Concavity.*

$$\begin{aligned}\mathcal{A}_t\big(\lambda\cdot X^{(t+1)} + (1-\lambda)Y^{(t+1)};\mathfrak{F}\big)&\\ &\le \lambda\cdot\mathcal{A}_t\big(X^{(t+1)};\mathfrak{F}\big) + (1-\lambda)\cdot\mathcal{A}_t\big(Y^{(t+1)};\mathfrak{F}\big) \text{ a.s.}\end{aligned}$$

for any $\lambda \in [0,1]$.

If we use the notion of the infimum for $\bar{L}_{p_t}\left(\Omega,\mathcal{F}_t,\mathbb{P}\right)$-spaces, it is again possible to analyse supergradients and conjugates for such mappings ([16]). It is also possible to generalise the notions of time consistency, recursivity and the weak projection property in this new framework:

Definition 3.3.2. A dynamic multi-period monotonic mapping has the *weak projection property* if

$$\mathcal{A}_t\left(X_t;\mathfrak{F}\right) = X_t$$

holds for any t and $X_t \in \bar{L}_{p_t}\left(\Omega,\mathcal{F}_t,\mathbb{P}\right)$.

Definition 3.3.3. A dynamic multi-period monotonic mapping is called *time consistent*, if for all $X, Y \in \times_{j=t+1}^{T} \bar{L}_{p_j}\left(\Omega,\mathcal{F}_j,\mathbb{P}\right)$ and all t the implication

$$\mathcal{A}_t(X^{(t+1)};\mathfrak{F}) \ge \mathcal{A}_t(Y^{(t+1)};\mathfrak{F}) \text{ a.s.} \wedge X_t \ge Y_t \Longrightarrow \mathcal{A}_{t-1}(X^{(t)};\mathfrak{F}) \ge \mathcal{A}_{t-1}(Y^{(t)};\mathfrak{F})$$

holds for any $X^{(t)} \in \times_{j=t}^{T} \bar{L}_{p_j}\left(\Omega,\mathcal{F}_j,\mathbb{P}\right)$.

Definition 3.3.4. A dynamic multi-period monotonic mapping is called *recursive*, if for all t

$$\mathcal{A}_{t-1}(X^{(t)};\mathfrak{F}) = \mathcal{A}_{t-1}\left((X_t, \mathcal{A}_t(X^{(t+1)}));\mathfrak{F}\right)$$

holds for any $X^{(t)} \in \times_{j=t}^{T} \bar{L}_{p_j}(\Omega, \mathcal{F}_j, \mathbb{P})$.

The idea for this definition goes back to [12]. Based on these definitions we can restate the equivalence of time consistency and recursivity for dynamic multi-period mappings, an assertion similar to Theorem 2.2.3.

Proposition 3.3.5. *Let $\mathcal{A}_t(\cdot;\mathfrak{F})$ be a dynamic multi-period mapping. If the weak projection property holds, the mapping is time consistent if and only if it is recursive.*

Proof. The proof is parallel to the proof of Theorem 2.2.3 but uses the generalised definitions of recursivity, time consistency and the weak projection property:

If $\mathcal{A}_t(X^{(t+1)};\mathfrak{F}) \geq \mathcal{A}_t(Y^{(t+1)};\mathfrak{F})$ a.s. $\wedge$ $X_t \geq Y_t$ holds, applying monotonicity we get

$$\mathcal{A}_{t-1}\left(X_t, \mathcal{A}_t(X^{(t+1)};\mathfrak{F});\mathfrak{F}\right) \geq \mathcal{A}_{t-1}\left(Y_t, \mathcal{A}_t(Y^{(t+1)};\mathfrak{F});\mathfrak{F}\right).$$

If recursivity is assumed, this means that

$$\mathcal{A}_{t-1}(X^{(t)};\mathfrak{F}) \geq \mathcal{A}_{t-1}(Y^{(t)};\mathfrak{F}),$$

which confirms time consistency.

For the other direction first remember that because of the projection property the equation $\mathcal{A}_t(\mathcal{A}_t(X^{(t+1)};\mathfrak{F});\mathfrak{F}) = \mathcal{A}_t(X^{(t+1)};\mathfrak{F})$ and hence also the inequalities $\mathcal{A}_t(\mathcal{A}_t(X^{(t+1)};\mathfrak{F});\mathfrak{F}) \leq \mathcal{A}_t(X^{(t+1)};\mathfrak{F})$ and $\mathcal{A}_t(\mathcal{A}_t(X^{(t+1)};\mathfrak{F});\mathfrak{F}) \geq \mathcal{A}_t(X^{(t+1)};\mathfrak{F})$ hold. In addition for any $\mathcal{F}_t$-measurable random variable X_t we have $X_t \leq X_t$ and $X_t \geq X_t$.

Assuming now time consistency, we can conclude the pair of inequalities

$$\mathcal{A}_{t-1}(X_t, \mathcal{A}_t(X^{(t+1)};\mathfrak{F});\mathfrak{F}) \leq \mathcal{A}_{t-1}(X^{(t)};\mathfrak{F})$$

and

$$\mathcal{A}_{t-1}(X_t, \mathcal{A}_t(X^{(t+1)};\mathfrak{F});\mathfrak{F}) \geq \mathcal{A}_{t-1}(X^{(t)};\mathfrak{F}).$$

Together this means equality and therefore recursivity holds. □

A special case of dynamic multi-period mappings are acceptability compositions ([21, 22]) which are defined in the following way:

Definition 3.3.6. Let $\mathcal{A}_t(\cdot|\mathcal{F}_t)$, $t = 1, \ldots, T$ be a sequence of conditional monotonic mappings, abbreviated as $\mathcal{A}_t(\cdot)$. An *additive monotonic composition* is a multi-period probability mapping $\mathcal{B}(\cdot|\mathfrak{F})$ which for any $X \in \times_{j=t}^{T} \bar{L}_{p_j}(\Omega, \mathcal{F}_j, \mathbb{P})$ is given by

$$\mathcal{B}_{T-1}(X_T|\mathcal{F}_{T-1}) = \mathcal{A}_{T-1}(X_T|\mathcal{F}_{T-1})$$

for $T-1$ and

$$\mathcal{B}_{t-1}(X^{(t)}|\mathcal{F}_{t-1}) = \mathcal{A}_{t-1}(X_t + \mathcal{B}_t(X^{(t+1)}|\mathcal{F}_t)|\mathcal{F}_{t-1})$$

for $t < T - 1$.

If the mappings $\mathcal{A}_t(\cdot|\mathcal{F}_t)$ are conditional acceptability mappings we call the composition a *additive acceptability composition.*

For additive acceptability compositions we have

$$\mathcal{B}_t(X^{(t+1)};\mathfrak{F}) = \mathcal{A}_t \circ \mathcal{A}_{t+1} \circ \cdots \circ \mathcal{A}_{T-1}\Bigg(\sum_{k=t+1}^{T} X_k\Bigg).$$

If we use a nontrivial σ-algebra for the first period we get dynamic multi-period monotonic mappings.

Such mappings are constructed in a recursive way and hence they are also recursive in the sense of Definition 3.3.4.

Corollary 3.3.7. *Additive monotonic compositions are recursive and hence time consistent multi-period mappings if their constituent mappings possess the weak projection property.*

Proof. This is true because of the recursive construction of the compositions $\mathcal{B}_t$. Time consistency follows from Proposition 3.3.5. □

For compositions of (conditional) acceptability mappings supergradients and supergradient representations have been derived ([13], Lemma 5.2.4, and Theorems 5.2.5 and 5.2.6).

It should be noted that compositions of acceptability mappings and compositions of certainty equivalents are time consistent in the same sense. For two periods, and for the case of only one final payment the main distinction was the self recursivity of certainty equivalents. Because certainty equivalents are not translation equivariant in general, the notion of self recursivity is not meaningful any more for compositions of certainty equivalents.

What is really needed for optimisation in a multi-period, multistage setting are functionals that evaluate the acceptability of a process unconditional - as one real number related to the actual period of decision. Such multi-period functionals can be described as mappings $\mathcal{A}(\cdot;\mathfrak{F})$ from product spaces $\times_{t=1}^{T}\bar{L}_p(\Omega,\mathcal{F}_t,\mathbb{P})$, which are Banach lattices together with the p-norms $\|X\|_p = \sum_{t=1}^{T}\mathbb{E}(|X_t|^p)^{\frac{1}{p}}$, $1 \le p \le \infty$, into the extended real line $\overline{\mathbb{R}}$.

Definition 3.3.8. We will call a multi-period functional $\mathcal{A}(X;\mathfrak{F})$ *multi-period acceptability functional*, if it is proper and satisfies

(MA0) *Information Monotonicity*: If $\mathfrak{F}^{(1)} = \big(\{\Omega,\emptyset\},\mathcal{F}_1^{(1)},\dots,\mathcal{F}_T^{(1)}\big)$, $\mathfrak{F}^{(2)} = \big(\{\Omega,\emptyset\},\mathcal{F}_1^{(2)},\dots,\mathcal{F}_T^{(2)}\big)$ are filtrations such that $\mathcal{F}_t^{(1)} \subseteq \mathcal{F}_t^{(2)}$ for all t, then

$$\mathcal{A}(X;\mathfrak{F}^{(1)}) \le \mathcal{A}(X;\mathfrak{F}^{(2)})$$

for any $X \in \operatorname{dom}\mathcal{A} \subseteq \times_{t=1}^{T}\bar{L}_p(\Omega,\mathcal{F}_t,\mathbb{P})$.

(MA1) *Translation Equivariance*: For all periods t

$$\mathcal{A}(X_1,\dots,X_t+c,\dots,X_T;\mathfrak{F}) = \mathcal{A}(X_1,\dots,X_t,\dots,X_T;\mathfrak{F}) + c$$

holds, if c is a real number.

(MA2) *Concavity*: The mapping $X \mapsto \mathcal{A}(X;\mathfrak{F})$ is concave.

(MA3) *Monotonicity*: $X_t \leq Y_t$ a.s. for all t implies $\mathcal{A}(X;\mathfrak{F}) \leq \mathcal{A}(Y;\mathfrak{F})$.

This definition goes back to [16], where a stronger version of (MA1) was used. For concave multi-period functionals $\mathcal{A}(X;\mathfrak{F})$ it is possible to define the concave conjugate

$$\mathcal{A}^+(Z;\mathcal{F}) = \inf\left\{\langle X, Z\rangle - \mathcal{A}(X;\mathcal{F}) :\ X \in \times_{t=1}^T \bar{L}_p(\Omega,\mathcal{F}_t,\mathbb{P})\right\}$$

and if $\mathcal{A}(X;\mathfrak{F})$ is proper and upper semicontinuous in addition the Fenchel–Moreau–Rockafellar Theorem ([19], Theorem 5) ensures again that the functional equals its biconjugate:

$$\mathcal{A}(X;\mathfrak{F}) = \mathcal{A}^{++}(X;\mathfrak{F}).$$

For multi-period acceptability functionals the supergradient representation is well known:

Proposition 3.3.9. *Let $\mathcal{A}(\cdot;\mathfrak{F})$ be an upper semicontinuous multi-period functional satisfying (MA1), (MA2) and (MA3). Then the representation*

$$\mathcal{A}(X;\mathcal{F}) = \inf_Z \left\{\sum_{t=1}^T \mathbb{E}(X_t\cdot Z_t) - \mathcal{A}^*(Z;\mathfrak{F}) :\ Z_t \geq 0;\ \mathbb{E}(Z_t) = 1\right\} \tag{3.1}$$

holds for every $X \in \times_{t=1}^T \bar{L}_p(\Omega,\mathcal{F}_t,\mathbb{P})$.

Conversely – if $\mathcal{A}(\cdot;\mathfrak{F})$ can be represented by a dual representation (3.1) and the conjugate $\mathcal{A}^*$ is proper, then $\mathcal{A}$ is proper, upper semicontinuous and satisfies (MA1), (MA2) and (MA3).

Proof. See Theorem 3.21 from [16]. □

There are many ways to construct multi-period functionals. One possibility uses dynamic multi-period mappings: Just define a dynamic multi-period mapping for the time periods $0,\dots,T$, where the filtration $\mathfrak{F}$ starts with the trivial σ-algebra $\mathcal{F}_0 = \{\Omega,\emptyset\}$. It is obvious to call such a multi-period functional time consistent, if the dynamic multi-period mapping used is time consistent itself.

In particular it is possible to construct a time consistent multi-period functional with properties (MA1), (MA2), (MA3) by additive-monotonic compositions of conditional acceptability mappings. By contrast additive-monotonic compositions of certainty equivalents – though meaningful functionals – will not fulfil conditions (MA1) – (MA3) in general.

Property (MA0) – information monotonicity – is even harder to achieve. In particular this property is not automatically compatible with a recursive construction of dynamic and multi-period mappings. We will see in section 4 that additive compositions of conditional versions of the same unconditional functional can cause severe troubles.

Another type of multi-period functionals are SEC-functionals ([16], Definition 3.27):

Definition 3.3.10. A dynamic multi-period acceptability mapping $\mathcal{B}_t(\cdot;\mathfrak{F})$ is called *separable expected conditional* (SEC) if it is of the form

$$\mathcal{B}_t\big(X^{(t+1)};\mathfrak{F}\big) = \sum_{j=t+1}^{T} \mathbb{E}\left(\mathcal{A}_{j-1}\left(X_j|\mathcal{F}_{j-1}\right)|\mathcal{F}_t\right),$$

where the $\mathcal{A}_j(\cdot|\mathcal{F}_{t-1})$ are conditional u.s.c. acceptability mappings.

Starting the sequence with $\mathcal{F}_0 = \{\Omega, \{\}\}$, a multi-period acceptability functional $\mathcal{B}(\cdot;\mathfrak{F})$ is called SEC if it has the form

$$\mathcal{B}(X;\mathfrak{F}) = \sum_{t=1}^{T} \mathbb{E}\left(\mathcal{A}_{t-1}\left(X_t|\mathcal{F}_{t-1}\right)\right)$$

SEC-functionals are not constructed as compositions, nevertheless, they are time-consistent in the sense of Definition 3.3.3.

Proposition 3.3.11. *With given acceptability mappings $\mathcal{A}_t(\cdot|\mathcal{F}_t)$ SEC-mappings are time consistent. If the constituting acceptability mappings possess the weak projection property they are recursive in addition.*

Proof. Consider sequences $X^{(t)}, Y^{(t)}$ such that $\mathcal{B}_t\left(X^{(t+1)};\mathfrak{F}\right) \geq \mathcal{B}_t\left(Y^{(t+1)};\mathfrak{F}\right)$ a.s. and $X_t \geq Y_t$ a.s. These assumptions imply $\mathcal{A}_{t-1}(X_t|\mathcal{F}_t) \geq \mathcal{A}_{t-1}(Y_t|\mathcal{F}_t)$ a.s. and $\mathbb{E}\left(\mathcal{B}_t\left(X^{(t+1)};\mathfrak{F}\right)|\mathcal{F}_{t-1}\right) \geq \mathbb{E}\left(\mathcal{B}_t\left(Y^{(t+1)};\mathfrak{F}\right)|\mathcal{F}_{t-1}\right)$ a.s. by monotonicity.

Summing those inequalities and applying the projection property for the conditional expectation we get

$$\begin{aligned}
&\mathcal{A}_{t-1}(X_t|\mathcal{F}_t) + \sum_{j=t+1}^{T} \mathbb{E}\left(\mathcal{A}_{t-1}\left(X_j|\mathcal{F}_{j-1}\right)|\mathcal{F}_{t-1}\right) \\
\geq\ &\mathcal{A}_{t-1}(Y_t|\mathcal{F}_t) + \sum_{j=t+1}^{T} \mathbb{E}\left(\mathcal{A}_{t-1}\left(Y_j|\mathcal{F}_{j-1}\right)|\mathcal{F}_{t-1}\right) \text{ a.s.}
\end{aligned}$$

or

$$\mathcal{B}_{t-1}\left(X^{(t)};\mathfrak{F}\right) \geq \mathcal{B}_{t-1}\left(Y^{(t)};\mathfrak{F}\right)$$

Recursivity then follows by Proposition 3.3.5. □

4 Information monotonicity

In the last section we want to discuss the notion of information monotonicity in more detail. Consider a stochastic (cash-flow) process $X = (X_1, X_2, \ldots, X_T)$, defined on some probability space $(\Omega, \mathcal{F}, \mathbb{P})$. If we want to assign an acceptability value $\mathcal{A}$ seemingly it would suffice to consider only the random variables Y_t, that means to define $\mathcal{A} = \mathcal{A}(Y_1, Y_2, \ldots, Y_T)$.

However in a multi-period setting not only the information generated by the process Y under consideration might be relevant. Typically there is other information (e.g. from the "economic environment") available and useful. This is particularly true in situations where decisions can be taken not only at the beginning but also at some intermediate points of time (multistage problems). A multi-period decision generally should prefer a process with a finer filtration over the same process with a coarser filtration. Otherwise it could be optimal to base a decision on outdated information or to create additional uncertainty in an attempt to get higher "acceptability". In addition it would not be possible to reduce risk by means of hedging.

This is the reason why we introduced the filtration $\mathfrak{F}$ into our definition of multi-period functionals and this justifies the introduction of property (MA0) in Definition 3.3.3.

Of course this property is not restricted to be used with acceptability functionals and can be applied to any (monotonic) multi-period functional. In addition we say that a valuation of a random process X is information monotonic if (MA0) holds at least for this concrete random process. We will also use the notion of information monotonicity with respect to subsets of the domain of a functional. An example would be the multi-period valuation of final cash-flows – in this case information monotonicity is restricted to processes of the form $(0, \ldots, 0, X_T)$.

If a multi-period functional is used as an objective function in a multi-stage decision problem, it should be information monotonic in the sense of (MA0), in order to use an information monotonic valuation on the whole space of possible decisions.

Sometimes we refer to a stronger version of time consistency:

Definition 4.0.12. A mapping $\mathcal{A}_t(\cdot; \mathfrak{F})$ is called *strictly information monotonic* if it is information monotonic and if, whenever $\mathcal{F}_t^{(1)} \subset \mathcal{F}_t^{(2)}$ holds for at least one t, the strict inequality

$$\mathcal{A}(X; \mathfrak{F}^{(1)}) < \mathcal{A}(X; \mathfrak{F}^{(2)})$$

is valid for at least one $X \in \times_{j=t}^{T} \bar{L}_{p_j}(\Omega, \mathcal{F}_j, \mathbb{P})$.

This notion is introduced to distinguish between functionals not responding in any way to additional information and functionals that respond positively at least for some processes.

As by now little is known definitely about the interplay of time consistency and information monotonicity the following facts and examples should make clear the questions arising.

First we look at final payoffs:

Fact 4.1. *Compositions of conditional certainty equivalents – applied to final payoffs – are self-recursive and time consistent. But this means that any intermediate information cancels out. This leads to information monotonicity, but strict information monotonicity does not hold.*

Fact 4.2. *The composition of conditional $AV@R_\alpha$-mappings with common parameter α is time consistent but not information monotonic.*

This can be seen from the following counterexample: Consider a final random payoff X_2 at time $t = 2$ under two alternative filtrations $\mathfrak{F}^{(1)}$ and $\mathfrak{F}^{(2)}$, represented by the following probability trees. Both possibilities are evaluated under the same measure given by the conditional probabilities on the arcs of the tree.

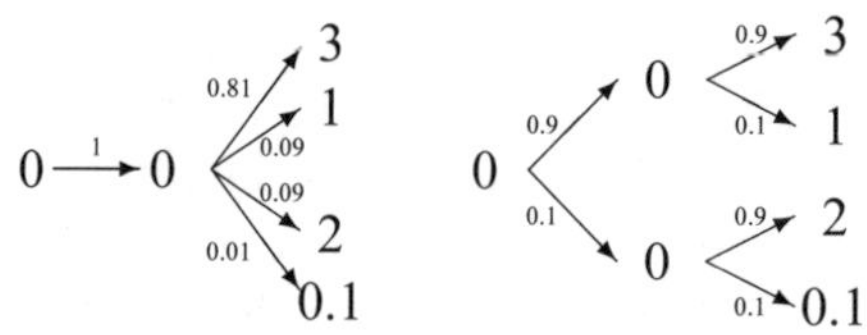

Of course the right tree ($\mathfrak{F}^{(2)}$) represents the more informative filtration.
If we calculate the composition of $AV@R_\alpha$ mappings we get

$$AV@R_{0.05}\left(AV@R_{0.05}(X_2|\mathcal{F}_1^{(1)})\right) = 0.82$$

and

$$AV@R_{0.05}\left(AV@R_{0.05}(X_2|\mathcal{F}_1^{(2)})\right) = 0.1,$$

which means that this pair can not be information monotonic.

Fact 4.3. *If the expectation is composed with the conditional $AV@R_\alpha$ the resulting expected conditional mapping $\mathbb{E}(AV@R_\alpha(\cdot|\mathcal{F}_1^{(1)}))$ will be time consistent and information monotonic ([16], Proposition 3.9). Because the information does not cancel out, strict information monotonicity is possible. This principle can hold also for expected conditional mappings, constructed with other conditional mappings.*

Fact 4.4. *Compositions of acceptability mappings can be information monotonic for some subset of $\times_{j=t}^{T} \bar{L}_{p_j}(\Omega, \mathcal{F}_j, \mathbb{P})$.*

If we use our tree above, an example would be the composition of the unconditional $AV@R_{0.6}$ with the conditional $AV@R_{0.05}$. In fact any unconditional $AV@R_\alpha$, with α high enough will suffice. The limiting case with $\alpha = 1$ is the expectation, which results in an information monotonic valuation by Fact 4.3.

Next we introduce intermediate payments and analyse additive monotonic compositions. In particular additive monotonic compositions of certainty equivalents seem to be a very reasonable concept, because the random variable

$$X_1 + u^{-1}\left(\mathbb{E}\left(u(X_2)|\mathcal{F}_1^{(1)}\right)\right)$$

sums quantities with consistent dimension.

Fact 4.5. *The additive composition of entropic mappings is time consistent and information monotonic but not in the strict sense.*

This can be seen from

$$-\ln\left(\mathbb{E}\left(\exp(-X_1)\cdot\mathbb{E}\left(\exp(-X_2|\mathcal{F}_1^{(1)})\right)\right)\right) = -\ln\left(\mathbb{E}\left(\exp(-X_1 - X_2)\right)\right).$$

Fact 4.6. *In general, additive compositions of other certainty equivalents are not information monotonic.*

If we introduce an intermediate payment X_1 at point in time $t = 1$ in our example trees the calculation of $\mathbb{E}\left[\sqrt{\left(X_1 + \mathbb{E}\left(\sqrt{X_2}|\mathcal{F}_1^{(1)}\right)^2\right)}\right]^2$ leads to a strange phenomenon: If X_1 is a positive payment, information monotonicity holds. But if X_1 is negative, the functional is information antitone.

Fact 4.7. *Monotonic compositions of $AV@R$ mappings with different but coordinated α-parameters will show information monotonicity and also time consistency for some concrete random variables.*

So far, none of the considered additive compositions jointly possesses all the favourable properties of time consistency and (strict) information monotonicity. But at least we can give one example that this is possible.

Fact 4.8. *As we saw in Proposition 3.3.11, SEC-functionals – although not constructed as additive compositions – are time consistent multi-period acceptability functionals. In addition, some of them – particularly if they are based on the expected average value at risk $\mathbb{E}\left(AV@R_\alpha(X_t|\mathcal{F}_{t-1})\right)$ – are information monotonic ([16], Proposition 3.9). Because the information does not cancel out, strict information monotonicity may hold.*

Bibliography

[1] K. J. Arrow, *Aspects of the theory of risk-bearing*, Helsinki, 1965.

[2] Ph. Artzner, F. Delbaen, J. M. Eber, and D. Heath, *Coherent Measures of Risk*, Mathematical Finance 9 (1999), pp. 203–228.

[3] Ph. Artzner, F. Delbaen, D. Heath, and H. Ku, *Coherent multiperiod risk adjusted values and Bellman's principle*, Annals of Operations Research 152 (2007), pp. 5–22.

[4] Ph. Artzner, Fr. Delbaen, and D. Heath, *Thinking Coherently*, Risk 10 (1997), pp. 203–228.

[5] J. Bion-Nadal, *Conditional risk measure and robust representation of convex conditional risk measures*, CMAP preprint 557, 2004.

[6] ______, *Dynamic risk measures: Time consistency and risk measures from BMO martingales*, Finance and Stochastics 12 (2008), pp. 219–244.

[7] K. Detlefsen and G. Scandolo, *Conditional and Dynamic Convex Risk Measures*, Finance and Stochastics 9 (2005), pp. 539–561.

[8] H. Föllmer and A. Schied, *Stochastic Finance – An Introduction in Discrete Time*, Studies in Mathematics, de Gruyter, 2002.

[9] H. Geman and Ohana S., *Time-consistency in managing a commodity portfolio: A dynamic risk measure approach*, Journal of Banking & Finance 32 (2008), pp. 1991–2005.

[10] Ch. Gollier, *The Economics of Risk and Time*, MIT Press, 2001.

[11] G. H. Hardy, J. E. Littlewood, and G. Pólya, *Inequalities*, Cambridge University Press, 1934.

[12] A. Jobert and L.C.G. Rogers, *Valuations and dynamic convex risk measures*, Mathematical Finance 18 (2008), pp. 1–22.

[13] R. Kovacevic, *Conditional Acceptability mappings: Convex Analysis in Banach Lattices*, Dissertation, University of Vienna, 2008.

[14] M. Kupper and W. Schachermayer, *Representation Results for Law Invariant Time Consistent Functions*, preprint, 2008.

[15] H. M. Markowitz, *Portfolio selection*, The Journal of Finance (1952).

[16] G. Ch. Pflug and W. Römisch, *Modeling, Measuring and Managing Risk*, World Scientific, August 2007.

[17] ———, *The role of information in multi-period risk measurement*, Stochastic Programming E Print Series (2009).

[18] F. Riedel, *Dynamic coherent risk measures*, Stochastic Processes and their Applications 112 (2004), pp. 185–200.

[19] R. T. Rockafellar, *Conjugate Duality and Optimization*, CBMS-NSF Regional Conference Series in Applied Mathematics, vol. 16, SIAM, Philadelphia, 1974.

[20] B. Roorda and J. M. Schumacher, *Time consistency conditions for acceptability measures, with an application to Tail Value at Risk*, Insurance: Mathematics and Economics 40 (2007), pp. 209–230.

[21] A. Ruszczyński and A. Shapiro, *Optimization of Risk Measures*, Probabilistic and Randomized Methods for Design under Uncertainty (G. Calafiore and F Dabbene, eds.), Springer, 2005, pp. 17–158.

[22] ———, *Conditional Risk Mappings*, Mathematics of Operations Research 31 (2006), pp. 544–561.

[23] A. Shapiro, *On a time consistency concept in risk averse multi-stage stochastic programming*, Preprint (eprints for the optimization community) (2008).

[24] J. Von Neumann and O. Morgenstern, *Theory of Games and Economic Behavior*, 3d. ed., Princeton University Press, 1953.

[25] S. Weber, *Distribution-invariant dynamic risk measures, information and dynamic consistency*, Mathematical Finance 16 (2006), pp. 419–441.

Author information

Raimund Kovacevic, Department of Statistics and Decision Support Systems, University of Vienna, Universitätsstraße 5/9, A-1010 Vienna, Austria.
Email: raimund.kovacevic@univie.ac.at

Georg Ch. Pflug, Department of Statistics and Decision Support Systems, University of Vienna and IIASA, Laxenburg, Austria.
Email: georg.pflug@univie.ac.at

Optimal investment and hedging under partial and inside information

Michael Monoyios

Abstract. This article concerns optimal investment and hedging for agents who must use trading strategies which are adapted to the filtration generated by asset prices, possibly augmented with some inside information related to the future evolution of an asset price. The price evolution and observations are taken to be continuous, so the partial (and, when applicable, inside) information scenario is characterised by asset price processes with an unknown drift parameter, which is to be filtered from price observations. With linear observation and signal process dynamics, this is done with a Kalman–Bucy filter. Using the dual approach to portfolio optimisation, we solve the Merton optimal investment problem when the agent does not know the drift parameter of the underlying stock. This is taken to be a random variable with a Gaussian prior distribution, which is updated via the Kalman filter. This results in a model with a stochastic drift process adapted to the observation filtration, and which can be treated as a full information problem, yielding an explicit solution. We also consider the same problem when the agent has noisy knowledge at time zero of the terminal value of the Brownian motion driving the stock. Using techniques of enlargement of filtration to accommodate the insider's additional knowledge, followed by filtering the asset price drift, we are again able to obtain an explicit solution. Finally we treat an incomplete market hedging problem. A claim on a non-traded asset is hedged using a correlated traded asset. We summarise the full information case, then treat the partial information scenario in which the hedger is uncertain of the true values of the asset price drifts. After filtering, the resulting problem with random drifts is solved in the case that each asset's prior distribution has the same variance, resulting in analytic approximations for the optimal hedging strategy.

Key words. Duality, filtering, incomplete information, optimal portfolios.

AMS classification. 49N30, 93E11, 93C41, 91B28

1 Introduction

This article examines some problems of optimal investment, and of optimal hedging of a contingent claim in an incomplete market, when the agent's information set is restricted to stock price observations, possibly augmented by some additional information related to the terminal value of a stock price.

In classical models of financial mathematics, one usually specifies a probability space $(\Omega, \mathcal{F}, P)$ equipped with a filtration $\mathbb{F} = (\mathcal{F}_t)_{0\leq t\leq T}$, and then writes down some stochastic process $S = (S_t)_{0\leq t\leq T}$ for an asset price, such that S is adapted to the filtration $\mathbb{F}$. A typical example would be the Black–Scholes (henceforth, BS) model of

Work presented during the *Special Semester on Stochastics with Emphasis on Finance*, September 3 – December 5, 2008, organised by RICAM (Johann Radon Institute for Computational and Applied Mathematics, Austrian Academy of Sciences) Linz, Austria

a stock price, following the geometric Brownian motion

$$dS_t = \sigma S_t(\lambda dt + dB_t), \tag{1.1}$$

where B is a $(P, \mathbb{F})$-Brownian motion and the volatility $\sigma > 0$ and the Sharpe ratio λ are assumed to be known constants. Of course, this is a strong assumption that an agent is assumed to be able to observe the Brownian motion process B, as well as the stock price process S. We refer to this as a *full information* scenario. In this case, an agent uses $\mathbb{F}$-adapted trading strategies in S, a process with known drift and diffusion coefficients.

We shall frequently relax the full information assumption in this article. We shall assume that the agent can only observe the stock price process, and not the Brownian motion B. The agent's trading strategies must therefore be adapted to the *observation filtration* $\widehat{\mathbb{F}} := (\widehat{\mathcal{F}}_t)_{0\le t\le T}$ generated by S. This is a *partial information* scenario. In recent years there has been a growing research activity in this area, as surveyed by Pham [28], for instance, who examines some different scenarios to the ones in this article.

With partial information, the parameter λ would be regarded as an unknown constant whose value needs to be determined from price data. In principle, one would also have to apply this philosophy to the volatility σ, but we shall make the approximation that price observations are continuous, so that σ can be computed from the quadratic variation $[S]_t$ of the stock price, since we have

$$[S]_t = \sigma^2 S_t^2 t, \quad 0 \le t \le T.$$

One way to model the uncertainty in our knowledge of the (supposed constant) parameter λ is to take a so-called Bayesian approach. This means we consider λ to be an $\mathcal{F}_0$-measurable random variable with a given initial distribution (the *prior* distribution). The prior distribution initialises the probability law of λ conditional on $\widehat{\mathcal{F}}_0$, and this is updated in the light of new price information, that is, as the observation filtration $\widehat{\mathbb{F}}$ evolves. (In the case that λ is some unknown process $(\lambda_t)_{0\le t\le T}$ (as opposed to an unknown constant), then we would consider it to be some $\mathbb{F}$-adapted process such that its starting value λ_0 has a given prior distribution conditional on $\widehat{\mathcal{F}}_0$.)

This is an example of a *filtering* problem. In the case of the BS model (1.1), where we model λ as an $\mathcal{F}_0$-measurable random variable, we are interested in computing the conditional expectation

$$\widehat{\lambda}_t := E[\lambda \,|\, \widehat{\mathcal{F}}_t], \quad 0 \le t \le T.$$

We shall see that the effect of filtering is that the model (1.1) may be replaced by a model specified on the filtered probability space $(\Omega, \widehat{\mathcal{F}}_T, \widehat{\mathbb{F}}, P)$ and written as

$$dS_t = \sigma S_t(\widehat{\lambda}_t dt + d\widehat{B}_t),$$

where $\widehat{B}$ is a $(P, \widehat{\mathbb{F}})$-Brownian motion. This model may now be treated as a full information model, since both $\widehat{B}$ and $\widehat{\lambda}$ are $\widehat{\mathbb{F}}$-adapted processes. The price we have paid for restoring a full information scenario is that the constant parameter λ has been replaced by a random process $\widehat{\lambda}$. The procedure by which a partial information model is replaced with a tractable full information model under the observation filtration is typically only achievable in special circumstances, such as Gaussian prior distributions and certain linearity properties in the relation between the observable and unobservable processes.

The rest of the article is as follows. In Section 2, we briefly introduce the innovations process of filtering theory and state the filtering algorithm that we shall use, the celebrated Kalman–Bucy filter [11]. In Section 3 we use the dual approach to portfolio optimisation (see Karatzas [13] for example), to solve the Merton problem [19, 20] of optimal investment, when the drift parameter of the stock must be filtered from price observations. In Section 4 we solve the Merton problem when the agent is again uncertain of the stock's drift, but is assumed to have some additional information in the form of knowledge of the value of a random variable I, representing noisy information on the underlying Brownian motion at time T. Further examples of optimal investment problems with inside information and parameter uncertainty are given in Danilova, Monoyios and Ng [2]. Finally, in Section 5 we consider the hedging of a claim in an incomplete market setting under partial information. Specifically, we consider a basis risk model involving the optimal hedging of a claim on a non-tradeable asset Y using a traded stock S, correlated with Y, when the hedger is restricted to trading strategies in S that are adapted to the observation filtration generated by the asset prices. A number of papers, such as Henderson [8], Monoyios [21, 22] and Musiela and Zariphopoulou [26], have used exponential indifference valuation methods to hedge the claim in an optimal manner in a full information scenario. We outline these results before moving on to the partial information case, where we assume the hedger does not know with certainty the drifts of S and Y. Analytic approximations for prices and hedging strategies are given. Further work on this topic can be found in Monoyios [23, 25].

2 Innovations and linear filtering

Filtering problems concern estimating something (in a manner to be made precise shortly) about an unobserved stochastic process Ξ given observations of a related process Λ. The problem was solved for linear systems in continuous time by Kalman and Bucy [11]. Subsequent work sought generalisations to systems with nonlinear dynamics, see Zakai [33] for instance. Kailath [10] developed the so-called innovations approach to linear filtering, which formulated the problem in the context of martingale theory. This approach to nonlinear filtering was given a definitive treatment by Fujisaki, Kallianpur and Kunita [7]. Textbook treatments can be found in Kallianpur [12], Lipster and Shiryaev [16, 17], Rogers and Williams [32], Chapter VI.8, and Fleming and Rishel [6].

The setting is a probability space $(\Omega, \mathcal{F}, P)$ equipped with a filtration $\mathbb{F} = (\mathcal{F}_t)_{0 \le t \le T}$. All processes are assumed to be $\mathbb{F}$-adapted. Note that $\mathbb{F}$ is not the observation filtration. Let us call $\mathbb{F}$ the *background filtration*. We consider two processes, both taken to be one-dimensional (for simplicity):

- a *signal process* $\Xi = (\Xi_t)_{0 \le t \le T}$ which is not directly observable;
- an *observation process* $\Lambda = (\Lambda_t)_{0 \le t \le T}$, which is observable and somehow correlated with Ξ, so that by observing Λ we can say something about the distribution of Ξ.

Let $\widehat{\mathbb{F}} := (\widehat{\mathcal{F}}_t)_{0 \le t \le T}$ denote the observation filtration generated by Λ. That is,

$$\widehat{\mathcal{F}}_t := \sigma(\Lambda_s; 0 \le s \le t), \quad 0 \le t \le T.$$

The filtering problem is to compute the conditional expectation

$$\widehat{\Xi}_t := E\big[\Xi_t \,|\, \widehat{\mathcal{F}}_t\big], \quad 0 \le t \le T. \tag{2.1}$$

To proceed further, we specify some particular model for the observation and signal processes. We shall focus on the linear case where both Λ and Ξ are solutions to linear stochastic differential equations (SDEs).

2.1 Linear observations and linear signal

Let $B = (B_t)_{0 \le t \le T}$ be an $\mathbb{F}$-Brownian motion. We assume the observation process Λ is of the form

$$\Lambda_t = \int_0^t G(s)\Xi_s \mathrm{d}s + B_t, \quad 0 \le t \le T, \quad \text{(linear observation)} \tag{2.2}$$

with $G(\cdot)$ a deterministic function such that $E\int_0^T G^2(t)\Xi_t^2 < \infty$.

We take the signal process to be of the form

$$\Xi_t = \Xi_0 + \int_0^t A(s)\Xi_s \mathrm{d}s + \int_0^t C(s)\mathrm{d}W_s, \quad 0 \le t \le T, \quad \text{(linear signal)}$$

for deterministic functions $A(\cdot), C(\cdot)$, with W a $(P, \mathbb{F})$-Brownian motion independent of the $\mathcal{F}_0$-measurable random variable Ξ_0, and correlated with B in the observation model (2.2) according to

$$[W, B]_t = \rho t, \quad 0 \le t \le T, \quad \rho \in [-1, 1].$$

Suppose further that the signal process has a Gaussian initial distribution, $\Xi_0 \sim \mathrm{N}(\mu, \mathrm{v})$, independent of B and of W, where $\mathrm{N}(\mu, \mathrm{v})$ denotes the normal probability law with mean μ and variance v. The two-dimensional process (Ξ, Λ) is then Gaussian, so the conditional distribution of Ξ_t given the sigma-field $\widehat{\mathcal{F}}_t$ will also be normal (and so, in particular, is completely characterised by its mean and variance), with mean given by (2.1) and variance

$$V_t := \mathrm{var}\big[\Xi_t \,|\, \widehat{\mathcal{F}}_t\big] = E\big[(\Xi_t - \widehat{\Xi}_t)^2 \,|\, \widehat{\mathcal{F}}_t\big] = \widehat{\Xi_t^2} - \Big(\widehat{\Xi}_t\Big)^2, \quad 0 \le t \le T.$$

Notice that the initial values are

$$\widehat{\Xi}_0 = E\big[\Xi_0 \,|\, \widehat{\mathcal{F}}_0\big] = E\Xi_0 = \mu,$$

and

$$V_0 = E\big[(\Xi_0 - \widehat{\Xi}_0)^2 \,|\, \widehat{\mathcal{F}}_0\big] = E\big[(\Xi_0 - \mu)^2\big] = \mathrm{var}(\Xi_0) = \mathrm{v}.$$

The problem then boils down to finding an algorithm for computing the sufficient statistics $\widehat{\Xi}_t, V_t$ from their initial values $\widehat{\Xi}_0 = \mu$, $V_0 = \mathrm{v}$. For linear systems it turns out that the conditional variance V_t is a *deterministic* function of t. Thus, there is in fact only one sufficient statistic, the conditional mean $\widehat{\Xi}_t$, which turns out to satisfies a linear SDE. This is the celebrated *Kalman–Bucy filter*, given in Theorem 2.1 shortly.

2.2 Innovations process

Define the $\widehat{\mathbb{F}}$-adapted *innovations process* $N = (N_t)_{0\leq t\leq T}$ by

$$N_t := \Lambda_t - \int_0^t G(s)\widehat{\Xi}_s \mathrm{d}s, \quad 0 \leq t \leq T.$$

We recall two crucial properties of the innovations process, which form the bedrock of filtering theory.

- The innovations process N is an $\widehat{\mathbb{F}}$-Brownian motion.
- Every local $\widehat{\mathbb{F}}$-martingale M admits a representation of the form

  $$M_t = M_0 + \int_0^t \Phi_s \mathrm{d}N_s, \quad 0 \leq t \leq T,$$

  where Φ is $\widehat{\mathbb{F}}$-adapted and $\int_0^T \Phi_t^2 \mathrm{d}t < \infty$ a.s.

For a proof of the above results, and of the following celebrated result for filtering of linear systems, see Rogers and Williams [32] or Lipster and Shiryaev [16], for instance.

Theorem 2.1 (One-dimensional Kalman–Bucy filter). *On a filtered probability space $(\Omega, \mathcal{F}, \mathbb{F}, P)$, with $\mathbb{F} = (\mathcal{F}_t)_{0\leq t\leq T}$, let $\Xi = (\Xi_t)_{0\leq t\leq T}$ be an $\mathbb{F}$-adapted signal process satisfying*

$$\mathrm{d}\Xi_t = A(t)\Xi_t \mathrm{d}t + C(t)\mathrm{d}W_t,$$

and let $\Lambda = (\Lambda_t)_{0\leq t\leq T}$ be an $\mathbb{F}$-adapted observation process satisfying

$$\mathrm{d}\Lambda_t = G(t)\Xi_t \mathrm{d}t + \mathrm{d}B_t, \quad \Lambda_0 = 0,$$

where W, B are $\mathbb{F}$-Brownian motions with correlation ρ, and the coefficients $A(\cdot)$, $C(\cdot)$, $G(\cdot)$ are deterministic functions satisfying

$$\int_0^T \left(|A(t)| + C^2(t) + G^2(t)\right) \mathrm{d}t < \infty.$$

Define the observation filtration $\widehat{\mathbb{F}} := (\widehat{\mathcal{F}}_t)_{0\leq t\leq T}$ *by*

$$\widehat{\mathcal{F}}_t := \sigma(\Lambda_s; 0 \leq s \leq t).$$

Suppose Ξ_0 *is an* $\mathcal{F}_0$*-measurable random variable, and that the distribution of* Ξ_0 *is Gaussian with mean* μ *and variance* v*, independent of* W *and* B*. Then the conditional expectation* $\widehat{\Xi}_t := E\big[\Xi_t \mid \widehat{\mathcal{F}}_t\big]$*, for* $0 \leq t \leq T$*, satisfies*

$$\mathrm{d}\widehat{\Xi}_t = A(t)\widehat{\Xi}_t\mathrm{d}t + [G(t)V_t + \rho C(t)]\,\mathrm{d}N_t, \quad \widehat{\Xi}_0 = \mu,$$

where $N = (N_t)_{0\leq t\leq T}$ *is the innovations process, an* $\widehat{\mathbb{F}}$*-Brownian motion satisfying*

$$\mathrm{d}N_t = \mathrm{d}\Lambda_t - G(t)\widehat{\Xi}_t\mathrm{d}t,$$

and $V_t = \mathrm{var}\big[\Xi_t \mid \widehat{\mathcal{F}}_t\big]$*, for* $0 \leq t \leq T$*, is the conditional variance, which is independent of* $\widehat{\mathcal{F}}_t$ *and satisfies the deterministic Riccati equation*

$$\frac{\mathrm{d}V_t}{\mathrm{d}t} = (1-\rho^2)C^2(t) + 2\big[A(t) - \rho C(t)G(t)\big]V_t - G^2(t)V_t^2, \quad V_0 = \mathrm{v}.$$

A multi-dimensional version of the Kalman–Bucy filter can be derived using similar techniques to the one-dimensional case. See Theorem V.9.2 in Fleming and Rishel [6], for instance.

3 Optimal investment problems with random drift

3.1 Portfolio optimisation via convex duality

We wish to apply the filtering results in the previous section to portfolio optimisation and optimal hedging problems when the agent does not know the drift parameters of the underlying assets. The filtering approach leads to portfolio problems in which the assets follow SDEs with random drift parameters. The dual approach to portfolio optimisation is now a classical technique, well suited to such problems. In this section we recall the main results of portfolio optimisation via convex duality. See Karatzas [13] for more details and further references.

Consider an agent with a continuous, differentiable, increasing, concave utility function $U : \mathbb{R}^+ \to \mathbb{R}$. Define the *convex conjugate* $\widetilde{U} : \mathbb{R}^+ \to \mathbb{R}$ of U by

$$\widetilde{U}(\eta) := \sup_{x\in\mathbb{R}^+} [U(x) - x\eta], \quad \eta > 0. \tag{3.1}$$

Then $\widetilde{U}$ is a decreasing, continuously differentiable, convex function given by

$$\widetilde{U}(\eta) = U\left(I(\eta)\right) - \eta I(\eta), \tag{3.2}$$

where I is the inverse of U'. Differentiating (3.2) gives

$$\widetilde{U}'(\eta) = -I(\eta). \tag{3.3}$$

We note that the defining duality relation (3.1) is equivalent to the bidual relation

$$U(x) = \inf_{\eta \in \mathbb{R}^+} [\widetilde{U}(\eta) + x\eta], \quad x > 0.$$

We are interested in solving an optimal portfolio problem for an agent in a complete market with a single stock whose price process is a continuous semimartingale. To be precise, on an a probability space $(\Omega, \mathcal{F}, P)$ equipped with a filtration $\mathbb{F} = (\mathcal{F}_t)_{0\le t\le T}$, suppose a stock price $S = (S_t)_{0\le t\le T}$ follows

$$\mathrm{d}S_t = \sigma_t S_t(\lambda_t \mathrm{d}t + \mathrm{d}B_t),$$

where $\sigma = (\sigma_t)_{0\le t\le T}$ and $\lambda = (\lambda_t)_{0\le t\le T}$ are $\mathbb{F}$-adapted processes, and $B = (B_t)_{0\le t\le T}$ is an $\mathbb{F}$-Brownian motion. For simplicity, we take the interest rate to be zero.

The wealth process $X = (X_t)_{0\le t\le T}$ associated with a self-financing portfolio involving S is given by

$$\mathrm{d}X_t = \sigma_t \theta_t X_t(\lambda_t \mathrm{d}t + \mathrm{d}B_t), \quad X_0 = x,$$

where the process $\theta = (\theta_t)_{0\le t\le T}$ represents the proportion of wealth placed in the stock, and constitutes the agent's trading strategy. Define the set $\mathcal{A}$ of admissible trading strategies as those satisfying $\int_0^T \sigma_t^2\theta_t^2 \mathrm{d}t < \infty$ a.s. and whose wealth process satisfies $X_t \ge 0$ a.s. for all $t \in [0, T]$.

The unique martingale measure $Q \sim P$ on $\mathcal{F}_T$ is defined by

$$\frac{\mathrm{d}Q}{\mathrm{d}P} = Z_T,$$

where $Z = (Z_t)_{0\le t\le T}$ is the exponential local martingale defined by

$$Z_t := \mathcal{E}(-\lambda \cdot B)_t, \quad 0 \le t \le T.$$

We assume that λ satisfies the Novikov condition

$$E \exp\left(\frac{1}{2}\int_0^T \lambda_t^2 \mathrm{d}t\right) < \infty,$$

so that Z is indeed a martingale and Q is indeed a probability measure equivalent to P.

Under Q, the process B^Q defined by

$$B_t^Q := B_t + \int_0^t \lambda_s \mathrm{d}s, \quad 0 \le t \le T,$$

is a Brownian motion. The Q-dynamics of S, X are

$$\mathrm{d}S_t = \sigma_t S_t \mathrm{d}B_t^Q, \quad \mathrm{d}X_t = \sigma_t\theta_t X_t \mathrm{d}B_t^Q.$$

In particular, the solution of the SDE for X, given $X_0 = x$, is

$$X_t = x\mathcal{E}(\sigma\theta \cdot B^Q)_t, \quad 0 \le t \le T.$$

We assume that

$$E^Q \exp\left(\frac{1}{2}\int_0^T \sigma_t^2\theta_t^2 \mathrm{d}t\right) < \infty,$$

so that X is a Q-martingale, satisfying $E^Q X_T = x$, or

$$E\big[Z_T X_T\big] = x, \tag{3.4}$$

which we shall regard as a *constraint* on the terminal wealth X_T. This is the foundation of the dual approach to portfolio optimisation, namely to enforce the martingale constraint on the wealth process.

The basic portfolio problem (the *primal* problem) is, given $X_0 = x$, to maximise expected utility of wealth at time T:

$$u(x) := \sup_{\theta\in\mathcal{A}} EU(X_T), \tag{3.5}$$

subject to (3.4).

The *dual* value function is $\tilde{u} : \mathbb{R}^+ \to \mathbb{R}$ defined by

$$\tilde{u}(\eta) := E\widetilde{U}\left(\eta\frac{\mathrm{d}Q}{\mathrm{d}P}\right), \quad \eta > 0.$$

The well-known result on portfolio optimisation via duality for this model is as follows.

Theorem 3.1. *1. The primal and dual value functions $u(x)$ and $\tilde{u}(\eta)$ are conjugate:*

$$\tilde{u}(\eta) = \sup_{x\in\mathbb{R}^+}[u(x) - x\eta], \qquad u(x) = \inf_{\eta>0}[\tilde{u}(\eta) + x\eta],$$

so that $u'(x) = \eta$ (equivalently, $\tilde{u}'(\eta) = -x$);

2. The optimal terminal wealth in (3.5) is X_T^* satisfying

$$U'(X_T^*) = \eta\frac{\mathrm{d}Q}{\mathrm{d}P}, \quad equivalently, \quad X_T^* = I\left(\eta\frac{\mathrm{d}Q}{\mathrm{d}P}\right).$$

A proof of this result can be found in Karatzas [13]. The idea behind the proof is to consider the maximisation of the objective functional $EU(X_T)$ subject to the constraint $E\big[Z_T X_T\big] = x$, via the Lagrangian

$$L(X_T,\eta) := EU(X_T) + \eta\left(x - E\big[Z_T X_T\big]\right).$$

The first order condition for an optimum then yields that the optimal terminal wealth is characterised by

$$U'(X_T^*) = \eta Z_T \Leftrightarrow X_T^* = I\left(\eta Z_T\right). \tag{3.6}$$

The value of the multiplier η is needed to fully determine X_T^*. We substitute (3.6) into the constraint $E\big[Z_T X_T^*\big] = x$, so that η is given by

$$E\big[Z_T I(\eta Z_T)\big] = x,$$

or, using the definition of $\tilde{u}(\eta)$ and (3.3),

$$\tilde{u}'(\eta) = -x.$$

This is precisely the relation we expect to hold when u and $\tilde{u}$ are conjugate.

3.1.1 Duality for incomplete markets

Similar duality theorems have been developed for incomplete market situations, and also when the agent has a random terminal endowment, possibly in the form of a contingent claim. For the incomplete market case, see the seminal paper by Karatzas et al. [14] for markets with continuous price processes, and Kramkov and Schachermayer [15] for the case with general semimartingale price processes. For problems involving a terminal random endowment in the form of an $\mathcal{F}_T$-measurable random variable, contributions have been made by (among others) Hugonnier and Kramkov [9], Owen [27] and by Delbaen et al. [5] for an agent with an exponential utility function. We shall use the results of [5] in Section 5, when we examine the exponential hedging of a contingent claim in a basis risk model.

For an incomplete market, in which the set $\mathcal{M}$ of martingale measures is no longer a singleton, the significant change is that the dual value function is then defined by

$$\tilde{u}(\eta) := \inf_{Q\in\mathcal{M}} E\widetilde{U}\left(\eta\frac{\mathrm{d}Q}{\mathrm{d}P}\right). \tag{3.7}$$

The form of the duality theorem for an incomplete market is similar to Theorem 3.1, but with the unique martingale measure Q of the complete market replaced by the optimal *dual minimiser* Q^* that achieves the infimum in (3.7). See [13], for instance, for details in an Itô process setting.

3.2 Optimal investment with Gaussian drift process

We wish to apply filtering theory and the martingale approach to portfolio optimisation to the classical optimal portfolio problem of Merton [19, 20], in the case that the agent does not know the drift parameter of the stock. As we shall see, this will involve a portfolio problem in which the market price of risk of the stock is a Gaussian process. Hence we first describe the solution to such a problem.

Suppose a stock price $S = (S_t)_{0\le t\le T}$ follows the process

$$\mathrm{d}S_t = \sigma S_t(\lambda_t \mathrm{d}t + \mathrm{d}B_t),$$

on a filtered probability space $(\Omega, \mathcal{F}, \mathbb{F} = (\mathcal{F}_t)_{0\le t\le T}, P)$, with B an $\mathbb{F}$-Brownian motion and λ an $\mathbb{F}$-adapted process following

$$\lambda_t = \lambda_0 + \int_0^t \mathrm{w}_s dB_s, \quad \mathrm{w}_t = \frac{\mathrm{w}_0}{1+\mathrm{w}_0 t}, \quad 0 \le t \le T, \tag{3.8}$$

for constants λ_0, w_0.

The self-financing wealth process X from trading S is given by

$$\mathrm{d}X_t = \sigma\theta_t X_t(\lambda_t \mathrm{d}t + \mathrm{d}B_t), \quad X_0 = x, \tag{3.9}$$

where the trading strategy $\theta = (\theta_t)_{0\le t\le T}$ is the proportion of wealth invested in stock. We define the set $\mathcal{A}$ of admissible strategies as those satisfying $\int_0^T \theta_t^2 \mathrm{d}t < \infty$ almost surely, such that $X_t \ge 0$ almost surely for all $t \in [0, T]$.

The value function is

$$u(x) := \sup_{\theta \in \mathcal{A}} E\big[U(X_T) \mid \mathcal{F}_0\big] \tag{3.10}$$

where $U(x)$ is the power utility function given by

$$U(x) = \frac{x^\gamma}{\gamma}, \quad 0 < \gamma < 1. \tag{3.11}$$

Theorem 3.2. *Assume that*

$$-1 < \mathrm{w}_0 T < \frac{1-\gamma}{\gamma}.$$

Then the value function (3.10) is given by

$$u(x) = \frac{x^\gamma}{\gamma} C^{1-\gamma}, \tag{3.12}$$

where C is given by

$$C = \left(\frac{(1+\mathrm{w}_0 T)^q}{1+q\mathrm{w}_0 T}\right)^{1/2} \exp\left(-\frac{1}{2}\frac{q(1-q)\lambda_0^2 T}{1+q\mathrm{w}_0 T}\right), \quad q = -\frac{\gamma}{1-\gamma}. \tag{3.13}$$

The optimal trading strategy θ^* achieving the supremum in (3.10) is given by

$$\theta_t^* = \frac{\lambda_t}{\sigma(1-\gamma)}\left(\frac{1}{1+q\mathrm{w}_t(T-t)}\right), \quad 0 \le t \le T. \tag{3.14}$$

Proof. Let Q denote the unique martingale measure for this market. The change of measure martingale $Z := (Z_t)_{0 \le t \le T}$ is given by

$$Z_t := \left.\frac{\mathrm{d}Q}{\mathrm{d}P}\right|_{\mathcal{F}_t} = \mathcal{E}(-\lambda \cdot B)_t, \quad 0 \le t \le T,$$

and satisfies the SDE

$$\mathrm{d}Z_t = -\lambda_t Z_t \mathrm{d}B_t, \quad Z_0 = 1. \tag{3.15}$$

Notice that

$$\lim_{\mathrm{w}_0 \to 0} Z_t = \mathcal{E}(-\lambda_0 B)_t = \exp\left(-\lambda_0 B_t - \frac{1}{2}\lambda_0^2 t\right). \tag{3.16}$$

We may write $Z_t = f(t, \lambda_t)$ where $f : [0,T] \times \mathbb{R} \to \mathbb{R}^+$ is a smooth function, and apply Itô's formula along with the SDE (3.8) for λ to give

$$\mathrm{d}Z_t = \left[f_t(t,\lambda_t) + \frac{1}{2}\mathrm{w}_t^2 f_{xx}(t,\lambda_t)\right]\mathrm{d}t + \mathrm{w}_t f_x(t,\lambda_t)\mathrm{d}B_t, \tag{3.17}$$

with subscripts of f denoting partial derivatives. Equating (3.15) and (3.17) yields the partial differential equations for f:

$$\begin{aligned} \mathrm{w}_t f_x(t,x) &= -x f(t,x), \\ f_t(t,x) + \frac{1}{2}\mathrm{w}_t^2 f_{xx}(t,x) &= 0, \end{aligned}$$

with $f(0,\cdot) = Z_0 = 1$. The solution to these equations gives Z_t in the form

$$Z_t = \left(\frac{\mathrm{w}_0}{\mathrm{w}_t}\right)^{1/2} \exp\left[-\frac{1}{2}\left(\frac{\lambda_t^2}{\mathrm{w}_t} - \frac{\lambda_0^2}{\mathrm{w}_0}\right)\right], \quad 0 \le t \le T. \tag{3.18}$$

Note that this function is actually well-defined even for $\mathrm{w}_0 \to 0$. It is not hard to check that (3.18) reduces to (3.16) in the limit $\mathrm{w}_0 \to 0$.

For power utility, the convex conjugate $\widetilde{U}$ of the utility function is given by

$$\widetilde{U}(\eta) = -\frac{\eta^q}{q}, \quad q = -\frac{\gamma}{1-\gamma}, \quad \eta > 0. \tag{3.19}$$

The dual value function is defined by

$$\tilde{u}(\eta) := E\big[\widetilde{U}(\eta Z_T) \mid \mathcal{F}_0\big], \quad \eta > 0.$$

Using (3.19) we obtain

$$\tilde{u}(\eta) = -\frac{\eta^q}{q} C,$$

where

$$C := E\big[Z_T^q \mid \mathcal{F}_0\big]. \tag{3.20}$$

From Theorem 3.1, the primal and dual value functions are conjugate, which yields that the primal value function is indeed given by (3.12), with C defined by (3.20). It therefore remains to show that C is indeed equal to the expression in (3.13) and that the optimal strategy is given by (3.14).

Once again using Theorem 3.1, the optimal terminal wealth X_T^*, attained by adopting the strategy that achieves the supremum in (3.10), is given by

$$X_T^* = -\widetilde{U}'(u'(x)Z_T).$$

Hence, using the form (3.12) for u, we obtain

$$X_T^* = \frac{x}{C}\,(Z_T)^{-(1-q)}.$$

The optimal wealth process X^* is a $(Q,\mathbb{F})$-martingale, so

$$X_t^* = E^Q\big[X_T^* | \mathcal{F}_t\big] = \frac{1}{Z_t} E\big[Z_T X_T^* \mid \mathcal{F}_t\big] = \frac{x}{CZ_t} E\big[Z_T^q | \mathcal{F}_t\big], \quad 0 \le t \le T. \tag{3.21}$$

So, to compute explicit formulae for $C = E\big[Z_T^q \mid \mathcal{F}_0\big]$ and the optimal wealth process (from which the optimal trading strategy will be derived), we need to evaluate the conditional expectation $E\big[Z_T^q | \mathcal{F}_t\big], 0 \le t \le T$.

From (3.8), for $t \le T$, and conditional on $\mathcal{F}_t$, λ_T is normally distributed according to

$$\mathrm{Law}(\lambda_T | \mathcal{F}_t) = \mathrm{N}(\lambda_t, \mathrm{w}_t - \mathrm{w}_T), \quad 0 \le t \le T.$$

For a normally distributed random variable $Y \sim \mathrm{N}(m, s^2)$, we have

$$E\exp(cY^2) = \frac{1}{\sqrt{1-2cs^2}} \exp\left(\frac{cm^2}{1-2cs^2}\right),$$

so that, given the explicit expression (3.18) for Z_t, both C and the right-hand side of (3.21) can be computed in closed form. We find that C is indeed given by (3.13). Notice that $1 + q\mathrm{w}_0 T > 0$ and $1 + \mathrm{w}_0 T > 0$ due to the conditions on $\mathrm{w}_0 T$, thus the solution is well defined.

For the optimal wealth process, we obtain the formula

$$X_t^* = x\left(\frac{\Psi_t}{\Psi_0}\right)^{1/2} \exp\left(\frac{1}{2}(1-q)(\Phi_t - \Phi_0)\right), \quad 0 \le t \le T, \tag{3.22}$$

where

$$\Psi_t := \frac{\mathrm{w}_t}{1 + q\mathrm{w}_t(T-t)}, \quad \Phi_t := \frac{\lambda_t^2}{\mathrm{w}_t(1 + q\mathrm{w}_t(T-t))}, \quad 0 \le t \le T.$$

To compute the optimal trading strategy θ^*, we apply the Itô formula to (3.22), using the SDE for λ and noting that the derivative of w_t is given by

$$\frac{\mathrm{d}\mathrm{w}_t}{\mathrm{d}t} = -\mathrm{w}_t^2.$$

We compare the coefficient of $\mathrm{d}B_t$ in $\mathrm{d}X_t^*$ with that in (3.9) for the case of the optimal wealth process. This gives (3.14). □

3.2.1 Classical Merton problem

In the limit $\mathrm{w}_0 \to 0$, the drift of the stock becomes the constant λ_0, and Theorem 3.2 gives the solution to the classical full information Merton optimal investment problem for a stock with constant market price of risk λ_0 and volatility σ. In this case it is easy to check that the value function (3.12) becomes

$$u(x) = \frac{x^\gamma}{\gamma} \exp\left(\frac{1}{2}\frac{\gamma}{1-\gamma}\lambda_0^2 T\right),$$

and the optimal trading strategy (3.14) becomes

$$\theta_t^* = \frac{\lambda_0}{\sigma(1-\gamma)}, \quad 0 \le t \le T.$$

That is, the Merton investor keeps a constant proportion of wealth invested in the stock, as is well known.

3.3 Merton problem with uncertain drift

We can now solve the Merton problem when the agent has uncertainty over the true value of the drift parameter. Optimal investment models under partial information have been considered by many authors. We refer the reader to Rogers [31], Björk, Davis and Landén [1], and Platen and Runggaldier [30], for example.

A stock price process $S = (S_t)_{0 \le t \le T}$ follows

$$dS_t = \sigma S_t(\lambda dt + dB_t), \tag{3.23}$$

on a complete probability space $(\Omega, \mathcal{F}, P)$ equipped with a filtration $\mathbb{F} := (\mathcal{F}_t)_{0 \le t \le T}$, with $B = (B_t)_{0 \le t \le T}$ an $\mathbb{F}$-Brownian motion.

Define the process $\xi = (\xi_t)_{0 \le t \le T}$, by

$$\xi_t := \frac{1}{\sigma} \int_0^t \frac{dS_u}{S_u} = \lambda t + B_t. \tag{3.24}$$

The process ξ will be considered as the observation process in a filtering framework, corresponding to noisy observations of λ, with B representing the noise. In a partial information model with continuous stock price observations, an agent must use $\widehat{\mathbb{F}}$-adapted trading strategies, where where $\widehat{\mathbb{F}} := (\widehat{\mathcal{F}}_t)_{0 \le t \le T}$ is the observation filtration, defined by

$$\widehat{\mathcal{F}}_t := \sigma(\xi_s; 0 \le s \le t) = \sigma(S_s; 0 \le s \le t).$$

Then σ is known from the quadratic variation of S, but λ is an unknown constant, and hence modelled as an $\mathcal{F}_0$-measurable random variable. We assume the distribution of λ is Gaussian, $\lambda \sim \mathrm{N}(\lambda_0, \mathrm{v}_0)$, independent of B.

We are faced with a Kalman–Bucy type filtering problem whose unobservable signal process is the market price of risk λ. The signal process SDE is

$$d\lambda = 0, \tag{3.25}$$

and the observation process SDE is (3.24).

We apply Theorem 2.1 to the signal process λ in (3.25) and observation process ξ in (3.24). Then the optimal filter

$$\widehat{\lambda}_t := E\big[\lambda \,|\, \widehat{\mathcal{F}}_t\big], \quad 0 \le t \le T,$$

satisfies

$$d\widehat{\lambda}_t = \mathrm{v}_t d\widehat{B}_t, \quad \widehat{\lambda}_0 = \lambda_0, \tag{3.26}$$

where

$$\mathrm{v}_t := E\big[(\lambda - \widehat{\lambda}_t)^2 | \widehat{\mathcal{F}}_t\big], \quad 0 \le t \le T,$$

is the conditional variance of λ. This satisfies the Riccati equation

$$\frac{d\mathrm{v}_t}{dt} = -\mathrm{v}_t^2, \tag{3.27}$$

with initial value v_0, so that

$$\mathrm{v}_t = \frac{\mathrm{v}_0}{1 + \mathrm{v}_0 t}, \quad 0 \le t \le T. \tag{3.28}$$

The process $\widehat{B}$ is an $\widehat{\mathbb{F}}$-Brownian motion, the innovations process, satisfying

$$d\widehat{B}_t = d\xi_t - \widehat{\lambda}_t dt. \tag{3.29}$$

Using this in (3.26), the optimal filter can also be written in terms of the observable ξ as

$$\widehat{\lambda}_t = \frac{\lambda_0 + \mathrm{v}_0 \xi_t}{1 + \mathrm{v}_0 t}, \quad 0 \le t \le T. \tag{3.30}$$

The effect of the filtering is that the agent is now investing in a stock with dynamics given by $\mathrm{d}S_t = \sigma S_t \mathrm{d}\xi_t$ which, using (3.29), becomes

$$\mathrm{d}S_t = \sigma S_t(\widehat{\lambda}_t \mathrm{d}t + \mathrm{d}\widehat{B}_t). \tag{3.31}$$

Our agent has a power utility function (3.11) and may invest a portion of his wealth in shares and the remaining wealth in a cash account with zero interest rate (for simplicity). The ($\widehat{\mathbb{F}}$-adapted) wealth process X^0 then follows

$$\mathrm{d}X^0_t = \sigma\theta^0_t X^0_t(\widehat{\lambda}_t \mathrm{d}t + \mathrm{d}\widehat{B}_t), \quad X^0_0 = x, \tag{3.32}$$

where θ^0_t is the proportion of wealth invested in shares at time $t \in [0,T]$, an $\widehat{\mathbb{F}}$-adapted process satisfying $\int_0^T \left(\theta^0_t\right)^2 \mathrm{d}t < \infty$ almost surely, and such that $X^0_t \ge 0$ almost surely for all $t \in [0,T]$. Denote by $\mathcal{A}_0$ the set of such admissible strategies.

The objective is to maximise expected utility of terminal wealth over the $\widehat{\mathbb{F}}$-adapted admissible strategies. The value function is

$$u_0(x) := \sup_{\theta \in \mathcal{A}_0} E\big[U(X^0_T) \mid \widehat{\mathcal{F}}_0\big].$$

This may now be treated as a full information problem, with state dynamics given by (3.32).

We see from equations (3.26), (3.28) and (3.31), that the solution to the partial information optimal portfolio problem is given by Theorem 3.2, when we replace the process λ of Theorem 3.2 by $\widehat{\lambda}$, and replace $(\mathrm{w}_t)_{0\le t\le T}$ by $(\mathrm{v}_t)_{0\le t\le T}$. We have therefore proved the following result.

Theorem 3.3 (Merton problem with uncertain drift). *In a complete market with stock price process S given by (3.23), suppose an agent is restricted to using stock price adapted strategies to maximise expected utility of terminal wealth, with power utility function given by (3.11). Suppose further that the agent's prior distribution for λ is Gaussian, according to*

$$\mathrm{Law}(\lambda \mid \widehat{\mathcal{F}}_0) = \mathrm{N}(\lambda_0, \mathrm{v}_0),$$

and assume that

$$-1 < \mathrm{v}_0 T < \frac{1-\gamma}{\gamma}.$$

Then the agent's value function is given by

$$u_0(x) = \frac{x^\gamma}{\gamma} C_0^{1-\gamma}.$$

where

$$C_0 = \left[\frac{(1+\mathrm{v}_0 T)^q}{1+q\mathrm{v}_0 T}\right]^{1/2} \exp\left[-\frac{1}{2}\frac{q(1-q)\lambda_0^2 T}{1+q\mathrm{v}_0 T}\right], \quad q = -\frac{\gamma}{1-\gamma}.$$

The optimal trading strategy is $\theta^{0,*} = (\theta_t^{0,*})_{0\le t\le T}$, *given by*

$$\theta_t^{0,*} = \frac{\widehat{\lambda}_t}{\sigma(1-\gamma)}\left(\frac{1}{1+q\mathrm{v}_t(T-t)}\right), \quad 0 \le t \le T,$$

where $\widehat{\lambda} = (\widehat{\lambda}_t)_{0\le t\le T}$ *satisfies (3.26) and* v_t *is given by (3.28).*

The classical Merton strategy is thus altered in two ways: the constant λ is replaced by its filtered estimate $\widehat{\lambda}_t$, and the risky asset proportion is decreased by the factor $(1 + q\mathrm{v}_t(T-t))^{-1}$. We note that the more risk averse the investor, the less likely he is to invest in shares, and as $t \to T$, the optimal strategy gets closer and closer to the Merton rule.

4 Investment with inside information and drift uncertainty

We again consider the Merton optimal investment problem in which the agent does not know the stock price drift, but now with the added feature that the agent has some additional information at time zero, represented by noisy knowledge of the terminal value B_T of the Brownian motion driving the stock. We refer the reader to Danilova, Monoyios and Ng [2] for further examples, such as when the additional information involves noisy knowledge of the terminal stock price. The work in this section and in [2] extends the classical inside information model of Pikovsky and Karatzas [29] by considering the situation where the insider does not know the stock's appreciation rate. The agent must use strategies that are adapted to the stock price filtration, but enlarged by the additional information. We must therefore utilise a filtering algorithm which computes the best estimate of the drift, given stock price observations and the additional information. The usual Kalman–Bucy equations hold in this scenario, but with modified initial conditions reflecting the additional information.

The market is the same one as in Section 3.3, with a single stock whose price process S follows (3.23), on a complete probability space $(\Omega, \mathcal{F}, P)$ equipped with a background filtration $\mathbb{F} := (\mathcal{F}_t)_{0\le t\le T}$, with B an $\mathbb{F}$-Brownian motion. We shall again allow for uncertainty in the value of λ, so consider it to be an $\mathcal{F}_0$-measurable random variable. Once again we take the interest rate to be zero.

As before, we define the observation process $\xi = (\xi_t)_{0\le t\le T}$ by (3.24), and the filtration generated by ξ is again denoted by $\widehat{\mathbb{F}} = (\widehat{\mathcal{F}}_t)_{0\le t\le T}$. Since the background filtration $\mathbb{F}$ contains the Brownian filtration and also the sigma-field generated by λ, we have $\widehat{\mathcal{F}}_t \subseteq \mathcal{F}_t$, for all $t \in [0, T]$.

Also as before, the uncertainty in the $\mathcal{F}_0$-measurable random variable λ is modelled by assuming that its prior distribution conditional on $\widehat{\mathcal{F}}_0$ is Gaussian, according to

$$\mathrm{Law}(\lambda \mid \widehat{\mathcal{F}}_0) = \mathrm{N}(\lambda_0, \mathrm{v}_0), \quad \text{independent of } B, \tag{4.1}$$

for given constants λ_0, v_0.[1]

In contrast to earlier, the utility-maximising agent will not only have access to $\widehat{\mathbb{F}}$ in order to estimate λ and implement an optimal strategy, but will be able to augment $\widehat{\mathbb{F}}$ with some additional information, represented by knowledge of a random variable I.

Our procedure in this section is to first enlarge the background filtration $\mathbb{F}$ with the information carried by the random variable I. Denote the enlarged filtration by $\mathbb{F}^{\sigma(I)} = (\mathcal{F}_t^{\sigma(I)})_{0\le t\le T}$, with

$$\mathcal{F}_t^{\sigma(I)} := \mathcal{F}_t \vee \sigma(I), \quad 0 \le t \le T.$$

By starting with an enlarged background filtration and then considering the optimal investment problem with uncertain drift, we aim to incorporate the insider's additional information in the estimation of the unknown market price of risk λ.

The next step is to write the stock price SDE (3.23) in terms of quantities adapted to $\mathbb{F}^{\sigma(I)}$. As $\mathbb{F}$ contains the Brownian filtration, we apply classical initial enlargement results (see, for instance, Mansuy and Yor [18]). There exists an $\mathbb{F}^{\sigma(I)}$-adapted process ν, the *information drift*, such that the Brownian motion B decomposes according to

$$B_t := B_t^I + \int_0^t \nu_s \mathrm{d}s, \quad 0 \le t \le T, \tag{4.2}$$

where B^I is an $\mathbb{F}^{\sigma(I)}$-Brownian motion. We shall characterise the information drift via Lemma 4.2 shortly.

Using (4.2), the stock price dynamics (3.23) is written in terms of $\mathbb{F}^{\sigma(I)}$-adapted processes, to give

$$\mathrm{d}S_t = \sigma S_t \left(\lambda_t^I \mathrm{d}t + \mathrm{d}B_t^I\right), \tag{4.3}$$

where

$$\lambda_t^I := \lambda + \nu_t, \quad 0 \le t \le T,$$

is $\mathbb{F}^{\sigma(I)}$-adapted. If the insider happened to know the value of λ, then we would interpret (4.3) as his stock price SDE, with a stochastic market price of risk λ^I, on the filtered probability space $(\Omega, \mathcal{F}_T^{\sigma(I)}, \mathbb{F}^{\sigma(I)}, P)$.

We study a problem where the inside information consists of noisy Brownian inside information. In other words, we take I to be given by

$$I := aB_T + (1-a)\epsilon, \quad 0 < a < 1, \tag{4.4}$$

and where ϵ is a standard normal random variable independent of B and λ.

Define the insider's observation filtration $\widehat{\mathbb{F}}^{\sigma(I)} = (\widehat{\mathcal{F}}_t^{\sigma(I)})_{0\le t\le T}$ by

$$\widehat{\mathcal{F}}_t^{\sigma(I)} := \sigma(I, \xi_s; 0 \le s \le t), \quad 0 \le t \le T.$$

We now incorporate the insider's uncertainty in the knowledge of λ by treating it as an $\mathcal{F}_0^{\sigma(I)}$-measurable Gaussian random variable with distribution conditional on $\widehat{\mathcal{F}}_0$ given

[1] One way to choose λ_0, v_0 would be to use past data before time zero to obtain a point estimate of λ, and to use the distribution of the estimator as the prior, as in Monoyios [23] and Section 5 of this article.

by (4.1). In this example, λ is independent of I, so its distribution conditional on $\widehat{\mathcal{F}}_0^{\sigma(I)}$ is unaltered from that in (4.1):

$$\mathrm{Law}(\lambda \mid \widehat{\mathcal{F}}_0^{\sigma(I)}) = \mathrm{Law}(\lambda \mid \widehat{\mathcal{F}}_0) = \mathrm{N}(\lambda_0, \mathrm{v}_0). \tag{4.5}$$

Treating λ^I as an unobservable signal process, we shall see that λ^I will satisfy a linear SDE with respect to $\mathbb{F}^{\sigma(I)}$. The Kalman–Bucy filter then allows the insider to infer the conditional expectation

$$\widehat{\lambda}_t^I := E\big[\lambda_t^I \mid \widehat{\mathcal{F}}_t^{\sigma(I)}\big], \quad 0 \le t \le T, \tag{4.6}$$

that is, the best estimate of the signal λ^I based on the insider's observation filtration $\widehat{\mathbb{F}}^{\sigma(I)}$, which turns out to be a Gaussian process, fully characterised by the filtering algorithm. The initial condition for the optimal filter incorporates the inside information, and the SDE for the filter augments this with the stock price observations. This will convert the partial information model (4.3) to a full information model on the filtered probability space $(\Omega, \widehat{\mathcal{F}}_T^{\sigma(I)}, \widehat{\mathbb{F}}^{\sigma(I)}, P)$ with the stock price following

$$\mathrm{d}S_t = \sigma S_t(\widehat{\lambda}_t^I \mathrm{d}t + \mathrm{d}\widehat{B}_t^I), \tag{4.7}$$

where $\widehat{B}^I$ is an $\widehat{\mathbb{F}}^{\sigma(I)}$-Brownian motion. Finally, once we have the full information model (4.7), we are able to compute the maximum utility via duality.

Denote the agent's $\widehat{\mathbb{F}}^{\sigma(I)}$-adapted wealth process by $X^I = (X_t^I)_{0\le t\le T}$, with trading strategy $\theta^I = (\theta_t^I)_{0\le t\le T}$, the proportion of wealth invested in the stock, an $\widehat{\mathbb{F}}^{\sigma(I)}$-adapted process satisfying $\int_0^T (\theta_t^I)^2\,\mathrm{d}t < \infty$ almost surely, such that $X_t^I \ge 0$ almost surely for all $t \in [0,T]$. Denote by $\mathcal{A}_I$ the set of such admissible strategies.

The value function for this problem is

$$u_I(x) := \sup_{\theta^I \in \mathcal{A}_I} E\big[U(X_T^I) \mid \widehat{\mathcal{F}}_0^{\sigma(I)}\big], \quad x > 0, \tag{4.8}$$

where U is the power utility function (3.11). We emphasise that the objective function in (4.8) is conditioned on $\widehat{\mathcal{F}}_0^{\sigma(I)}$.

Define the modulated terminal time T_a by

$$T_a := T + \left(\frac{1-a}{a}\right)^2, \tag{4.9}$$

which will appear in our results. Then the solution to this problem is as follows.

Theorem 4.1. *Assume that*

$$\frac{T}{T_a} - 1 < \mathrm{v}_0 T < \frac{T}{T_a} + \frac{1-\gamma}{\gamma}.$$

Define the function $\mathrm{v}^I : [0,T] \to \mathbb{R}$ *by*

$$\mathrm{v}_t^I := \frac{\mathrm{v}_0^I}{1+\mathrm{v}_0^I t}, \quad \mathrm{v}_0^I := \mathrm{v}_0 - \frac{1}{T_a}, \quad 0 \le t \le T. \tag{4.10}$$

Then the process $\widehat{\lambda}^I$ *in (4.6) is given by*

$$\widehat{\lambda}^I_t = \lambda_0 + \frac{I}{aT_a} + \int_0^t \mathrm{v}^I_s \mathrm{d}\widehat{B}^I_s, \quad 0 \le t \le T, \tag{4.11}$$

where I *is defined in (4.4) and* T_a *in (4.9). The value function of the insider with knowledge of* I *at time zero is given by*

$$u_I(x) = \frac{x^\gamma}{\gamma} C_I^{1-\gamma}, \tag{4.12}$$

where C_I *is the* $\widehat{\mathcal{F}}^I_0$*-measurable random variable given by*

$$C_I = \left(\frac{(1+\mathrm{v}^I_0 T)^q}{1+q\mathrm{v}^I_0 T} \right)^{1/2} \exp\left(-\frac{1}{2} \frac{q(1-q)(\widehat{\lambda}^I_0)^2 T}{1+q\mathrm{v}^I_0 T} \right), \quad q = -\frac{\gamma}{1-\gamma}.$$

The insider's optimal trading strategy is $\theta^{I,*} = (\theta^{I,*}_t)_{0\le t\le T}$*, given by*

$$\theta^{I,*}_t = \frac{\widehat{\lambda}^I_t}{\sigma(1-\gamma)} \left(\frac{1}{1+q\mathrm{v}^I_t(T-t)} \right), \quad 0 \le t \le T.$$

Of course, the value function (4.12) depends explicitly on I, through its dependence on $\widehat{\lambda}^I_0$. We note the similarity in the structure of the solution to this problem with that of the Merton problem with uncertain drift and no inside information. The function v^I plays a similar role to the function v in the conventional partial information problem. It turns out that v^I is related to (but not identical to) the variance of λ^I conditional on $\widehat{\mathbb{F}}^I$, as we shall see.

4.1 Computing the information drift

The first result we need in order to prove Theorem 4.1 is a lemma that gives an explicit formula for the information drift in (4.2). Recall that we begin with a background filtration $\mathbb{F} = (\mathcal{F}_t)_{0\le t\le T}$ that includes the Brownian filtration and the sigma-field generated by λ. We enlarge $\mathbb{F}$ with the information carried by the random variable I. Define, for a bounded Borel function $f : \mathbb{R} \to \mathbb{R}$, the process $(\pi_t(f))_{0\le t\le T}$ as the continuous version of the martingale $(E[f(I) \mid \mathcal{F}_t])_{0\le t\le T}$:

$$\pi_t(f) := E[f(I) \mid \mathcal{F}_t], \quad 0 \le t \le T.$$

There then exists a predictable family of measures $(\mu_t(\mathrm{d}x))_{0\le t\le T}$ such that

$$\pi_t(f) = \int_{\mathbb{R}} f(x)\mu_t(\mathrm{d}x).$$

For fixed $t \in [0,T]$, the measure $\mu_t(\mathrm{d}x)$ is the conditional distribution of I given $\mathcal{F}_t$. Suppose I is such that there exists a density function $g(t,x,y)$ for each $t \in [0,T]$, and such that

$$\pi_t(f) = \int_{\mathbb{R}} f(x)\mu_t(\mathrm{d}x) = \int_{\mathbb{R}} f(x) g(t,x,B_t)\mathrm{d}x. \tag{4.13}$$

The enlargement decomposition formula is given by the following lemma.

Lemma 4.2. *Suppose that I is continuous random variable with conditional (on $\mathcal{F}_t$) distribution given by $g(t, x, B_t)$. Assume also that this distribution satisfies the following conditions:*

$$\int_{\mathbb{R}} |g_y(t,x,y)|\,\mathrm{d}x < \infty, \qquad \int_{\mathbb{R}} \left|\frac{g_y(t,x,y)}{g(t,x,y)}\right| \mathrm{d}x < \infty,$$

for a.e. $t \in [0,T]$ and a.e. $y \in \mathbb{R}$. Then the $\mathbb{F}$-Brownian motion B decomposes with respect to the enlarged filtration $\mathbb{F}^{\sigma(I)}$ according to

$$B_t = B_t^I + \int_0^t \nu_s \mathrm{d}s, \quad 0 \le t \le T,$$

where B^I is an $\mathbb{F}^{\sigma(I)}$-Brownian motion. The information drift ν is given by

$$\nu_t = \frac{g_y(t,I,B_t)}{g(t,I,B_t)}, \quad 0 \le t \le T.$$

Proof. Let f be a test function. Introduce the $\mathbb{F}$-predictable process $(\dot{\pi}_t(f))_{0\le t\le T}$ such that

$$\pi_t(f) = Ef(I) + \int_0^t \dot{\pi}_s(f)\mathrm{d}B_s,$$

which exists by the representation property of Brownian martingales as stochastic integrals with respect to B. There exists a predictable family of measures $(\dot{\mu}_t(\mathrm{d}x))_{0\le t\le T}$ such that

$$\dot{\pi}_t(f) = \int_{\mathbb{R}} f(x)\dot{\mu}_t(\mathrm{d}x),$$

and such that for each $t \in [0,T]$ the measure $\dot{\mu}_t(\mathrm{d}x)$ is absolutely continuous with respect to $\mu_t(\mathrm{d}x)$. Define $\alpha(t,x)$ by

$$\dot{\mu}_t(\mathrm{d}x) = \alpha(t,x)\mu_t(\mathrm{d}x).$$

Now suppose we have a continuous $\mathbb{F}$-martingale M given by

$$M_t = \int_0^t m_s \mathrm{d}B_s, \qquad 0 \le t \le T.$$

By Theorem 1.6 in Mansuy and Yor [18], there exists an $\mathbb{F}^{\sigma(I)}$-local martingale M^I such that

$$M_t = M_t^I + \int_0^t \alpha(s,I)\,\mathrm{d}[M,B]_s,$$

provided that, almost surely,

$$\int_0^t |\alpha(s,I)|\,\mathrm{d}[M,B]_s < \infty.$$

In particular, if $\int_0^t |\alpha(s,I)|\,\mathrm{d}s < \infty$ almost surely, then B decomposes as

$$B_t = B_t^I + \int_0^t \alpha(s,I)\,\mathrm{d}s, \quad 0 \le t \le T,$$

with B^I an $\mathbb{F}^{\sigma(I)}$-Brownian motion.

From the definition of $\alpha(t,x)$ we have

$$\dot{\pi}_t(f) = \int_{\mathbb{R}} f(x)\alpha(t,x)\mu_t(\mathrm{d}x) = \int_{\mathbb{R}} f(x)\alpha(t,x)g(t,x,B_t)\mathrm{d}x.$$

Hence,

$$\mathrm{d}\pi_t(f) = \dot{\pi}_t(f)\mathrm{d}B_t = \left(\int_{\mathbb{R}} f(x)\alpha(t,x)g(t,x,B_t)\mathrm{d}x\right)\mathrm{d}B_t,$$

so that

$$\mathrm{d}[\pi(f),M]_t = \left(\int_{\mathbb{R}} f(x)\alpha(t,x)g(t,x,B_t)\mathrm{d}x\right)\mathrm{d}[B,M]_t. \tag{4.14}$$

But from the defining representation (4.13), the right-hand side of which is a smooth function of B_t, the Itô formula gives

$$\mathrm{d}[\pi(f),M]_t = \left(\int_{\mathbb{R}} f(x)g_y(t,x,B_t)dx\right)\mathrm{d}[B,M]_t, \tag{4.15}$$

and comparing (4.14) with (4.15) yields the result. □

Proof of Theorem 4.1. For I given by (4.4), the conditional distribution of I given $\mathcal{F}_t$, for $t \leq T$, is

$$\mathrm{N}(aB_t, a^2(T-t)+(1-a)^2) = \mathrm{N}(aB_t, a^2(T_a-t)),$$

where T_a is defined in (4.9). Hence the conditional density is

$$g(t,x,B_t) = \frac{1}{a\sqrt{2\pi(T_a-t)}}\exp\left[-\frac{1}{2}\frac{(x-aB_t)^2}{a^2(T_a-t)}\right].$$

So by Lemma 4.2, the information drift is

$$\nu_t = \frac{I-aB_t}{a(T_a-t)}, \quad 0 \leq t \leq T. \tag{4.16}$$

Using the information drift in (4.16) we write the stock price SDE (3.23) in terms of $\mathbb{F}^{\sigma(I)}$-adapted processes, to obtain (4.3), where the $\mathbb{F}^{\sigma(I)}$-adapted market price of risk λ^I is given by

$$\lambda^I_t := \lambda + \nu_t = \lambda + \frac{I-aB_t}{a(T_a-t)} =: h(t,B_t), \quad 0 \leq t \leq T,$$

and where $h : [0,T]\times\mathbb{R} \to \mathbb{R}$ is defined by

$$h(t,x) := \lambda + \frac{I-ax}{a(T_a-t)}.$$

Applying the Itô's formula and using $\mathrm{d}B_t = \nu_t\mathrm{d}t + \mathrm{d}B^I_t$, we obtain

$$\mathrm{d}\lambda^I_t = -\frac{1}{T_a-t}\mathrm{d}B^I_t, \quad \lambda^I_0 = \lambda + \frac{I}{aT_a}. \tag{4.17}$$

With ξ being the returns process in (3.24), we have

$$\mathrm{d}\xi_t = \lambda_t^I \mathrm{d}t + \mathrm{d}B_t^I. \tag{4.18}$$

We now regard λ as an unknown constant, and hence a random variable, whose distribution conditional on $\widehat{\mathcal{F}}_0^{\sigma(I)}$ is given by (4.5). Then we regard $(\lambda_t^I)_{0\le t\le T}$ as an unobservable signal process following (4.17), and ξ as an observation process following (4.18), in a filtering framework to estimate of λ_t^I conditional on $\widehat{\mathcal{F}}_t^{\sigma(I)}$.

Using (4.5), we can write down the initial distribution of λ_0^I given $\widehat{\mathcal{F}}_0^{\sigma(I)}$:

$$\mathrm{Law}(\lambda_0^I|\widehat{\mathcal{F}}_0^{\sigma(I)}) = \mathrm{Law}\left(\lambda + \frac{I}{aT_a}\middle|\widehat{\mathcal{F}}_0^{\sigma(I)}\right) = \mathrm{N}\left(\lambda_0 + \frac{I}{aT_a}, \mathrm{v}_0\right).$$

This defines the prior distribution of the signal process λ^I. Of course, since I is $\widehat{\mathcal{F}}_0^{\sigma(I)}$-measurable, it does not contribute to the initial variance.

The Kalman–Bucy filter, Theorem 2.1, is directly applicable, and yields that the optimal filter

$$\widehat{\lambda}_t^I := E[\lambda_t^I \mid \widehat{\mathcal{F}}_t^{\sigma(I)}], \quad 0\le t\le T,$$

satisfies the SDE

$$\mathrm{d}\widehat{\lambda}_t^I = \left(V_t^I - \frac{1}{T_a - t}\right)\mathrm{d}\widehat{B}_t^I, \quad \widehat{\lambda}_0^I = \lambda_0 + \frac{I}{aT_a}, \tag{4.19}$$

where $\widehat{B}^I$ is the innovations process, an $\widehat{\mathbb{F}}^{\sigma(I)}$-Brownian motion defined by

$$\widehat{B}_t^I := \xi_t - \int_0^t \widehat{\lambda}_s^I \mathrm{d}s, \quad 0\le t\le T, \tag{4.20}$$

and V_t^I is the conditional variance of λ_t^I:

$$V_t^I := E\left[\,(\lambda_t^I - \widehat{\lambda}_t^I)^2\middle|\,\widehat{\mathcal{F}}_t^{\sigma(I)}\right], \quad 0\le t\le T,$$

which satisfies

$$\frac{\mathrm{d}V_t^I}{\mathrm{d}t} = \frac{2}{T_a - t}V_t^I - \left(V_t^I\right)^2, \quad V_0^I = \mathrm{v}_0.$$

If we define

$$\mathrm{v}_t^I := V_t^I - \frac{1}{T_a - t}, \quad 0\le t\le T,$$

then (4.19) becomes

$$\mathrm{d}\widehat{\lambda}_t^I = \mathrm{v}_t^I \mathrm{d}\widehat{B}_t^I, \quad \widehat{\lambda}_0^I = \lambda_0 + \frac{I}{aT_a}. \tag{4.21}$$

Note that (4.21) is of the same form as (3.8) with w_t replaced by v_t^I and with B_t replaced by $\widehat{B}_t^I$. Indeed, v_t^I plays the role of an 'effective variance', satisfying the Riccati equation (3.27), with a modified initial condition:

$$\frac{\mathrm{d}\mathrm{v}_t^I}{\mathrm{d}t} = -\left(\mathrm{v}_t^I\right)^2, \quad \mathrm{v}_0^I = \mathrm{v}_0 - \frac{1}{T_a}.$$

The solution to this equation is then given by (4.10), and the solution to (4.21) is then (4.11).

Using (4.20) in the SDE (4.21), the optimal filter may also be written explicitly in terms of the observable ξ, as

$$\widehat{\lambda}_t^I = \frac{\widehat{\lambda}_0^I + \mathrm{v}_0^I \xi_t}{1 + \mathrm{v}_0^I t}, \quad 0 \le t \le T.$$

This is of the same form as (3.30), with λ_0 replaced by $\widehat{\lambda}_0^I$ and v_0 replaced by v_0^I.

The effect of the filtering is that the agent is now investing in a stock with dynamics given by $\mathrm{d}S_t = \sigma S_t \mathrm{d}\xi_t$ which, using (4.20), becomes (4.7). The $\widehat{\mathbb{F}}^{\sigma(I)}$-adapted wealth process X^I then follows

$$\mathrm{d}X_t^I = \sigma\theta_t^I X_t^I (\widehat{\lambda}_t^I \mathrm{d}t + \mathrm{d}\widehat{B}_t^I), \quad X_0^I = x,$$

where θ^I is the $\widehat{\mathbb{F}}^{\sigma(I)}$-adapted trading strategy. The theorem then follows immediately from making the replacements

$$\mathrm{w} \to \mathrm{v}^I, \quad \lambda \to \widehat{\lambda}^I,$$

in Theorem 3.2. □

It can be shown that the additional information increases the insider's utility over the regular agent: see [2] for this and other effects of the inside information.

5 Optimal hedging of basis risk with partial information

In this section we analyse the hedging of a contingent claim in a basis risk model, a tractable example of an incomplete market, first under a full information assumption, and then under a partial information scenario. Basis risk models involve a claim on a non-traded asset, which is hedged using a correlated traded asset. They were first studied systematically by Davis [4] (whose preprint on the subject originated in 2000) who used a dual approach to derive approximations for indifference prices. Subsequently, Henderson [8], and Musiela and Zariphopoulou [26] derived an expectation representation (given in Theorem 5.3) for the value function of the utility maximisation problem involving a random endowment of the claim. This was used by Monoyios [21] to derive accurate analytic approximations for indifference prices and hedging strategies. In simulation experiments, Monoyios showed that exponential indifference hedging could outperform the BS approximation of taking the traded asset as a good proxy for the non-traded asset. Unfortunately, the utility-based hedge requires knowledge of the drift parameters of the assets. These are hard to estimate accurately, as shown by Rogers [31] and Monoyios [22], who showed that drift parameter mis-estimation could ruin the effectiveness of the optimal hedge. Finally, in [23, 25] Monoyios developed a filtering algorithm to deal with the drift parameter uncertainty, and showed that with this added ingredient, utility-based hedging was indeed effective, even in the face of parameter uncertainty. We shall describe some of these results in this section.

5.1 Basis risk model: full information case

In a full information model, the setting is a filtered probability space $(\Omega, \mathcal{F}, \mathbb{F} := (\mathcal{F}_t)_{0\leq t\leq T}, P)$, where the filtration $\mathbb{F}$ is the P-augmentation of that generated by a two-dimensional Brownian motion $(B, B^{\perp})$. A traded stock price $S := (S_t)_{0\leq t\leq T}$ follows a log-Brownian process given by

$$dS_t = \sigma S_t(\lambda dt + dB_t) =: \sigma S_t d\xi_t, \tag{5.1}$$

where $\sigma > 0$ and λ are known constants. For simplicity, the interest rate is taken to be zero. The process ξ in (5.1) defined by $d\xi_t := \lambda dt + dB_t$ will subsequently play a role as one component of an observation process in a partial information model, when λ will be treated as a random variable rather than as a known constant.

A non-traded asset price $Y := (Y_t)_{0\leq t\leq T}$ follows the correlated log-Brownian motion

$$dY_t = \beta Y_t(\theta dt + dW_t) =: \beta Y_t d\zeta_t, \tag{5.2}$$

with $\beta > 0$ and θ known constants. The Brownian motion W is correlated with B according to

$$[B, W]_t = \rho t, \quad W = \rho B + \sqrt{1-\rho^2}B^{\perp}, \quad \rho \in [-1, 1],$$

and the process ζ, given by $d\zeta_t := \theta dt + dW_t$, will act as the second component of an observation process in a partial information model, when θ will be considered a random variable. We shall henceforth refer to the Sharpe ratios λ (respectively, θ) as the drift of S (respectively, Y), for brevity.

A European contingent claim pays the non-negative random variable $h(Y_T)$ at time T, where $h : \mathbb{R}^+ \to \mathbb{R}^+$. In what follows we shall consider utility maximisation problems with the additional random terminal endowment $nh(Y_T)$, for $n \in \mathbb{R}$. We assume the random endowment $nh(Y_T)$ is continuous and bounded below, with finite expectation under any martingale measure.

An agent may trade the stock in a self-financing fashion, leading to the portfolio wealth process $X = (X_t)_{0\leq t\leq T}$ satisfying

$$dX_t = \sigma\pi_t(\lambda dt + dB_t),$$

where $\pi := (\pi_t)_{0\leq t\leq T}$ is the wealth in the stock, representing the agent's trading strategy, satisfying $\int_0^T \pi_t^2 dt < \infty$ almost surely.

5.1.1 Perfect correlation case

This market is incomplete for $|\rho| \neq 1$. If the correlation is perfect, however, the market becomes complete and perfect hedging is possible, as we now show.

The minimal martingale measure Q^M has density process with respect to P given by

$$\left.\frac{dQ^M}{dP}\right|_{\mathcal{F}_t} = \mathcal{E}(-\lambda \cdot B)_t, \quad 0 \leq t \leq T.$$

Under Q^M, (S, Y) follow

$$\begin{aligned} \mathrm{d}S_t &= \sigma S_t \mathrm{d}B_t^{Q^M}, \\ \mathrm{d}Y_t &= \beta(\theta - \rho\lambda) Y_t \mathrm{d}t + \beta Y_t \mathrm{d}W_t^{Q^M}, \end{aligned} \tag{5.3}$$

where B^{Q^M}, W^{Q^M} are correlated Brownian motions under Q^M. The stock price S is a local Q^M-martingale, but this is not the case for the non-traded asset, unless we have the perfect correlation case, $\rho = 1$. In this case Y is effectively a traded asset (as Y_t is then a deterministic function of S_t), so the Q^M-drift of Y vanishes. Therefore, given σ, β, when $\rho = 1$ the Sharpe ratios λ, θ are equal:

$$\theta = \lambda.$$

In this case the market becomes complete, and perfect hedging is possible. It is easy to show that with $\rho = 1$, so that $W = B$, we have

$$Y_t = Y_0 \left(\frac{S_t}{S_0}\right)^{\beta/\sigma} e^{ct}, \quad c = \frac{1}{2}\sigma\beta\left(1 - \frac{\beta}{\sigma}\right).$$

Let the claim price process be $v(t, Y_t), 0 \le t \le T$, where $v : [0, T] \times \mathbb{R}^+ \to \mathbb{R}^+$ is smooth enough to apply the Itô formula, so that

$$\mathrm{d}v(t, Y_t) = \left[v_t(t, Y_t) + \mathcal{A}^Y v(t, Y_t)\right] \mathrm{d}t + \beta Y_t v_y(t, Y_t) \mathrm{d}W_t,$$

where $\mathcal{A}^Y$ is the generator of the process Y in (5.2). The replication conditions are

$$X_t = v(t, Y_t), \quad 0 \le t \le T, \quad \mathrm{d}X_t = \mathrm{d}v(t, Y_t).$$

Standard arguments then show that to perfectly hedge the claim one must hold Δ_t shares of S at $t \in [0, T]$, given by

$$\Delta_t = \frac{\beta}{\sigma}\frac{Y_t}{S_t}\frac{\partial v}{\partial y}(t, Y_t), \quad 0 \le t \le T, \tag{5.4}$$

and the claim pricing function $v(t, y)$ satisfies

$$v_t(t, y) + \beta(\theta - \lambda) y v_y(t, y) + \frac{1}{2}\beta^2 y^2 v_{yy}(t, y) = 0, \quad v(T, y) = h(y).$$

But with $\rho = 1$, $\theta = \lambda$, so we get the BS partial differential equation (PDE), and hence

$$v(t, Y_t) = \mathrm{BS}(t, Y_t), \quad 0 \le t \le T,$$

where $\mathrm{BS}(t, y)$ denotes the BS option pricing formula at time t, with underlying asset price y.

Therefore, a position in n claims is hedged by $\Delta_t^{(\mathrm{BS})}$ units of S at $t \in [0, T]$, where

$$\Delta_t^{(\mathrm{BS})} = -n\frac{\beta}{\sigma}\frac{Y_t}{S_t}\frac{\partial}{\partial y}\mathrm{BS}(t, Y_t; \beta), \quad 0 \le t \le T, \tag{5.5}$$

and where $\mathrm{BS}(t, y; \beta)$ denotes the BS formula at time t for underlying asset price y and volatility β. From our perspective, the salient feature of (5.5) is that the perfect hedge does not require knowledge of the values of the drifts λ, θ.

5.1.2 Incomplete case

Now suppose the correlation is not perfect, so that the market is incomplete. We embed the problem in a utility maximisation framework in a manner that is by now classical. Let the agent have risk preferences expressed via the exponential utility function

$$U(x) = -\exp(-\alpha x), \quad x \in \mathbb{R}, \quad \alpha > 0.$$

The agent maximises expected utility of terminal wealth at time T, with a random endowment of n units of claim payoff:

$$J(t,x,y;\pi) = E\big[U(X_T + nh(Y_T)) \mid X_t = x, Y_t = y\big].$$

The value function is $u^{(n)}(t,x,y) \equiv u(t,x,y)$, defined by

$$u(t,x,y) := \sup_{\pi \in \mathcal{A}} J(t,x,y;\pi), \quad u(T,x,y) = U(x + nh(y)), \tag{5.6}$$

where $\mathcal{A}$ is the set of admissible strategies. This is composed of S-integrable processes whose gains process is a Q-martingale for any martingale measure with finite relative entropy with respect to P. Denote the optimal trading strategy that achieves the supremum in (5.6) by $\pi^* \equiv \pi^{*,n}$, and denote the optimal wealth process by $X^* \equiv X^{*,n}$.

The following definitions of utility-based price and hedging strategy are now standard.

Definition 5.1 (Indifference price). The indifference price per claim at $t \in [0,T]$, given $X_t = x, Y_t = y$, is $p(t,x,y) \equiv p^{(n)}(t,x,y)$, defined by

$$u^{(n)}(t, x - np^{(n)}(t,x,y), y) = u^{(0)}(t,x,y).$$

We allow for possible dependence on t,x,y of $p^{(n)}$ in the above definition, but with exponential preferences it turns out that there is no dependence on x.

Definition 5.2 (Optimal hedging strategy). The optimal hedging strategy for n units of the claim is $\pi^H := (\pi_t^H)_{0 \le t \le T}$ given by

$$\pi_t^H := \pi_t^{*,n} - \pi_t^{*,0}, \quad 0 \le t \le T.$$

We have the following representation for the value function and indifference price.

Theorem 5.3. *The value function $u^{(n)}$ and indifference price $p^{(n)}$, given $X_t = x, Y_t = y$ for $t \in [0,T]$, are given by*

$$\begin{aligned} u^{(n)}(t,x,y) &= -e^{-\alpha x - \frac{1}{2}\lambda^2 (T-t)} \left[F(t,Y_t)\right]^{1/(1-\rho^2)}, \\ F(t,y) &= E^{Q^M}\left[\exp\left(-\alpha(1-\rho^2)nh(Y_T)\right)\middle| Y_t = y\right], \qquad (5.7) \\ p^{(n)}(t,y) &= -\frac{1}{\alpha(1-\rho^2)n}\log F(t,y). \qquad (5.8) \end{aligned}$$

Proof. The Hamilton–Jacobi–Bellman (HJB) equation for the value function $u^{(n)}$ is

$$u_t^{(n)} + \sigma \sup_{\pi} \left(\lambda \pi u_x^{(n)} + \frac{1}{2}\sigma \pi^2 u_{xx}^{(n)} + \rho\beta\pi y u_{xy}^{(n)} \right) + \mathcal{A}^Y u^{(n)} = 0.$$

Performing the maximisation gives the optimal feedback control as $\Pi^{*,n}(t,x,y)$, where the function $\Pi^{*,n} : [0,T] \times \mathbb{R} \times \mathbb{R}^+$ is given by

$$\Pi^{*,n}(t,x,y) := -\left(\frac{\lambda u_x^{(n)} + \rho\beta y u_{xy}^{(n)}}{\sigma u_{xx}^{(n)}} \right). \tag{5.9}$$

The optimal trading strategy $\pi^{*,n}$ is then given by $\pi_t^{*,n} = \Pi^*(t, X_t^*, Y_t)$. Substituting the optimal Markov control back into the Bellman equation gives the HJB PDE

$$u_t^{(n)} + \mathcal{A}^Y u^{(n)} - \frac{\left(\lambda u_x^{(n)} + \rho\beta y u_{xy}^{(n)}\right)^2}{2u_{xx}^{(n)}} = 0.$$

The function $F(t,y)$ in (5.7) satisfies the linear PDE

$$F_t + \beta(\theta - \rho\lambda)F_y + \frac{1}{2}\beta^2 y^2 F_{yy} = 0, \quad F(T,y) = \exp(-\alpha(1-\rho^2)nh(y)),$$

by virtue of the Feynman–Kac theorem. It is then straightforward to verify that $u^{(n)}$ as given in the theorem solves the above HJB equation, and the definition of the indifference price gives the formula (5.8). □

This leads to the following representation for the optimal hedging strategy.

Theorem 5.4. *The optimal hedging strategy for a position in n claims is to hold Δ_t^H shares at $t \in [0,T]$, given by*

$$\Delta_t^H = -n\rho \frac{\beta}{\sigma} \frac{Y_t}{S_t} \frac{\partial p^{(n)}}{\partial y}(t, Y_t), \quad 0 \le t \le T. \tag{5.10}$$

Proof. From Theorem 5.3 the value function may be written in terms of the indifference price as

$$u^{(n)}(t,x,y) = -\exp\left(-\alpha(x + np^{(n)}(t,y)) - \frac{1}{2}\lambda^2(T-t) \right). \tag{5.11}$$

The optimal trading strategy is $\pi_t^{*,n} = \Pi^*(t, X_t^*, Y_t)$, where the function $\Pi^{*,n}(t,x,y)$ is given in (5.9), in terms of derivatives of the value function. Using (5.11) we obtain

$$\pi_t^{*,n} = \frac{\lambda - \rho\alpha n\beta Y_t p_y^{(n)}(t, Y_t)}{\alpha\sigma}, \quad 0 \le t \le T.$$

The optimal trading strategy for the problem with no claims, $\pi_t^{*,0}$ is obtained trivially by setting $n = 0$ in this result, and then applying Definition 5.2 proves the theorem. □

Notice that, given the PDE satisfied by F, the indifference pricing function $p(t,y) \equiv p^{(n)}(t,y)$ satisfies

$$p_t + \beta(\theta - \rho\lambda)yp_y + \frac{1}{2}\beta^2y^2p_{yy} - \frac{1}{2}\beta^2y^2\alpha n(1-\rho^2)(p_y)^2 = 0.$$

So for $\rho = 1$, in which case $\theta = \lambda$, the claim price then satisfies the BS PDE and we recover the perfect delta hedge (5.4).

In [21, 22] the hedging strategy in (5.10) is shown to be superior to the BS-style hedge (5.5), in terms of the terminal hedging error distribution produced by selling the claim at the appropriate price (the indifference price or the BS price) and investing the proceeds in the corresponding hedging portfolio. But from (5.3) we see that the exponential hedge requires knowledge of λ, θ, which are impossible to estimate accurately (see Rogers [31] or Monoyios [22]). This can ruin the effectiveness of indifference hedging, as shown in [22]. It is therefore dubious to draw any meaningful conclusions on the effectiveness of utility-based hedging in this model without relaxing the assumption that the agent knows the true values of the drifts.

5.2 Partial information case

Now we assume the hedger does not know the values of the return parameters λ, θ, so these are considered to be random variables. Equivalently, the agent cannot observe the Brownian motions B, W driving the asset prices, so is required to use strategies adapted to the observation filtration $\widehat{\mathbb{F}}$ generated by asset returns.

5.2.1 Choice of prior

We take the the two-dimensional random variable

$$\Xi := \begin{pmatrix} \lambda \\ \theta \end{pmatrix}$$

to have a Gaussian distribution which will be updated as the agent attempts to filter the values of the drifts from asset observations during the hedging interval $[0,T]$.

The choice of Gaussian prior is motivated by the idea that the agent has some past observations of S, Y before time zero, uses these to obtain classical point estimates of the drifts, and the joint distribution of the estimators is used as the prior in a Bayesian framework. Ultimately, in order to obtain explicit solutions, we shall assume that the agent uses observations before time zero of equal length for both assets. In setting the prior this way, we make the approximation that the asset price observations are continuous, so that σ, β, ρ are known from the quadratic variation and co-variation of S, Y. This is because our goal here is to focus on the severest problem of drift parameter uncertainty.

So, consider, for the moment, an observer with data for S over a time interval of length t_S, and for Y over a window of length t_Y, who considers λ and θ as *constants*, and records the returns $\mathrm{d}S_t/S_t$ and $\mathrm{d}Y_t/Y_t$ in order to estimate the values of the drifts.

An unbiased estimator of λ is $\bar{\lambda}(t_S)$ given by

$$\bar{\lambda}(t_S) = \frac{1}{t_S}\int_{t_0}^{t_0+t_S} \frac{\mathrm{d}S_u}{\sigma S_u} = \lambda + \frac{B_{t_0+t_S}}{t_S} \sim \mathrm{N}\left(\lambda, \frac{1}{t_S}\right).$$

The estimator of λ is normally distributed, with a similar computation for the estimator of θ. The estimator, $(\bar{\lambda}, \bar{\theta})$, of the (supposed constant) vector (λ, θ) is bivariate normal. Defining $\mathrm{v}_0 := 1/t_S$ and $\mathrm{w}_0 := 1/t_Y$ it is easily checked that

$$\begin{pmatrix} \bar{\lambda} \\ \bar{\theta} \end{pmatrix} \sim \mathrm{N}(M, C_0),$$

where the mean vector M and covariance matrix C_0 are given by

$$M = \begin{pmatrix} \lambda \\ \theta \end{pmatrix}, \qquad C_0 = \begin{pmatrix} \mathrm{v}_0 & \rho\min(\mathrm{v}_0, \mathrm{w}_0) \\ \rho\min(\mathrm{v}_0, \mathrm{w}_0) & \mathrm{w}_0 \end{pmatrix}. \tag{5.12}$$

With this in mind, we shall suppose that (λ, θ), now considered as a *random variable*, is bivariate normal according to

$$\lambda \sim \mathrm{N}(\lambda_0, \mathrm{v}_0), \quad \theta \sim \mathrm{N}(\theta_0, \mathrm{w}_0), \quad \mathrm{cov}(\lambda, \theta) = \mathrm{c}_0 := \rho\min(\mathrm{v}_0, \mathrm{w}_0),$$

for some chosen values λ_0, θ_0, typically obtained from past data prior to time zero. This distribution will be updated via subsequent observations of

$$\xi_t := \frac{1}{\sigma}\int_0^t \frac{\mathrm{d}S_u}{S_u} = \lambda t + B_t, \qquad \zeta_t := \frac{1}{\beta}\int_0^t \frac{\mathrm{d}Y_u}{Y_u} = \theta t + W_t,$$

over the hedging interval $[0, T]$.

5.2.2 Two-dimensional Kalman–Bucy filter

We are firmly within the realm of a two-dimensional Kalman filtering problem, which we treat as follows. Define the observation filtration by

$$\widehat{\mathbb{F}} := (\widehat{\mathcal{F}}_t)_{0\le t\le T}, \qquad \widehat{\mathcal{F}}_t = \sigma(\xi_s, \zeta_s; 0 \le s \le t).$$

The observation process, Λ, and unobservable signal process, Ξ, are defined by

$$\Lambda := \begin{pmatrix} \xi_t \\ \zeta_t \end{pmatrix}_{0\le t\le T}, \qquad \Xi := \begin{pmatrix} \lambda \\ \theta \end{pmatrix},$$

satisfying the stochastic differential equations

$$\mathrm{d}\Lambda_t = \Xi\mathrm{d}t + \begin{pmatrix} 1 & 0 \\ \rho & \sqrt{1-\rho^2} \end{pmatrix} \mathrm{d}\begin{pmatrix} B_t \\ B_t^{\perp} \end{pmatrix}, \qquad \mathrm{d}\Xi = \begin{pmatrix} 0 \\ 0 \end{pmatrix}.$$

The optimal filter is $\widehat{\Xi}_t := E[\Xi \mid \widehat{\mathcal{F}}_t], 0 \le t \le T$, a two-dimensional process defining the best estimates of λ and θ given observations up to time $t \in [0, T]$:

$$\widehat{\Xi}_t \equiv \begin{pmatrix} \widehat{\lambda}_t \\ \widehat{\theta}_t \end{pmatrix} := \begin{pmatrix} E[\lambda \mid \widehat{\mathcal{F}}_t] \\ E[\theta \mid \widehat{\mathcal{F}}_t] \end{pmatrix}, \qquad \begin{pmatrix} \widehat{\lambda}_0 \\ \widehat{\theta}_0 \end{pmatrix} = \begin{pmatrix} \lambda_0 \\ \theta_0 \end{pmatrix}. \tag{5.13}$$

The solution to this filtering problem converts the partial information model to a full information model with random drifts, given in the following proposition. To avoid a proliferation of symbols, we abuse notation and write $\widehat{\lambda}_t \equiv \widehat{\lambda}(t, S_t)$ and $\widehat{\theta} \equiv \widehat{\theta}(t, Y_t)$ for processes $\widehat{\lambda}, \widehat{\theta}$ that will turn out to be functions of time and current asset price.

Proposition 5.5. *The partial information model is equivalent to a full information model in which the asset price dynamics in the observation filtration $\widehat{\mathbb{F}}$ are*

$$\begin{aligned} \mathrm{d}S_t &= \sigma S_t(\widehat{\lambda}_t \mathrm{d}t + \mathrm{d}\widehat{B}_t), &(5.14)\\ \mathrm{d}Y_t &= \beta Y_t(\widehat{\theta}_t \mathrm{d}t + \mathrm{d}\widehat{W}_t), &(5.15) \end{aligned}$$

where $\widehat{B}, \widehat{W}$ are $\widehat{\mathbb{F}}$-Brownian motions with correlation ρ, and the random drifts $\widehat{\lambda}, \widehat{\theta}$ are $\widehat{\mathbb{F}}$-adapted processes.

If λ and θ have common initial variance v_0, then $\widehat{\lambda}, \widehat{\theta}$ are given by

$$\begin{pmatrix} \widehat{\lambda}_t \\ \widehat{\theta}_t \end{pmatrix} = \begin{pmatrix} \lambda_0 \\ \theta_0 \end{pmatrix} + \int_0^t \mathrm{v}_s \begin{pmatrix} \mathrm{d}\widehat{B}_s \\ \mathrm{d}\widehat{W}_s \end{pmatrix}, \quad 0 \le t \le T, \tag{5.16}$$

where $(\mathrm{v}_t)_{0 \le t \le T}$ is the deterministic function

$$\mathrm{v}_t := \frac{\mathrm{v}_0}{1 + \mathrm{v}_0 t}, \qquad 0 \le t \le T.$$

Equivalently, $\widehat{\lambda}, \widehat{\theta}$ are given as functions of time and current asset price by

$$\widehat{\lambda}_t = \widehat{\lambda}(t, S_t) = \frac{\lambda_0 + \mathrm{v}_0 \xi_t}{1 + \mathrm{v}_0 t}, \qquad \widehat{\theta}_t = \widehat{\theta}(t, Y_t) = \frac{\theta_0 + \mathrm{v}_0 \zeta_t}{1 + \mathrm{v}_0 t}, \tag{5.17}$$

with

$$\xi_t = \frac{1}{\sigma} \log\left(\frac{S_t}{S_0}\right) + \frac{1}{2}\sigma t, \qquad \zeta_t = \frac{1}{\beta} \log\left(\frac{Y_t}{Y_0}\right) + \frac{1}{2}\beta t. \tag{5.18}$$

Proof. Using a two-dimensional Kalman–Bucy filter (see, for example, Theorem V.9.2 in Fleming and Rishel [6]), $\widehat{\Xi}$ satisfies the stochastic differential equation

$$\mathrm{d}\widehat{\Xi}_t = C_t \left(DD^T\right)^{-1} (\mathrm{d}\Lambda_t - \widehat{\Xi}_t \mathrm{d}t) =: C_t \left(DD^T\right)^{-1} dN_t, \tag{5.19}$$

where $(N_t)_{0 \le t \le T}$ is the innovations process, defined by

$$N_t := \Lambda_t - \int_0^t \widehat{\Xi}_s \mathrm{d}s = \begin{pmatrix} \xi_t - \int_0^t \widehat{\lambda}_s \mathrm{d}s \\ \zeta_t - \int_0^t \widehat{\theta}_s \mathrm{d}s \end{pmatrix} =: \begin{pmatrix} \widehat{B}_t \\ \widehat{W}_t \end{pmatrix}, \tag{5.20}$$

and $\widehat{B}, \widehat{W}$ are $\widehat{\mathbb{F}}$-Brownian motions with correlation ρ. The deterministic matrix function C_t is the conditional variance-covariance matrix defined by

$$C_t := E\left[(\Xi - \widehat{\Xi}_t)(\Xi - \widehat{\Xi}_t)^{\mathrm{T}}\middle|\widehat{\mathcal{F}}_t\right] = E\left[(\Xi - \widehat{\Xi}_t)(\Xi - \widehat{\Xi}_t)^{\mathrm{T}}\right],$$

(T denoting transpose) where the last equality follows because the error $\Xi - \widehat{\Xi}_t$ is independent of $\widehat{\mathcal{F}}_t$ (Theorem V.9.2 in [6] again).

Using (5.20), and writing $\mathrm{d}S_t$ in terms of $\mathrm{d}\xi_t$, as in (5.1), gives the dynamics (5.14) of S in the observation filtration; (5.15) is established similarly.

The matrix $C = (C_t)_{0 \le t \le T}$ satisfies the Riccati equation

$$\frac{\mathrm{d}C_t}{\mathrm{d}t} = -C_t \left(DD^T\right)^{-1} C_t,$$

with C_0 given in (5.12). Then $R_t := C_t^{-1}$ satisfies the Lyapunov equation

$$\frac{\mathrm{d}R_t}{\mathrm{d}t} = \left(DD^T\right)^{-1}.$$

Define the elements of the conditional covariance matrix by

$$C_t =: \begin{pmatrix} \mathrm{v}_t & \mathrm{c}_t \\ \mathrm{c}_t & \mathrm{w}_t \end{pmatrix}.$$

Then the filtering equation (5.19) is a pair of coupled stochastic differential equations:

$$\begin{aligned}\begin{pmatrix}\mathrm{d}\widehat{\lambda}_t \\ \mathrm{d}\widehat{\theta}_t\end{pmatrix} &= \frac{1}{1-\rho^2}\begin{pmatrix} \mathrm{v}_t - \rho\mathrm{c}_t & \mathrm{c}_t - \rho\mathrm{v}_t \\ \mathrm{c}_t - \rho\mathrm{w}_t & \mathrm{w}_t - \rho\mathrm{c}_t \end{pmatrix}\begin{pmatrix}\mathrm{d}\xi_t - \widehat{\lambda}_t\mathrm{d}t \\ \mathrm{d}\zeta_t - \widehat{\theta}_t\mathrm{d}t\end{pmatrix} \\ &= \frac{1}{1-\rho^2}\begin{pmatrix} \mathrm{v}_t - \rho\mathrm{c}_t & \mathrm{c}_t - \rho\mathrm{v}_t \\ \mathrm{c}_t - \rho\mathrm{w}_t & \mathrm{w}_t - \rho\mathrm{c}_t \end{pmatrix}\begin{pmatrix}\mathrm{d}\widehat{B}_t \\ \mathrm{d}\widehat{W}_t\end{pmatrix}.\end{aligned}$$

Solving the Lyapunov equation yields three equations for $\mathrm{v}_t, \mathrm{w}_t, \mathrm{c}_t$:

$$\begin{aligned}\frac{\mathrm{v}_t}{\mathrm{v}_t\mathrm{w}_t - \mathrm{c}_t^2} - \frac{\mathrm{v}_0}{\mathrm{v}_0\mathrm{w}_0 - \mathrm{c}_0^2} &= \frac{t}{1-\rho^2}, \\ \frac{\mathrm{w}_t}{\mathrm{v}_t\mathrm{w}_t - \mathrm{c}_t^2} - \frac{\mathrm{w}_0}{\mathrm{v}_0\mathrm{w}_0 - \mathrm{c}_0^2} &= \frac{t}{1-\rho^2}, \\ \frac{\mathrm{c}_t}{\mathrm{v}_t\mathrm{w}_t - \mathrm{c}_t^2} - \frac{\mathrm{c}_0}{\mathrm{v}_0\mathrm{w}_0 - \mathrm{c}_0^2} &= \frac{\rho t}{1-\rho^2},\end{aligned} \tag{5.21}$$

where we have written $\mathrm{c}_0 \equiv \rho\min(\mathrm{v}_0, \mathrm{w}_0)$ for brevity.

Now make the simplification $\mathrm{w}_0 = \mathrm{v}_0$. From the discussion in Section 5.2.1, we see that this corresponds to using past observations over the same length of time, $t_S = t_Y$, for both S and Y in fixing the prior. Then $\mathrm{c}_0 = \rho\mathrm{v}_0$, and the solution to the system of equations (5.21) gives the entries of the matrix C_t as

$$\mathrm{v}_t = \frac{\mathrm{v}_0}{1 + \mathrm{v}_0 t}, \quad \mathrm{w}_t = \mathrm{v}_t, \quad \mathrm{c}_t = \rho\mathrm{v}_t.$$

With this simplification, the equation for the optimal filter simplifies to

$$\begin{pmatrix} d\widehat{\lambda}_t \\ d\widehat{\theta}_t \end{pmatrix} = \mathrm{v}_t \begin{pmatrix} d\xi_t - \widehat{\lambda}_t dt \\ d\zeta_t - \widehat{\theta}_t dt \end{pmatrix} = \mathrm{v}_t \begin{pmatrix} d\widehat{B}_t \\ d\widehat{W}_t \end{pmatrix},$$

which, along with the initial condition in (5.13), yields (5.16) and (5.17).

Finally, the expressions in (5.18) for ξ_t, ζ_t follow directly from the solutions of (5.1) and (5.2) for S and Y. □

Armed with Proposition 5.5 we may now treat the model as a full information model with random drift parameters $(\widehat{\lambda}_t, \widehat{\theta}_t)$, and this is done in the next section.

5.2.3 Optimal hedging with random drifts

On the stochastic basis $(\Omega, \widehat{\mathcal{F}}_T, \widehat{\mathbb{F}}, P)$, the wealth process associated with trading strategy $\pi := (\pi_t)_{0\le t\le T}$, an $\widehat{\mathbb{F}}$-adapted process satisfying $\int_0^T \pi_t^2 dt < \infty$ a.s., is $X = (X_t)_{0\le t\le T}$, satisfying

$$dX_t = \sigma\pi_t(\widehat{\lambda}_t dt + d\widehat{B}_t). \tag{5.22}$$

The class $\mathcal{M}$ of local martingale measures for this model consists of measures Q with density processes defined by

$$Z_t := \left.\frac{dQ}{dP}\right|_{\widehat{\mathcal{F}}_t} = \mathcal{E}(-\widehat{\lambda}\cdot\widehat{B} - \psi\cdot\widehat{B}^{\perp})_t, \quad 0 \le t \le T, \tag{5.23}$$

for integrands ψ satisfying $\int_0^t \psi_s^2 ds < \infty$ a.s., for all $t \in [0, T]$ (it is not hard to show that $\int_0^t \widehat{\lambda}_s^2 ds < \infty, 0 \le t \le T$). For $\psi = 0$ we obtain the minimal martingale measure Q^M.

Under $Q \in \mathcal{M}$, $(\widehat{B}^Q, \widehat{B}^{\perp,Q})$ is two-dimensional Brownian motion, where

$$d\widehat{B}_t^Q := d\widehat{B}_t^Q + \widehat{\lambda}_t dt, \qquad d\widehat{B}_t^{\perp,Q} := d\widehat{B}_t^{\perp} + \psi_t dt,$$

and the asset prices and random drifts satisfy

$$\begin{aligned} dS_t &= \sigma S_t d\widehat{B}_t^Q, \\ dY_t &= \beta Y_t[(\widehat{\theta}_t - \rho\widehat{\lambda}_t - \sqrt{1-\rho^2}\psi_t)dt + d\widehat{W}_t^Q], \\ d\widehat{\lambda}_t &= \mathrm{v}_t[-\widehat{\lambda}_t dt + d\widehat{B}_t^Q], \\ d\widehat{\theta}_t &= \mathrm{v}_t[-(\rho\widehat{\lambda}_t + \sqrt{1-\rho^2}\psi_t)dt + d\widehat{W}_t^Q], \end{aligned} \tag{5.24}$$

where $\widehat{W}^Q = \rho\widehat{B}^Q + \sqrt{1-\rho^2}\widehat{B}^{\perp,Q}$.

The relative entropy between $Q \in \mathcal{M}$ and P is defined by

$$\begin{aligned} \mathcal{H}(Q, P) &:= E\left[\frac{dQ}{dP}\log\frac{dQ}{dP}\right] \\ &= E^Q\left[-\int_0^T \widehat{\lambda}_t d\widehat{B}_t^Q - \int_0^T \psi_t d\widehat{B}_t^{\perp,Q} + \frac{1}{2}\int_0^T \left(\widehat{\lambda}_t^2 + \psi_t^2\right) dt\right]. \end{aligned}$$

Using the Q-dynamics of $\widehat{\lambda}_t$ it is straightforward to establish that $E^Q \int_0^t \widehat{\lambda}_s^2 \mathrm{d}s < \infty$ for all $t \in [0,T]$. If, in addition, we have the integrability condition

$$E^Q \int_0^t \psi_s^2 \mathrm{d}s < \infty, \quad 0 \le t \le T, \tag{5.25}$$

then

$$\mathcal{H}(Q,P) = E^Q \left[\frac{1}{2} \int_0^T \left(\widehat{\lambda}_t^2 + \psi_t^2 \right) \mathrm{d}t \right] < \infty. \tag{5.26}$$

In this case we write $Q \in \mathcal{M}_f$, where $\mathcal{M}_f$ denotes the set of martingale measures Q with finite relative entropy with respect to P, and we define $\mathcal{H}(Q,P) := \infty$ otherwise. From (5.26) we note that the minimal entropy measure Q^E is characterised by

$$\mathcal{H}(Q^E,P) = E^{Q^E} \left[\frac{1}{2} \int_0^T \widehat{\lambda}_t^2 \mathrm{d}t \right],$$

corresponding to $\psi \equiv 0$ in (5.26). This means that the minimal martingale measure and the minimal entropy measure in this model coincide: $Q^E = Q^M$.

For an initial time $t \in [0,T]$, we define the conditional entropy between $Q \in \mathcal{M}$ and P by

$$H_t(Q,P) := E \left[\frac{Z_T}{Z_t} \log \left(\frac{Z_T}{Z_t} \right) \middle| \widehat{\mathcal{F}}_t \right], \quad 0 \le t \le T, \tag{5.27}$$

satisfying $H_0(Q,P) \equiv \mathcal{H}(Q,P)$. Provided the integrability condition (5.25) is satisfied, then

$$H_t(Q,P) = E^Q \left[\frac{1}{2} \int_t^T \left(\widehat{\lambda}_u^2 + \psi_u^2 \right) \mathrm{d}u \middle| \widehat{\mathcal{F}}_t \right],$$

and we define $H_t(Q,P) := \infty$ otherwise. In particular, therefore, recalling that $\widehat{\lambda}_t \equiv \widehat{\lambda}(t,S_t)$ is a smooth and Lipschitz function of time and current stock price, and that the Q-dynamics of $\widehat{\lambda}_t$ do not depend on ψ_t for any $Q \in \mathcal{M}$, the minimal conditional entropy $(H_t(Q^E,P))_{0 \le t \le T}$ will be a deterministic function of time and stock price, given by $H_t(Q^E,P) \equiv H^E(t,S_t)$ for a $C^{1,2}([0,T] \times \mathbb{R}^+)$ function H^E defined by

$$H^E(t,s) := E^{Q^E} \left[\frac{1}{2} \int_t^T \widehat{\lambda}^2(u,S_u) \mathrm{d}u \middle| S_t = s \right]. \tag{5.28}$$

5.2.4 The primal problem

We use an exponential utility function, $U(x) = -\exp(-\alpha x)$, $x \in \mathbb{R}$, $\alpha > 0$. The primal value function $u^{(n)}$ is defined as the maximum expected utility of wealth at T from trading S and receiving n units of the claim on Y, when starting at time $t \in [0,T]$:

$$u^{(n)}(t,x,s,y) := \sup_{\pi \in \mathcal{A}} E\big[U(X_T + nh(Y_T)) \,|\, X_t = x, S_t = s, Y_t = y \big], \tag{5.29}$$

where $\mathcal{A}$ denotes the set of admissible trading strategies. The dynamics of the state variables X, S, Y are given by (5.22) and (5.14, 5.15). For starting time zero we write $u^{(n)}(x) \equiv u^{(n)}(0, x, \cdot, \cdot)$.

The set of admissible strategies is defined as follows. Denote by $\Delta := \pi/S$ be the adapted S-integrable process for the number of shares held. The space of permitted strategies is

$$\mathcal{A} = \{\Delta : (\Delta \cdot S) \text{ is a } (Q, \widehat{\mathbb{F}})\text{-martingale for all } Q \in \mathcal{M}_f\},$$

where $(\Delta \cdot S)_t = \int_0^t \Delta_u \mathrm{d}S_u$ is the gain from trading over $[0, t], t \in [0, T]$.

Denote the optimal trading strategy by $\pi^* \equiv \pi^{*,n}$, and the optimal wealth process by $X^* \equiv X^{*,n}$. The utility-based price $p^{(n)}$ and optimal hedge for a position in n claims are defined along the lines of Definitions 5.1 and 5.2. The indifference price per claim at $t \in [0, T]$, given $X_t = x, S_t = s, Y_t = y$, is $p^{(n)}$ given by

$$u^{(n)}(t, x - np^{(n)}(t, x, s, y), s, y) = u^{(0)}(t, x, s).$$

The optimal hedging strategy is to hold $(\Delta_t^H)_{0 \le t \le T}$ shares of stock at time t, where $\Delta_t^H S_t =: \pi_t^H S_t$, and $\pi^H := \left(\pi_t^H\right)_{0 \le t \le T}$, is defined by

$$\pi_t^H := \pi_t^{*,n} - \pi_t^{*,0}, \quad 0 \le t \le T. \tag{5.30}$$

It is well known that with exponential utility the indifference price is independent of the initial cash wealth x, so we shall write $p^{(n)}(t, x, s, y) \equiv p^{(n)}(t, s, y)$ from now on.

For small positions in the claim (or, equivalently, for small risk aversion), we shall later approximate the indifference price by the marginal utility-based price introduced by Davis [3]. This is the indifference price for infinitesimal diversions of funds into the purchase or sale of claims, and is equivalent (as is well-known, see for example Monoyios [24]) to the limit of the indifference price as $n \to 0$.

Definition 5.6 (Marginal price). The marginal utility-based price of the claim at $t \in [0, T]$ is $\widehat{p}(t, s, y)$ defined by

$$\widehat{p}(t, s, y) := \lim_{n \to 0} p^{(n)}(t, s, y).$$

It is well known that with exponential utility the marginal price is also equivalent to the limit of the indifference price as risk aversion goes to zero. Under appropriate conditions (satisfied in this model) it is given by the expectation of the payoff under the optimal measure of the dual problem without the claim. For exponential utility this measure is the minimal entropy measure Q^E and, as we have already seen, in our model $Q^E = Q^M$, giving the representation $\widehat{p}(t, s, y) = E^{Q^M}\left[h(Y_T) \mid S_t = s, Y_t = y\right]$, as we shall see in the next section.

5.2.5 Dual problem and optimal hedge

We attack the primal utility maximisation problem (5.29) using classical duality results. For a problem with the random terminal endowment of a European claim, and with exponential utility, as in this paper, Delbaen et al. [5] establish the required duality relations between the primal and dual problems in a semimartingale setting. We shall use these results below to establish a simple algebraic relation (Lemma 5.7) between the primal value function and the indifference price, which we shall then exploit to derive the representation for the optimal hedging strategy.

The dual problem with starting time zero has value function defined by

$$\tilde{u}^{(n)}(\eta) := \inf_{Q\in\mathcal{M}} E\big[\widetilde{U}(\eta Z_T) + \eta Z_T nh(Y_T)\big],$$

where Z is the density process in (5.23) and $\widetilde{U}$ is the convex conjugate of the utility function. For exponential utility $\widetilde{U}$ is given by

$$\widetilde{U}(\eta) = \frac{\eta}{\alpha}\left[\log\left(\frac{\eta}{\alpha}\right) - 1\right].$$

Hence the dual value function has the well-known entropic representation

$$\tilde{u}^{(n)}(\eta) = \widetilde{U}(\eta) + \frac{\eta}{\alpha}\inf_{Q\in\mathcal{M}}\left[\mathcal{H}(Q,P) + \alpha n E^Q h(Y_T)\right].$$

Denoting the dual minimiser that attains the above infimum by $Q^{*,n}$, we observe that $Q^{*,n} \in \mathcal{M}_f$.

For a starting time $t \in [0,T]$ the dual value function is defined by

$$\tilde{u}^{(n)}(t,\eta,s,y) := \inf_{Q\in\mathcal{M}} E\left[\widetilde{U}\left(\eta\frac{Z_T}{Z_t}\right) + \eta\frac{Z_T}{Z_t}nh(Y_T)\middle|\, S_t = s, Y_t = y\right], \tag{5.31}$$

and we write $\tilde{u}^{(n)}(\eta) \equiv \tilde{u}^{(n)}(0,\eta,\cdot,\cdot)$.

Lemma 5.7. *The primal value function and indifference price are related by*

$$u^{(n)}(t,x,s,y) = u^{(0)}(t,x,s)\exp\left(-\alpha n p^{(n)}(t,s,y)\right), \tag{5.32}$$

where the value function without the claim is given by

$$u^{(0)}(t,x,s) = -\exp\left(-\alpha x - H^E(t,s)\right), \tag{5.33}$$

and $H^E(t,s)$ is the conditional minimal entropy function defined in (5.28).

Proof. For brevity, we give the proof for $t = 0$. The proof for a general starting time follows similar lines, and we make some comments on how to adapt the following argument for that case at the end of the proof.

The fundamental duality linking the primal and dual problems in Delbaen et al. [5] implies that the value functions $u^{(n)}(x)$ and $\tilde{u}^{(n)}(\eta)$ are conjugate:

$$\tilde{u}^{(n)}(\eta) = \sup_{x\in\mathbb{R}}[u^{(n)}(x) - x\eta], \qquad u^{(n)}(x) = \inf_{\eta>0}[\tilde{u}^{(n)}(\eta) + x\eta].$$

The value of η attaining the above infimum is η^*, given by $\tilde{u}^{(n)}_\eta(\eta^*) = -x$, so that

$$u^{(n)}(x) = \tilde{u}^{(n)}(\eta^*) + x\eta^*,$$

which translates to

$$u^{(n)}(x) = -\exp\left(-\alpha x - \inf_{Q\in\mathcal{M}}\left[\mathcal{H}(Q,P) + \alpha n E^{Q} h(Y_T)\right]\right). \tag{5.34}$$

So, in particular,

$$u^{(0)}(x) = -\exp\left[-\alpha x - \mathcal{H}(Q^E, P)\right], \tag{5.35}$$

where Q^E is the minimal entropy measure: $Q^E = Q^{*,0}$.

Combining the dual representations (5.34) and (5.35) for the primal problems with and without the claim, with the definition of the indifference price, gives the dual representation for the utility-based price in the form

$$p^{(n)} = \frac{1}{\alpha n}\left[\inf_{Q\in\mathcal{M}}\left[\mathcal{H}(Q,P) + \alpha n E^{Q} h(Y_T)\right] - \mathcal{H}(Q^E, P)\right], \tag{5.36}$$

which is the representation found in Delbaen et al. [5], modified slightly as we have a random endowment of n claims ([5] considered the case $n = -1$).

In particular, for $n \to 0$ or $\alpha \to 0$, we obtain the marginal price of Davis [3]:

$$\widehat{p} := \lim_{n\to 0} p^{(n)} = E^{Q^E} h(Y_T) = E^{Q^M} h(Y_T), \tag{5.37}$$

the last equality following from the equality of Q^M and Q^E, as implied by (5.26).

From (5.34)–(5.36), the relation between the primal value functions and indifference price then follows immediately, as

$$\begin{aligned} u^{(n)}(x) &= -\exp\left(-\alpha x - \mathcal{H}(Q^E, P) - \alpha n p^{(n)}\right) \\ &= u^{(0)}(x)\exp\left(-\alpha n p^{(n)}\right). \end{aligned}$$

Similarly, a corresponding relation for a starting time $t \in [0,T]$ may also be derived. This is achieved using the definition (5.31) of the dual value function for an initial time $t \in [0,T]$, the conjugacy of $u^{(n)}(t,x,s,y)$ and $\tilde{u}^{(n)}(t,\eta,s,y)$ and the definitions (5.27) and (5.28) of the conditional entropy and conditional minimal entropy. □

Using Lemma 5.7 we obtain the following representation for the optimal hedging strategy associated with the indifference price. In what follows we assume that the indifference price is a suitably smooth function of (t,s,y), so that (given Lemma 5.7) we may assume the primal value function is smooth enough to be a classical solution of the associated Hamilton–Jacobi–Bellman (HJB) equation. This smoothness property is confirmed in [23].

Theorem 5.8. *The optimal hedge for a position in n claims is to hold Δ_t^H units of S at $t \in [0,T]$, where*

$$\Delta_t^H = -n\left(p_s^{(n)}(t,S_t,Y_t) + \rho\frac{\beta}{\sigma}\frac{Y_t}{S_t}p_y^{(n)}(t,S_t,Y_t)\right).$$

Remark 5.9. We note the extra term in the hedging formula compared with the corresponding full information result (5.10). The drift parameter uncertainty results in additional risk, manifested as dependence of the indifference price on the stock price, and hence the derivative with respect to the stock price appears in the theorem.

Proof. The HJB equation associated with the primal the value function is

$$u_t^{(n)} + \max_{\pi} \mathcal{A}_{X,S,Y} u^{(n)} = 0,$$

where $\mathcal{A}_{X,S,Y}$ is the generator of (X, S, Y) under P. Performing the maximisation over π yields the optimal Markov control as $\pi_t^{*,n} = \pi^{*,n}(t, X_t^{*,n}, S_t, Y_t)$, where

$$\pi^{*,n}(t,x,s,y) = -\left(\frac{\widehat{\lambda} u_x^{(n)} + \sigma s u_{xs}^{(n)} + \rho\beta y u_{xy}^{(n)}}{\sigma u_{xx}^{(n)}}\right),$$

and where the arguments of the functions on the right-hand side are omitted for brevity. For the case $n = 0$ there is no dependence on y in the value function $u^{(0)}$, and we have $\pi_t^{*,0} = \pi^{*,0}(t, X_t^{*,0}, S_t)$, where

$$\pi^{*,0}(t,x,s) = -\left(\frac{\widehat{\lambda} u_x^{(0)} + \sigma s u_{xs}^{(0)}}{\sigma u_{xx}^{(0)}}\right).$$

Applying the definition (5.30) of the optimal hedging strategy along with the representations (5.32) and (5.33) from Lemma 5.7 for the value functions, gives the result. □

5.2.6 Stochastic control representation of the indifference price

Using the expression (5.26) for the relative entropy between measures in $Q \in \mathcal{M}_f$ and P in the dual representation (5.36) of $p^{(n)}$, we obtain the indifference price of the claim at time zero as the value function of a control problem:

$$p^{(n)} = \inf_{\psi} E^Q \left[\frac{1}{2\alpha n} \int_0^T \psi_t^2 \mathrm{d}t + h(Y_T)\right],$$

to be minimised over control processes $(\psi_t)_{0 \le t \le T}$, such that $Q \in \mathcal{M}_f$. Of course, we need only consider measures with finite relative entropy since a martingale measure with $\mathcal{H}(Q, P) = \infty$ will not attain the infimum in (5.36). The dynamics for S, Y are given in the system of equations (5.24). Equivalently, since $\widehat{\lambda}, \widehat{\theta}$ may be expressed as functions of time and current asset price by (5.17), we may write the state dynamics of the control problem for the indifference price as

$$\begin{aligned} \mathrm{d}S_t &= \sigma S_t \mathrm{d}\widehat{B}_t^Q, \\ \mathrm{d}Y_t &= \beta Y_t[(\widehat{\theta}(t,Y_t) - \rho\widehat{\lambda}(t,S_t) - \sqrt{1-\rho^2}\psi_t)\mathrm{d}t + \mathrm{d}\widehat{W}_t^Q]. \end{aligned}$$

Adopting a dynamic programming approach, we consider a starting time $t \in [0, T]$. Then we have

$$p^{(n)}(t,s,y) = \inf_{\psi} E^Q \left[\frac{1}{2\alpha n} \int_t^T \psi_u^2 \mathrm{d}u + h(Y_T) \middle| S_t = s, Y_t = y \right].$$

The HJB dynamic programming PDE associated with $p^{(n)}(t,s,y)$ is

$$p_t^{(n)} + \mathcal{A}_{S,Y}^{Q^M} p^{(n)} + \inf_{\psi} \left[\frac{1}{2\alpha n}\psi^2 - \beta\sqrt{1-\rho^2}\psi y p_y^{(n)} \right] = 0, \quad p(T,s,y) = h(y),$$

where $\mathcal{A}_{S,Y}^{Q^M}$ is the generator of (S,Y) under the minimal measure:

$$\mathcal{A}_{S,Y}^{Q^M} f(t,s,y) = \beta(\widehat{\theta}(t,y) - \rho\widehat{\lambda}(t,s)) y f_y + \frac{1}{2}s^s f_{ss} + \frac{1}{2}\beta^2 y^2 f_{yy} + \rho\sigma\beta s y f_{sy}.$$

Performing the minimisation in the HJB equation, the optimal Markov control is $\psi_t^{*,n} \equiv \psi^{*,n}(t, S_t, Y_t)$, where

$$\psi^{*,n}(t,s,y) = \alpha n\sqrt{1-\rho^2}\beta y p_y^{(n)}(t,s,y),$$

and note that $\psi^{*,0} = 0$. Substituting back into the HJB equation, we find that $p^{(n)}$ solves the semi-linear PDE

$$p_t^{(n)} + \mathcal{A}_{S,Y}^{Q^M} p^{(n)} - \frac{1}{2}\alpha n(1-\rho^2)\beta^2 y^2 \left(p_y^{(n)}\right)^2 = 0, \quad p^{(n)}(T,s,y) = h(y).$$

We note that for $n = 0$ this becomes a linear PDE for the marginal price $\widehat{p}$, so that by the Feynman–Kac theorem we have

$$\widehat{p}(t,s,y) = E_{t,s,y}^{Q^M} h(Y_T), \tag{5.38}$$

consistent with the general result (5.37). We shall see that in this case the marginal price is given by a BS-type formula.

5.2.7 Analytic approximation for the indifference price

To obtain analytic results we approximate the indifference price by the marginal price in (5.38). The marginal price (and hence the associated trading strategy) can be computed in analytic form since, under Q^M, $\log Y_T$ is Gaussian. We have the following result.

Proposition 5.10. *Under Q^M, conditional on $S_t = s, Y_t = y$, $\log Y_T \sim \mathrm{N}(m, \Sigma^2)$, where $m \equiv m(t,s,y)$ and $\Sigma^2 \equiv \Sigma^2(t)$ are given by*

$$\begin{aligned} m(t,s,y) &= \log y + \beta\left(\widehat{\theta}(t,y) - \rho\widehat{\lambda}(t,s) - \frac{1}{2}\beta\right)(T-t) \\ \Sigma^2(t) &= \left[1 + (1-\rho^2)\mathrm{v}_t(T-t)\right]\beta^2(T-t). \end{aligned}$$

Proof. This is established by computing the SDEs for Y and for $\widehat{\theta}_t - \rho\widehat{\lambda}_t$ under Q^M. Indeed, applying the Itô formula to $\log Y_t$ under Q^M, we obtain, for $t < T$,

$$\log Y_T = \log Y_t + \beta\int_t^T \left(\widehat{\theta}_u - \rho\widehat{\lambda}_u\right) du - \frac{1}{2}\beta^2(T-t) + \beta\int_t^T d\widehat{W}_u^{Q^M}, \tag{5.39}$$

where $\widehat{W}^{Q^M}$ is a Brownian motion under Q^M. The dynamics of $\widehat{\theta}_t - \rho\widehat{\lambda}_t$ under Q^M are

$$\mathrm{d}(\widehat{\theta}_t - \rho\widehat{\lambda}_t) = \sqrt{1-\rho^2}\mathrm{v}_t \mathrm{d}\widehat{B}_t^{\perp,Q^M},$$

where $\widehat{B}^{\perp,Q^M}$ is a Q^M-Brownian motion perpendicular to that driving the stock, related to $\widehat{W}^{Q^M}$ by $\widehat{W}^{Q^M} = \rho\widehat{B}^{Q^M} + \sqrt{1-\rho^2}\widehat{B}^{\perp,Q^M}$, and where $\widehat{B}^{Q^M}$ is the Brownian motion driving S. Hence, for $u > t$, after changing the order of integration in a double integral, we obtain

$$\int_t^T \left(\widehat{\theta}_u - \rho\widehat{\lambda}_u\right)\mathrm{d}u = \left(\widehat{\theta}_t - \rho\widehat{\lambda}_t\right)(T-t) + \sqrt{1-\rho^2}\int_t^T \mathrm{v}_u(T-u)\mathrm{d}\widehat{B}_u^{\perp,Q^M}.$$

This can be inserted into (5.39) to yield the desired result. □

We are thus able to obtain BS-style formulae for the price and hedge. For a put option of strike K we easily obtain the following explicit formulae for the marginal price and the associated optimal hedging strategy, where Φ denotes the standard cumulative normal distribution function.

Corollary 5.11. *With m and Σ as in Proposition 5.10, define $b \equiv b(t,s,y)$ by*

$$m = \log y + b - \frac{1}{2}\Sigma^2.$$

Then the marginal price at time $t \in [0,T]$ of a put option with payoff $(K - Y_T)^+$ is $\widehat{p}(t, S_t, Y_t)$, where

$$\begin{aligned} \widehat{p}(t,s,y) &= K\Phi(-d_1+\Sigma) - ye^b\Phi(-d_1), \\ d_1 &= \frac{1}{\Sigma}\left[\log\left(\frac{y}{K}\right) + b + \frac{1}{2}\Sigma^2\right]. \end{aligned}$$

The optimal hedging strategy given by Theorem 5.8 with $\widehat{p}$ as an approximation to the indifference price is $\widehat{\Delta}_t \equiv \widehat{\Delta}(t, S_t, Y_t)$, where

$$\widehat{\Delta}(t,s,y) = n\rho\frac{\beta}{\sigma}\frac{y}{s}e^b\Phi(-d_1).$$

In [23] these results are used to conduct a simulation study of the effectiveness of the optimal hedge under partial information (that is, with Bayesian learning about the drift parameters of the assets), compared with the BS-style hedge and the optimal hedge without learning. The results show that optimal hedging combined with a filtering algorithm to deal with drift parameter uncertainty can indeed give improved hedging performance over methods which take S as a perfect proxy for Y, and over methods which do not incorporate learning via filtering.

Bibliography

[1] T. Björk, M. H. A. Davis and C. Landén, *Optimal investment under partial information*, preprint (2008)

[2] A. Danilova, M. Monoyios and A. Ng, *Optimal investment with inside information and parameter uncertainty*, preprint (2009).

[3] M. H. A. Davis, *Option pricing in incomplete markets*, Mathematics of Derivative Securities (Cambridge) (M. A. H. Dempster and S. R. Pliska, eds.), Cambridge University Press, 1997, pp. 216–226.

[4] M. H. A. Davis, *Optimal hedging with basis risk* From Stochastic Calculus to Mathematical Finance : The Shiryaev Festschrift (Berlin) (Y. Kabanov, R. Lipster and J. Stoyanov, eds.), Springer, 2006, pp. 169–188.

[5] F. Delbaen, P. Grandits, T. Rheinländer, D. Samperi, M. Schweizer and C. Stricker, *Exponential hedging and entropic penalties*, Mathematical Finance 12 (2002), pp. 99–123.

[6] W. H. Fleming and R. W. Rishel, *Deterministic and Stochastic Optimal Control*, Springer-Verlag, New York, 1975.

[7] M. Fujisaki, G. Kallianpur and H. Kunita, *Stochastic differential equations for the nonlinear filtering problem*, Osaka J. Math. 9 (1972), pp. 19–40.

[8] V. Henderson, *Valuation of claims on nontraded assets using utility maximization*, Mathematical Finance 12 (2002), pp. 351–373.

[9] J. Hugonnier and D. Kramkov, *Optimal investments with random endowments in incomplete markets*, Annals of Applied Probability 14 (2004), pp. 845–864.

[10] T. Kailath, *An innovations approach to least-squares estimation, Part I: Linear Filtering in Additive Noise*, IEEE Trans. Automatic Control 13 (1968), pp. 646–655.

[11] R. E. Kalman and R. S. Bucy, *New results in linear filtering and prediction theory*, Trans. ASME Ser. D. J. Basic Engineering 83 (1961), pp. 95–108.

[12] G. Kallianpur, *Stochastic Filtering Theory*, Springer, 1980.

[13] I. Karatzas, I *Lectures on the Mathematics of Finance*, CRM Monographs 8 (1996), American Mathematical Society.

[14] I. Karatzas, J. P. Lehoczky, S. E. Shreve and G-L. Xu, 1991 *Martingale and duality methods for utility maximization in an incomplete market*, SIAM Journal on Control and Optimization 29 (1991), pp. 702–730.

[15] D. Kramkov and W. Schachermayer, *The asymptotic elasticity of utility functions and optimal investment in incomplete markets*, Annals of Applied Probability 9 (1999), pp. 904–950.

[16] R. S. Lipster and A. N. Shiryaev, *Statistics of Random Processes I: General Theory*, Springer, 2001.

[17] R. S. Lipster and A. N. Shiryaev, *Statistics of Random Processes II: Applications*, Springer, 2001.

[18] R. Mansuy and M. Yor, *Random times and enlargements of filtrations in a Brownian setting*, Lecture Notes in Mathematics 1873 (2006), Springer.

[19] R. C. Merton, *Lifetime portfolio selection under uncertainty: the continuous-time case*, Rev. Econom. Statist. 51 (1969), pp. 247–257.

[20] R. C. Merton, *Optimum consumption and portfolio rules in a continuous-time model*, J. Econom. Theory 3 (1971), pp. 373–413. *Erratum:* ibid. 6 (1973) pp. 213–214.

[21] M. Monoyios, *Performance of utility-based strategies for hedging basis risk*, Quantitative Finance 4 (2004), pp. 245–255.

[22] M. Monoyios, *Optimal hedging and parameter uncertainty*, IMA Journal of Management Mathematics 18 (2007), pp. 331–351.

[23] M. Monoyios, *Marginal utility-based hedging of claims on non-traded assets with partial information*, preprint (2008).

[24] M. Monoyios, *Utility indifference pricing with market incompleteness*, Nonlinear Models in Mathematical Finance: New Research Trends in Option Pricing (New York) (M. Ehrhardt, ed.), Nova Science Publishers, 2008.

[25] M. Monoyios, *Asymptotic expansions for optimal hedging of basis risk with partial information*, preprint (2009).

[26] M. Musiela and T. Zariphopoulou, *An example of indifference prices under exponential preferences*, Finance & Stochastics 8 (2004), pp. 229–239.

[27] M. P. Owen, *Utility based optimal hedging in incomplete markets*, Annals of Applied Probability 12 (2002), pp. 691–709.

[28] H. Pham, *Portfolio optimization under partial observation: theoretical and numerical aspects*, Handbook of Nonlinear Filtering (Oxford) (D. Crisan and B. Rozovsky, eds.), Oxford University Press, to appear.

[29] I. Pikovsky and I. Karatzas, *Anticipative portfolio optimization*, Adv. in App. Prob. 28 (1996), pp. 1095–1122.

[30] E. Platen and W. J. Runggaldier, *A benchmark approach to portfolio optimization under partial information*, Asia Pacific Financial Markets 14 (2007), pp. 25–43.

[31] L. C. G. Rogers, *The relaxed investor and parameter uncertainty*, Finance & Stochastics 5 (2001), pp. 131–154.

[32] L. C. G. Rogers and D. Williams, *Diffusions, Markov Processes, and Martingales, Vol. 2: Itô Calculus*, 2nd. ed., Cambridge University Press, Cambridge, UK, 2000.

[33] M. Zakai, *On the optimal filtering of diffusion processes*, Z. Wahrscheinlichkeitstheorie und Verw. Gebiete 11 (1969), pp. 230–243.

Author information

Michael Monoyios, Mathematical Institute, University of Oxford, UK.
Email: monoyios@maths.ox.ac.uk

Investment/consumption choice in illiquid markets with random trading times

Huyên Pham

Abstract. We consider a portfolio/consumption selection problem in a liquidity risk model introduced in [11], and further investigated in [12] and [4]. This survey paper summarises the main results in these works. In this illiquidity market modelling, stock prices are quoted and observed only at exogenous random times corresponding to the arrivals of buy/sell orders. The investor trades the stock at these random times, while consuming continuously from his cash holdings, and the goal is to maximise the expected utility from consumption. This mixed discrete/continuous stochastic control problem is solved by a dynamic programming approach, which leads to a coupled system of Integro–Partial Differential Equations (IPDE). Analytic characterisation of the value functions and of the optimal strategies are derived, and we provide a convergent numerical algorithm for the resolution to this coupled system of IPDE. Several numerical experiments illustrate the impact of the restricted liquidity trading opportunities, and we measure in particular the utility loss with respect to the classical Merton consumption problem.

Key words. Liquidity, random trading times, portfolio/consumption problem, cost of liquidity, integro-partial differential equations, viscosity solutions.

AMS classification. 93E20, 49K22, 91B28

1 Introduction

Liquidity risk is one of the most significant risk factors in financial economy, yet a lot remains to be done at the theoretical level for designing appropriate measures of liquidity and understanding the mechanisms which underly it. In general, the market liquidity is the ability to quickly liquidate big volumes to low costs when assets have to be converted into cash. Therefore, the liquidity is a three-dimensional measure composed of: (i) *volume*: the size of traded position, (ii) *price*: the costs which are caused by trading the position, (iii) *time*: the point in time when one has to trade or execute the position. There have been numerous approaches to modelling liquidity over the years, mostly focusing on the volume and price measures of market liquidity. In this direction, the recent papers [1], [2], [3] or [14] studied the price impact, that is the correlation betwen an incoming order (to buy or sell) and the subsequent price change.

The temporal dimension of market liquidity is related to the restriction on asset price observation, trade or execution times, and is a crucial determinant for liquidity measure. It has been largely studied in the econometrics of high-frequency data, especially for volatility estimation. The next important issue in this modelling is to understand the implications for pricing and risk management. In this perspective, Schwartz and Tebaldi [15] and Longstaff [8] assumed in their model that illiquid assets could only be traded at the starting date and at a fixed terminal horizon. In a less extreme modelling, Rogers and Zane [13] and Matsumoto [9] consider random trading dates with continuous-time observation, by assuming that trade succeed only at the jump times of a Poisson process, and study the impact on the portfolio choice problem.

In this paper, we consider a description of liquidity risk introduced in Pham and Tankov [11], which is consistent with the situation often viewed by practitioners where their ability to trade assets is limited or restricted to the times when a quote comes into the market. The price of the risky asset can be observed and trade orders can be passed only at random times of an exogenous Poisson process . These times model the arrival of buy/sell orders in an illiquid market, or the dates on which the results of a hedge fund are published. This setup was inspired by recent papers of Frey and Runggaldier [6] and Cvitanic, Liptser and Rozovskii [5], who assume in addition that there is an unobservable stochastic volatility, and are interested in the estimation of this volatility. In our liquidity risk context, we suppose that the investor is also allowed to consume (or distribute dividends) continuously from the bank account, and the objective is to maximise the expected discounted utility of consumption. The resulting optimisation problem is a nonstandard mixed discrete/continuous time stochastic control problem, which leads via the dynamic programming principle to a coupled system of nonlinear integro-partial differential equations (IPDE). The aim of this paper is to summarise the main results recently obtained in [11], [12] and [4] about the IPDE characterisation and regularity of the value functions, the existence and representation of the optimal strategies, and the numerical resolution of this investment/consumption problem in a liquidity risk context with computational illustrations compared to the classical Merton problem.

The rest of the paper is structured as follows. In Section 2, we describe the liquidity risk model, and we formulate in Section 3 the optimal investment/consumption problem. Section 4 contains the main results of the paper, by stating the IPDE viscosity characterisation and regularity of the value function, which is then used for deriving the optimal portfolio and consumption policies. Finally, we describe in Section 5 a convergent numerical algorithm and give some numerical tests for measuring the impact of our liquidity trading constraints.

2 The liquidity risk model

We consider a market model in which the bids and offers on a risky asset are not available at any time. The arrivals of buy/sell orders occur at the jumps $(\tau_k)_k$, $\tau_0 = 0 < \tau_1 < \ldots < \tau_k$ of a Poisson process with constant intensity λ, independent of the asset price process S. For simplicity, we assume that the continuous time price process S follows a Black–Scholes dynamics:

$$dS_t = S_t(bdt + \sigma dW_t), \tag{2.1}$$

where W is a standard brownian motion on a probability space $(\Omega, \mathcal{G}, \mathbb{P})$, b, $\sigma > 0$ are positive constants, and we denote by $\mathbb{F} = (\mathbb{F}_t)_{t\geq 0}$ the natural filtration of W, which is also the filtration generated by the asset price S.

In this illiquid market, the investor can observe and trade S only at the random times $(\tau_k)_{k\geq 0}$. We denote by

$$Z_{t,s} \quad = \quad \frac{S_s - S_t}{S_t}, \quad 0 \leq t \leq s,$$

and by

$$Z_k \quad = \quad Z_{\tau_{k-1},\tau_k}, \quad k \geq 1,$$

the observed return process valued in $(-1, \infty)$. We set by convention Z_0 to some fixed constant. The investor may also consume continuously from the bank account (interest rate is assumed w.l.o.g. to be zero) between two trading dates. We introduce the continuous observation filtration $\mathbb{G}^c = (\mathcal{G}_t)_{t\geq 0}$ with :

$$\mathcal{G}_t \quad = \quad \sigma\left\{(\tau_k, Z_k) \; : \tau_k \leq t\right\},$$

and the discrete observation filtration $\mathbb{G}^d = (\mathcal{G}_{\tau_k})_{k\geq 0}$.

A portfolio/consumption policy is a mixed discrete-continuous process (α, c), where $\alpha = (\alpha_k)_{k\geq 0}$ is real-valued $\mathbb{G}^d$-adapted, and $c = (c_t)_{t\geq 0}$ is a nonnegative $\mathbb{G}^c$-adapted process: α_k represents the amount of stock invested for the period $(\tau_k, \tau_{k+1}]$ after observing the stock price at time τ_k, and c_t is the consumption rate at time t based on the available information. Starting from an initial capital $x \geq 0$, and given a control policy (α, c), we denote by $X^x_{\tau_k}$ the wealth of the investor at time τ_k given by:

$$X^x_{\tau_k} \quad = \quad x - \int_0^{\tau_k} c_t dt + \sum_{i=0}^{k} \alpha_i Z_{i+1}, \quad k \geq 0. \tag{2.2}$$

Given $x \geq 0$, we say that a control policy (α, c) is admissible, and we denote it by (α, c) $\in \mathcal{A}(x)$ if :

$$X^x_{\tau_k} \quad \geq \quad 0, \quad \text{a.s.} \quad \forall\, k \geq 0. \tag{2.3}$$

Remark 2.1. For all $k \geq 0$, conditionally on the interarrival times $\tau_{k+1} - \tau_k = t \geq 0$, we see from (2.1) that Z_{k+1} is independent of $\mathcal{G}_{\tau_k}$, and has distribution $p(t, dz)$ of support $(-1, \infty)$, with

$$p(t, dz) \quad = \quad \mathbb{P}\Big[e^{(b-\frac{\sigma^2}{2})t+\sigma W_t} - 1 \in dz\Big].$$

Notice that

$$\int zp(t, dz) \quad = \quad \mathbb{E}\big[Z_{k+1}\big|\mathcal{G}_{\tau_k}, \tau_{k+1} - \tau_k = t\big] = e^{bt} - 1 \;\geq\; 0, \; k \geq 0, t \geq 0. \tag{2.4}$$

Remark 2.2. *Constrained policies.* Since $X^x_{\tau_{k+1}} = X^x_{\tau_k} - \int_{\tau_k}^{\tau_{k+1}} c_u du + \alpha_k Z_{k+1}$, and the support of Z_{k+1} is $(-1, \infty)$, we see that the admissibility condition (2.3) for (α, c) $\in \mathcal{A}(x)$ is written as:

$$X^x_{\tau_k} - \int_{\tau_k}^{s} c_u du + \alpha_k z \quad \geq \quad 0, \quad \forall k \geq 0, \; \forall s \geq \tau_k, \; \forall z \in (-1, \infty)$$

almost surely. This means that we have a no-short sale constraint (both on the risky asset and bank account):

$$0 \leq \alpha_k \quad \leq \quad X^x_{\tau_k}, \quad \forall k \geq 0, \tag{2.5}$$

together with the consumption constraint:

$$\int_{\tau_k}^{\infty} c_u du \quad \leq \quad X^x_{\tau_k} - \alpha_k, \quad \forall k \geq 0. \tag{2.6}$$

Remark 2.3. *Embedding in a continuous-time wealth process.* Let us introduce the continuous time filtration $\mathbb{H} = (\mathcal{H}_t)_{t\geq 0} = \mathbb{F} \vee \mathbb{G}^c$. In other words, $\mathcal{H}_t$ corresponds to the path observation of the asset price and of the random times up to time t. Notice that W is still a Brownian motion under $\mathbb{H}$, and the dynamics of S under $(\mathbb{P}, \mathbb{H})$ is still governed by (2.1). Given $x \geq 0$, and $(\alpha, c) \in \mathcal{A}(x)$ with corresponding discrete time wealth process $(X^x_{\tau_k})_{k\geq 0}$, let us define the continuous time process $(X^x_t)_{t\geq 0}$ by

$$\begin{aligned} X^x_t &= X^x_{\tau_k} - \int_{\tau_k}^t c_u du + \alpha_k Z_{\tau_k, t}, \quad \tau_k < t \leq \tau_{k+1} \;\; k \geq 0, \\ &= x - \int_0^t c_u du + \int_0^t H_u dS_u, \quad t \geq 0, \end{aligned} \tag{2.7}$$

where H is the simple integrand process

$$H_t \quad = \quad \sum_{k=0}^{\infty} \frac{\alpha_k}{S_{\tau_k}} 1_{\tau_k < t \leq \tau_{k+1}}, \quad t \geq 0,$$

representing the number of shares invested in the risky asset. Notice that H is $\mathbb{H}$-predictable (hence adapted), and c is $\mathbb{H}$-adapted. Moreover, from (2.5) and (2.6), we easily see that H satisfies the no-short sale constraint:

$$0 \leq H_t S_t \quad \leq \quad X^x_t, \quad t \geq 0. \tag{2.8}$$

The continuous time process X has the meaning of a shadow wealth process: it is not observed except for at times τ_k.

3 The optimal investment/consumption problem

We investigate an optimal investment/consumption problem in the illiquid market described in the previous section. We are given a utility function defined on $\mathbb{R}_+$, with $U(0) = 0$, strictly increasing, strictly concave and C^1 on $(0, \infty)$, satisfying the Inada conditions $U'(0^+) = \infty$ and $U'(\infty) = 0$. We also assume the following growth condition on U: there exists $\gamma \in (0, 1)$ s.t.

$$U(x) \quad \leq \quad K_1 x^\gamma, \quad x \geq 0,$$

for some positive constant K_1.

We consider the optimal investment/consumption problem:

$$v(x) = \sup_{(\alpha,c)\in\mathcal{A}(x)} \mathbb{E}\left[\int_0^\infty e^{-\rho t}U(c_t)dt\right], \quad x \geq 0, \tag{3.1}$$

where $\rho > 0$ is a positive discount factor. Actually, it is proved in [12] that for ρ large enough, namely

$$\rho > b\gamma,$$

(which we shall assume in the sequel), then the nonnegative value function v is finite, and satisfies the growth condition

$$v(x) \leq Kx^\gamma, \quad x \geq 0, \tag{3.2}$$

for some positive constant K.

Remark 3.1. Given $x \geq 0$, denote by $\mathcal{A}_{\mathbb{H}}(x)$ (resp. $\mathcal{A}_{\mathbb{F}}(x)$) the set of pairs of $\mathbb{H}$-adapted (resp. $\mathbb{F}$-adapted) processes (H, c) with c nonnegative, and corresponding wealth process given in (2.7), such that the no-short sale constraint (2.8) holds. Consider the associated continuous time optimal investment/consumption problems

$$v_{\mathbb{H}}(x) = \sup_{(H,c)\in\mathcal{A}_{\mathbb{H}}(x)} \mathbb{E}\left[\int_0^\infty e^{-\rho t}U(c_t)dt\right], \quad x \geq 0,$$

and

$$v_M(x) = \sup_{(H,c)\in\mathcal{A}_{\mathbb{F}}(x)} \mathbb{E}\left[\int_0^\infty e^{-\rho t}U(c_t)dt\right], \quad x \geq 0. \tag{3.3}$$

Problem (3.3) is the classical Merton portfolio/consumption choice problem under no-short sale constraints, and based on the continuous time observation of the stock price. It is not hard to check that independent information provided by the random times τ_k does not increase the maximal expected utility of consumption; in other words, the value functions $v_{\mathbb{H}}$ and v_M coincide, and in view of Remark 2.3, we have

$$v \leq v_M \ (= v_{\mathbb{H}}). \tag{3.4}$$

We use a dynamic programming approach for solving the control problem (3.1). The starting point is the following version of the dynamic programming principle (DPP) adapted to our context, and proved rigorously in [11]:

$$v(x) = \sup_{(\alpha,c)\in\mathcal{A}(x)} \mathbb{E}\left[\int_0^{\tau_1} e^{-\rho t}U(c_t)dt + e^{-\rho\tau_1}v(X^x_{\tau_1})\right]. \tag{3.5}$$

From the expression (2.2) of the wealth, and the measurability conditions on the control, the above dynamic programming relation is written as

$$v(x) = \sup_{(a,c)\in A_d(x)} \mathbb{E}\left[\int_0^{\tau_1} e^{-\rho t}U(c_t)dt + e^{-\rho\tau_1}v\left(x - \int_0^{\tau_1} c_t dt + a\, Z_1\right)\right], \tag{3.6}$$

where $A_d(x)$ is the set of pairs (a, c) with a deterministic constant, and c a deterministic nonnegative process s.t. (see Remark 2.2) $a \in [0, x]$ and

$$\int_0^\infty c_u du \quad \le \quad x - a \tag{3.7}$$

Given $a \in [0, x]$, we denote by $\mathcal{C}_a(0, x)$ the set of deterministic nonnegative processes satisfying (3.7). The r.h.s. of (3.6) is then written explicitly in:

$$v(x) = \sup_{\substack{a \,\in\, [0,\, x] \\ c \,\in\, \mathcal{C}_a(0,\, x)}} \int_0^\infty e^{-(\rho+\lambda)t} \left[U(c_t) + \lambda \int v \left(x - \int_0^t c_s ds + a\, z \right) p(t, dz) \right] dt. \tag{3.8}$$

Denote by

$$\mathcal{D} = \mathbb{R}_+ \times \mathcal{X} \quad \text{with } \mathcal{X} \; = \; \{(x, a) \in \mathbb{R}_+ \times \mathbb{R}_+ : a \; \le x\},$$

and let us introduce the dynamic auxiliary control problem: for $(t, x, a) \in \mathcal{D}$,

$$\hat{v}(t, x, a) = \sup_{c \in \mathcal{C}_a(t,x)} \int_t^\infty e^{-(\rho+\lambda)(s-t)} \left[U(c_s) + \lambda \int v \left(Y_s^{t,x} + a\, z \right) p(s, dz) \right] ds, \tag{3.9}$$

where $\mathcal{C}_a(t, x)$ is the set of deterministic nonnegative processes $c = (c_s)_{s \ge t}$ s.t.

$$\int_t^\infty c_u du \quad \le \quad x - a \tag{3.10}$$

and $Y^{t,x}$ is the deterministic controlled process by $c \in \mathcal{C}_a(t, x)$:

$$Y_s^{t,x} \quad = \quad x - \int_t^s c_u du, \quad s \ge t. \tag{3.11}$$

From (3.8)–(3.9), the original value function is then related to this auxiliary optimisation problem by:

$$v(x) \quad = \quad \sup_{a \in [0,x]} \hat{v}(0, x, a). \tag{3.12}$$

The Hamilton–Jacobi (in short HJ) equation associated to the deterministic control problem (3.9) is the following Integro Partial Differential Equation (in short IPDE):

$$(\rho + \lambda)\hat{v} - \frac{\partial \hat{v}}{\partial t} - \tilde{U}\left(\frac{\partial \hat{v}}{\partial x} \right) - \lambda \int v(x + a\, z) p(t, dz) \quad = \quad 0, (t, x, a) \in \mathcal{D}, \tag{3.13}$$

where $\tilde{U}$ is the Legendre transform of U, i.e. $\tilde{U}(y) = \sup_{x \ge 0}[U(x) - xy]$.

To sum up, the dynamic programming principle for our original stochastic optimisation problem (3.1) leads to a first-order coupled IPDE (3.12)–(3.13): Problem (3.9) is a family over $a \in \mathbb{R}_+$ of standard deterministic control problems on infinite horizon, associated to the HJ equation (3.13), and the coupling comes from the fact that the reward function appearing in the definition of problem (3.9) or in its IPDE (3.13) depends on the value function of problem (3.12) and vice-versa. The next section provides a rigorous analytic characterisation of the value functions through their dynamic programming (in short DP) equations (3.12)–(3.13), by showing the regularity properties of the value functions, and then as a byproduct the existence (and uniqueness) of the optimal control feedback.

4 Analytic characterisation of the value functions and optimal strategies

We first recall from [12] some basic properties on the value functions $(v, \hat{v})$ defined in the previous section. The value function v is strictly increasing, concave on $\mathbb{R}_+$, and lies in $C_+(\mathbb{R}_+)$, the set of nonnegative continuous functions on $\mathbb{R}_+$. The value function $\hat{v}$ lies in $C_+(\mathcal{D})$, the set of nonnegative continuous functions on $\mathcal{D}$, and satisfies the boundary condition

$$\lim_{x \downarrow a} \hat{v}(t,x,a) \quad = \quad \lambda \int_t^\infty e^{-(\rho+\lambda)(s-t)} \int v(a \, + a \, z) p(s,dz) ds, \quad \forall t \geq 0. \quad (4.1)$$

It satisfies the growth estimate

$$\hat{v}(t,x,a) \quad \leq \quad K(e^{bt}x)^\gamma, \quad (t,x,a) \in \mathcal{D}, \tag{4.2}$$

for some positive constant K. Moreover, $\hat{v}$ is strictly increasing in $x \geq a$, given (t,a) $\in \mathbb{R}_+ \times \mathbb{R}_+$, and is concave in $(x,a) \in \mathcal{X}$, given $t \in \mathbb{R}_+$.

We now provide a characterisation of the value functions to the DP equation (3.12)–(3.13) by means of the notion of viscosity solution adapted to our text.

Definition 4.1. A pair of functions $(w, \hat{w}) \in C_+(\mathbb{R}_+) \times C_+(\mathcal{D})$ is a viscosity solution to (3.12)–(3.13) if the two following properties hold simultaneously:
(i) *viscosity supersolution property*: $w(x) \geq \sup_{a \in [0,x]} \hat{w}(0,x,a)$, for all $x \geq 0$, and

$$(\rho+\lambda)\hat{w}(\bar{t},\bar{x},a) - \frac{\partial \varphi}{\partial t}(\bar{t},\bar{x}) - \tilde{U}\left(\frac{\partial \varphi}{\partial x}(\bar{t},\bar{x})\right) - \lambda \int w(\bar{x} + a \, z) p(\bar{t},dz) \quad \geq \quad 0,$$

for all $a \in \mathbb{R}_+$, for any test function $\varphi \in C^1(\mathbb{R}_+ \times (a,\infty))$, and $(\bar{t},\bar{x}) \in \mathbb{R}_+ \times (a,\infty)$, which is a local minimum of $(\hat{w}(\cdot,\cdot,a) - \varphi)$.
(ii) *viscosity subsolution property* : $w(x) \leq \sup_{a \in [0,x]} \hat{w}(0,x,a)$, for all $x \geq 0$, and

$$(\rho+\lambda)\hat{w}(\bar{t},\bar{x},a) - \frac{\partial \varphi}{\partial t}(\bar{t},\bar{x}) - \tilde{U}\left(\frac{\partial \varphi}{\partial x}(\bar{t},\bar{x})\right) - \lambda \int w(\bar{x} + a \, z) p(\bar{t},dz) \quad \leq \quad 0,$$

for all $a \in \mathbb{R}_+$, for any test function $\varphi \in C^1(\mathbb{R}_+ \times (a,\infty))$, and $(\bar{t},\bar{x}) \in \mathbb{R}_+ \times (a,\infty)$, which is a local maximum of $(\hat{w}(\cdot,\cdot,a) - \varphi)$.

We reformulate the viscosity characterisation result proved in [12].

Theorem 4.2. *The pair of value functions* $(v, \hat{v})$ *defined in* (3.1), (3.9) *is the unique viscosity solution to* (3.12)–(3.13) *in the sense of Definition 4.1, satisfying the growth conditions* (3.2), (4.2), *and the boundary condition* (4.1).

The above characterisation makes the computation of the value functions possible (see the next section) but does not yield the optimal policies in explicit form. We need to go beyond the viscosity property, and focus on the regularity of the value functions. By using arguments of (semi)concavity and the strict convexity of the Hamiltonian for the IPDE in connection with viscosity solutions, it is proved in [4] that the value functions are continuously differentiable.

Theorem 4.3. *(1) The value function* v *lies in* $C^1(0,\infty)$, *and any maximum point in* (3.12) *is interior for every* $x > 0$. *Moreover,* $v'(0^+) = \infty$.
(2) For all $a \in \mathbb{R}_+$, *we have* $\hat{v}(\cdot,\cdot,a) \in C^1([0,\infty) \times (a,\infty))$, *and*

$$\lim_{x \downarrow a} \frac{\partial \hat{v}}{\partial x}(t,x,a) \quad = \quad \infty, \quad t \geq 0.$$

In particular, $\hat{v}$ *satisfies the IPDE* (3.13) *in the classical sense.*

From the regularity of the value functions, we derive the existence of an optimal control through a verification theorem, and the optimal consumption strategy is produced in feedback form in terms of the classical derivatives of the value functions. We denote by $I = (U')^{-1} : (0,\infty) \to (0,\infty)$ the inverse function of the derivative U', and we consider for each $a \in \mathbb{R}_+$ the nonnegative measurable function

$$\hat{c}(t,x,a) \quad = \quad \arg\max_{c \geq 0} \Big[U(c) - c\frac{\partial \hat{v}}{\partial x}(t,x,a) \Big] \; = \; I\Big(\frac{\partial \hat{v}}{\partial x}(t,x,a) \Big), \quad (t,x,a) \in \mathcal{D}.$$

Theorem 4.4. *(1) Let* $(x,a) \in \mathcal{X}$, *i.e.* $x \geq a \geq 0$. *There exists a unique solution, denoted by* $\hat{Y}(x,a)$, *to the equation*

$$Y_t \quad = \quad x - \int_0^t \hat{c}(s, Y_s, a) ds, \quad t \geq 0, \tag{4.3}$$

and the pair $(\hat{Y}(x,a), a)$ *lies in* $\mathcal{X}$, *i.e.* $\hat{Y}_t(x,a) \geq a$, *for all* $t \geq 0$. *The feedback control*

$$\{\hat{c}(t, \hat{Y}_t(x,a), a), t \geq 0\} \quad \text{is optimal for} \quad \hat{v}(0,x,a).$$

(2) For any $x \geq 0$, *there exists an optimal control policy* $(\alpha^*, c^*) \in \mathcal{A}(x)$ *for* $v(x)$, *given by*

$$\alpha_k^* \quad = \quad \arg\max_{a \in [0, X^x_{\tau_k}]} \hat{v}(0, X^x_{\tau_k}, a), \quad k \geq 0, \tag{4.4}$$

$$c_t^* \quad = \quad \hat{c}(t - \tau_k, Y_t^{(k)}, \alpha_k^*), \quad \tau_k \leq t < \tau_{k+1}, \quad k \geq 0, \tag{4.5}$$

where $X^x_{\tau_k}$ is the wealth of the investor at time τ_k given in (2.2) with the feedback control (α^*, c^*), and $Y_t^{(k)} = \hat{Y}_{t-\tau_k}(X^x_{\tau_k}, \alpha^*_k)$, $t \geq \tau_k$, solution to

$$Y_t^{(k)} = X^x_{\tau_k} - \int_{\tau_k}^t c^*_s ds, \quad t \geq \tau_k,$$

represents the wealth between two trading dates τ_k and τ_{k+1}.

Proof. (1) Let $c \in \mathcal{C}_a(0,x)$ and $Y^x = x - \int_0 c dt$ the corresponding wealth process. By applying standard differential calculus to $e^{-(\rho+\lambda)t}\hat{v}(t, Y_t^x, a)$ between $t = 0$ and $t = T$, we have

$$\begin{aligned} & e^{-(\rho+\lambda)T}\hat{v}(T, Y_T^x, a) - \hat{v}(0,x,a) \\ = & \int_0^T e^{-(\rho+\lambda)t}\Big[-(\rho+\lambda)\hat{v} + \frac{\partial \hat{v}}{\partial t} - c_t\frac{\partial \hat{v}}{\partial x}\Big](t, Y_t^x, a)dt \\ = & -\int_0^T e^{-(\rho+\lambda)t}\Big[U(c_t) + \lambda\int v(Y_t^x + az)\Big]p(t,dz) \\ & + \int_0^T e^{-(\rho+\lambda)t}\Big[U(c_t) - c_t\frac{\partial \hat{v}}{\partial x}(t, Y_t^x, a) - \tilde{U}\Big(\frac{\partial \hat{v}}{\partial x}(t, Y_t^x, a)\Big)\Big]dt, \end{aligned} \tag{4.6}$$

where we used in the last equality the property that $\hat{v}$ satisfies the IPDE (3.13). From the growth estimate (4.2) for $\hat{v}$, the increasing monotonicity of $\hat{v}(T,.,a)$, and since $\rho > b\gamma$, we see that $\lim_{T\to\infty} e^{-(\rho+\lambda)T}\hat{v}(T, Y_T^x, a) = 0$, and thus by sending T to infinity into (4.6):

$$\begin{aligned} \hat{v}(0,x,a) = & \int_0^\infty e^{-(\rho+\lambda)t}\Big[U(c_t) + \lambda\int v(Y_t^x + az)\Big]p(t,dz) \qquad (4.7) \\ & + \int_0^\infty e^{-(\rho+\lambda)t}\Big[U(c_t) - c_t\frac{\partial \hat{v}}{\partial x}(t, Y_t^x, a) - \tilde{U}\Big(\frac{\partial \hat{v}}{\partial x}(t, Y_t^x, a)\Big)\Big]dt. \end{aligned}$$

The existence and uniqueness of a solution $\hat{Y}(x,a)$ to (4.3), which satisfies $\hat{Y}_t(x,a) \geq a$ for all $t \geq 0$ is proved in [4]. The wealth process $\hat{Y}(x,a)$ is associated to the admissible control $\hat{c}_t = \hat{c}(t, \hat{Y}_t(x,a), a)$, $t \geq 0$, and by definition of the function $\hat{c}$, we have

$$\tilde{U}\Big(\frac{\partial \hat{v}}{\partial x}(t, \hat{Y}_t(x,a), a)\Big) = U(\hat{c}_t) - \hat{c}_t\frac{\partial \hat{v}}{\partial x}(t, \hat{Y}_t(x,a), a), \quad t \geq 0,$$

so that from (4.7)

$$\hat{v}(0,x,a) = \int_0^\infty e^{-(\rho+\lambda)t}\Big[U(\hat{c}_t) + \lambda\int v(\hat{Y}_t(x,a) + az)\Big]p(t,dz), \tag{4.8}$$

which shows the optimality of the control $\hat{c}$.

(2) We first show that for any $(\alpha, c) \in \mathcal{A}(x)$, and $k \geq 0$,

$$\begin{aligned} & \mathbb{E}\Big[\int_{\tau_k}^{\tau_{k+1}} e^{-\rho(t-\tau_k)}U(c_t)dt + e^{-\rho(\tau_{k+1}-\tau_k)}v(X^x_{\tau_{k+1}})\Big| \mathcal{G}_{\tau_k}\Big] \qquad (4.9) \\ = & \int_{\tau_k}^\infty e^{-(\rho+\lambda)(t-\tau_k)}\Big[U(c_t) + \lambda\int v\Big(X^x_{\tau_k} - \int_{\tau_k}^t c_u du + \alpha_k z\Big)p(t-\tau_k, dz)\Big]dt. \end{aligned}$$

Indeed, since $X^x_{\tau_{k+1}} = X^x_{\tau_k} - \int_{\tau_k}^{\tau_{k+1}} c_u du + \alpha_k Z_{k+1}$, we have by the law of conditional toy expectations:

$$\mathbb{E}\left[\int_{\tau_k}^{\tau_{k+1}} e^{-\rho(t-\tau_k)}U(c_t)dt + e^{-\rho(\tau_{k+1}-\tau_k)}w(X^x_{\tau_{k+1}})\Big|\mathcal{G}_{\tau_k}\right]$$

$$= \mathbb{E}\Bigg[\int_{\tau_k}^{\tau_{k+1}} e^{-\rho(t-\tau_k)}U(c_t)dt +$$

$$e^{-\rho(\tau_{k+1}-\tau_k)}\mathbb{E}\left[w\left(X^x_{\tau_k} - \int_{\tau_k}^{\tau_{k+1}} c_u du + \alpha_k Z_{k+1}\right)\Big|\mathcal{G}_{\tau_k}, \tau_{k+1}-\tau_k\right]\Bigg|\mathcal{G}_{\tau_k}\Bigg]$$

$$= \mathbb{E}\Bigg[\int_{\tau_k}^{\tau_{k+1}} e^{-\rho(t-\tau_k)}U(c_t)dt +$$

$$e^{-\rho(\tau_{k+1}-\tau_k)}\int w\left(X^x_{\tau_k} - \int_{\tau_k}^{\tau_{k+1}} c_u du + \alpha_k z\right)p(\tau_{k+1}-\tau_k, dz)\Bigg|\mathcal{G}_{\tau_k}\Bigg]$$

$$= \int_0^\infty \Bigg[\int_{\tau_k}^{\tau_k+s} e^{-\rho(t-\tau_k)}U(c_t)dt +$$

$$e^{-\rho s}\int w\left(X^x_{\tau_k} - \int_{\tau_k}^{\tau_k+s} c_u du + \alpha_k z\right)p(s, dz)\Bigg]\lambda e^{-\lambda s}ds,$$

where we used Remark 2.1 in the second equality and the fact that $\tau_{k+1} - \tau_k$ follows an exponential law of parameter λ, in the last one. We obtain (4.9) with Fubini's theorem and the change of variable $s \to s + \tau_k$.

We next prove that for any $(\alpha, c) \in \mathcal{A}(x)$, and $n \geq 0$,

$$\mathbb{E}\big[e^{-\rho\tau_n}(X^x_{\tau_n})^\gamma\big] \;\leq\; x^\gamma\delta^n, \quad \text{with } \delta \;=\; \frac{\lambda}{\rho - b\gamma + \lambda} < 1 \tag{4.10}$$

Indeed, from Jensen's inequality

$$\begin{aligned}\mathbb{E}\big[(X^x_{\tau_{n-1}} + \alpha_{n-1}Z_n)^\gamma\big|\mathcal{G}_{\tau_{n-1}}, \tau_n - \tau_{n-1}\big] \;&\leq\; \left(X^x_{\tau_{n-1}} + \alpha_n\int z\, p(\tau_n - \tau_{n-1}, dz)\right)^\gamma \\ &\leq\; \left(X^x_{\tau_{n-1}} + X^x_{\tau_{n-1}}(e^{b(\tau_n-\tau_{n-1})} - 1)\right)^\gamma \\ &=\; (X^x_{\tau_{n-1}})^\gamma e^{b\gamma(\tau_n-\tau_{n-1})},\end{aligned}$$

where we used (2.4) and (2.5). Thus, by writing that $X^x_{\tau_n} \leq X^x_{\tau_{n-1}} + \alpha_{n-1}Z_n$, and by the law of iterated conditional expectations, we get:

$$\begin{aligned}\mathbb{E}\big[e^{-\rho\tau_n}(X^x_{\tau_n})^\gamma\big] \;&\leq\; \mathbb{E}\big[e^{-(\rho-b\gamma)(\tau_n-\tau_{n-1})}e^{-\rho\tau_{n-1}}(X^x_{\tau_{n-1}})^\gamma\big] \\ &=\; \mathbb{E}\Big[e^{-\rho\tau_{n-1}}(X^x_{\tau_{n-1}})^\gamma\int_0^\infty \lambda e^{-(\rho-b\gamma+\lambda)t}dt\Big] \\ &=\; \delta\,\mathbb{E}\big[e^{-\rho\tau_{n-1}}(X^x_{\tau_{n-1}})^\gamma\big].\end{aligned}$$

We obtain the required inequality (4.10) by induction on n.

Consider the control policy in (4.4)–(4.5). By definition of $\hat{Y}$ in (4.3), the associated wealth process satisfies for all $k \geq 0$

$$\begin{aligned} X^x_{\tau_{k+1}} &= X^x_{\tau_k} - \int_{\tau_k}^{\tau_{k+1}} c^*_t \, dt + \alpha^*_k Z_{k+1} \\ &= \hat{Y}_{t-\tau_k}(X^x_{\tau_k}, \alpha^*_k) + \alpha^*_k Z_{k+1} \geq \alpha^*_k + \alpha^*_k Z_{k+1} \geq 0, \text{ a.s.,} \end{aligned}$$

and thus $(\alpha^*, c^*) \in \mathcal{A}(x)$. From (3.12), definition of α^* and (4.8), we have

$$\begin{aligned} v(X^x_{\tau_k}) &= \hat{v}(0, X^x_{\tau_k}, \alpha^*_k) \\ &= \int_{\tau_k}^{\infty} e^{-(\rho+\lambda)(t-\tau_k)} \Big[U(c^*_t) + \lambda \int v(Y^{(k)}_t + \alpha^*_k z) \Big] p(t-\tau_k, dz) \\ &= \mathbb{E}\Big[\int_{\tau_k}^{\tau_{k+1}} e^{-\rho(t-\tau_k)} U(c^*_t) dt + e^{-\rho(\tau_{k+1}-\tau_k)} v(X^x_{\tau_{k+1}}) \Big| \mathcal{G}_{\tau_k} \Big], \end{aligned}$$

where we used (4.9) in the last equality. By iterating these relations for all k, and using the law of iterated conditional expectations, we obtain

$$v(x) = \mathbb{E}\Big[\int_0^{\tau_n} e^{-\rho t} U(c^*_t) dt + e^{-\rho \tau_n} v(X^x_{\tau_n}) \Big].$$

From the growth estimate (3.2), relation (4.10), and sending n to infinity, we conclude that

$$v(x) = \mathbb{E}\Big[\int_0^{\infty} e^{-\rho t} U(c^*_t) dt \Big].$$

□

Furthermore, by using maximum principle, additional properties on the consumption policy between two trading dates are derived in [4], as solution of an Euler–Lagrange ordinary differential equation.

Proposition 4.5. *Suppose that $U \in C^2(0, \infty)$. Given an investment $a \in \mathbb{R}_+$ at time t in the stock, and starting from an initial capital $Y_t(t, x, a) = x \geq a$, the optimal wealth process $\hat{Y}(t, x, a)$ between two trading dates is twice differentiable, satisfies the second-order ordinary differential equation*

$$\begin{aligned} \frac{d^2 \hat{Y}_s(t,x,a)}{ds^2} &= \frac{\lambda \int v'(\hat{Y}_s(t,x,a) + az) p(s, dz) - (\rho+\lambda) U'(c_s)}{U''(c_s)}, \quad s \geq t, \\ c_s &= -\frac{d\hat{Y}_s(t,x,a)}{ds}, \end{aligned}$$

and we have $\lim_{s \to \infty} \hat{Y}_s(t, x, a) = a$.

5 Numerical solution and illustrations

In this section, we focus on the resolution of the DP equation (3.12)–(3.13), and we give some numerical tests for illustrating the impact of liquidity risk induced by the random trading times.

5.1 A numerical decoupling algorithm

The main difficulty in the numerical resolution of the IPDE (3.13) for $\hat{v}$ comes from the coupling in the integral term involving $\hat{v}$ via v . To overcome this problem, we suggest the following iterative procedure. We start from an initial function v_0 defined on $\mathbb{R}_+$, as the value function of the consumption problem without trading:

$$v_0(x) \quad = \quad \sup_{c\in\mathcal{C}(x)} \int_0^\infty e^{-\rho t} U(c_t) dt,$$

where $\mathcal{C}(x)$ is the set of nonnegative (deterministic) processes $c = (c_t)_t$ s.t. $x - \int_0^t c_s ds \geq 0$ for all $t \geq 0$. v_0 is the unique solution with linear growth condition to the first-order differential equation

$$\rho v_0 - \tilde{U}\left(\frac{\partial v_0}{\partial x}\right) \quad = \quad 0, \quad x > 0,$$

together with the boundary condition $v_0(0^+) = 0$. We then construct by induction a sequence of functions $(\hat{v}_n)_{n\geq 1}$ defined on $\mathcal{D}$ and $(v_n)_{n\geq 0}$ defined on $\mathbb{R}_+$ by:

$$\begin{aligned} \hat{v}_{n+1}(t,x,a) &= \sup_{c\in\mathcal{C}_a(t,x)} \int_t^\infty e^{-(\rho+\lambda)(s-t)} \left[U(c_s) + \lambda \int v_n(Y_s^{t,x} + a\ z) p(s,dz) \right] ds \\ v_{n+1}(x) &= \sup_{a\in[0,x]} \hat{v}_{n+1}(0,x,a), \qquad n \geq 0. \end{aligned} \tag{5.1}$$

By the dynamic programming principle, the function $\hat{v}_{n+1}$ satisfies the first-order PDE

$$-(\rho+\lambda)\hat{v}_{n+1} + \frac{\partial \hat{v}_{n+1}}{\partial t} + \tilde{U}\left(\frac{\partial \hat{v}_{n+1}}{\partial x}\right) + \lambda \int v_n(x + a\ z) p(t,dz) = 0, \ (t,x,a) \in \mathcal{D},$$

and we have an approximate trading policy by taking:

$$\alpha_k^{(n)} \quad \in \quad \arg \max_{a\in[0,X_{\tau_k}^x]} \hat{v}_n(0, X_{\tau_k}^x, a), \quad k \geq 0.$$

The convergence of this iterative decoupling algorithm was studied in [11], where it is proved that the sequence of functions $(v_n, \hat{v}_n)_n$ converges uniformly on any compact subset of $\mathcal{D}$ and $\mathbb{R}_+$ to $(v, \hat{v})$. More precisely, for any compact subset F and G of $\mathcal{D}$ and $\mathbb{R}_+$, there exist some positive constants C_F and C_G s.t.

$$0 \ \leq \ \sup_F(\hat{v} - \hat{v}_n) \ \leq \ C_F\, \delta^n, \quad \text{and} \quad 0 \ \leq \ \sup_G(v - v_n) \ \leq \ C_G\, \delta^n,$$

where $0 < \delta < 1$ is defined in (4.10).

5.2 Numerical illustrations

We now provide simulations for illustrating the impact of liquidity constraints on the attainable utility level and on the investment strategy. We shall compare our numerical experiments with the original Merton problem with no-short sale constraints, and defined in (3.3). We consider the case of power utility functions $U(x) = x^\gamma/\gamma$, and we recall that the value function and the optimal trading strategy (in amount) are explicitly given by

$$v_{_M}(x) = K_{_M} x^\gamma, \qquad \bar{\alpha}_t^M = \min\Big[\frac{b}{(1-\gamma)\sigma^2}, 1\Big] X_t^x,$$

with

$$K_{_M} = \frac{1}{\gamma}\Big(\frac{1-\gamma}{\rho-\eta}\Big)^{1-\gamma}, \qquad \eta = \gamma \max_{\pi\in[0,1]} \big[\pi b - \frac{1}{2}\pi^2(1-\gamma)\sigma^2\big].$$

We know from (3.4) that $v \le v_{_M}$. On the other hand, the value function v is always bounded from below by the value function of the consumption problem without trading v_0, given in our present setting by $v_0(x) = K_0 x^\gamma$, with $K_0 = \frac{1}{\gamma}\big(\frac{1-\gamma}{\rho}\big)^{1-\gamma}$.

Given $(t, x, a) \in \mathcal{D}$, notice that for any $\beta > 0$, we have $c \in \mathcal{C}_a(t, x)$ if and only if $\beta c \in \mathcal{C}_{\beta a}(t, \beta x)$. We then easily deduce from (3.9) and (3.12) a scaling relation for the value function v and the auxiliary value function $\hat{v}$:

$$\hat{v}(t, \beta x, \beta a) = \beta^\gamma \hat{v}(t, x, a), \qquad v(\beta x) = \beta^\gamma v(x), \quad \forall \beta > 0.$$

The scaling relation for v shows that it is of power type: $v(x) = v(1)x^\gamma$, hence of the same form as in the Merton model (see Figure 5.1). The scaling relation for $\hat{v}$ implies that for all $\beta > 0$,

$$\hat{a} \in \arg\max_{a\in[0,x]} \hat{v}(0, x, a) \quad \text{if and only if} \quad \beta a \in \arg\max_{a\in[0,\beta x]} \hat{v}(0, \beta x, a).$$

From the feedback form (4.4), this shows that α_k^* is linear in $X_{\tau_k}^x$, or in other words the optimal investmet strategy consists in investing a fixed proportion of the wealth into the risky asset. Moreover, we can reduce the dimension of the problem and denote by

$$v(x) = \vartheta_1 x^\gamma, \quad \hat{v}(t, x, a) = a^\gamma \bar{v}(t, \xi), \quad \xi = x/a,$$

where ϑ_1 and $\bar{v}$ are solution to

$$(\rho+\lambda)\bar{v} - \frac{\partial \bar{v}}{\partial t} - \tilde{U}\Big(\frac{\partial \bar{v}}{\partial \xi}\Big) - \lambda\vartheta_1 \int (\xi+z)^\gamma p(t, dz) = 0,$$
$$\vartheta_1 = \sup_{\xi\ge 1} \xi^{-\gamma}\bar{v}(0, \xi),$$

In the sequel, for the numerical experiments, we consider a power utility function with $\gamma = 0.5$. We choose parameters for which $\frac{b}{(1-\gamma)\sigma^2} < 1$, and such that $K_{_M}$ is substantially different from K_0. These two requirements on the model parameters

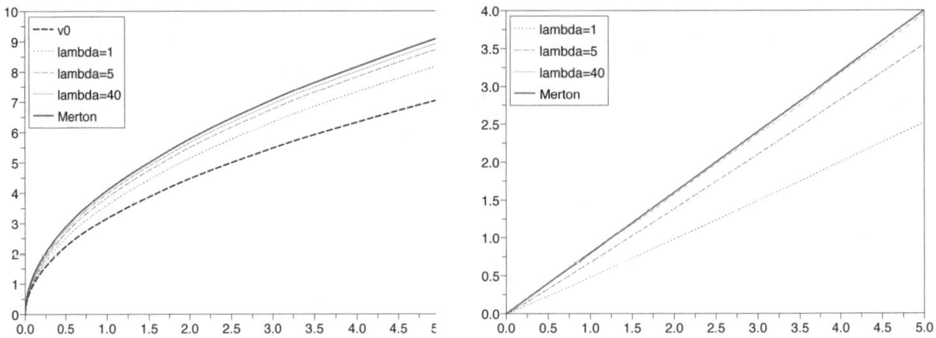

Figure 5.1. Behavior of the value function in an illiquid market (left) and of the optimal investment policy (right) for different values of the Poisson parameter λ.

correspond to a high-risk return market, where the economic agent can considerably increase her utility with relatively little investment. In addition, the discount factor ρ must satisfy $\rho > b\gamma$. To satisfy all these conditions, we take $b = 0.4$, $\sigma = 1$ and $\rho = 0.2$, yielding $K_0 = 3.16$, $K_M = 4.08$ and $\frac{b}{(1-\gamma)\sigma^2} = 0.8$. The intensity λ is a free parameter that can be changed to adjust the "illiquidity" of the market.

A first series of tests computed in [11] studied the performance of the decoupling algorithm in a strongly illiquid market ($\lambda = 1$). In Figure 5.2, the left graph shows the form of the value function and the right graph that of the optimal investment strategy obtained at different iterations of the numerical decoupling algorithm. As expected, the limiting value function lies between the solution corresponding to the model without trading v_0 and the value function of the Merton problem v_M.

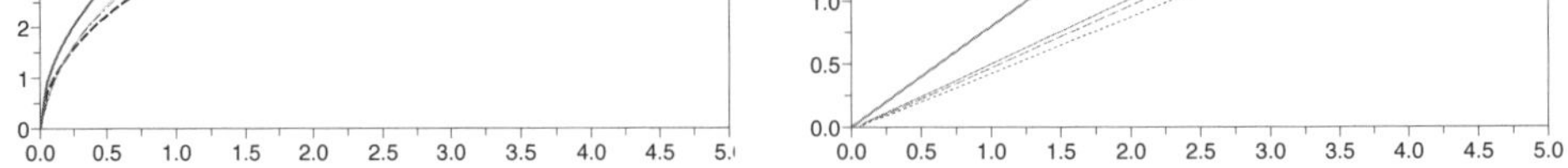

Figure 5.2. Left : Convergence of the iterative algorithm for computing the value function in an illiquid market with $\lambda = 1$. Right : Convergence of the iterative algorithm for computing the optimal investment policy (the amount to invest in stock as a function of the total wealth at the trading date).

In the second experiment, we vary the Poisson parameter λ governing the trading frequency, to study the convergence of the illiquid market to the Merton portfolio problem. Figure 5.1 presents the behaviour of the value functions $v(x)$ and the associated optimal trading strategies. From these graphs we observe, empirically, that

(i) for a fixed value of x, both the value function and the optimal investment policy are increasing in λ and

(ii) as $\lambda \to \infty$, the value function and the optimal investment policy seem to converge to the corresponding functions in the Merton portfolio problem.

Next, we would like to study the utility loss due to liquidity constraints. Following the utility-indifference pricing approach introduced in [7], we define the utility loss in monetary terms (which can also be called cost of liquidity) as the extra amount of initial wealth $\pi(x)$ needed to reach the same level of expected utility as an investor without trading restrictions and initial capital x. This cost of liquidity is then computed as the solution to $v(x + \pi(x)) = v_M(x)$. In our setting (power utility), the cost of liquidity $\pi(x)$ is roughly proportional to x. We therefore study the cost of liquidity per unit of initial wealth $\pi(1)$. Table 5.1 reproduces the values $\pi(1)$ for different values of the Poisson parameter λ. As expected, the cost of liquidity decreases to zero as $\lambda \to \infty$.

λ	0 (No trading)	1	5	40
$\pi(1)$	0.6671	0.2749	0.1214	0.0539

Table 5.1. Cost of liquidity $\pi(1)$ as a function of the parameter λ.

Bibliography

[1] Almgren R. and N. Criss (2001): Optimal execution of portfolio transaction, *Journal of Risk*, **3**, 5–39.

[2] Bank P. and D. Baum (2004): Hedging and portfolio optimisation in illiquid financial markets with a large trader, *Mathematical Finance*, **14**, 1–18.

[3] Cetin U., Jarrow R. and P. Protter (2004): Liquidity risk and arbitrage pricing theory, *Finance and Stochastics*, **8**, 311–341.

[4] Cretarola A., Gozzi F., Pham H. and P. Tankov (2008): Optimal consumption policies in illiquid markets, to appear in *Finance and Stochastics*.

[5] Cvitanic J., Liptser R. and B. Rozovskii (2006): A filtering approach to tracking volatility from prices observed at random times, *Annals of Applied Probability*, **16**, 1633–1652.

[6] Frey R. and W. Runggaldier (2001): A nonlinear filtering approach to volatility estimation with a view towards high frequency data, *International Journal of Theoretical and Applied Finance*, **4**, 199–210.

[7] Hodges S. and A. Neuberger (1989): Optimal replication of contingent claims under transaction costs, *Review of Futures Markets*, **8**, 222–239.

[8] Longstaff F. (2005): Asset pricing in markets with illiquid assets, Preprint UCLA.

[9] Matsumoto K. (2006): Optimal portfolio of low liquid assets with a log-utility function, *Finance and Stochastics*, **10**, 121–145.

[10] Merton R. (1971) : Optimum consumption and portfolio rules in a continuous-time model, *Journal of Economic Theory*, **3**, 373–413.

[11] Pham H. and Tankov P. (2008): A Model of Optimal Consumption under Liquidity Risk with Random Trading Time, *Mathematical Finance*, **18** (4), 613–627.

[12] Pham H. and Tankov P. (2007): A Coupled System of Integrodifferential Equations Arising in Liquidity Risk Model, *Applied Mathematics and Optimisation*, **59** (2), 147–173.

[13] Rogers C. and O. Zane (2002): A simple model of liquidity effects, in *Advances in Finance and Stochastics: Essays in Honour of Dieter Sondermann*, eds. K. Sandmann and P. Schoenbucher, pp 161–176.

[14] Schied A. and T. Schöneborn (2009): Risk aversion and the dynamics of optimal liquidation strategies in illiquid markets, *Finance and Stochastics*, **13**, 181–204.

[15] Schwartz E. and C. Tebaldi (2004): Illiquid assets and optimal portfolio choice, Preprint UCLA.

Author information

Huyên Pham, Laboratoire de Probabilités et Modèles Aléatoires, CNRS, UMR 7599, Université Paris 7, Crest and Institut Universitaire de France, France.
Email: pham@math.jussieu.fr

Radon Series Comp. Appl. Math 8, 427–453

Optimal asset allocation in a stochastic factor model – an overview and open problems

Thaleia Zariphopoulou

Abstract. This paper provides an overview of the optimal investment problem in a market in which the dynamics of the risky security are affected by a correlated stochastic factor. The performance of investment strategies is measured using two criteria. The first criterion is the traditional one, formulated in terms of expected utility from terminal wealth while the second is based on the recently developed forward investment performance approach.

Key words. Excess hedging demand, forward performance process, Hamilton–Jacobi–Bellman equation, market participation puzzle, myopic portfolio, utility maximisation.

AMS classification. 91B16, 91B28

1 Introduction

The aim herein is to present an overview of results and open problems arising in optimal investment models in which the dynamics of the underlying stock depend on a correlated stochastic factor. Stochastic factors have been used in a number of academic papers to model the time-varying predictability of stock returns, the volatility of stocks as well as stochastic interest rates (see, for example, [1], [15], [42] and other references discussed in the next section). The performance of the investment decisions is, typically, measured via an expected utility criterion which is often formulated in a finite trading horizon.

From the technical point of view, a stochastic factor model is the simplest and most direct extension of the celebrated Merton model ([66] and [67]), in which stock dynamics are taken to be lognormal. However, as it is discussed herein, very little is known about the maximal expected utility as well as the form and properties of the optimal policies once the lognormality assumption is relaxed and correlation between the stock and the factor is introduced. This is despite the Markovian nature of the problem at hand, the advances in the theories of fully nonlinear pdes and stochastic control, and the computational tools that exist today. Specifically, results on the validity of the Dynamic Programming Principle, regularity of the value function, existence and verification of optimal feedback controls, representation of the value function and numerical approximations are still lacking. The only cases that have been extensively analysed are the ones of special utilities, namely, the exponential, power and logarithmic. In these

The author would like to thank the organisers of the special semester on "Stochastics with emphasis on Finance" at RICAM for their hospitality. She would also like to thank S. Malamud, H. Pham, N. Touzi and G. Zitkovic for their fruitful comments. Special thanks go to M. Sirbu for his help and suggestions. This work was partially supported by the National Science Foundation (NSF grants: DMS-FRG-0456118 and DMS-RTG-0636586).

cases, convenient scaling properties reduce the associated Hamilton–Jacobi–Bellman (HJB) equation to a quasilinear one. The analysis, then, simplifies considerably both from the analytic as well as the probabilistic points of view.

The lack of rigorous results for the value function when the utility function is general limits our understanding of the optimal policies. Informally speaking, the first-order conditions in the HJB equation yield that the optimal feedback portfolio consists of two components. The first is the so-called myopic portfolio and has the same functional form as the one in the classical Merton problem. The second component, usually referred to as the excess hedging demand, is generated by the stochastic factor. Conceptually, very little is understood about this term. In addition, the sum of the two components may become zero which implies that it is optimal for a risk averse investor not to invest in a risky asset with positive risk premium. A satisfactory explanation for this counter intuitive phenomenon – related to the so-called market participation puzzle – is also lacking.

Besides these difficulties, there are other issues that limit the development of an optimal investment theory in complex market environments. One of them is the "static" choice of the utility function at the specific investment horizon. Indeed, once the utility function is chosen, no revision of risk preferences is possible at any earlier trading time. In addition, once the horizon is chosen, no investment performance criteria can be formulated for horizons longer than the initial one. These limitations have been partly addressed by allowing infinite horizon, long-term growth criteria, random horizon, recursivity and others.

Herein, we discuss a new approach that complements the existing ones. The alternative criterion has the same fundamental requirements as the classical value function process but allows for both revision of preferences and arbitrary trading horizons. It is given by a stochastic process, called the forward investment performance, defined for all times. A stochastic partial differential equation emerges which is the "forward" analogue of the HJB equation. The key new element is the performance volatility process which, in contrast to the classical formulation, is not a priori given.

The special case of zero-volatility deserves special attention as it yields useful insights for the optimal portfolios. It turns out that for this class of risk preferences, the non-myopic component always disappears, independently of the dynamics of the stochastic factor. This result might give an answer to the market participation puzzle mentioned earlier. In addition, closed form solutions can be found for the performance process as well as the associated optimal wealth and portfolio processes for general preferences and arbitrary factor dynamics.

Two classes of non-zero volatility processes and their associated optimal portfolios are, also, discussed. While from the technical point of view these cases reduce to the zero-volatility case, they provide useful results on the structure of optimal investments when the investor has alternative views for the upcoming market movements or wishes to measure performance in reference to a different numeraire/benchmark.

We finish this section mentioning that there is a very rich body of research for the analysis of the classical expected utility models based on duality techniques. This powerful approach is applicable to general market models and yields elegant results for the value function and the optimal wealth. The optimal portfolios can be then characterised via martingale representation results for the optimal wealth process (see, among others, [48], [57], [58], [80] and [81]). However, little can be said about the structure and properties of the optimal investments. Because of their volume as well as their different nature and focus, these results are not discussed herein.

The paper is organised as follows. In section 2, we present the market model. In section 3, we discuss the existing results in the classical (backward) formulation. We present some examples and state some open problems. In section 4, we present the alternative (forward) investment performance criterion and analyse, in some detail, the zero-volatility case. We also present the non-zero volatility cases, concrete examples and some open problems.

2 The model

The market consists of a risky and a riskless asset. The risky asset is a stock whose price S_t, $t \geq 0$, is modelled as a diffusion process solving

$$dS_t = \mu(Y_t) S_t dt + \sigma(Y_t) S_t dW_t^1, \tag{2.1}$$

with $S_0 > 0$.The stochastic factor Y_t, $t \geq 0$, satisfies

$$dY_t = b(Y_t) dt + d(Y_t) \left(\rho dW_t^1 + \sqrt{1-\rho^2} dW_t^2\right), \tag{2.2}$$

with $Y_0 = y$, $y \in \mathbb{R}$. The process $W_t = \left(W_t^1, W_t^2\right)$, $t \geq 0$, is a standard $2-$dimensional Brownian motion, defined on a filtered probability space $(\Omega, \mathcal{F}, \mathbb{P})$. The underlying filtration is $\mathcal{F}_t = \sigma(W_s : 0 \leq s \leq t)$. It is assumed that $\rho \in (-1, 1)$.

The market coefficients $f = \mu, \sigma, b$ and d satisfy the global Lipschitz and linear growth conditions

$$|f(y) - f(\bar{y})| \leq K |y - \bar{y}| \quad \text{and} \quad f^2(y) \leq K\left(1 + y^2\right), \tag{2.3}$$

for $y, \bar{y} \in \mathbb{R}$. Moreover, it is assumed that the non degeneracy condition $\sigma(y) \geq l > 0$, $y \in \mathbb{R}$, holds.

The riskless asset, the savings account, offers constant interest rate $r > 0$.

We introduce the process

$$\lambda(Y_t) = \frac{\mu(Y_t) - r}{\sigma(Y_t)}. \tag{2.4}$$

We will occasionally refer to it as the market price of risk.

Starting with an initial endowment x, the investor invests at future times in the riskless and risky assets. The present value of the amounts allocated in the two accounts are denoted, respectively, by π_t^0 and π_t. The present value of her investment is, then, given by $X_t^\pi = \pi_t^0 + \pi_t$, $t > 0$. We will refer to X_t^π as the discounted wealth. Using (2.1) we easily deduce that it satisfies

$$dX_t^\pi = \sigma(Y_t) \pi_t \left(\lambda(Y_t) dt + dW_t^1\right). \tag{2.5}$$

The investment strategies will play the role of control processes and are taken to satisfy the standard assumption of being self-financing. Such a portfolio, π_t, is deemed admissible if, for $t > 0$, $\pi_t \in \mathcal{F}_t$, $E_{\mathbb{P}}\big(\int_0^t \sigma^2(Y_s)\,\pi_s^2 ds\big) < \infty$ and the associated discounted wealth satisfies the state constraint $X_t^\pi \in \mathbb{D}$, $t \geq 0$, for some acceptability domain $\mathbb{D} \subseteq \mathbb{R}$. We will denote the set of admissible strategies by $\mathcal{A}$.

The form of the spatial domain $\mathbb{D}$ and the consequences of this choice to the structure of the optimal portfolios are subjects of independent interest and will not be discussed herein. Frequently, portfolio constraints are also present which complicate the analysis further. For the model at hand, we will not allow for such generality as the focus is mainly on the choice and impact of risk preferences on investment decisions. To ease the notation, however, we will carry out the $\mathbb{D}-$notation and make it more specific when appropriate.

Stochastic factors have been used in portfolio choice to model asset predictability and stochastic volatility. The predictability of stock returns was first discussed in [34], [35] and [38]; see also [13], [14], [17] and [18]. More complex models were analysed in [1] and [12]. The role of stochastic volatility in investment decisions was studied in [3], [22], [38], [39], [42], [76], [84] and others. Models that combine predictability and stochastic volatility, as the one herein, were analysed, among others, in [51], [56], [64], [77] and [93].

In a different modelling direction, stochastic factors have been incorporated in asset allocation models with stochastic interest rates (see, for example, [15], [16], [19], [24], [25], [28], [29], [79] and [89]). From the technical point of view, the analysis is not much different as long as the model remains Markovian. However, various technically interesting questions arise (see, for example, [54], [56] and [87]).

3 The backward formulation

The traditional criterion for optimal portfolio choice has been based on maximal expected utility[1] (see, for example, [66] and [67]). The key ingredients are the choices of the trading horizon, $[0, T]$, and the investor's utility, u_T, at terminal time T. The utility function reflects the risk attitude of the investor at time T and is an increasing and concave function of his wealth[2]. It is important to observe that once these choices are made, the risk preferences cannot be revised. In addition, no investment decisions can be assessed for times beyond T.

The objective is to maximise the expected utility of terminal wealth over the set of admissible strategies. The solution, known as the value function, is defined as

$$V(x, y, t; T) = \sup_{\mathcal{A}} E_{\mathbb{P}}\left(u_T(X_T)\right| X_t = x, Y_t = y), \tag{3.1}$$

for $(x, y, t) \in \mathbb{D} \times \mathbb{R} \times [0, T]$ and $\mathcal{A}$ being the set of admissible strategies. For conditions on the asymptotic behaviour of u_T in infinite and semi-infinite domains see [80] and [81].

[1] See, for example, the review article [96].

[2] The quadratic utility represents an exception as it is not globally increasing. This utility, albeit popular for tractability reasons, yields non intuitive optimal portfolios and is not discussed herein.

As solution of a stochastic optimisation problem, the value function is expected to satisfy the Dynamic Programming Principle (DPP), namely,

$$V(x,y,t;T) = \sup_{\mathcal{A}} E_{\mathbb{P}}\left(V(X_s,Y_s,s;T)\right| X_t = x, Y_t = y), \tag{3.2}$$

for $t \leq s \leq T$. This is a fundamental result in optimal control and has been proved for a wide class of optimisation problems. For a detailed discussion on the validity (and strongest forms) of the DPP in problems with controlled diffusions, we refer the reader to [37] (see, also, [8], [32], [60] and [62]). Key issues are the measurability and continuity of the value function process as well as the compactness of the set of admissible controls. It is worth mentioning that a proof specific to the problem at hand has not been produced to date. Recently, a weak version of the DPP was proposed in [11] where conditions related to measurable selection and boundness of controls are relaxed.

Besides its technical challenges, the DPP exhibits two important properties of the value function process. Specifically, $V(x,Y_s,s;T)$, $s \in [t,T]$, is a supermartingale for an arbitrary investment strategy and becomes a martingale at an optimum (provided certain integrability conditions hold). One may, then, view $V(x,Y_s,s;T)$ as the intermediate (indirect) utility in the relevant market environment. It is worth noticing, however, that the notions of utility and risk aversion for times $t \in [0,T)$ are tightly connected to the investment opportunities the investor has in the specific market. Observe that the DPP yields a backward in time algorithm for the computation of the maximal utility, starting at expiration with u_T and using the martingality property to compute the solution for earlier times. For this, we refer to this formulation of the optimal portfolio choice problem as *backward.*

The Markovian assumptions on the stock price and stochastic factor dynamics allow us to study the value function via the associated HJB equation, stated in (3.3) below. Fundamental results in the theory of controlled diffusions yield that if the value function is smooth enough then it satisfies the HJB equation. Moreover, optimal policies may be constructed in a feedback form from the first-order conditions in the HJB equation, provided that the candidate feedback process is admissible and the wealth SDE has a strong solution when the candidate control is used. The latter usually requires further regularity on the value function. In the reverse direction, a smooth solution of the HJB equation that satisfies the appropriate terminal and boundary (or growth) conditions may be identified with the value function, provided the solution is unique in the appropriate sense. These results are usually known as the "verification theorem" and we refer the reader to [37], [60] and [92] for a general exposition on the subject.

In maximal expected utility problems, it is rarely the case that the arguments in either direction of the verification theorem can be established. Indeed, it is very difficult to show a priori regularity of the value function, with the main difficulties coming from the lack of global Lipschitz regularity of the coefficients of the controlled process with respect to the controls and from the non-compactness of the set of admissible policies. It is, also, very difficult to establish existence, uniqueness and regularity of the solutions to the HJB equation. This is caused primarily by the presence of the control policy in the volatility of the controlled wealth process which makes the classical assumptions of global Lipschitz conditions of the equation with regards to the non linearities fail.

Additional difficulties come from state constraints and the non-compactness of the admissible set.

To our knowledge, regularity results for the value function (3.1) for general utility functions have not been obtained to date except for the special cases of homothetic preferences (see, for example, [36], [56], [68], [77] and [93]). The most general result in this direction, and in a much more general market model, was recently obtained in [59] where it is shown that the value function is twice differentiable in the spatial argument but without establishing its continuity.

Because of lack of general rigorous results, we proceed with an informal discussion about the optimal feedback policies.

For the model at hand, the associated HJB equation turns out to be

$$\begin{aligned} V_t &+ \max_{\pi}\left(\frac{1}{2}\sigma^2(y)\,\pi^2 V_{xx} + \pi\left(\mu(y)\,V_x + \rho\sigma(y)\,d(y)\,V_{xy}\right)\right) \qquad (3.3)\\ &+ \frac{1}{2}d^2(y)\,V_{yy} + b(y)\,V_y = 0\,, \end{aligned}$$

with $V(x,y,T;T) = u_T(x)\,,\,(x,y,t)\in\mathbb{D}\times\mathbb{R}\times[0,T]\,.$

The verification results would yield that under appropriate regularity and growth conditions, the feedback policy

$$\pi_s^* = \pi^*\left(X_s^*, Y_s, s; T\right), \quad t\le s\le T,$$

with $\pi^* : \mathbb{D}\times\mathbb{R}\times[0,T]$ given by

$$\pi^*(x,y,t;T) = -\frac{\lambda(y)}{\sigma(y)}\frac{V_x(x,y,t;T)}{V_{xx}(x,y,t;T)} - \rho\frac{d(y)}{\sigma(y)}\frac{V_{xy}(x,y,t;T)}{V_{xx}(x,y,t;T)} \qquad (3.4)$$

and X_s^*, $t\le s\le T$, solving

$$dX_s^* = \sigma(Y_s)\,\pi\left(X_s^*, Y_s, s; T\right)\left(\lambda(Y_s)\,ds + dW_s^1\right), \qquad (3.5)$$

is admissible and optimal.

Some answers to the questions related to the characterisation of the solutions to the HJB equation may be given if one relaxes the requirement to have classical solutions. An appropriate class of weak solutions turns out to be the so called viscosity solutions ([26], [62], [63] and [88]). The analysis and characterisation of the value function in the viscosity sense has been carried out for the special cases of power and exponential utility (see, for example, [93]). However, proving that the value function is the unique viscosity solution of (3.3) has not been addressed.

A key property of viscosity solutions is their robustness (see [63]). If the HJB has a unique viscosity solution (in the appropriate class), robustness is used to establish convergence of numerical schemes for the value function and the optimal feedback laws. Such numerical studies have been carried out successfully for a number of applications. However, for the model at hand, no such studies are available. Numerical results using Monte Carlo techniques have been obtained in [30] for a model more general than the one herein.

Besides the technically challenging issues that problem (3.1) gives rise to, there is a number of very interesting questions on the economic properties of the optimal portfolios. From (3.4) one sees that the optimal feedback portfolio functional consists of two terms, namely,

$$\pi^{*,m}(x,y,t;T) = -\frac{\lambda(y)}{\sigma(y)}\frac{V_x(x,y,t;T)}{V_{xx}(x,y,t;T)} \tag{3.6}$$

and

$$\pi^{*,h}(x,y,t;T) = -\rho\frac{d(y)}{\sigma(y)}\frac{V_{xy}(x,y,t;T)}{V_{xx}(x,y,t;T)}. \tag{3.7}$$

The first component, $\pi^{*,m}(x,y,t;T)$, is known as the *myopic* investment strategy. It corresponds functionally to the investment policy followed by an investor in markets in which the investment opportunity set remains constant through time. The myopic portfolio is always positive for a nonzero market price of risk.

The second term, $\pi^{*,h}(x,y,t;T)$, is called the *excess hedging demand.* It represents the additional investment caused by the presence of the stochastic factor. It does not have a constant sign, for the signs of the correlation coefficient ρ and the mixed derivative V_{xy} are not definite. The excess risky demand vanishes in the uncorrelated case, $\rho = 0$, and when the volatility of the stochastic factor process is zero, $d(y) = 0$, $y \in \mathbb{R}$. In the latter case, using a simple deterministic time-rescaling argument reduces the problem to the classical Merton one. Finally, $\pi^{*,h}(x,y,t;T)$ vanishes for the case of logarithmic utility (see (3.8)). Despite the nomenclature "hedging demand", a rigorous study for the precise characterisation and quantification of the risk that is not hedged has not been carried out. Indeed, in contrast to derivative valuation where the notion of imperfect hedge is well defined, such a notion has not been established in the area of investments (see [85] for a special case).

The total allocation in the risky asset might become zero even if the risk premium is not zero. This phenomenon, related to the so called market participation puzzle, appears at first counter intuitive, for classical economic ideas suggest that a risk averse investor should always retain nonzero holdings in an asset that offers positive risk premium. We refer the reader to, among others, [4], [20] and [43].

Important questions arise on the dependence, sensitivity and robustness of the optimal feedback portfolio in terms of the market parameters, the wealth, the level of the stochastic factor and the risk preferences. Such questions are central in financial economics and have been studied, primarily in simpler models in which intermediate consumption is also incorporated (see, among others, [2], [52], [61], [75] and [78]). For diffusion models with and without a stochastic factor qualitative results can be found in [30], [51], [53], [64], [90] and, recently, in [9] (see, also, [65] for a general incomplete market discrete model). However, a qualitative study for general utility functions and/or arbitrary factor dynamics has not been carried out to date.

Some open problems

Problem 1: What are the weakest conditions on the market coefficients and the utility function so that the Dynamic Programming Principle holds?

Problem 2: What are the weakest conditions on the market coefficients and the utility function so that existence and uniqueness of viscosity solutions to the HJB equation hold?

Problem 3: Study the regularity of the value function and establish the associated verification theorem.

Problem 4: Develop numerical schemes for the value function and the optimal feedback policies for general utility functions.

Problem 5: Study the behaviour of the optimal portfolio in terms of market inputs, the horizon length and risk preferences for general utility functions and arbitrary stochastic factor dynamics. Compute and analyse the distribution of the optimal wealth and portfolio processes as well as their moments.

3.1 The CARA, CRRA and logarithmic cases

We provide examples for the most frequently used utilities, namely, the exponential, power and logarithmic ones. They have convenient homogeneity properties which, in combination with the linearity of the wealth dynamics in the control policies, enable us to reduce the HJB equation to a quasilinear one. Under a "distortion" transformation (see, for example, [93]) the latter can be linearised and solutions in closed form can be produced using the Feynman–Kac formula. The smoothness of the value function and, in turn, the verification of the optimal feedback policies follows easily.

Multi-factor models for these preferences have been analysed by various authors. The theory of BSDE has been successfully used to characterise and represent the solutions of the reduced HJB equation (see [33]). The regularity of its solutions has been studied using PDE arguments by [77] and [68], for power and exponential utilities, respectively. Finally, explicit solutions for a three factor model can be found in [64].

Exponential case: We have $u_T(x) = -e^{-\gamma x}$, $x \in \mathbb{R}$ and $\gamma > 0$.

This case has been extensively studied not only in optimal investment models but, also, in indifference pricing where valuation is done primarily under exponential preferences (see [21] for a concise collection of relevant references). The value function is multiplicatively separable and given, for $(x, y, t) \in \mathbb{R} \times \mathbb{R} \times [0, T]$, by

$$V(x, y, t; T) = -e^{-\gamma x} h(y, t; T)^{\delta}, \qquad \delta = \frac{1}{1-\rho^2},$$

where $h : \mathbb{R} \times [0, T] \to \mathbb{R}$ solves

$$h_t + \frac{1}{2}\sigma^2(y) h_{yy} + \left(b(y) - \rho \frac{d(y)}{\sigma(y)} \right) h_y = \frac{1}{2}\left(1-\rho^2\right)\lambda^2(y) h,$$

with $h(x, y, T; T) = 1$. The optimal feedback investment strategy is independent of the wealth level and given by

$$\pi^*(x, y, t; T) = \frac{\lambda(y)}{\sigma^2(y)} + \frac{\rho}{1-\rho^2} \frac{d(y)}{\sigma(y)} \frac{h_y(y, t; T)}{h(y, t; T)}.$$

The optimal wealth and portfolio processes follow directly from (3.4) and (3.5). Namely, for $t \leq s \leq T$,

$$\pi_s^* = \pi^*(x, Y_s, s; T) = \frac{\lambda(Y_s)}{\sigma^2(Y_s)} + \frac{\rho}{1-\rho^2} \frac{d(Y_s)}{\sigma(Y_s)} \frac{h_y(Y_s, s; T)}{h(Y_s, s; T)}$$

and

$$X_s^* = x + \int_t^s \sigma(Y_u)\lambda(Y_u)\pi_u^* du + \int_t^s \sigma(Y_u)\pi_u^* dW_u^1.$$

A well-known criticism of the exponential utility is that the optimal portfolio does not depend on the investor's wealth. While this property might be desirable in asset equilibrium pricing, it appears to be problematic and counter intuitive for investment problems. We note, however, that this property is directly related to the choice of the savings account as the numeraire. If the benchmark changes, the optimal portfolio ceases to be independent of wealth (see (4.44)).

The next two utilities are defined on the half-line and the stochastic optimisation problem is a state-constraint one. We easily deduce from the form of the optimal portfolios that the non-negativity wealth constraint is always satisfied.

Power case: We have $u_T(x) = \frac{1}{\gamma}x^\gamma,\ \ 0 < \gamma < 1, \gamma \neq 0.$

The value function is multiplicatively separable and given, for $(x, y, t) \in \mathbb{R}^+ \times \mathbb{R} \times [0, T]$, by

$$V(x, y, t; T) = \frac{1}{\gamma} x^\gamma f(y, t; T)^\delta, \qquad \delta = \frac{1-\gamma}{1-\gamma+\rho^2\gamma},$$

where $f : \mathbb{R} \times [0, T] \to \mathbb{R}^+$ solves the linear parabolic equation

$$f_t + \frac{1}{2}d^2(y) f_{yy} + \left(b(y) + \rho\frac{\gamma}{1-\gamma}\lambda(y) d(y)\right) f_y + \frac{\gamma}{2(1-\gamma)} \frac{\lambda^2(y)}{\delta} f = 0,$$

with $f(x, y, T; T) = 1$. The optimal policy feedback function is linear in wealth,

$$\pi^*(x, y, t; T) = \frac{1}{1-\gamma} \frac{\lambda(y)}{\sigma(y)} x + \frac{\rho}{(1-\gamma)+\rho^2\gamma} \frac{d(y)}{\sigma(y)} \frac{f_y(y, t; T)}{f(y, t; T)} x.$$

The optimal investment and wealth processes are, in turn, given by

$$\pi_s^* = m_s X_s^*$$

and

$$X_s^* = x \exp\left(\int_t^s \left(\sigma^2(Y_u)\lambda(Y_u) m_u - \frac{1}{2}\sigma^2(Y_u) m_u^2\right) du + \int_t^s \sigma(Y_u) m_u dW_u^1\right),$$

with

$$m_s = \frac{1}{1-\gamma} \frac{\lambda(Y_s)}{\sigma(Y_s)} + \frac{\rho}{(1-\gamma)+\rho^2\gamma} \frac{d(Y_s)}{\sigma(Y_s)} \frac{f_y(Y_s, s; T)}{f(Y_s, s; T)}.$$

The range of the risk aversion parameter can be relaxed to include negative values. Its choice plays an important role in the boundary and asymptotic behaviour of the value function as well as the long-term behaviour of the optimal wealth and portfolio processes (see [51] and [64]). Verification results for weak conditions on the risk premium can be found, among others, in [55] and [56].

Logarithmic utility: We have $u_T(x) = \ln x$, $x > 0$.
The value function is additively separable, namely,

$$V(x, y, t; T) = \ln x + h(y, t; T),$$

with $h : \mathbb{R}\times [0, T] \to \mathbb{R}^+$ solving

$$h_t + \frac{1}{2} d^2(y) h_{yy} + b(y) h_y + \frac{1}{2}\lambda^2(y) h = 0$$

and $h(y, T; T) = 1$. The optimal portfolio takes the simple linear form

$$\pi^*(x, y, t; T) = \frac{\lambda(y)}{\sigma(y)} x. \tag{3.8}$$

In turn, the optimal investment and wealth processes are given, for $t \le s \le T$, by

$$\pi_s^* = \frac{\lambda(Y_s)}{\sigma(Y_s)} X_s^* \quad \text{and} \quad X_s^* = x \exp\left(\int_t^s \frac{1}{2}\lambda^2(Y_u)\, du + \int_t^s \lambda(Y_u)\, dW_u^1\right).$$

The logarithmic utility plays a special role in portfolio choice. Because of the additively separable form of the value function, the optimal portfolio is always myopic. It is known as the "growth optimal portfolio" and has been extensively studied in general market settings (see, for example, [6] and [50]). The associated optimal wealth is the so-called "numeraire portfolio". It has also been extensively studied, for it is the numeraire with regards to which all wealth processes are supermartingales under the historical measure (see, among others, [40] and [41]).

4 The forward formulation

As discussed in the previous section, the main feature of the expected utility approach is the a priori choice of the utility at the end of the trading horizon. Direct consequences of this choice are, from one hand, the lack of flexibility to revise the risk preferences at other times and, from the other, the inability to assess the performance of investment strategies beyond the prespecified horizon.

Addressing these limitations has been the subject of a number of studies and various approaches have been proposed. With regards to the horizon length, the most popular alternative has been the formulation of the investment problem in $[0, +\infty)$ and incorporating either intermediate consumption or optimising the investor's long-term optimal behaviour (see, among others, [47], [48] and [86]). Investment models with random horizon have also been examined ([23]). The revision of risk preferences has been partially addressed by recursive utilities (see, for example, [31], [82] and [83]).

Next, we present another alternative approach which addresses both shortcomings of the expected utility approach. The associated criterion is developed in terms of a family of stochastic processes defined on $[0, \infty)$ and indexed by the wealth argument. It will be called *forward performance process*. Its key properties are the martingality at an optimum and supermartingality away from it. These are in accordance with the analogous properties of the value function process that stem out from the Dynamic Programming Principle (cf. (3.2)). However, in contrast to the existing framework, the risk preferences are specified for today[1] and not for a (possibly remote) future time.

We recall that $\mathcal{F}_t$, $t \geq 0$, is the filtration generated by $W_t = \left(W_t^1, W_t^2\right), t \geq 0$, and $\mathcal{A}$ the set of admissible policies. As in the previous section, we use $\mathbb{D}$ to denote the generic admissible space domain.

Definition 4.1. An $\mathcal{F}_t$–adapted process $U(x,t)$ is a forward performance if for $t \geq 0$ and $x \in \mathbb{D}$:

i) the mapping $x \to U(x,t)$ is concave and increasing,

ii) for each portfolio process $\pi \in \mathcal{A}$, $E_{\mathbb{P}}\left(U\left(X_t^\pi, t\right)\right)^+ < \infty$, and

$$E_{\mathbb{P}}\left(U\left(X_s^\pi, s\right)|\mathcal{F}_t\right) \leq U\left(X_t^\pi, t\right), \qquad s \geq t, \tag{4.1}$$

iii) there exists a portfolio process $\pi^* \in \mathcal{A}$, for which

$$E_{\mathbb{P}}\Big(U\big(X_s^{\pi^*}, s\big)|\mathcal{F}_t\Big) = U\left(X_t^{\pi^*}, t\right), \qquad s \geq t, \tag{4.2}$$

and

iv) at $t = 0$, $U(x,0) = u_0(x)$, where $u_0 : \mathbb{D} \to \mathbb{R}$ is increasing and concave.

The concept of forward performance process was introduced in [69] (see, also, [70]). The model therein is incomplete binomial and the initial data is taken to be exponential. The exponential case was subsequently and extensively analysed in [71] and [95].

Ideas related to the forward approach can also be found in [23] where the authors consider random horizon choices, aiming at alleviating the dependence of the value function on a fixed deterministic horizon. Their model is more general in terms of the assumptions on the price dynamics but the focus in [23] is primarily on horizon effects. Horizon issues were also considered in [44] for the special case of lognormal stock dynamics.

It is worth observing the following differences and similarities between the forward performance process and the traditional value function. Namely, the process $U(x,t)$ is defined for all $t \geq 0$, while the value function $V(x,y,t;T)$, is defined only on $[0,T]$. In the classical set up discussed in the previous section, $V(x,y,T;T) \in \mathcal{F}_0$, due to the deterministic choice of the terminal utility u_T. If the terminal utility is taken to be

[1] The choice of the initial condition gives rise to interesting mathematical and modelling questions (see, for example, [73] and references therein).

state-dependent, $V(x,y,T;T) \in \mathcal{F}_T$, (see, for example, [49], [81] as well as [10], [27] and [46]), the traditional and new formulations are, essentially, identical in $[0,T]$.

Recently, it was shown in [74] that a sufficient condition for a process $U(x,t)$ to be a forward performance is that it satisfies a stochastic partial differential equation (see (4.5) below). For completeness, we state the result for a general incomplete market model with k risky stocks whose prices are modelled as Ito processes driven by a d-dimensional Brownian motion. We use σ_t, $t \geq 0$, to denote their $d \times k$ random volatility matrix and μ_t the k-dim vector with coordinates the mean rate of return of each stock. It is assumed that the volatility vectors are such that $\mu_t - r_t\mathbf{1} \in Lin\left(\sigma_t^T\right)$, where $Lin\left(\sigma_t^T\right)$ denotes the linear space generated by the columns of σ_t^T. This implies that $\sigma_t^T\left(\sigma_t^T\right)^+ (\mu_t - r_t\mathbf{1}) = \mu_t - r_t\mathbf{1}$ and, therefore, the market price of risk vector

$$\lambda_t = \left(\sigma_t^T\right)^+ (\mu_t - r_t\mathbf{1}) \tag{4.3}$$

is a solution to the equation $\sigma_t^T x = \mu_t - r_t\mathbf{1}$. The matrix $\left(\sigma_t^T\right)^+$ is the Moore–Penrose pseudo-inverse of the matrix σ_t^T. It easily follows that, for $t \geq 0$,

$$\sigma_t \sigma_t^+ \lambda_t = \lambda_t. \tag{4.4}$$

It is assumed from now on that there exists a deterministic constant $c \geq 0$ such that, for $t \geq 0$, $\lambda(Y_t) \leq c$.

Proposition 4.2. *Let $U(x,t) \in \mathcal{F}_t$ be such that the mapping $x \to U(x,t)$ is increasing and concave. Let, also, $U(x,t)$ be a solution to the stochastic partial differential equation*

$$dU(x,t) = \frac{1}{2}\frac{\left|U_x(x,t)\lambda_t + \sigma_t\sigma_t^+ a_x(x,t)\right|^2}{U_{xx}(x,t)}dt + a(x,t)\cdot dW_t, \tag{4.5}$$

where $a(x,t) \in \mathcal{F}_t$. Then $U(x,t)$ is a forward performance process.

It might seem that all Definition 4.1 produces is a criterion that is dynamically consistent across time. Indeed, internal consistency is an ubiquitous requirement and needs to be ensured in any proposed criterion. It is satisfied, for example, by the traditional value function. However, the new criterion allows for much more flexibility as it is manifested by the volatility process $a(x,t)$ introduced above. Characterising the appropriate class of admissible volatility processes is, in our view, an interesting and challenging question.

The forward performance SPDE (4.5) poses several challenges. It is fully nonlinear and *not* (degenerate) elliptic; the latter is a direct consequence of the "forward in time" nature of the involved stochastic optimisation problem. Thus, existing results of existence, uniqueness and regularity of weak (viscosity) solutions are not directly applicable. An additional difficulty comes from the fact that the volatility coefficient may depend on the second order derivative of U. In such cases, it might not be possible to reduce the SPDE, using the method of stochastic characteristics, into a PDE with random coefficients.

For the model at hand, the coefficients appearing in (4.5) take the form

$$\sigma_t = (\sigma(Y_t), 0)^T, \quad \sigma_t^+ = \left(\frac{1}{\sigma(Y_t)}, 0\right) \quad \text{and} \quad \lambda_t = \left(\frac{\mu(Y_t) - r}{\sigma(Y_t)}, 0\right)^T.$$

We easily see that (4.4) is trivially satisfied.

Proposition 4.3. *i) Let $U(x,t) \in \mathcal{F}_t$ be such that the mapping $x \to U(x,t)$ is increasing and concave. Let, also, $U(x,t)$ be a solution to the stochastic partial differential equation*

$$dU(x,t) = \frac{1}{2} \frac{\left(\lambda(Y_t) U_x(x,t) + a_x^1(x,t)\right)^2}{U_{xx}(x,t)} dt + a^1(x,t)\, dW_t^1 + a^2(x,t)\, dW_t^2,$$

where $a(x,t) = \left(a^1(x,t), a^2(x,t)\right)^T$, with $a^i(x,t) \in \mathcal{F}_t$, $i = 1,2$. Then $U(x,t)$ is a forward performance process.

ii) Let $U(x,t)$ be a solution to the SPDE (4.5) such that, for each $t \geq 0$, the mapping $x \to U(x,t)$ is increasing and concave. Consider the process π_t^*, $t \geq 0$, given by

$$\pi_t^* = -\frac{\lambda(Y_t)}{\sigma(Y_t)} \frac{U_x(X_t^*, t)}{U_{xx}(X_t^*, t)} - \frac{a_x^1(X_t^*, t)}{\sigma(Y_t) U_{xx}(X_t^*, t)} \tag{4.6}$$

where X_t^*, $t \geq 0$, solves

$$dX_t^* = \sigma(Y_t)\, \pi_t^* \left(\lambda(Y_t)\, dt + dW_t^1\right), \tag{4.7}$$

with $X_0^* = x$. If $\pi_t^* \in \mathcal{A}$ and (4.7) has a strong solution, then π_t^* and X_t^* are optimal.

Remark: The same stochastic partial differential equation emerges in the classical formulation of the optimal portfolio choice problem. Indeed, assuming for the moment that the appropriate regularity assumptions hold, expanding the process $V(x, Y_t, t; T)$ (cf. (2.2) and (3.1)), yields,

$$\begin{aligned} dV(x, Y_t, t) = &\left(V_t(x, Y_t, t) + \frac{1}{2} d^2(Y_t) V_{yy}(x, Y_t, t) + b(Y_t) V_y(x, Y_t, t)\right) dt \\ &+ \rho d(Y_t) V_y(x, Y_t, t)\, dW_t^1 + \sqrt{1-\rho^2} d(Y_t) V_y(x, Y_t, t)\, dW_t^2. \end{aligned}$$

Using that $V(x, y, t; T)$ solves the HJB equation and rearranging terms, we deduce that

$$\begin{aligned} dV(x, Y_t, t) = &\frac{1}{2} \frac{\left(\lambda(Y_t) V_x(x, Y_t, t) + \rho d(Y_t) V_{xy}(x, Y_t, t)\right)^2}{V_{xx}(x,t)} dt \\ &+ \rho d(Y_t) V_y(x, Y_t, t)\, dW_t^1 + \sqrt{1-\rho^2} d(Y_t) V_y(x, Y_t, t)\, dW_t^2. \end{aligned}$$

The above SPDE corresponds to the volatility choice, for $0 \leq t < T$,

$$a^1(x,t) = \rho d(Y_t) V_y(x, Y_t, t) \quad \text{and} \quad a^2(x,t) = \sqrt{1-\rho^2} d(Y_t) V_y(x, Y_t, t).$$

Notice that in the backward optimal investment model, there is *no* freedom in choosing the volatility coefficients, for they are uniquely obtained from the Ito decomposition of the value function process.

4.1 The zero volatility case

An important class of forward performance processes are the ones that are *decreasing* in time. They yield an intuitively rich family of performance criteria which compile in a transparent way the dynamic risk profile of the investor and the information coming from the evolution of the investment opportunity set. This section is dedicated to the representation of these processes and the construction of the associated optimal wealth and portfolios. These issues have been extensively studied in [72] and [73], and we refer the reader therein for the proofs of the results that follow.

The *local risk tolerance* function $r(x,t)$, $t \geq 0$, defined below, plays a crucial role in the representation of the optimal investment and wealth processes. It represents the dynamic counterpart of the static risk tolerance function, $r_T(x) = -\frac{u'_T(x)}{u''_T(x)}$. Observe that similarly to its static analogue, it is chosen exogenously to the market. However, now it is time-dependent and solves the autonomous fast diffusion equation (4.12)[2]. The reciprocal of the risk tolerance, the *local risk aversion,* $\gamma = r^{-1}$ solves the porous medium equation (4.13). We recall that $u_0(x)$ is the initial condition of the forward performance process. It is assumed that $u_0 \in \mathcal{C}^4(\mathbb{D})$.

Theorem 4.4. *Let λ be as in (2.4) and define the time-rescaling process*

$$A_t = \int_0^t \lambda(Y_s)^2 \, ds, \quad t \geq 0. \tag{4.8}$$

Let, also, $u \in \mathcal{C}^{4,1}(\mathbb{D} \times (0,+\infty))$ be a concave and increasing in the spatial argument function satisfying

$$u_t = \frac{1}{2} \frac{u_x^2}{u_{xx}}, \tag{4.9}$$

and $u(x,0) = u_0(x)$. Then, the time-decreasing process

$$U_t(x) = u(x, A_t) \tag{4.10}$$

is a forward performance.

Proposition 4.5. *Let the local risk tolerance function $r : \mathbb{D} \times [0,+\infty) \to \mathbb{R}_0^+$ be defined by*

$$r(x,t) = -\frac{u_x(x,t)}{u_{xx}(x,t)}, \tag{4.11}$$

with u solving (4.9). Then, r satisfies

$$r_t + \frac{1}{2} r^2 r_{xx} = 0, \tag{4.12}$$

with $r(x,0) = -\frac{u'_0(x)}{u''_0(x)}$. Its reciprocal, $\gamma = r^{-1}$, solves

$$\gamma_t + \frac{1}{2} \left(\frac{1}{\gamma} \right)_{xx} = 0, \tag{4.13}$$

[2] See [7] and [45] for a similar equation arising in the traditional Merton problem.

with $\gamma(x,0) = -\frac{u_0''(x)}{u_0'(x)}$.

An analytically explicit construction of the function u was recently developed in [73]. A strictly increasing space-time harmonic function, $h : \mathbb{R} \times [0,+\infty) \to \mathbb{D}$, solving the backward heat equation

$$h_t + \frac{1}{2}h_{xx} = 0, \tag{4.14}$$

plays a key role. This function is always globally defined but its range varies as $Range\,(h) = \mathbb{D}$, with $\mathbb{D}$ being the domain of u.

It was shown in [73] that there is a one-to-one correspondence between strictly increasing solutions of (4.14) and strictly increasing and concave solutions of (4.9) (see, Propositions 9, 13 and 14 therein). Pivotal role in the analysis is played by a positive Borel measure, ν, through which the function h is represented in an integral form (see (4.19) and (4.25) below). This representation stems from classical results of Widder for the solutions of the (backward) heat equation (see [91]). We note that in the applications at hand, these results are not directly applicable, for the range of h is not always constrained to the positive semi-axis. Indeed, we will see that h is used to represent the optimal wealth (cf. (4.30)), which, in unconstrained problems, may take arbitrary values.

The results that follow correspond to the infinite domain case, $\mathbb{D} = \mathbb{R}$. To ease the presentation we introduce the following sets,

$$\begin{aligned}
\mathcal{B}^+(\mathbb{R}) &= \Big\{\nu \in \mathcal{B}(\mathbb{R}) : \forall B \in \mathcal{B},\ \nu(B) \geq 0 \\
&\qquad \text{and } \int_{\mathbb{R}} e^{yx}\nu(dy) < \infty,\ x \in \mathbb{R}\Big\}, && (4.15)\\
\mathcal{B}_0^+(\mathbb{R}) &= \left\{\nu \in \mathcal{B}^+(\mathbb{R}) \text{ and } \nu(\{0\}) = 0\right\}, && (4.16)\\
\mathcal{B}_+^+(\mathbb{R}) &= \left\{\nu \in \mathcal{B}_0^+(\mathbb{R}) : \nu((-\infty,0)) = 0\,\right\} \text{ and} && (4.17)\\
\mathcal{B}_-^+(\mathbb{R}) &= \left\{\nu \in \mathcal{B}_0^+(\mathbb{R}) :\ \nu((0,+\infty)) = 0\,\right\}. && (4.18)
\end{aligned}$$

We start with representation results for strictly increasing solutions of (4.14) with unbounded range.

Proposition 4.6. *i) Let* $\nu \in \mathcal{B}^+(\mathbb{R})$ *and* $C \in \mathbb{R}$. *Then, the function* h *defined, for* $(x,t) \in \mathbb{R}\times[0,+\infty)$, *by*

$$h(x,t) = \int_{\mathbb{R}} \frac{e^{yx-\frac{1}{2}y^2t} - 1}{y}\nu(dy) + C, \tag{4.19}$$

is a strictly increasing solution to (4.14).

Moreover, if $\nu(\{0\}) > 0$, *or* $\nu \in \mathcal{B}_+^+(\mathbb{R})$ *and* $\int_{0^+}^{+\infty} \frac{\nu(dy)}{y} = +\infty$, *or* $\nu \in \mathcal{B}_-^+(\mathbb{R})$ *and* $\int_{-\infty}^{0^-} \frac{\nu(dy)}{y} = -\infty$, *then* $Range\,(h) = (-\infty,+\infty)$, *for* $t \geq 0$. *On the other hand, if* $\nu \in \mathcal{B}_+^+(\mathbb{R})$ *with* $\int_{0^+}^{+\infty} \frac{\nu(dy)}{y} < +\infty$ *(resp.* $\nu \in \mathcal{B}_-^+(\mathbb{R})$ *with* $\int_{-\infty}^{0^-} \frac{\nu(dy)}{y} > -\infty$*), then*

$Range(h) = (C - \int_{0+}^{+\infty} \frac{\nu(dy)}{y}, +\infty)$ (resp. $Range(h) = (-\infty, C - \int_{-\infty}^{0^-} \frac{\nu(dy)}{y})$), for $t \geq 0$.

ii) Conversely, let $h : \mathbb{R} \times [0, +\infty) \to \mathbb{R}$ be a strictly increasing solution to (4.14). Then, there exists $\nu \in \mathcal{B}^+(\mathbb{R})$ such that h is given by (4.19).

Moreover, if $Range(h) = (-\infty, +\infty)$, $t \geq 0$, then it must be either that $\nu(\{0\}) > 0$, or $\nu \in \mathcal{B}_+^+(\mathbb{R})$ and $\int_{0+}^{+\infty} \frac{\nu(dy)}{y} = +\infty$, or $\nu \in \mathcal{B}_-^+(\mathbb{R})$ and $\int_{-\infty}^{0^-} \frac{\nu(dy)}{y} = -\infty$. On the other hand, if $Range(h) = (x_0, +\infty)$ (resp. $Range(h) = (-\infty, x_0)$), $t \geq 0$ and $x_0 \in \mathbb{R}$, then it must be that $\nu \in \mathcal{B}_+^+(\mathbb{R})$ with $\int_{0+}^{+\infty} \frac{\nu(dy)}{y} < +\infty$ (resp. $\nu \in \mathcal{B}_-^+(\mathbb{R})$ with $\int_{-\infty}^{0^-} \frac{\nu(dy)}{y} > -\infty$).

The next proposition yields the one-to-one correspondence between the solutions h and u. Without loss of generality, we will normalise the values

$$h(0,0) = 0, \tag{4.20}$$

choosing $C = 0$, and[3]

$$u(0,0) = 0 \text{ and } u_x(0,0) = 1. \tag{4.21}$$

Proposition 4.7. *i) Let $\nu \in \mathcal{B}^+(\mathbb{R})$ and $h : \mathbb{R} \times [0, +\infty) \to \mathbb{R}$ be as in (4.19) with the measure ν being used. Assume that h is of full range, for each $t \geq 0$, and let $h^{(-1)} : \mathbb{R} \times [0, +\infty) \to \mathbb{R}$ be its spatial inverse. Then, the function u defined for $(x,t) \in \mathbb{R} \times [0, +\infty)$ and given by*

$$u(x,t) = -\frac{1}{2} \int_0^t e^{-h^{(-1)}(x,s) + \frac{s}{2}} h_x\left(h^{(-1)}(x,s), s\right) ds + \int_0^x e^{-h^{(-1)}(z,0)} dz, \tag{4.22}$$

is an increasing and strictly concave solution of (4.9) satisfying (4.21).

Moreover, for $t \geq 0$, the Inada conditions,

$$\lim_{x \to -\infty} u_x(x,t) = +\infty \quad \text{and} \quad \lim_{x \to +\infty} u_x(x,t) = 0, \tag{4.23}$$

are satisfied.

ii) Conversely, let u be an increasing and strictly concave function satisfying, for $(x,t) \in \mathbb{R} \times [0, +\infty)$, (4.9) and (4.21), and the Inada conditions (4.23), for $t \geq 0$. Then, there exists $\nu \in \mathcal{B}^+(\mathbb{R})$, such that u admits representation (4.22) with h given by (4.19), for $(x,t) \in \mathbb{R} \times [0, +\infty)$. Moreover, h is of full range, for each $t \geq 0$, and satisfies (4.20).

The cases of semi-finite domain, $\mathbb{D} = \mathbb{R}^+, \mathbb{R}_0^+, \mathbb{R}^-$ and $\mathbb{R}_0^-$ deserve special attention as they are used in the popular choices of power and logarithmic risk preferences. In these cases, the support of the measure is constrained to the half-line. The representation results above need to be modified for semi-infinite domains. Various cases emerge, depending on certain characteristics of the measure ν which affect the boundary behaviour of the solution u. The arguments are both computationally cumbersome

[3] The first equality is imposed in an ad hoc way. The second one, however, is in accordance with (4.20). For details see the proof of Proposition 9 in [73].

and long. For completeness we state one of these cases and we refer the reader to [73] for the others. To this end, we assume that

$$\nu \in \mathcal{B}_+^+ (\mathbb{R}) \quad \text{and} \quad \int_{0+}^{+\infty} \frac{\nu(dy)}{y} < +\infty, \tag{4.24}$$

with $\mathcal{B}_+^+(\mathbb{R})$ given in (4.17). Choosing for convenience $C = \int_{0+}^{+\infty} \frac{1}{y}\nu(dy)$ in (4.19) yields[4] the solution to (4.14)

$$h(x,t) = \int_{0+}^{+\infty} \frac{e^{yx-\frac{1}{2}y^2 t}}{y} \nu(dy), \tag{4.25}$$

with *Range* $(h) = (0, +\infty)$.

Proposition 4.8. *i) Let* ν *satisfy (4.24) and, in addition,* $\nu((0,1]) = 0$. *Let, also,* $h : \mathbb{R} \times [0,+\infty) \to (0,+\infty)$ *be as in (4.25) and* $h^{(-1)} : (0,+\infty) \times [0,+\infty) \to \mathbb{R}$ *be its spatial inverse. Then, the function* u *defined, for* $(x,t) \in (0,+\infty) \times [0,+\infty)$, *by*

$$u(x,t) = -\frac{1}{2}\int_0^t e^{-h^{(-1)}(x,s)+\frac{s}{2}} h_x\left(h^{(-1)}(x,s), s\right) ds + \int_0^x e^{-h^{(-1)}(z,0)} dz, \tag{4.26}$$

is an increasing and strictly concave solution of (4.9) with

$$\lim_{x\to 0} u(x,t) = 0, \quad \text{for } t \geq 0. \tag{4.27}$$

Moreover, for $t \geq 0$, *the Inada conditions*

$$\lim_{x\to 0} u_x(x,t) = +\infty \quad \text{and} \quad \lim_{x\to+\infty} u_x(x,t) = 0 \tag{4.28}$$

are satisfied.

ii) Conversely, let u, *defined for* $(x,t) \in (0,+\infty) \times [0,+\infty)$, *be an increasing and strictly concave function satisfying (4.9), (4.27) and the Inada conditions (4.28). Then, there exists* $\nu \in \mathcal{B}^+(\mathbb{R})$ *satisfying (4.24) and* $\nu((0,1]) = 0$, *such that* u *admits representation (4.26) with* h *given by (4.25), for* $(x,t) \in \mathbb{R} \times [0,+\infty)$.

Note that the above results yield implicit representation constraints for the initial condition u_0. For example, from (4.22) we must have $u_0'(x) = e^{-h^{(-1)}(x,0)}$, $x \in \mathbb{R}$, with the integrand $e^{-h^{(-1)}(x,0)}$ specified from $h(x,0) = \int_{\mathbb{R}} \frac{e^{yx}-1}{y}\nu(dy)$. This, in turn, yields that the inverse of u_0' must be represented as

$$\left(u_0'\right)^{(-1)}(x) = \int_{\mathbb{R}} \frac{e^{-y\ln x}-1}{y}\nu(dy), \quad x > 0.$$

Characterising the set of admissible initial data and providing an intuitively meaningful interpretation is, in our view, an interesting question.

We continue with the construction of the optimal wealth and portfolio processes for the class of time decreasing performance processes. As the theorem below shows, the optimal processes can be calculated in closed form.

[4] One may alternatively represent h as $h(x,t) = \int_0^{+\infty} e^{yx-\frac{1}{2}y^2 t}\mu(dy)$ with $\mu(dy) = \frac{\nu(dy)}{y}$. Note that $\mu \in \mathcal{B}^+(\mathbb{R})$. Such a representation was used in [5].

Theorem 4.9. *i) Let h be a strictly increasing solution to (4.14), for $(x,t) \in \mathbb{R}\times [0,+\infty)$, and assume that the associated measure ν satisfies, for $t>0$,*

$$\int_{\mathbb{R}} e^{yx+\frac{1}{2}y^2t}\nu\,(dy) < \infty. \tag{4.29}$$

Let also A_t be as in (4.8) and M_t, $t \geq 0$, be given by

$$M_t = \int_0^t \lambda\,(Y_s)\,dW_s^1.$$

Define the processes X_t^* and π_t^* by

$$X_t^* = h\left(h^{(-1)}\,(x,0) + A_t + M_t, A_t\right) \tag{4.30}$$

and

$$\pi_t^* = \frac{\lambda\,(Y_t)}{\sigma\,(Y_t)} h_x\left(h^{(-1)}\,(x,0) + A_t + M_t, A_t\right), \tag{4.31}$$

$t \geq 0$, $x \in \mathbb{R}$, with h as above and $h^{(-1)}$ standing for its spatial inverse. Then, the portfolio π_t^* is admissible and generates X_t^*, i.e.,

$$X_t^* = x + \int_0^t \sigma\,(Y_s)\,\pi_s^*\left(\lambda\,(Y_s)\,ds + dW_s^1\right). \tag{4.32}$$

ii) Let u be the associated with h increasing and strictly concave solution to (4.9). Then, the process $u\,(X_t^*, A_t)$, $t \geq 0$, satisfies

$$du\,(X_t^*, A_t) = u_x\,(X_t^*, A_t)\,\sigma(Y_t)\pi_t^* dW_t^1, \tag{4.33}$$

with X_t^* and π_t^* as in (4.30) and (4.31). Therefore, the processes X_t^* and π_t^* are optimal.

The optimal portfolio π_t^* may be also represented in terms of the risk tolerance process, R_t^*, defined as

$$R_t^* = r\,(X_t^*, A_t), \tag{4.34}$$

with X_t^* solving (4.32) and r as in (4.11). Indeed, one can show that the local risk tolerance function satisfies, for $(x,t) \in \mathbb{D} \times [0,+\infty)$,

$$r\,(x,t) = h_x\big(h^{(-1)}\,(x,t)\,, t\big). \tag{4.35}$$

Therefore, (4.31) yields

$$\pi_t^* = \frac{\lambda\,(Y_t)}{\sigma\,(Y_t)} R_t^*. \tag{4.36}$$

One then sees that under the investment performance criterion (4.10), the investor will *always* follow a myopic strategy. The excess hedging demand component disappears as long as the volatility performance process remains zero.

4.2 The CARA, CRRA and generalised CRRA cases

Case 1: Let $\nu = \delta_0$, where δ_0 is a Dirac measure at 0. Then, from (4.19) we obtain $h(x,t) = x$ and, thus, (4.22) yields $u(x,t) = 1 - e^{-x+\frac{t}{2}}$. The optimal performance process is

$$U(x,t) = 1 - e^{-x+\frac{A_t}{2}}.$$

Formulae (4.32) and (4.31) yield, respectively,

$$X_t^* = x + A_t + M_t \quad \text{and} \quad \pi_t^* = \frac{\lambda(Y_t)}{\sigma(Y_t)}.$$

This class of forward performance processes is analysed in detail in [71] (see, also, [95]).

Case 2: Let $\nu = \delta_\gamma$, $\gamma > 1$. Then (4.25) yields $h(x,t) = \frac{1}{\gamma} e^{\gamma x - \frac{1}{2}\gamma^2 t}$. Since $\nu((0,1]) = 0$, u is given by (4.26) and, therefore, $u(x,t) = \frac{\gamma^{\frac{\gamma-1}{\gamma}}}{\gamma-1} x^{\frac{\gamma-1}{\gamma}} e^{-\frac{\gamma-1}{2}t}$. The forward performance process is

$$U_t(x) = \frac{\gamma^{\frac{\gamma-1}{\gamma}}}{\gamma-1} x^{\frac{\gamma-1}{\gamma}} e^{-\frac{\gamma-1}{2}A_t}, \; t \geq 0.$$

The optimal wealth and portfolio processes are given, respectively, by

$$X_t^* = x \exp\left(\gamma\left(1 - \frac{\gamma}{2}\right) A_t + \gamma M_t\right) \quad \text{and} \quad \pi_t^* = \gamma \frac{\lambda(Y_t)}{\sigma(Y_t)} X_t^*.$$

For the cases $\nu = \delta_\gamma$ with $\gamma = 1$, $\gamma \in (0,1)$ and $\gamma = -\frac{1}{2k+1}$, $k > 0$, see [94].

Case 3: Let $\nu = \frac{b}{2}(\delta_a + \delta_{-a})$, $a, b > 0$, and $\delta_{\pm a}$ are Dirac measures at $\pm a$, $a \neq 1$. We, then, have $h(x,t) = \frac{b}{a} e^{-\frac{1}{2}a^2 t} \sinh(ax)$ and, from (4.22),

$$u(x,t)$$

$$= \frac{\sqrt[a]{\alpha}}{\alpha^2 - 1} e^{\frac{1-a}{2}t} \frac{b^2 e^{-\alpha t} + a(1+\alpha)\left(\alpha x^2 + x\sqrt{\alpha^2 x^2 + b^2 e^{-\alpha^2 t}}\right)}{\left(\alpha x + \sqrt{\alpha^2 x^2 + b^2 e^{-\alpha^2 t}}\right)^{1+\frac{1}{\alpha}}} - \frac{\sqrt[a]{\alpha}}{\alpha^2 - 1} b^{1-\frac{1}{a}}.$$

Equalities (4.30) and (4.31) yield the optimal wealth and portfolio processes

$$X_t^* = \frac{b}{a} e^{-\frac{1}{2}a^2 A_t} \sinh\left(a\left(h^{(-1)}(x,0)\right) + A_t + M_t\right)$$

and

$$\pi_t^* = b \frac{\lambda(Y_t)}{\sigma(Y_t)} e^{-\frac{1}{2}a^2 A_t} \cosh\left(a\left(h^{(-1)}(x,0)\right) + A_t + M_t\right).$$

The case $a = 1$ deserves special attention as it corresponds to the generalised logarithmic case (see [94] for details).

4.3 Two special cases of volatilities

We focus on the case that the volatility coefficient a is a local affine function of U and xU_x. These examples can be reduced to the zero-volatility case but in markets with modified risk premia.

The "market-view" case: $\left(\alpha^1(x,t), a^2(x,t)\right) = U(x,t)\left(\varphi_t^1, \varphi_t^2\right), \varphi_t^1, \varphi_t^2 \in \mathcal{F}_t$

We assume that the processes φ_t^1, φ_t^2 are bounded by a (deterministic) constant.

The forward performance SPDE, (4.5), becomes

$$dU(x,t) = \frac{1}{2}\left(\lambda(Y_t) + \varphi_t^1\right)^2 \frac{\left(U_x(x,t)\right)^2}{U_{xx}(x,t)} dt + U(x,t)\left(\varphi_t^1 dW_t^1 + \varphi_t^2 dW_t^2\right). \quad (4.37)$$

We introduce the process

$$U(x,t) = u(x, A_t^{\varphi}) M_t, \quad (4.38)$$

with u as in (4.9), the process $A_t^{\varphi},\ t \geq 0$, defined as

$$A_t^{\varphi} = \int_0^t \left(\lambda(Y_s) + \varphi_s^1\right)^2 ds \quad (4.39)$$

and the exponential martingale $M_t, t \geq 0$, solving

$$dM_t = M_t\left(\varphi_t^1 dW_t^1 + \varphi_t^2 dW_t^2\right) \quad \text{with} \quad M_0 = 1.$$

One may interpret M_t as a device that offers the flexibility to modify our views on asset returns, changing the original market risk premium, $\lambda(Y_t)$, to $\lambda_t^M = \lambda(Y_t) + \varphi_t^1$.

The optimal allocation vector, $\pi_t^*, t > 0$, has the same functional form as (3.4) but for a different time-rescaling process, namely,

$$\pi_t^* = -\frac{\lambda_t^M}{\sigma(Y_t)} \frac{u_x(X_t^*, A_t^{\varphi})}{u_{xx}(X_t^*, A_t^{\varphi})} = \frac{\lambda_t^M}{\sigma(Y_t)} r(X_t^*, A_t^{\varphi}),$$

with A^{φ} as in (4.39) and r as in (4.11). The optimal wealth process solves

$$dX_t^* = r(X_t^*, A_t)\lambda_t^M\left(\lambda(Y_t)dt + dW_t^1\right).$$

It is worth noticing that if we choose $\varphi_t^1 = -\lambda(Y_t), t \geq 0$, solutions become static, independently of the choice of the second volatility component. Indeed, the time-rescaling process vanishes, $A_t^{\varphi} = 0, t > 0$. In turn, the forward performance process becomes constant, $U(x,t) = u_0(x), t > 0$ and the optimal investment and wealth processes degenerate,

$$\pi_t^* = 0 \quad \text{and} \quad X_t^* = x, \quad t \geq 0.$$

An optimal policy is to allocate zero wealth in the risky asset.

The "benchmark" case: $\left(a^1(x,t), a^2(x,t)\right) = \left(-\delta_t x U_x(x,t), 0\right), \delta_t \in \mathcal{F}_t$

It is assumed that δ_t, $t \geq 0$, is bounded by a deterministic constant. The forward performance SPDE, (4.5), becomes

$$dU(x,t) = \frac{1}{2}\frac{\left(U_x(x,t)\left(\lambda(Y_t) - \delta_t\right) - xU_{xx}(x,t)\delta_t\right)^2}{U_{xx}(x,t)}dt - xU_x(x,t)\delta_t dW_t^1. \quad (4.40)$$

Let A_t^δ, $t \geq 0$, be

$$A_t^\delta = \int_0^t \left(\lambda(Y_s) - \delta_s\right)^2 ds, \quad (4.41)$$

and consider the process N_t, $t \geq 0$, solving

$$dN_t = N_t\delta_t\left(\lambda(Y_t)\,dt + dW_t^1\right) \quad \text{with} \quad N_0 = 1. \quad (4.42)$$

One can then show that the process

$$U(x,t) = u\left(\frac{x}{N_t}, A_t^\delta\right), \quad (4.43)$$

with u as in (4.9), is a forward performance.

One may interpret the auxiliary process N_t, $t \geq 0$, as a benchmark with respect to which the performance of investment policies is measured. It is, then, natural to look at the benchmarked optimal portfolio and wealth processes, $\tilde{\pi}_t^*$ and $\tilde{X}_t^*$, $t \geq 0$, defined, respectively, as

$$\tilde{\pi}_t^* = \frac{\pi_t^*}{N_t} \quad \text{and} \quad \tilde{X}_t^* = \frac{X_t^*}{N_t}.$$

Using (4.6) and (4.43) we obtain, setting $\lambda_t^N = \lambda(Y_t) - \delta_t$,

$$\begin{aligned}\tilde{\pi}_t^* &= \frac{\delta_t}{\sigma(Y_t)}\tilde{X}_t^* - \frac{\lambda_t^N}{\sigma(Y_t)}\frac{u_x\left(\tilde{X}_t^*, A_t^\delta\right)}{u_{xx}\left(\tilde{X}_t^*, A_t^\delta\right)} \\ &= \frac{\delta_t}{\sigma(Y_t)}\tilde{X}_t^* + \frac{\lambda_t^N}{\sigma(Y_t)}r\left(\tilde{X}_t^*, A_t^\delta\right),\end{aligned} \quad (4.44)$$

with A_t^δ, $t \geq 0$ as in (4.41), r as in (4.11) and $\tilde{X}_t^*$ solving

$$d\tilde{X}_t^* = \tilde{R}_t^*\lambda_t^N\left(\lambda(Y_t)dt + dW_t^1\right).$$

The optimal portfolio process is represented as the sum of two funds, say $\tilde{\pi}^{*,X}$ and $\tilde{\pi}_t^{*,R}$, defined as

$$\tilde{\pi}^{*,X} = \frac{\delta_t}{\sigma(Y_t)}\tilde{X}_t^* \quad \text{and} \quad \tilde{\pi}_t^{*,R} = \frac{\lambda_t^N}{\sigma(Y_t)}r\left(\tilde{X}_t^*, A_t^\delta\right).$$

The first component is independent of the risk preferences, depends linearly on wealth and vanishes if $\delta_t = 0$. The situation is reversed for the other component in that it depends only on the investor's risk preferences and vanishes when $\delta_t = \lambda(Y_t)$. The latter condition corresponds to the case when the stock becomes the benchmark itself. Note that, even for exponential preferences, the optimal portfolio may depend on the wealth if performance is measured in terms of a benchmark different than the savings account.

Some open problems

Problem 1: Characterise the class of volatility processes for which the SPDE (4.5) has a solution which satisfies the requirements of a forward performance process.

Problem 2: Prove a verification theorem for the forward stochastic optimisation problem (4.5).

Problem 3: Characterise the family of initial risk preferences $u_0(x)$ for which a forward performance process exists.

Problem 4: Infer the investor's initial risk preferences from his desirable investment targets.

Problem 5: Study the invariance and consistency of the forward performance process and the associated optimal portfolios in terms of different numeraires and benchmarks.

Bibliography

[1] Ait-Sahalia, Y. and M. Brandt: Variable selection for portfolio choice, *Journal of Finance,* 56, 1297–1351 (2001).

[2] Arrow, K.: Aspects of the theory of risk bearing, Helsinki, Hahnson Foundation (1965).

[3] Bates, D.S.: Post-87 crash fears and S&P futures options, *Journal of Econometrics,* 94, 181–238 (2000).

[4] Benzoni, L.P., Collin-Dufrense, C. and R.S. Goldstein: Portfolio Choice over the Life-Cycle when the Stock and Labor Markets Are Cointegrated, *The Journal of Finance,* 62(5), 2123–2167 (2007).

[5] Barrier F., Rogers L.C. and M. Tehranchi: A characterization of forward utility functions, preprint (2007).

[6] Becherer, D.: The numeraire portfolio for unbounded semimartingales, *Finance and Stochastics,* 5, 327–344 (2001).

[7] Black, F.: Investment and consumption through time, Financial Note 6B (1968).

[8] Borkar, V.S.: Optimal control of diffusion processes, *Pitman Research Notes*, 203 (1983).

[9] Borrell, C.: Monotonicity properties of optimal investment strategies for log-Brownian asset prices, *Mathematical Finance,* 17(1), 143–153 (2007).

[10] Bouchard, B. and H. Pham: Wealth-path dependent utility maximization in incomplete markets, *Finance and Stochastics,* 8, 579–603 (2004).

[11] Bouchard, B. and N. Touzi: Weak Dynamic Programming Principle for viscosity solutions, submitted for publication (2009).

[12] Brandt, M.: Estimating portfolio and consumption choice: A conditional Euler equation approach, *Journal of Finance,* 54, 1609–1645 (1999).

[13] Brennan, M.J., Schwartz, E.S. and R. Lagnado: Strategic asset allocation, *Journal of Economic Dynamics and Control,* 21, 1377–1402 (1997).

[14] Brennan, M.J.: The role of learning in dynamic portfolio decisions, *European Finance Review,* 1, 295–306 (1998).

[15] Brennan, M. and Y. Xia: Stochastic interest rates and the bond-stock mix, *European Finance Review,* 4, 197–210 (2000).

[16] Brennan, M. and Y. Xia: Dynamic asset allocation under inflation, *Journal of Finance,* 57, 1201–1238 (2002).

[17] Campbell, J.Y. and L. Viceira: Consumption and portfolio decisions when expected returns are time varying, *Quarterly Journal of Economics,* 114, 433–495 (1999).

[18] Campbell, J.Y. and J. Cochrane: By force of habit: A consumption-based explanation of aggregate stock market behavior, *Journal of Political Economy,* 107, 205–251 (1999).

[19] Campbell, J.Y. and L.M. Viceira: Who should buy long-term bonds?, *The American Economic Review,* 91, 99–127 (2001).

[20] Canner, N., Mankiw, N.G. and D. N. Weil: An asset allocation puzzle, *The American Economic Review,* 87, 181–191 (1997).

[21] Carmona, R. (Ed.): Indifference pricing, Princeton University Press (2009).

[22] Chacko, G. and L. M. Viceira: Dynamic consumption and portfolio choice with stochastic volatility in incomplete markets, *Review of Financial Studies,* 18, 1369–1402 (2005).

[23] Choulli, T., Stricker, C. and J. Li: Minimal Hellinger martingale measures of order q, *Finance and Stochastics,* 11(3), 399–427 (2007).

[24] Constantinides, G.: A theory of the nominal term structure of interest rates, *Review of Financial Studies,* 5, 531–552 (1992).

[25] Cox, J.C., Ingersoll, J.E. and S.A. Ross: A theory of the term structure of interest rates, *Econometrica,* 53, 385–407 (1985).

[26] Crandall, M., Ishii, H. and P.-L. Lions: User's guide to viscosity solutions of second order partial differential equations, *Bulletin of the American Mathematical Society,* 27, 1–67 (1992).

[27] Cvitanic, J., Schachermayer, W. and H. Wang: Utility maximization in incomplete markets with random endowment, *Finance and Stochastics,* 5, 259–272 (2001).

[28] Deelstra, G., Grasselli, M. and P.-F. Koehl: Optimal investment strategies in a CIR framework, *Journal of Applied Probability,* 37, 936–946 (2000).

[29] Detemple, J. and M. Rindisbacher: Closed-form solutions for optimal portfolio selection with stochastic interest rate and investment constraints, *Mathematical Finance,* 15(4), 539–568 (2005).

[30] Detemple, J., Garcia, R. and M. Rindisbacher: A Monte Carlo method for optimal portfolios, *The Journal of Finance,* 58(1), 401–446 (2003).

[31] Duffie, D. and P.-L. Lions: PDE solutions of stochastic differential utility, *Journal of Mathematical Economics,* 21, 577–606 (1992).

[32] El Karoui, N., Nguyen, D.H. and M. Jeanblanc: Compactification methods in the control of degenerate diffusions: existence of an optimal control, *Stochastics,* 20, 169–220 (1987).

[33] El Karoui, N., Peng, S. and M.C. Quenez: Backward stochastic differential equations in finance, *Mathematical Finance,* 7(1), 1–71 (1997).

[34] Fama, W.E. and G.W. Schwert: Asset returns and inflation, *Journal of Financial Economics,* 5, 115–146 (1977).

[35] Ferson, W.E. and C. R. Harvey: The risk and predictability of international equity returns, *Review of Financial Studies,* 6, 527–566 (1993).

[36] Fleming, W. and D. Hernandez-Hernandez: An optimal consumption model with stochastic volatility, *Finance and Stochastics,* 7, 245–262 (2003).

[37] Fleming, W.H. and M.H. Soner: Controlled Markov processes and viscosity solutions, Springer-Verlag, 2nd edition (2005).

[38] French, K.R., Schwert, G.W. and R.F. Stambaugh: Expected stock returns and volatility, *Journal of Financial Economics,* 19, 3–29 (1987).

[39] Glosten, L.R., Jagannathan, R. and D.E. Runkle, On the relation between the expected value and the volatility of the nominal excess return of stocks, *Journal of Finance,* 48, 1779–1801 (1993).

[40] Goll, T. and J. Kallsen: Optimal portfolios for logarithmic utility, *Stochastic Processes and their Applications,* 89, 31–48 (2000).

[41] Goll, T. and J. Kallsen: A complete explicit solution to the log-optimal portfolio problem, *The Annals of Applied Probability,* 12(2), 774–799 (2003).

[42] Harvey, C.R.: Time-varying conditional covariances in tests of asset pricing models, *Journal of Financial Economics,* 24, 289–317 (1989).

[43] Heaton, J. and D. Lucas, Market frictions, savings behavior and portfolio choice, *Macroeconomic Dynamics,* 1, 76–101 (1997).

[44] Henderson, V. and D. Hobson: Horizon-unbiased utility functions, *Stochastic processes and their applications,* 117(11), 1621–1641 (2007).

[45] Huang, C.-F. and T. Zariphopoulou: Turnpike behavior of long-term investments, *Finance and Stochastics* 3(1), 15–34 (1999).

[46] Huggonier, J. and D. Kramkov: Optimal investment with random endowments in incomplete markets, *Annals of Applied Probability,* 14, 845–864 (2004).

[47] Karatzas, I.: Lectures on the Mathematics of Finance, *CRM Monograph Series,* American Mathematical Society (1997).

[48] Karatzas, I., Lehoczky, J.P., Shreve S. E. and G.-L. Xu: Martingale and duality methods for utility maximization in an incomplete market, *SIAM Journal on Control and Optimization,* 25, 1157–1586 (1987).

[49] Karatzas, I. and G. Zitkovic: Optimal consumption from investment and random endowment in incomplete semimartingale markets, *Annals of Applied Probability,* 31(4), 1821–1858 (2003).

[50] Karatzas, I. and Kardaras, C.: The numeraire portfolio in semimartingale financial models, *Finance and Stochastics,* 11, 447–493 (2007).

[51] Kim, T.S. and E. Omberg: Dynamic nonmyopic portfolio behavior, *Review of Financial Studies,* 9, 141–161 (1996).

[52] Kimball, M.S.: Precautionary saving in the Small and in the Large, *Econometrica,* 58, 53–73 (1990).

[53] Korn, R and H. Kraft: On the stability of continuous-time portfolio problems with stochastic opportunity set, *Mathematical Finance* 14, 403–414 (2003).

[54] Korn, R and H. Kraft: A stochastic control approach to portfolio problems with stochastic interest rates, *SIAM Journal on Control and Optimization,* 40, 1250–1269 (2001).

[55] Korn, R. and E. Korn: Option pricing and portfolio optimization – Modern methods of Financial Mathematics, *American Mathematical Society* (2001).

[56] Kraft, H.: Optimal portfolios and Heston's stochastic volatility model, *Quantitative Finance* 5, 303–313 (2005).

[57] Kramkov, D. and W. Schachermayer: The asymptotic elasticity of utility functions and optimal investment in incomplete markets, *The Annals of Applied Probability,* 9(3), 904–950 (1999).

[58] Kramkov, D. and W. Schachermayer: Necessary and sufficient conditions in the problem of optimal investment in incomplete markets, *The Annals of Applied Probability,* 13(4), 1504–1516 (2003).

[59] Kramkov, D. and M. Sirbu: On the two times differentiability of the value functions in the problem of optimal investment in incomplete market, *The Annals of Applied Probability,* 16(3), 1352–1384 (2006).

[60] Krylov, N.: Controlled diffusion processes, Springer-Verlag (1987).

[61] Landsberger, M. and I. Meilijnson: Demand for risky assets: A portfolio analysis, *Journal of Economic Theory,* 50, 204–213 (1990).

[62] Lions, P.-L.: Optimal control of diffusion processes and Hamilton–Jacobi–Bellman equations. Part I: The Dynamic Programming Principle and applications, *Communications in Partial Differential Equations*, 8, 1101–1174 (1983).

[63] Lions, P.-L.: Optimal control of diffusion processes and Hamilton–Jacobi–Bellman equations. Part II: Viscosity solutions and uniqueness, *Communications in Partial Differential Equations*, 8, 1229–1276 (1983).

[64] Liu, J.: Portfolio selection in stochastic environments, *Review of Financial Studies,* 20(1), 1–39 (2007).

[65] Malamud, S. and E. Trubowitz: The structure of optimal consumption streams in general incomplete markets, *Mathematics and Financial Economics,* 1, 129–161 (2007).

[66] Merton, R.: Lifetime portfolio selection under uncertainty: the continuous-time case, *The Review of Economics and Statistics,* 51, 247–257 (1969).

[67] Merton, R.: Optimum consumption and portfolio rules in a continuous-time model, *Journal of Economic Theory,* 3, 373–413 (1971).

[68] Mnif, M.: Portfolio Optimization with Stochastic Volatilities and Constraints: An Application in High Dimension, *Applied Mathematics and Optimization*, 56, 243–264 (2007).

[69] Musiela M. and T. Zariphopoulou: The backward and forward dynamic utilities and their associated pricing systems: The case study of the binomial model, preprint (2003).

[70] Musiela M. and T. Zariphopoulou: The single period binomial model, *Indifference Pricing,* R. Carmona (ed.), Princeton University Press (2009).

[71] Musiela, M. and Zariphopoulou, T.: Optimal asset allocation under forward exponential criteria, *Markov Processes and Related Topics: A Festschrift for Thomas. G. Kurtz,* IMS Collections, Institute of Mathematical Statistics, 4, 285–300 (2008).

[72] Musiela M. and T. Zariphopoulou: Portfolio choice under dynamic investment performance criteria, *Quantitative Finance,* in press.

[73] Musiela M. and T. Zariphopoulou: Portfolio choice under space-time monotone performance criteria, submitted for publication (2008).

[74] Musiela M. and T. Zariphopoulou: Stochastic partial differential equations in portfolio choice, preprint (2007).

[75] Neave, E.H.: Multi-period consumption-investment decisions and risk preferences, *Journal of Economic Theory,* 3, 40–53 (1971).

[76] Pagan, A.R. and G.W. Schwert: Alternative models for conditional stock volatility, *Journal of Econometrics,* 45, 267–290 (1990).

[77] Pham, H: Smooth solutions to optimal investment models with stochastic volatilities and portfolio constraints, *Applied Mathematics and Optimization*, 46, 1–55 (2002).

[78] Ross, S.A.: Some stronger measures of risk aversion in the small and in the large with applications, *Econometrica,* 49(3), 621–639 (1981).

[79] Sangvinatsos, A. and J. Wachter: Does the failure of the expectations hypothesis matter for long-term investors?, *Journal of Finance,* 60, 179–230 (2005).

[80] Schachermayer, W.: Optimal investment in incomplete markets when wealth may become negative, *Annals of Applied Probability* 11(3), 694–734 (2001).

[81] Schachermayer, W.: A super-martingale property of the optimal portfolio process, *Finance and Stochastics,* 7(4), 433–456 (2003).

[82] Schroder, M. and C. Skiadas: Optimal lifetime consumption-portfolio strategies under trading constraints and generalized recursive preferences, *Stochastic Processes and their Applications,* 108, 155–202 (2003).

[83] Schroder, M. and C. Skiadas: Lifetime consumption-portfolio choice under trading constraints, recursive preferences and nontradeable income, *Stochastic Processes and their Applications,* 115, 1–30 (2005).

[84] Scruggs, J.T.: Resolving the puzzling intertemporal relation between the market: risk premium and conditional market variance: a two-factor approach, *Journal of Finance,* 53, 575–603 (1998).

[85] Stoikov, S. and T. Zariphopoulou: Optimal investments in the presence of unhedgeable risks and under CARA preferences, *IMA Volume Series*, Institute for Mathematics and its Applications, in press (2009).

[86] Stutzer, M.: Portfolio choice with endogenous utility: A large deviations approach, *Journal of Econometrics*, 116, 365–386 (2003).

[87] Tehranchi, M. and N. Ringer: Optimal portfolio choice in the bond market, *Finance and Stochastics,* 10(4), 553–573 (2006).

[88] Touzi, N.: Stochastic control problems, viscosity solutions and application to finance, Lecture Notes, Scuola Normale Superiore, Pisa (2002).

[89] Wachter, J.: Risk aversion and allocation to long term bonds, *Journal of Economic Theory,* 112, 325–333 (2003).

[90] Wachter, J.: Portfolio and consumption decisions under mean-reverting returns: An exact solution for complete markets, *Journal of Financial and Quantitative Analysis,* 37, 63–91 (2002).

[91] Widder, D.V.: *The heat equation*, Academic Press (1975).

[92] Yong, J. and X. Y. Zhou, Stochastic Controls: Hamiltonian Systems and HJB Equations, Springer, New York (1999).

[93] Zariphopoulou, T.: A solution approach to valuation with unhedgeable risks, *Finance and Stochastics,* 5, 61–82 (2001).

[94] Zariphopoulou, T. and T. Zhou: Investment performance measurement under asymptotically linear local risk tolerance, *Handbook of Numerical Analysis,* A. Bensoussan (Ed.), in print (2009).

[95] Zitkovic, G.: A dual characterization of self-generation and log-affine forward performances, *Ann. of Appl. Probab.*, in press.

[96] Zitkovic, G.: Utility theory-historical perspectives, *Encyclopedia of Quantitative Finance,* in press (2009).

Author information

Thaleia Zariphopoulou, The University of Texas at Austin, Department of Mathematics, 1 University Station – C1200, Austin, TX 78712-0257, U.S.A.
Email: zariphop@math.utexas.edu